Bacterial Resistance to Antimicrobials

SECOND EDITION

Bacterial Resistance
to Antimicrobials

SECOND EDITION

Edited by
Richard G. Wax • Kim Lewis
Abigail A. Salyers • Harry Taber

CRC Press
Taylor & Francis Group
Boca Raton London New York

CRC Press is an imprint of the
Taylor & Francis Group, an **informa** business

CRC Press
Taylor & Francis Group
6000 Broken Sound Parkway NW, Suite 300
Boca Raton, FL 33487-2742

First issued in paperback 2019

ISBN-13: 978-0-8493-9190-3 (hbk)
ISBN-13: 978-0-367-38807-2 (pbk)

Library of Congress Cataloging-in-Publication Data

Bacterial resistance to antimicrobials / editors, Richard G. Wax ... [et al.]. -- 2nd ed.
    p. ; cm.
"A CRC title."
Includes bibliographical references and index.
ISBN 978-0-8493-9190-3 (hardcover : alk. paper)
   1. Drug resistance in microorganisms. I. Wax, Richard G. II. Title.
   [DNLM: 1. Drug Resistance, Bacterial. 2. Bacteria--genetics. QW 52 B13144 2008]

QR177.B33 2008
616.9'201--dc22   2007020179

Visit the Taylor & Francis Web site at
http://www.taylorandfrancis.com

and the CRC Press Web site at
http://www.crcpress.com

Contents

Preface

On June 9, 1999, the *New York Times* published a lengthy obituary for Anne Miller. Ms. Miller, who was 90 when she died, was not a celebrity or a high-profile politician. Her claim to fame was that, at the age of 33, she had been one of the first people to be given the new and largely untested antibiotic penicillin. The transformation in her condition, which occurred within days, from a young woman slipping into death to a woman who could sit up in bed, eat meals, and chat with visitors was a stunning demonstration of what was to become commonplace in a new era of medicine. Such seemingly miraculous cures soon led physicians and the public to call antibiotics "miracle drugs."

Since then, antibiotics have not only saved people with pneumonia and other dreaded diseases, such as tuberculosis, but also have become the foundation on which much of modern medicine rests. Antibiotics make routine surgery feasible. They protect cancer patients whose chemotherapy had rendered them temporarily susceptible to a variety of infections. They even cure diseases like ulcers that had been considered uncurable chronic conditions. In recent years, antibiotic use has been extended to agriculture, where it plays an important role in preventing infections and in promoting animal growth.

The success of antibiotics in so many areas has, ironically, led antibiotics to become an endangered category of drugs. Bacteria have once again demonstrated their enormous genetic flexibility by becoming resistant to one antibiotic after another. At first, bacterial resistance to antibiotics, such as penicillin, did not seem very alarming because new antibiotics were regularly being discovered and introduced into clinical use. In the 1970s, however, a scant two decades after the introduction of the first antibiotics, the number of new antibiotics entering the pipeline from laboratory to clinic began to decrease. Antibiotic discovery and development are expensive, especially considering the speed with which bacterial resistance can arise. And they are becoming more and more difficult to discover and develop. These factors have led pharmaceutical companies to be less and less interested in antibiotic production. One company after another has shut down or cut back on its antibiotic discovery program.

Finally, the medical community has begun to take antibiotic-resistant bacteria seriously. The public has also become alarmed. This alarm is reflected in the number of articles in the popular press anguishing about the new "superbugs." Agricultural use of antibiotics has been called into question as a possible threat to human health. There is also the potential fallout if antibiotics were to be "lost." Medical researchers have failed to cure many diseases, and the public accepts these failures with grumbling stoicism. But what if overuse of antibiotics caused physicians to lose a cure, an event that would be a first in history? How would this affect public confidence in the medical community?

This book explores many of the aspects of the growing problem posed by antibiotic-resistant bacteria. What is unique about this book is that it is a blend of the purely scientific and the practical, an approach that is essential because antibiotic resistance is a social and economic problem as well as a scientific problem. Chapter 1 explores the history of antibiotics and how bacteria became resistant to them. Understanding the forces leading to the overuse and abuse of antibiotics that have sped the appearance of ever more resistant bacteria is important because it impresses on people the need for rapid and effective future action. The speed with which resistance has arisen is something that everyone needs to appreciate.

Chapter 2 discusses the ecology of antibiotic resistance genes. In recent years, scientists have realized that there is more to the epidemiology of resistance than the transmission of resistant strains of bacteria. Resistance genes are also moving from one bacterium to another, across species and genus lines. Bacteria do not have to spend years mutating their way to resistance; they can become resistant within hours by obtaining genes from other bacteria. Also clear from this chapter, however, is how primitive and inadequate our understanding of resistance ecology still is.

Chapters 3 through 14 describe the means by which bacteria become resistant to antibiotics, methods of detecting resistance genes, and the latest findings on resistance or susceptibility specific to particular groups of bacteria. The bacteria that cause human and animal disease exhibit a staggering diversity. There is no one answer to the question of how bacteria become resistant to antibiotics. Understanding resistance mechanisms is the foundation for more rational design of new antibiotics that are themselves resistant to resistance mechanisms.

A complementary approach, exemplified by combination of a compound that inhibits bacterial β-lactamases with a β-lactam antibiotic, offers great promise. More such successes are needed. To take such an approach, however, is necessary to understand the mechanisms of resistance at a very basic level. Even in the case of the β-lactamase inhibitors, variations in the mechanisms of resistance have foiled this approach in some bacteria that do not use β-lactamases as a resistance mechanism. These chapters pull together all of the information on resistance mechanisms in different groups of bacteria in a way that should help future efforts to develop such combination therapies.

Chapters 15 and 16 examine the public health aspects of the resistance problem. Science alone is not going to solve the resistance problem. Communicating scientific advances and new understandings of forces that promote the rapid development of resistance is essential if the public is to join in the effort to slow the increase in bacterial resistance to antibiotics. Taking antibiotics is a personal matter for most people, a decision made by them and their physicians. As long as antibiotic use remains a personal matter and is not put in the context of public welfare, it is unlikely that progress will be made toward saving antibiotics.

Chapter 17 addresses the problem of finding and developing new antibiotics. This chapter is written by an "insider," a scientist who runs an antibiotic discovery program and thus knows the industry side of the problem. Since the resistance genie is out of the bottle and it will not be easy to put him back in, the continued discovery of new antibiotics is going to be a critical part of the effort to combat resistant bacterial strains. This effort is a critical legacy that we owe our children, who are the ones most likely to bear the consequences of the crisis we have precipitated.

This book is one-stop shopping for anyone interested in all of the facets of bacterial resistance to antibiotics. The breadth of the topics covered reflects the input of a diversity of editors, some of whom have spent their careers in the ivory tower of academic research, some who have had an interest in the public health issues involving the resistance problem, and some who have had direct experience with antibiotic discovery and development. The book represents a unique contribution to the continuing discussion of the best ways to respond to the challenge posed by resistant bacteria. Victory in this battle is not going to be easy. After all, our bacterial adversaries have had a 3-billion-year evolutionary head start. Their diversity and ability to respond to adversity are amazing and frightening. Disseminating information and thus stimulating more scientists to become part of the solution to the problem of resistant bacteria is our best strategy for victory.

Richard G. Wax
Kim Lewis
Abigail A. Salyers
Harry Taber

About the Editors

Richard G. Wax, Ph.D., was an Associate Research Fellow at Pfizer Global Research until his retirement in 2005. He received his B.S. in Chemical Engineering from the Polytechnic University of New York and his M.S. in Biophysics from Yale University. He followed his mentor, Professor Ernest Pollard, to The Pennsylvania State University, University Park, where he received his Ph.D. in Biophysics. He was a Staff Fellow at the U.S. National Institutes of Health (NIH), Bethesda, Maryland, and an NIH Special Fellow at the Weizmann Institute, Rehovot, Israel.

Dr. Wax's career has focused on secondary metabolism. In early research he developed a medium, AGFK, which is now the primary means used for germinating *Bacillus subtilis* spores. At the Merck Research Laboratories his laboratory created high-yielding mutants that allowed commercially feasible antibiotic production. He was co-discoverer of efrotomycin, an antibiotic that acts specifically on bacterial elongation factor Tu. Prior to joining Pfizer he served as Section Head of the Fermentation Group at the Frederick Cancer Research Facility, Frederick, Maryland.

Dr. Wax's avocation is a study of the roles of microbes in altering human history, and he has published and lectured on this subject.

Kim Lewis, Ph.D., is Professor of Biology and Director of the Antimicrobial Discovery Center at Northeastern University in Boston. He is also Director of NovoBiotic Pharmaceuticals, a biotechnology company focused on discovery of antibiotics from previously uncultured bacteria. Dr. Lewis received his B.S. in Biochemistry and his Ph.D. in Microbiology from Moscow University. After moving to the United States, he was a faculty member at the Massachusetts Institute of Technology, the University of Maryland, and Tufts University.

Dr. Lewis has worked in the field of multidrug pumps and established a program for studying antimicrobial tolerance of biofilms and persister cells.

Abigail A. Salyers, Ph.D., is Arends Professor for Molecular and Cellular Biology in the Department of Microbiology at the University of Illinois (Urbana-Champaign). She received her B.A. and Ph.D. from The George Washington University. She spent several years at the Virginia Polytechnic Institute Anaerobe Laboratory, where she began to work on human colonic *Bacteroides* spp.

From 1995 to 1999 Dr. Salyers was a Co-Director of the Microbial Diversity summer course at the Marine Biological Laboratory, Woods Hole, Massachusetts. She was President of the American Society for Microbiology from 2001 to 2002. Her current research focuses on the mechanisms and ecology of antibiotic resistance gene transfer in the human colon, with particular emphasis on *Bacteroides* species.

She is the author of *Revenge of the Microbes*, a book on antibiotics and antibiotic-resistant bacteria that is directed at the general public.

Harry Taber, Ph.D., is Director of the Division of Laboratory Quality Certification at the Wadsworth Center of the New York State Department of Health, Albany, New York. He received his B.A. in Chemistry from Reed College in Portland, Oregon and his Ph.D. in Biochemistry from the University of Rochester School of Medicine and Dentistry in Rochester, New York. He received postdoctoral training at Rochester, the National Institutes of Health in Bethesda, Maryland, and the Centre National de la Recherche Scientifique in Gif-sur-Yvette, France. He was on the Microbiology faculties of the University of Rochester and Albany Medical College before joining the Wadsworth Center as a Research Scientist. He is Past Director of the Division of Infectious Disease at Wadsworth.

Dr. Taber's research has been in the area of genetic regulation of bacterial respiratory systems, particularly as this regulation affects sporulation of *Bacillus subtilis* and the uptake of aminoglycoside antibiotics. He has broad interests in the public health aspects of bacterial antibiotic resistance and in the use of genotyping technologies for tuberculosis control.

Contributors

Keith A. Bostian
Mpex Pharmaceuticals Inc.
San Diego, Connecticut

Jooyoung Cha
Department of Chemistry and
    Biochemistry
University of Notre Dame
Notre Dame, Indiana

Stewart T. Cole
Unité de Génétique Moléculaire
    Bactérienne
Institut Pasteur
Paris, France

Christopher Gerard Dowson
Department of Biological Sciences
University of Warwick
Coventry, United Kingdom

George M. Eliopoulos
Division of Infectious Diseases
Beth Israel Deaconess Medical
    Center
Boston, Massachusetts

Ad C. Fluit
Eijkman-Winkler Institute
University Medical Center Utrecht
Utrecht, the Netherlands

Cindy R. Friedman
Centers for Disease Control and
    Prevention
Atlanta, Georgia

Nafsika H. Georgopapadakou
MethylGene Inc.
Montreal, Quebec, Canada

Keeta S. Gilmore
Schepens Eye Research Institute
Boston, Massachusetts

Michael S. Gilmore
Schepens Eye Research Institute
Boston, Massachusetts

David C. Hooper
Division of Infectious Diseases
Massachusetts General Hospital
Boston, Massachusetts

Lakshmi P. Kotra
Departments of Pharmaceutical
    Sciences and Chemistry
University of Toronto
and
Division of Cell and Molecular Biology
Toronto General Research Institute
University Health Network
Toronto, Ontario, Canada

Olga Lomovskaya
Mpex Pharmaceuticals Inc.
San Diego, Connecticut

Paul F. Miller
Infectious Diseases Therapeutic Area
Pfizer Global Research and
    Development
Groton, Connecticut

Shahriar Mobashery
Departments of Chemistry and
Biochemistry
University of Notre Dame
Notre Dame, Indiana

Steven J. Projan
Wyeth Research
Cambridge, Massachusetts

Alex S. Pym
Unit for Clinical and Biomedical
Tuberculosis Research
South African Medical Research
Council
Durban, South Africa

Philip N. Rather
Departments of Microbiology and
Immunology
Emory University School of Medicine
Atlanta, Georgia

Daniel F. Sahm
Eurofins Medinet
Herndon, Virginia

David Schlesinger
Department of Microbiology
University of Illinois
Urbana, Illinois

Paul Shears
Sheffield Teaching Hospitals
NHS Trust
Sheffield, United Kingdom

Nadja Shoemaker
Department of Microbiology
University of Illinois
Urbana, Illinois

Arjun Srinivasan
Centers for Disease Control and
Prevention
Atlanta, Georgia

William C. Summers
Yale University School of
Medicine
New Haven, Connecticut

Krzysztof Trzcinski
Departments of Epidemiology,
Immunology, and Infectious
Diseases
Harvard School of Public
Health
Boston, Massachusetts

Gerard D. Wright
Department of Biochemistry and
Biomedical Sciences
McMaster University
Hamilton, Ontario, Canada

Helen I. Zgurskaya
Department of Chemistry and
Biochemistry
University of Oklahoma
Norman, Oklahoma

1 Microbial Drug Resistance: A Historical Perspective

William C. Summers

CONTENTS

Almost as soon as it was known that microorganisms could be killed by certain substances, it was recognized that some microbes could survive normally lethal doses and were described as "drug-fast" (German: *-fest* = -proof, as in *feuerfest* = fire-proof; hence "drug-proof," in common usage by at least 1913). These early studies [1–3] conceived of microbial resistance in terms of "adaptation" to the toxic agents. By 1907, Ehrlich [4] more clearly focused on the concept of resistant organisms in his discussion of the development of resistance of *Trypanosoma brucei* to *p*-roseaniline, and in 1911 Morgenroth and Kaufmann [5] reported that pneumococci could develop resistance to ethylhydrocupreine. For every new agent that killed or inhibited microorganisms, resistance became an interest as well.

While we think of antibiotic resistance as a phenomenon of recent concern, the basic conceptions of the problems, the controversies, and even the fundamental mechanisms were well developed in the early decades of the twentieth century. These principles were, of course, elaborated in terms of resistance to anti-microbial toxins, such as the arsenicals, dyes, such as trypan red, and disinfectants, such as acid, phenols, and the like. However, by the time the first antibiotics were employed in the 1940s and resistance was first observed, the framework for understanding this phenomenon was already in place.

DRUG-FASTNESS

Drug-fastness became a topic of importance as microbiologists sought understanding of the growth, metabolism, and pathogenicity of bacteria, protozoa, and fungi. In 1913, Paul Ehrlich clearly described the basic mechanisms of drug action on microbes [6]: "parasites are only killed by those materials to which they have a certain relationship, by means of which they are fixed by them." He went on to describe specific drug binding (fixation) to specific organisms and elaborated "The principle of fixation in chemotherapy."

Once this principle was accepted, one could investigate how drugs are fixed by microbes, what kinds of cross-sensitivities existed, and what happened when organisms became resistant to chemotherapeutic agents. Ehrlich noted that both trypanosomes and spirochaetes, his favorite experimental organisms, exhibited different chemoreceptors that were specific for drugs of a given chemical class. Thus, there seemed to be a chemoreceptor for arsenic compounds (arsenious acid, arsanilic acid, and arsenophenylglycine) that differed from the receptor for azo-dyes (trypan red and trypan blue) as well as from the receptor for certain basic triphenylmethane dyes, such as fuchsin and methyl violet.

Drug-fastness, therefore, was readily explained as "a reduction of their (the chemoreceptors) affinity for certain chemical groupings connected with the remedy (the drug), which can only be regarded as purely chemical" [6]. Clearly, Ehrlich's approach was an outgrowth of his earlier work on histological staining and dye chemistry and reflected his strong chemical thinking.

Already in 1913, the problem of clinical drug resistance was confronting the physician and microbiologist. Ehrlich discussed the problem of "relapsing crops" of parasites as a result of the parasites' biological properties. His views were mildly selectionist, but he also held the common view that microbes had great adaptive power and that the few that managed to escape destruction by drugs (or immune serum) could subsequently change into new varieties that were drug-fast or serum-proof.

One corollary of the specific chemoreceptor hypothesis was that combined chemotherapy was best carried out with agents that attack entirely different chemo-receptors of the microbes. Ehrlich, who frequently resorted to military metaphors, wrote: "It is clear that in this manner a simultaneous and varied attack is directed at the parasites, in accordance with the military maxim: 'March apart but fight combined'" [6]. He also allowed for the possibility of drug synergism so that in favorable cases the effects of the drugs may be multiplied rather than simply additive. From the earliest days of chemotherapy, it appears that multiple drug therapy with agents with different mechanisms was seen as a way to circumvent the problem of "relapsing crops" or emergence of resistant organisms.

Ehrlich, too, realized the relationship between evolution of resistant variants and the dose of the agent used to treat the infection. Clinical practice often used remedies in increasing dosages, perhaps a therapeutic principle derived from empirical treatment practice of long tradition. He noted that these were precisely the conditions likely to lead to emergence of drug-fast organisms and developed the idea of "*therapia sterilisans magna*" (total sterilization) in which he advocated the maximum microbicidal dose that was non-toxic to the host [7]. Indeed, by 1916, there was experimental confirmation in controlled *in vitro* laboratory studies that gradual increases in drug concentration would lead to outgrowth of resistant spirochetes, while exposure to initial high concentrations of antitreponemal agents (arsenicals, mercuric, and iodide compounds) would not [7].

DISINFECTION

Often early research on antimicrobial agents was directed to problems of "disinfection" and related matters of public health, and the origins and properties of resistant organisms became of concern in the "fight against germs" [8]. Protocols for inducing drug-resistance *in vivo* were elaborated, and the relevance of *in vitro* resistance to "natural" *in vivo* resistance was debated in the literature of the 1930s and 1940s. One interesting aspect, now forgotten, was the widespread belief in bacterial life cycles as an explanation for the changing properties of bacterial cultures under what we would now call "selection." This theory of bacterial life cycles [9–11], called "cyclogeny," held that bacteria had definite phases of growth, and that properties of bacteria, such as shape, nutritional requirements, pathogenicity, antigenic reactivities, and chemical resistances, were variable properties of the organism that simply reflected the growth phase of the culture. This cyclogenic variation revived an old nineteenth century controversy in bacteriology, namely that of Koch's monomorphism versus Cohn's polymorphism. Ferdinand Cohn believed that bacterial forms were highly variable so that one "species" of bacteria could exist in many shapes and with many different properties, while Robert Koch held that specific bacterial "species" had unique morphologies and properties that were unchanging. This debate, of course, had far-reaching implications both for problems of bacterial classification and for understanding variation and mutation of bacterial characteristics.

MICROBIAL METABOLISM AND ADAPTATION

The basic issue, as we would see it today, that faced microbiologists in the early days of antimicrobial research is one of "adaptation versus mutation." It was passionately debated and contested by leading microbiologists from the mid-1930s until the early 1960s. Even those who viewed most microbial resistance as some sort of heritable change, or mutation, were divided on the basic problem of whether the mutations arose in response to the agent, or occurred spontaneously and were simply observed after selection against the sensitive organisms. This problem was unresolved until the 1940s and 1950s, but has returned in a new form recently, as will be discussed subsequently.

As early as the 1920s, the ability of bacterial cells to undergo infrequent abrupt and permanent changes in characteristics was interpreted as a manifestation of the phenomenon of mutation as had been described in higher organisms [12]. The relation of these mutations to the growth conditions where they could be observed, was, however, unclear. In the 1930s, this question was confronted directly by I.M. Lewis [13], who studied the mutation of a lactose-negative strain of "*Bacillus coli mutabile*" (*Escherichia coli*) to lactose-utilizing proficiency. Lewis laboriously isolated colonies and found that even in the absence of growth in lactose, the ability to ferment this sugar arose spontaneously in about one cell in 10^5. This work was the beginning of a long line of investigations that quite conclusively showed that mutation is (almost always) independent of selection.

The second kind of adaptation, that "due to chemical environment," is of special historical interest. As early as 1900, Frédéric Dienert [14] found that yeast that were grown for some time in galactose-containing medium became adapted to this medium and would grow rapidly without a lag when subcultured into fresh galactose medium, but that this "adaptation" was lost after a period of growth in glucose-containing medium. By 1930, Hennig Karström in Helsinki had found several instances of such adaptation [15]. For example, he found that a strain of *Bacillus aerogenes* could grow on ("ferment" to use the older term) xylose if "adapted" to do so, but that this strain could ferment glucose "constitutively" without the need for adaptation. When he examined the enzyme content of these adapted and unadapted cells, he found that there were some enzymes that were "constitutive" and some that were "adaptive." Thus, the metabolic properties of the culture mirrored the intracellular chemistry. By experiments in which the medium was changed in various ways, Karström and others showed that metabolic adaptation could sometimes take place even without measurable increase in cell numbers in the culture.

Marjory Stephenson, a leading mid-twentieth century bacterial physiologist, described these variations in her influential book, *Bacterial Metabolism* [16], as "Adaptation by Natural Selection" and "Adaptation due to Chemical Environment." The former included the phenomenon that is now termed mutation.

Between 1931 and the start of World War II, Stephenson and her students, John Yudkin and Ernest Gale, investigated bacterial metabolic variation in detail, often exploiting the lactose-fermenting system in enteric bacteria to study it. The mechanism of chemical adaptation, however, eluded them. The final paragraph of her monograph expressed her belief in the importance of the study of bacterial metabolism: "It (the bacterial cell) is immensely tolerant of experimental meddling and offers material for the study of processes of growth, variation and development of enzymes without parallel in any other biological material" [16].

In 1934, another research group on "bacterial chemistry" consisting of Paul Fildes and B.C.J.G. Knight was established at Middlesex Hospital in London [17]. Fildes and Knight investigated bacterial nutrition and established vitamin B1 (thiamine) as a growth factor for *Staphylococcus aureus*. Their work on bacterial growth factors suggested a unity of metabolic biochemistry at the cellular level, and they investigated the variations in growth factor requirements. One recurrent theme in their early work was the finding that they could "train" bacteria to grow on media deficient in some essential metabolite. For example, they could train *Bact. typhosum* (modern name *Salmonella typhi*) to grow on medium without tryptophan or without indole. Fildes noted that "during this time little attention was given to the mechanism of the training process, but it was certainly supposed that the enzyme make-up of the bacteria became altered as a result of a stimulus produced by the deficiency of the metabolite" [18].

By the mid-1940s, however, Fildes and his colleagues undertook a study of the mechanism of this ubiquitous "training." Was it another example of enzyme adaptation or was it something else? Using only simple growth curves, viable colony counts on agar plates, and ingenious experimental designs, they concluded "that 'training' bacteria to dispense with certain nutritive substances normally essential may be looked upon as a cumbersome method for selecting genetic mutants" [18]. Little by little, the underlying mechanisms of the different kinds of biochemical variations seen in bacteria were becoming clear, and little by little, genetics was joining biochemistry as a powerful approach to study bacterial physiology. This understanding, of course, was central to discovering the underlying mechanisms involved in the variation of microbial behavior related to drug resistance.

This approach, however, was not uncontested and matters were not so easily settled as Arthur Koch pointed out in an important review of the field in 1981 [19]. A more extreme view of cellular metabolism was proposed by Cyril Hinshelwood, a Nobel Prize winner, no less, who argued that all variations in cellular functions, such as enzyme inductions, changes in nutritional requirements, and drug resistances, were but readjustments of complex multiple equilibria of chemical reactions already active in the cell [20].

ADAPTATION OR MUTATION?

With the discovery and development of antibiotics and their medical applications, drug resistance took on new relevance and new approaches became possible. No sooner were new antibiotics announced than reports of drug resistance appeared: sulfonamide resistance in 1939 [21], penicillin resistance in 1941 [22], and streptomycin resistance in 1946 [23], to cite a few early reports in the widely read literature. Research on resistance focused on three major problems: (*i*) cross-resistance to other agents, that is, was resistance to one agent accompanied by resistance to another agent? (*ii*) distribution of resistance in nature, that is, what was the prevalence of resistance in naturally occurring strains of the same organism from different sources? (*iii*) induction of resistance, that is, what regimens of drug exposure led to the induction or selection of resistant organisms?

While many practically useful results came from such research, two lines of investigation emerged that were later to prove scientifically interesting. Rare nutritional markers were somewhat limited and such mutations often resulted in loss of function, usually recessive traits that were difficult to manipulate experimentally. Drug resistance, on the other hand, provided a potent experimental tool to microbiologists who were studying bacterial genes and mutations because it allowed the analysis of events that took place at extremely low frequencies. For example, in 1936, Lewis [13] tested for preexisting, spontaneous mutations to lactose utilization in a previously lactose-negative strain of *E. coli*, but his results gave only indirect evidence for the random, spontaneous nature of bacterial mutation (as did the statistical approach of Luria and Delbrück in 1943 [24]). However, Lederberg and Lederberg [25] were able to use both streptomycin resistance and their newly devised replica plating technique to provide direct and convincing evidence to support the belief that mutations to drug resistance occurred even in the absence of the selective agent. Not only did such work on drug resistance clarify the nature of microbe–drug interactions, but it provided a much-needed tool to the nascent field of microbial genetics [26].

Just as Paul Ehrlich's 1913 summary of the principles of chemotherapy provided a window on early understanding of drug resistance, we can find a similar succinct presentation of the mid-twentieth century state of the field in a review by Bernard Davis in 1952 [27]. By this time, genetics of microbes had replaced microbial biochemistry as the fashionable mode of explanation for bacterial drug resistance. Although bacteria did not have a cytologically visible nucleus with stainable chromosomes, it was recognized that they had "nucleoid bodies" and that the material in this structure appeared to behave in a way similar to the chromosomes of higher organisms. Davis boldly (for the time) asserted that bacteria have nuclei, and that "within these nuclei are chromosomes that appear to undergo mitosis." He went even further to note that "some bacterial strains can inherit features (including acquired drug resistance) from two different parents, as in the sexual process of higher organisms." Thus, by the mid-twentieth century, bacteria had become "real" cells, with conventional genetic properties. If bacteria were like higher organisms, and since "almost all the inherited properties of animals or plants are transmitted by their genes," it was only logical, Davis argued, to consider genetic mutations as the basis for inherited drug resistance.

Davis, however, gave a fair consideration to the possible neo-Lamarckian hypothesis that single-cell organisms, where there is no separation between somatic and germ cells, might behave differently from higher sexually dimorphic organisms. To his mind, however, the recent work in microbial genetics by Luria and Delbrück [24], by Lederberg and Lederberg [25], and by Newcombe [28], settled the matter: the mutations to drug resistance were already present, having originated by some "spontaneous" process, and were simply selected by the application of the drug.

A very important clinical correlate of this new understanding of the nature of bacterial drug resistance was its application to combination chemotherapy. Since it became clear that mutations to resistance to different agents were independent events, the concept of multiple drug therapy, initially envisioned by Ehrlich [6], was refined and made precise. It was realized that adequate dosages and lengths of treatment were necessary if the emergence of resistant organisms was to be avoided [27,29].

DRUG DEPENDENCE

The second observation of basic significance was the odd phenomenon of drug dependence, which was first noted for streptomycin in 1947 by Miller and Bohnhoff [30]. This finding seemed to be restricted to streptomycin, but was extensively investigated at the time, and was thought to offer clues to the problems of antibiotic resistance in general. Later, however, this puzzling finding would be fundamental to understanding the functioning of the ribosome, and rather specific to the mode of action of streptomycin. The history of this aspect of drug resistance emphasizes our inability to predict the future course of research and our failure to identify, beforehand, just where the likely advances will take us.

MULTIPLE DRUG RESISTANCE AND CROSS RESISTANCE

In the 1950s, in the era of many new antibiotics and the emphasis on surveys of both cross resistance and distributions of resistance in natural microbial populations, especially in Japan, it was recognized that many strains with multiple drug resistances were emerging. The appearance of such multiple drug resistance could not be adequately explained on the basis of random, independent mutational events. Also, the patterns of resistance were complex and did not fit a simple mutational model. For example, resistance to chloramphenicol was rarely, if ever, observed alone, but it was common in multiply-resistant strains. Careful epidemiological and bacteriological studies of drug-resistant strains in Japan led Akiba et al. [31] and Ochiai et al. [32] to suggest that multiple drug resistance may be transmissible both *in vivo* and *in vitro* between bacterial strains by so-called resistance transfer factors (RTFs) [33].

Genetic analysis of this phenomenon showed that the genes for these antibiotic resistance properties resided on the bacterial genome, yet were transmissible between strains albeit at low frequency. Further study showed that the transfer of these genes was mediated by a conjugal plasmid and that the resistance genes could associate with the conjugal plasmid; it was suggested that the resistance gene could be horizontally transmitted to other strains in a fashion similar to that for the integrative recombination for the temperate phage lambda [34]. It soon became clear, however, that the F-episome/F-lac system in *E. coli* was a better analogous genetic system. In some cases, the resistance genes and the transfer genes could be separated both genetically and physically [35]. Because of the promiscuous nature of the RTF, once a gene for drug resistance evolves, it can rapidly spread to other organisms. Additionally, because the R-factor plasmids replicate to high copy number, probably as a way to provide high levels of the drug-resistant protein, these plasmids have become the molecule of choice for molecular cloning technology.

With the better understanding of the genetics of drug resistance and the classification of the types of resistance, the biochemical bases for resistance were elucidated. Knowledge of the mechanism of action of an agent led to understanding of possible mechanisms of resistance. The specific role of penicillin in blocking cell wall biosynthesis, coupled with the knowledge of the structures of bacterial cell walls, could explain the sensitivity of Gram-positive organisms and the resistance of Gram-negative organisms to this antibiotic. Likewise, understanding of its metabolic fate led to the finding that penicillin was often inactivated by degradation by β-lactamase, which provides one mechanism of bacterial drug resistance. Detailed biochemical studies of the actions of antimicrobials have led to the understanding of the many ways in which microbes evolve to become resistant to such agents.

NEWLY FOUND MODES OF RESISTANCE

Not all voices for the adaptation hypothesis of drug resistance were drowned by the din of the genetic and conjugal mechanists. In the 1970s, mainly through the work of Samson and Cairns [36] and their colleagues, a variant of the adaptative model was revived and new mechanisms for bacterial drug resistance were discovered. Cairns and his colleagues observed that in accord with some of the older work, indeed, bacteria could be "trained" to resist certain agents by prior exposure to small, sublethal concentrations of the agent. They found that alkylating agents could induce the expression of specific genes whose products react with the alkylators, thus acting as a sink for further alkylating damage and rendering the cell hyper-resistant. While this phenomenon seems to represent a specialized pathway for dealing with alkylation damages, it suggests that a century after its first observation, microbial drug resistance is still a fruitful and surprising area of research.

REFERENCES

1. Kossiakoff MG. De la propriété que possèdent les microbes de s'accomoder aux milieux antiseptiques. *Ann Inst Pasteur* 1887; 1:465–476.
2. Effront J. Koch's Jahresber Gärungorganisimen 1891; 2:154 (quoted in Schnitzer RJ, Grunberg E. Drug Resistance of Microorganisms. New York: Academic Press, 1957. p 1).
3. Davenport CB, Neal HV. On the acclimatization of organisms to poisonous chemical substances. *Arch Entwicklungsmech Organ* 1895–1896; 2:564–583.
4. Ehrlich P. Chemotherapie Trypanosomen-Studien. *Berl Klin Wochenschr* 1907; 44:233–238.
5. Morgenroth J, Levy R. Chemotherapie der Pneumokokkeninfektion. *Berl Klin Wochenschr* 1911; 48:1560.
6. Ehrlich P. Chemotherapy: scientific principles, methods, and results. *Lancet* 1913; 2:445–451.
7. Akatsu S, Noguchi H. The drug-fastness of spirochetes to arsenic, mercurial, and iodide compounds in vitro. *J Expt Med* 1917; 25(3):349–362.
8. Tomes N. The Gospel of Germs: Men, Women and Microbes in American Life. Cambridge: Harvard University Press, 1998.
9. Enderlein G. Bakterien Cyclogenie: Prologomena zu Untersuchungen über Bau, geschlechtliche und ungeschlechtliche Fortpflanzung und Entwicklung der Bakterien. Berlin: W. de Guyter, 1925.
10. Hadley P. Microbic dissociation. *J Infect Dis* 1927; 40:1–312.
11. Amsterdamska O. Stabilizing instability: the controversy over cyclogenic theories of bacterial variation during the interwar period. *J Hist Biol* 1991; 24:191–222.
12. Summers WC. From culture as organism to organism as cell: Historical origins of bacterial genetics. *J Hist Biol* 1991; 24:171–190.
13. Lewis IM. Bacterial variation with special reference to behavior of some mutabile strains of colon bacteria in synthetic media. *J Bacteriol* 1934; 28:619–639.
14. Dienert F. Sur la fermentation du galactose et sur l'accountumance de levures à ce sucre. *Ann Inst Pasteur* 1900; 14:139–189.
15. Karström H. Über die Enzymbildung in Bakterien. Thesis. Helsingfors (Helsinki), Finland, 1930.
16. Stephenson M. Bacterial Metabolism. London: Longmans, Green and Co., 1930.
17. Fildes P. André Lwoff: An appreciation. In: *Of Microbes and Life*, J Monod and E Borek eds, New York: Columbia University Press, 1971.
18. Fildes P, Whitaker K. "Training" or mutation of bacteria. *Br J Exp Path* 1948; 29:240–248.
19. Koch AL. Evolution of antibiotic resistance gene function. *Microbiol Rev* 1981; 45:355–378.
20. Hinshelwood C. The Chemical Kinetics of the Bacterial Cell. Oxford: Oxford University Press, 1946.
21. Maclean IH, Rogers KB, Fleming A. M. & B. 693 and pneumococci. *Lancet* 1939; i:562–568.
22. Abraham EP, Chain E, Fletcher CM, et al. Further observations on penicillin. *Lancet* 1941; ii:177–189.
23. Murray R, Kilham L, Wilcox C, Finland M. Development of streptomycin resistance of Gram-negative bacilli in vitro and during treatment. *Proc Soc Exptl Biol Med* 1946; 63:470–474.

24. Luria S, Delbrück M. Mutations of bacteria from virus sensitivity to virus resistance. *Genetics* 1943; 28:491–511.
25. Lederberg J, Lederberg EM. Replica plating and indirect selection of bacterial mutants. *J Bacteriol* 1952; 63:399–406.
26. Brock TD. The Emergence of Bacterial Genetics. Cold Spring Harbor, NY: Cold Spring Harbor Laboratory Press, 1991.
27. Davis BD. Bacterial genetics and drug resistance. *Public Health Reports* 1952; 67(4):376–379.
28. Newcombe HB. Origin of bacterial variants. *Nature* 1949; 164:150.
29. Szybalski W. Theoretical basis of multiple chemotherapy. *Tuberculology* 1956; 15:82–85.
30. Miller CP, Bohnhoff M. Development of streptomycin-resistant variants of meningococcus. *Science* 1947; 105:620–621.
31. Akiba T, Koyama K, Ishiki Y, Kimura S, Fukushima T. On the mechanism of the development of multiple-drug-resistant clones of *Shigella*. *Japanese J Microbiol* 1960; 4:219–227.
32. Ochiai K, Yamanaka T, Kimura K, Sawada O. Inheritance of drug resistance (and its transfer) between *Shigella* strains and between *Shigella* and *E. coli* strains. *Nihon Iji Shimpo* 1959: 1861:34–46 (in Japanese).
33. Watanabe T. Infective heredity of multiple drug resistance in bacteria. *Bact Rev* 1963; 27:87–115.
34. Campbell A. in Bacterial Episomes and Plasmids, Ciba Foundation Symp. Boston: Little Brown and Co, 1969:117.
35. Cohen SN, Miller CA. Non-chromosomal antibiotic resistance in bacteria, III. Isolation of the discrete transfer unit of the R-factor R1. *Proc Natl Acad Sci* 1970; 67:510–516.
36. Samson L, Cairns J. A new pathway for DNA repair in *Escherichia coli*. *Nature* 1977; 267:281–283.

2 Ecology of Antibiotic Resistance Genes

Abigail A. Salyers, Nadja Shoemaker, and David Schlesinger

CONTENTS

The movement of antibiotic resistance genes, as opposed to the movement of resistant bacterial strains, has become an issue of interest in connection with clinical and agricultural antibiotic use patterns. Evidence to date suggests that extensive DNA transfer is occurring in natural settings, such as the human intestine. This transfer activity, especially transfers that cross genus lines, is probably being mediated mainly by conjugative transfer of plasmids and conjugative transposons. Natural transformation and phage transduction probably contribute mainly to transfers within species or groups of closely related species, but the extent of this contribution is not clear. A considerable amount of information is available about the mechanisms of resistance gene transfer. The goal of future work on resistance ecology will focus on new approaches to detecting gene transfer events in nature and incorporating this information into a framework that explains and predicts the effects of human antibiotic use patterns on resistance development.

INTRODUCTION

For many years, surveillance systems designed to monitor patterns of bacterial resistance to antibiotics focused exclusively on antibiotic-resistant strains of bacteria. Moreover, of necessity, these surveillance efforts had to focus on a limited number of clinically important bacterial species such as *Staphylococcus aureus* [1–4] and *Salmonella* spp. [5]. A limitation of this approach is not just that it can monitor only a limited number of species but also that it does not take into account the dynamic nature of the bacterial genome. In theory, DNA is constantly flowing into and out of bacterial cells located in a natural setting. Thus, the pattern of resistance gene distribution could be as important, if not more so, than the distribution of resistant strains of a particular species. This is especially true if resistance genes from one species can move to another species. Even if a newly acquired resistance gene is not expressed initially in a bacterial host, selective pressures imposed by the widespread clinical and agricultural use of antibiotics could select for promoter or codon usage mutations that allow the resistance gene to be expressed [6,7].

The importance of understanding the flow of resistance genes became particularly evident in discussions of possible impacts of agricultural use of antibiotics. In this case, initial attention focused on *Salmonella* and *Campylobacter* spp., types of bacteria that could cause human disease. Attention soon expanded, however, to include a broader question. Was it possible that even non-pathogenic bacteria, moving through the food supply from farm to the consumer, could transfer resistance genes to human intestinal bacteria [8–10]? Since human intestinal bacteria are a common cause of post-surgical infections [11,12], increased resistance due to acquisition of genes from swallowed bacteria passing through the intestinal tract could indeed have a direct impact on human health [13,14].

Assertions such as this prompted an old idea, called the "reservoir hypothesis" to resurface [15–17]. The reservoir hypothesis as it applies to human colonic bacteria is illustrated in Figure 2.1, but similar sorts of gene flows could occur almost anywhere in nature. According to the reservoir hypothesis, commensal bacteria in the colon, including those that could act as opportunistic pathogens and those that were truly non-pathogenic, exchange DNA with one another. They can also acquire DNA from or donate DNA to swallowed bacteria that cannot colonize the human colon, but spend enough time in the colon for DNA transfer to occur [18,19].

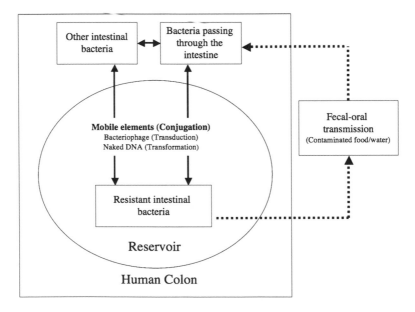

FIGURE 2.1   The reservoir hypothesis. Bacteria in the human colon serve as "reservoirs" for resistance genes that can be acquired from ingested bacteria.

But how likely are such exchanges to occur, especially broad host range transfers between members of different species and genera? This is the type of transfer that could be most problematic because it would allow resistance genes to move into bacteria capable of causing human disease. In trying to answer this question, attention has focused on conjugative gene transfer because this is the type of transfer known to be capable of crossing genus and phylum lines [20]. Initially, however, the focus was somewhat larger because early studies sought examples in which the same gene, with "same" defined as DNA sequence identity of more than 95%, was found in two very distantly related species of bacteria. That is, the only criterion was evidence that some sort of DNA transfer had occurred, without specifying the mechanism. The 95% cutoff was arbitrary but was motivated by the need to eliminate the possibility of convergent evolution. In convergent evolution, the same amino acid sequence might arise by selection from two different genes. Since two genes can differ by as much as 20% at the DNA sequence level and still have the same amino acid sequence, the requirement for 95% or higher DNA sequence identity seemed to be a good way to restrict attention to recent horizontal transfers of resistance genes.

In fact, the cutoff could have been 98%, because it proved all too easy to find resistance genes in different genera and species that were 98% to 100% identical at the DNA sequence level. Some examples are shown in Figure 2.2, where the resistance gene designation is shown inside the oval at the center and the names of Gram-positive and Gram-negative bacterial species found to have that gene are shown on either side of the oval. What is striking about this figure is that not only has the same gene been found in widely divergent species, but also in species commonly found in different locations. That is, the same genes were found not only in human colonic bacteria, but also in bacteria from other sites, such as soil, the intestinal tracts of

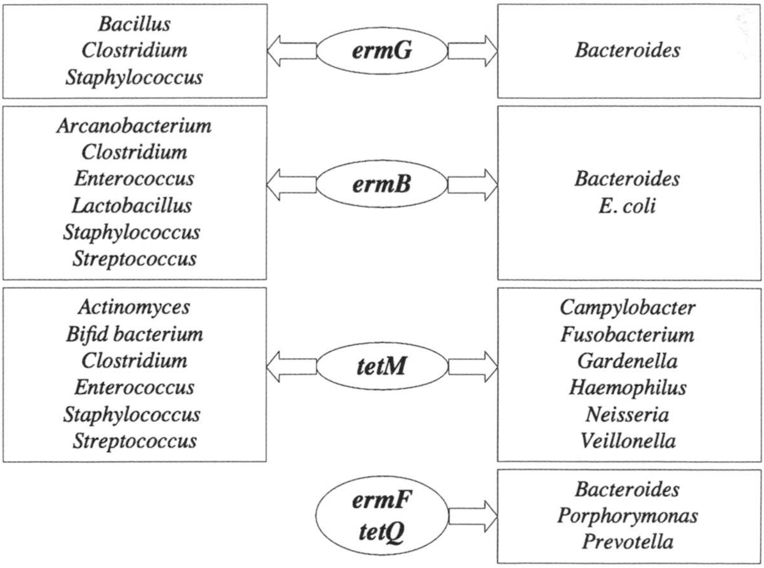

FIGURE 2.2 Example of genes with more than 95% sequence identity that have been found in distantly related bacteria from different sites.

non-human animals, and the human mouth. Most of these genes are genes that confer resistance to tetracycline (*tetM, tetQ*) or to macrolides (*ermB, ermF, ermG*). The tetracycline resistance genes are not the ones that encode efflux pumps, but encode a cytoplasmic protein that protects the bacterial ribosome from tetracycline. Why these two types of genes seem to be the ones most commonly found in different species and genera is not clear but could have something to do with the fact that they have a cytoplasmic location and thus do not need to be coupled with the proton motive force in membranes or to be secreted through the cytoplasmic membrane, requiring localization functions that could be species-specific.

A striking feature of all of the genes shown in Figure 2.2 is that they have been found almost exclusively on a type of integrated conjugative element called a conjugative transposon (CTn). CTns normally reside in the chromosome, but can excise to form a non-replicating circular intermediate, which transfers similarly to a plasmid. That is, there is a single stranded nick in the circular form, followed by transfer of a single strand of the DNA through a multi-protein complex that joins the cytoplasms of the donor and recipient. Once in the recipient, the circular copy of the CTn becomes double stranded and integrates into the recipient chromosome. Presumably the copy of the CTn in the donor has the same fate. Even if the copy of the CTn in the donor is sometimes lost, this affects only a small fraction of donor cells and the outcome of the process is a net increase in the number of bacteria carrying the CTn, especially if there is antibiotic selection for resistant cells.

CTns were first discovered in the Gram-positive bacteria and in the *Bacteroides* group of Gram-negative bacteria, but now that their existence is known, scientists are discovering CTns in other types of Gram-negative bacteria, such as *Vibrio cholerae, Salmonella* spp., and *Rhizobium* spp. [21–23]. There is no consistent nomenclature for this type of integrated transmissible element. They have also been called integrating, conjugative elements (ICE) elements and constins as well as CTns [24,25]. These alternative terms have the advantage that they avoid the word "transposon." Calling the CTns "transposons" is misleading, because their excision and integration is quite different from that of transposons, such as Tn*5* and Tn*10*. In fact, the enzyme that catalyzes the integration reaction, the CTn integrase, has most often proved to be a member of the tyrosine recombinase family, a family associated with many lambdoid phages. In some ways, the CTns resemble "phage" that travel from cell to cell through a multi-protein "capsid," similarly to the fusigenic viruses of mammalian cells. We will use the nomenclature CTn, because for better or worse, this nomenclature has been the one most commonly used in the literature.

Just as there are mobilizable plasmids that are transferred with the help of self-transmissible plasmids, there are also mobilizable transposons (MTns). The first of these to be discovered was NBU1, an MTn that is mobilized by a *Bacteroides* CTn, CTnDOT [26]. CTns can also mobilize plasmids [27,28].

MOVEMENT OF CTNs BETWEEN SPECIES OF HUMAN COLONIC *BACTEROIDES* SPP.

Figure 2.1 posits that gene transfer events occur between different species of colonic bacteria. What is the evidence that such transfers can occur and that if they do occur, they are common? A first attempt to answer this question was made in a 2001 publication by Shoemaker et al. [29]. In this study, two sets of human colonic *Bacteroides* strains were screened. One set had been isolated prior to 1970 and was obtained from the culture collection of the now defunct Virginia Polytechnic Institute Anaerobe Laboratory (Blacksburg, Virginia, U.S.A.). The second set included isolates obtained after 1990. The two sets of strains were further divided into clinical and community isolates. The community isolates were derived from healthy people. The clinical isolates were obtained from patients with *Bacteroides* infections. The reason for looking at these two groups separately was that if the reservoir hypothesis is correct, both sets should follow the same pattern of gene acquisition, rather than clinical isolates exhibiting a different ecology as might be expected if events happened primarily in a clinical setting.

The patterns of antibiotic resistance genes seen in the clinical and community isolates were indeed similar. A striking difference was apparent, however, when the pre-1970 and post-1990 strains were compared. The older strains had a much lower rate of carriage of *tetQ* and the *erm* genes than the strains isolated after 1990. So, something had happened in the two-decade period that separated the two sets of strains, a period characterized by extensive use of antibiotics, such as tetracycline and the macrolides [30]. It is also surprising how high the carriage rate was in the isolates obtained prior to 1970, before the onset of intensive use of antibiotics in the treatment of human disease. This type of anomaly has been seen in other cases, such as detection of antibiotic-resistant bacteria in "pristine" environments [31,32]. This raises the question of whether antibiotics are the only force selecting for antibiotic-resistant bacteria, a still-unanswered question to which we will return at the end of this chapter.

The high number of strains in the post-1990 period that carry *tetQ*, even in the community isolates obtained from people who were not taking antibiotics, indicates that once acquired, *tetQ* is maintained very stably. Since, as already indicated, *tetQ* is found almost exclusively on a type of CTn exemplified by CTnDOT, a human *Bacteroides* CTn, this indicates that the CTn itself is also maintained very stably. It is interesting to note another characteristic of CTnDOT: its excision and transfer are stimulated 100- to 1000-fold by exposure of the bacteria to tetracycline [33–35]. Tetracycline is used not only to treat acute human infections, but also in dermatology and agriculture. In the treatment of acne, tetracycline is administered orally in relatively low doses over a period that can extend from months to years [36,37]. In agriculture, tetracycline has been used to stimulate growth of some animals [38]. Thus, long dosage regimens for tetracycline have been widespread and could have been responsible for the increased carriage of *tetQ* between 1970 and 1990.

The *tetQ* gene is not the only gene whose carriage has increased over the past few decades. Carriage of some of the *erm* genes, principally *ermB*, *ermF*, and *ermG*, increased dramatically between the pre-1970 and post-1990 period. A particularly interesting aspect of this increase in carriage by human colonic *Bacteroides* strains is that *ermB* and *ermG* were previously thought to be "Gram-positive" resistance genes, because they were found primarily in Gram-positive bacteria. These genes seem to have entered *Bacteroides* spp. only very recently [29]. Could they be coming in from Gram-positive bacteria? The largest population of bacteria in the human colon is that of the Gram-positive anaerobes, a little studied and poorly understood group of bacteria [39,40]. Similarly, Gram-positive bacteria are the predominant population of bacteria in the human mouth and in the intestines of farm animals [41,42].

CHARTING THE MOVEMENT OF RESISTANCE GENES INTO *BACTEROIDES* SPP.

Given that the *ermB* and *ermG* genes had been found previously exclusively in the Gram-positive bacteria, is it possible that these genes were obtained from Gram-positive bacteria? Recently, it became possible to ask this question, because a CTn that carries *ermB*, CTnBST, was found in *Bacteroides* spp: It has been sequenced. The results of this analysis are both revealing and confusing [43]. We had hoped that the answer would be a simple one, that is, that a single CTn of Gram-positive origin would be revealed as having moved into *Bacteroides*. What we found was that the *ermB* gene was carried on a segment of DNA that is at least 7 kbp in size, and has integrated into a CTn that has been found previously in *Bacteroides fragilis*. The CTn is now clearly a chimera of Gram-positive and *Bacteroides*-like DNA. The *Bacteroides*-like DNA may not be from *Bacteroides* after all, however, because the percentage G+C content of the CTn outside the *ermB* region is higher than the percentage G+C content of *Bacteroides* spp. The chimeric nature of CTns and plasmids is becoming an old story. Recently, *tetM*, a Gram-positive tetracycline resistance gene, has been found in *Escherichia coli* [44]. Whether this gene is on a transmissible element remains to be seen. In the Gram-positive bacteria, *tetM* is usually found on CTns.

Some of the same resistance genes seen in human oral and colonic bacteria are also found in animal feces. The *ermB* and *tetQ* genes are examples of this. The *tetQ* gene was reported in a bacterium isolated from the rumen of cattle in the 1990s. This gene was not on a CTn but on a plasmid. Nonetheless, its DNA sequence was more than 95% identical to the sequences of the *tetQ* genes we were finding in human colonic bacteria [10]. More recently, the *ermB* gene has been found in isolates, mostly Gram positive, from a below-barn pig manure collection tank.

The overwhelming majority of reports of antibiotic resistance genes in bacteria isolated from animals and humans have focused on such foodborne pathogens as *Salmonella* and *Campylobacter*. Since these pathogens can colonize humans as well as animals, it is perhaps not surprising that they would move as resistant strains between human and animal reservoirs. More surprising is the apparent movement of genes, such as *tetQ* and *ermB* between members of the normal microflora of humans and animals, populations of bacteria that differ in species composition [10,29]. In these cases, it is almost certainly the genes that are moving rather than just the bacterial strains.

THE TRICLOSAN QUANDRY

A cause for concern is the widespread use of antimicrobial agents in products ranging from soaps to cutting boards. The story of triclosan is a good example of marketing gone wild. Triclosan is an antibacterial compound that has been added for years to plastic products to maintain the integrity of the products. One day, some marketing genius realized that by adding the label "antibacterial" to the product, the product suddenly gained added value in the public eye. Soon, triclosan was being added to soaps, toothpaste, and mouthwash, among other products.

Initially, triclosan was thought to be a disinfectant, but it has since been found to have a specific mode of action. It inhibits fatty acid biosynthesis. In 1998, Stuart Levy and co-workers first showed that *E. coli* strains resistant to triclosan could be isolated and that these strains had a specific defect in fatty acid synthesis [45]. Since then, many studies of the mechanism of triclosan action have been published, but the question that is still hanging fire is the question of how widespread triclosan use might affect the distribution of antibiotic-resistant strains. Fortunately, obtaining approval to use other antibacterial compounds in personal products is not easy, so there may be time to evaluate the impact of triclosan before decisions on newer antibacterials are made. How best to evaluate the impact of triclosan? The most obvious approach is to assess the ease with which triclosan-resistant mutants are selected, but this is not the critical question. The critical question is whether triclosan use could cross-select for strains resistant to other antibiotics. This question remains to be answered.

Whatever the impact of triclosan use on antibiotic resistance patterns, the sudden popularity of "antibacterial" products is a cautionary tale. Public health officials were unprepared for the sudden advent of such products, and it remains unclear what the appropriate response to such changes in public consumption patterns is and how best the implications of such usage changes can be evaluated for safety.

THE ECOLOGY OF THE FUTURE

Although the ecology of antibiotic resistance genes is still a relatively new area, some problems and challenges are evident. First, very few systematic studies of the distribution and movement of resistance genes in nature have been done. Comparisons of the incidence of resistant strains in farms that do or do not use antibiotics are misleading if variables such as the proximity of water supplies that might be contaminated with antibiotics or the movement of wild birds and rodents between the farms are not taken into account. The finding of significant concentrations of antibiotics in some water sources has demonstrated what should have been obvious all along: antibiotics do not necessarily stay in the location where they are used. Antibiotics used in the hospital or in agriculture can appear later in water released from sewage treatment plants [46,47]. Or water recovered from animal manure and used to irrigate vegetable crops can spread antibiotics to locations where antibiotics are not being used intentionally [48–50].

An unanswered question is how widely distributed antibiotic resistance genes are in nature outside the human body. Our finding that even in strains isolated from humans prior to 1970, the *tetQ* gene was already present in nearly one-third of *Bacteroides* strains is perplexing since tetracycline use only became widespread in the 1960s and 1970s. Is it possible that there are non-antibiotic selections for antibiotic-resistant strains? Production of antibiotics by antibiotic-producing bacteria is very low in natural settings, but plant compounds that mimic antibiotics may be more abundant. Also, it is important to keep in mind that resistance genes are often linked on the same element. Integrons are an excellent example of this phenomenon. If a set of genes in an operon includes, for example, a cadmium resistance gene as well as several antibiotic resistance genes, cadmium may select for maintenance of the antibiotic resistance genes. In the case of the *Bacteroides* CTns in our studies of human colonic bacteria, the CTnDOT type element contained both a tetracycline resistance gene, *tetQ*, and a macrolide resistance gene, *ermF*, so that selection for either resistance gene tends to select for maintenance of the other type [51,52].

Relatively few studies have been done to evaluate the distribution of antibiotic resistance patterns in environmental bacteria, especially bacteria in sites outside of farms or areas of human settlement. It would be informative to conduct a study similar to those often done by marine microbiologists, in which sites around the perimeter of an island that differ in the amount of human pollution are sampled and evaluated. Unfortunately, none of the major funding agencies regards this type of survey as part of its mission. Thus, the question of whether there is such a thing as a truly pristine site, free of antibiotic-resistant bacteria, remains unanswered. Also, surprisingly, the question of the extent to which animal or human pollution affects the incidence of antibiotic resistance genes is also unanswered.

Clearly, systematic surveys of antibiotic resistance gene distribution are needed, and ideally surveys should be guided by the principles developed by environmental microbiologists who have had long experience in ecology. An interesting approach to this type of analysis has been taken by Randall Singer and his associates. The approach is called landscape ecology [53]. It is a form of mathematical modeling that assesses correlations between antibiotic use patterns and the incidence of resistant strains. Proving association is not the same as proving cause and effect, but the fact that scientists are beginning to explore mathematical modeling of resistance patterns as a means of seeking possible cause-and-effect connections is encouraging.

REFERENCES

1. Adkin, D. A., S. S. Davis, R. A. Sparrow, P. D. Huckle, A. J. Phillips, and I. R. Wilding. 1995. The effects of pharmaceutical excipients on small intestinal transit. *Br J Clin Pharmacol* 39:381–7.
2. Akinbowale, O. L., H. Peng, and M. D. Barton. 2006. Antimicrobial resistance in bacteria isolated from aquaculture sources in Australia. *J Appl Microbiol* 100:1103–13.
3. Aubert, D., T. Naas, C. Heritier, L. Poirel, and P. Nordmann. 2006. Functional characterization of IS1999, an IS4 family element involved in mobilization and expression of beta-lactam resistance genes. *J Bacteriol* 188:6506–14.
4. Baldwin, H. E. 2006. Oral therapy for rosacea. *J Drugs Dermatol* 5:16–21.
5. Balis, E., A. C. Vatopoulos, M. Kanelopoulou, E. Mainas, G. Hatzoudis, V. Kontogianni, H. Malamou-Lada, S. Kitsou-Kiriakopoulou, and V. Kalapothaki. 1996. Indications of *in vivo* transfer of an epidemic R plasmid from *Salmonella enteritidis* to *Escherichia coli* of the normal human gut flora. *J Clin Microbiol* 34:977–9.
6. Barbosa, T. M. and S. B. Levy. 2000. The impact of antibiotic use on resistance development and persistence. *Drug Resist Updat* 3:303–11.
7. Bryan, A., N. Shapir, and M. J. Sadowsky. 2004. Frequency and distribution of tetracycline resistance genes in genetically diverse, nonselected, and nonclinical *Escherichia coli* strains isolated from diverse human and animal sources. *Appl Environ Microbiol* 70:2503–7.
8. Bucking, C. and C. M. Wood. 2006. Water dynamics in the digestive tract of the freshwater rainbow trout during the processing of a single meal. *J Exp Biol* 209:1883–93.

9. Burrus, V., J. Marrero, and M. K. Waldor. 2006. The current ICE age: biology and evolution of SXT-related integrating conjugative elements. *Plasmid* 55:173–83.

10. Casadevall, A. and L. A. Pirofski. 2000. Host-pathogen interactions: basic concepts of microbial commensalism, colonization, infection, and disease. *Infect Immun* 68:6511–8.

11. Chee-Sanford, J. C., R. I. Aminov, I. J. Krapac, N. Garrigues-Jeanjean, and R. I. Mackie. 2001. Occurrence and diversity of tetracycline resistance genes in lagoons and groundwater underlying two swine production facilities. *Appl Environ Microbiol* 67:1494–502.

12. Cheng, Q., B. J. Paszkiet, N. B, Shoemaker, J. F. Gardner, and A. A. Salyers. 2000. Integration and excision of a *Bacteroides* conjugative transposon, CTnDOT. *J Bacteriol* 182:4035–43.

13. Courvalin, P. and C. Carlier. 1987. Tn1545: a conjugative shuttle transposon. *Mol Gen Genet* 206:259–64.

14. Courvalin, P. and C. Carlier. 1986. Transposable multiple antibiotic resistance in *Streptococcus pneumoniae*. *Mol Gen Genet* 205:291–7.

15. Dargatz, D. A. and J. L. Traub-Dargatz. 2004. Multidrug-resistant *Salmonella* and nosocomial infections. *Vet Clin North Am Equine Pract* 20:587–600.

16. Davies, J. 1994. Inactivation of antibiotics and the dissemination of resistance genes. *Science* 264:375–82.

17. Eckburg, P. B., E. M. Bik, C. N. Bernstein, E. Purdom, L. Dethlefsen, M. Sargent, S. R. Gill, K. E. Nelson, and D. A. Relman. 2005. Diversity of the human intestinal microbial flora. *Science* 308:1635–8.

18. Edlund, C. and C. E. Nord. 1991. A model of bacterial-antimicrobial interactions: the case of oropharyngeal and gastrointestinal microflora. *J Chemother* 3 Suppl 1:196–200.

19. Grillot-Courvalin, C., S. Goussard, F. Huetz, D. M. Ojcius, and P. Courvalin. 1998. Functional gene transfer from intracellular bacteria to mammalian cells. *Nat Biotechnol* 16:862–6.

20. Iwanaga, M., C. Toma, T. Miyazato, S. Insisiengmay, N. Nakasone, and M. Ehara. 2004. Antibiotic resistance conferred by a class I integron and SXT constin in *Vibrio cholerae* O1 strains isolated in Laos. *Antimicrob Agents Chemother* 48:2364–9.

21. Jensen, L. B., Y. Agerso, and G. Sengelov. 2002. Presence of erm genes among macrolide-resistant Gram-positive bacteria isolated from Danish farm soil. *Environ Int* 28:487–91.

22. Kang, J. G., S. H. Kim, and T. Y. Ahn. 2006. Bacterial diversity in the human saliva from different ages. *J Microbiol* 44:572–6.

23. Klare, I., C. Konstabel, D. Badstubner, G. Werner, and W. Witte. 2003. Occurrence and spread of antibiotic resistances in *Enterococcus faecium*. *Int J Food Microbiol* 88:269–90.

24. Kruse, H. and H. Sorum. 1994. Transfer of multiple drug resistance plasmids between bacteria of diverse origins in natural microenvironments. *Appl Environ Microbiol* 60:4015–21.

25. Kummerer, K. 2001. Drugs in the environment: emission of drugs, diagnostic aids and disinfectants into wastewater by hospitals in relation to other sources—a review. *Chemosphere* 45:957–69.

26. Kuroda, M., T. Ohta, I. Uchiyama, T. Baba, H. Yuzawa, I. Kobayashi, L. Cui, A. Oguchi, K. Aoki, Y. Nagai, J. Lian, T. Ito, M. Kanamori, H. Matsumaru, A. Maruyama, H. Murakami, A. Hosoyama, Y. Mizutani-Ui, N. K. Takahashi, T. Sawano, R. Inoue, C. Kaito, K. Sekimizu, H. Hirakawa, S. Kuhara, S. Goto, J. Yabuzaki, M. Kanehisa, A. Yamashita, K. Oshima, K. Furuya, C. Yoshino, T. Shiba, M. Hattori, N. Ogasawara, H. Hayashi, and K. Hiramatsu. 2001. Whole genome sequencing of meticillin-resistant *Staphylococcus aureus*. *Lancet* 357:1225–40.

27. Li, L. Y., N. B. Shoemaker, and A. A. Salyers. 1993. Characterization of the mobilization region of a *Bacteroides* insertion element (NBU1) that is excised and transferred by *Bacteroides* conjugative transposons. *J Bacteriol* 175:6588–98.

28. Luna, V. A., D. B. Jernigan, A. Tice, J. D. Kellner, and M. C. Roberts. 2000. A novel multiresistant *Streptococcus pneumoniae* serogroup 19 clone from Washington State identified by pulsed-field gel electrophoresis and restriction fragment length patterns. *J Clin Microbiol* 38:1575–80.

29. Mackie, R. I., S. Koike, I. Krapac, J. Chee-Sanford, S. Maxwell, and R. I. Aminov. 2006. Tetracycline residues and tetracycline resistance genes in groundwater impacted by swine production facilities. *Anim Biotechnol* 17:157–76.

30. Martel, A., A. Decostere, E. D. Leener, M. Marien, E. D. Graef, M. Heyndrickx, H. Goossens, C. Lammens, L. A. Devriese, and F. Haesebrouck. 2005. Comparison and transferability of the erm (B) genes between human and farm animal streptococci. *Microb Drug Resist* 11:295–302.

31. McMurry, L. M., M. Oethinger, and S. B. Levy. 1998. Triclosan targets lipid synthesis. *Nature* 394:531–2.

32. Moon, K., N. B. Shoemaker, J. F. Gardner, and A. A. Salyers. 2005. Regulation of excision genes of the *Bacteroides* conjugative transposon CTnDOT. *J Bacteriol* 187:5732–41.

33. Musher, D. M., M. E. Dowell, V. D. Shortridge, R. K. Flamm, J. H. Jorgensen, P. Le Magueres, and K. L. Krause. 2002. Emergence of macrolide resistance during treatment of pneumococcal pneumonia. *N Engl J Med* 346:630–1.

34. Nikolich, M. P., G. Hong, N. B. Shoemaker, and A. A. Salyers. 1994. Evidence for natural horizontal transfer of tetQ between bacteria that normally colonize humans and bacteria that normally colonize livestock. *Appl Environ Microbiol* 60:3255–60.

35. Patrick, D. M., F. Marra, J. Hutchinson, D. L. Monnet, H. Ng, and W. R. Bowie. 2004. Per capita antibiotic consumption: how does a North American jurisdiction compare with Europe? *Clin Infect Dis* 39:11–7.

36. Pruden, A., R. Pei, H. Storteboom, and K. H. Carlson. 2006. Antibiotic resistance genes as emerging contaminants: studies in northern Colorado. *Environ Sci Technol* 40:7445–50.

37. Schmidt, A. S., M. S. Bruun, I. Dalsgaard, and J. L. Larsen. 2001. Incidence, distribution, and spread of tetracycline resistance determinants and integron-associated antibiotic resistance genes among motile aeromonads from a fish farming environment. *Appl Environ Microbiol* 67:5675–82.

38. Shoemaker, N. B., C. Getty, E. P. Guthrie, and A. A. Salyers. 1986. Regions in *Bacteroides* plasmids pBFTM10 and pB8-51 that allow *Escherichia coli-Bacteroides* shuttle vectors to be mobilized by IncP plasmids and by a conjugative *Bacteroides* tetracycline resistance element. *J Bacteriol* 166:959–65.

39. Shoemaker, N. B., H. Vlamakis, K. Hayes, and A. A. Salyers. 2001. Evidence for extensive resistance gene transfer among *Bacteroides* spp. and among *Bacteroides* and other genera in the human colon. *Appl Environ Microbiol* 67:561–8.

40. Singer, R. S., M. P. Ward, and G. Maldonado. 2006. Can landscape ecology untangle the complexity of antibiotic resistance? *Nat Rev Microbiol* 4:943–52.

41. Su, L. H., H. L. Chen, J. H. Chia, S. Y. Liu, C. Chu, T. L. Wu, and C. H. Chiu. 2006. Distribution of a transposon-like element carrying bla(CMY-2) among *Salmonella* and other *Enterobacteriaceae*. *J Antimicrob Chemother* 57:424–9.

42. Tan, A. W. and H. H. Tan. 2005. Acne vulgaris: a review of antibiotic therapy. *Expert Opin Pharmacother* 6:409–18.

43. Schlesinger, D. J., N. B. Shoemaker, and A. A. Salyers. 2007. Possible origins of CTnBST, a conjugative transposon found recently in a human colonic *Bacteroides* strain. *Appl Environ Microbiol* 73:4226–33.

44. Tancrede, C. 1992. Role of human microflora in health and disease. *Eur J Clin Microbiol Infect Dis* 11:1012–5.

45. Trieu-Cuot, P., M. Arthur, and P. Courvalin. 1987. Origin, evolution and dissemination of antibiotic resistance genes. *Microbiol Sci* 4:263–6.

46. Ulrich, A. and A. Puhler. 1994. The new class II transposon Tn163 is plasmid-borne in two unrelated *Rhizobium leguminosarum* biovar viciae strains. *Mol Gen Genet* 242:505–16.

47. Valentine, P. J., N. B. Shoemaker, and A. A. Salyers. 1988. Mobilization of *Bacteroides* plasmids by *Bacteroides* conjugal elements. *J Bacteriol* 170:1319–24.

48. Valenzuela, J. K., L. Thomas, S. R. Partridge, T. van der Reijden, L. Dijkshoorn, and J. Iredell. 2007. Horizontal gene transfer in a polyclonal outbreak of carbapenem-resistant *Acinetobacter baumannii*. *J Clin Microbiol* 45:453–60.

49. Waldor, M. K., H. Tschape, and J. J. Mekalanos. 1996. A new type of conjugative transposon encodes resistance to sulfamethoxazole, trimethoprim, and streptomycin in *Vibrio cholerae* O139. *J Bacteriol* 178:4157–65.

50. Wang, Y., E. R. Rotman, N. B. Shoemaker, and A. A. Salyers. 2005. Translational control of tetracycline resistance and conjugation in the *Bacteroides* conjugative transposon CTnDOT. *J Bacteriol* 187:2673–80.

51. Werner, G., B. Hildebrandt, and W. Witte. 2003. Linkage of erm(B) and aadE-sat4-aphA-3 in multiple-resistant *Enterococcus faecium* isolates of different ecological origins. *Microb Drug Resist* 9 Suppl 1:S9–16.

52. Whitford, M. F., R. J. Forster, C. E. Beard, J. Gong, and R. M. Teather. 1998. Phylogenetic analysis of rumen bacteria by comparative sequence analysis of cloned 16S rRNA genes(ss). *Anaerobe* 4:153–63.

53. Wilson, K. H., J. S. Ikeda, and R. B. Blitchington. 1997. Phylogenetic placement of community members of human colonic biota. *Clin Infect Dis* 25 Suppl 2:S114–6.

3 Global Response Systems That Confer Resistance

Paul F. Miller and Philip N. Rather

CONTENTS

The majority of attention on antibiotic resistance mechanisms has been justifiably focused on those factors that are highly transmissible among species and that lead to high levels of resistance to a specific class of antibiotics. Less is known about the ability of bacteria to alter their susceptibility to noxious agents by modulating their own intrinsic physiological systems. In this chapter, we describe two of the better-studied examples of this latter situation, both of which occur in Gram-negative species.

In the first example, the *mar/sox* regulatory network found in *Escherichia coli* is described. This system acts to modulate factors that limit the accumulation of a wide range of noxious agents, including several clinically important antibiotics. As such, we discuss a network of sensory and regulatory factors that operate to control the expression of genes whose products either actively extrude antibiotics or enhance the effectiveness of external permeability barriers. Because Chapter 4 specifically addresses efflux pumps, our discussion focuses on the structure and function of the *marRAB* and *soxRS* regulatory loci. We review evidence describing the high degree of molecular redundancy shared by these two regulatory systems, leading toward the concept that these are two semi-independent sensory systems that control a nearly identical set of target genes, although in quantitatively different ways. These differences may reflect the distinct types of signals that are sensed by the two systems, such that a protective response to inducers of one (e.g., superoxide generating compounds for *soxRS*) may require a slightly different gene expression pattern than would the response to inducers of the second (phenolic agents and antibiotics for *marRAB*).

In the second example, we describe the regulatory mechanisms controlling the *aac(2')-Ia* gene in *Providencia stuartii*. The *aac(2')-Ia* gene is a member of a growing family of chromosomally encoded aminoglycoside acetyltransferases that are intrinsic to certain bacterial species. Although the role of these acetyltransferases is largely unknown, the AAC(2')-Ia enzyme in *P. stuartii* functions as a peptidoglycan *O*-acetyltransferase. Given the possibility of diverse functions for these enzymes, we anticipate that the regulation of these genes will involve distinct mechanisms. However, the information on *aac(2')-Ia* expression that has been compiled to date may serve as a useful preliminary model for other systems.

INTRODUCTION

Microorganisms live in intimate proximity to their environment. For free-living species, this situation equates to the constant threat of exposure to a wide variety of potentially toxic agents produced either deliberately (e.g., by other organisms for defense against microbial encroachment) or as a consequence of normal organic turnover. Similarly, commensal and pathogenic organisms must protect themselves from both specific and non-specific agents elicited by the host. Not surprisingly, then, unicellular species have evolved an elaborate array of defenses designed to reduce or prevent the accumulation of unwanted toxic substances. There is, for example, a remarkable inventory of efflux systems that can be identified in the genomes of almost all bacteria. The mechanisms by which efflux pumps operate are discussed in Chapter 4 of this volume.

   With such a genetic investment in defense systems, it also makes sense that these organisms would possess similarly intricate regulatory mechanisms, which allow them to control the deployment of these systems. In this chapter, we highlight our understanding of a few of the better-characterized regulatory systems, including global resistance systems and intrinsic modifying enzymes. Although the systems described in this chapter have been studied primarily in *E. coli* and *P. stuartii*, it is reasonable to expect that these systems will serve as formal paradigms for as yet undiscovered control networks in other bacterial species.

GLOBAL REGULATORS OF ANTIBIOTIC RESISTANCE IN *ESCHERICHIA COLI*

THE *MAR* REGULATORY LOCUS

Undoubtedly, the best-characterized global antibiotic resistance regulatory system is the *mar* (multiple antibiotic resistance) system in *E. coli*. An excellent review of the molecular genetics of this system has been published [1]. Much of the detailed work described in that review is only summarized here, and the reader is encouraged to look to that source for additional detailed information. The *mar* locus was first described in 1983 in the pioneering studies of George and Levy. As a component of an ongoing effort to understand the mechanisms contributing to tetracycline resistance, these investigators identified a locus on the *E. coli* chromosome that was associated with the frequent emergence of low-level resistant strains [2]. Moreover, it was shown that these tetracycline-resistant (tet^r) strains had also acquired a concomitant resistance to other structurally unrelated antibiotics including chloramphenicol, rifampicin, and fluoroquinolones [2]; mechanistically this phenotype was associated with reduced accumulation and efflux of the affected agents [2–4]. The substrate spectrum for this system was later expanded to include certain organic solvents and disinfectants [5,6]. A Tn5 insertion at the 34 min region of the chromosome reversed the resistance phenotype for all of these agents, and identified the genetic locus, which was designated as *mar* [7]. DNA sequence analysis of cloned genetic segments that could complement the Mar phenotype associated with either the Tn5 insertion or a larger chromosomal deletion encompassing this region revealed a three-gene regulatory operon, designated *marRAB* [8–11]. The Tn5 insertion originally isolated by George and Levy was located in the second gene, *marA*. Overexpression of this gene by itself was shown to be sufficient to confer the Mar phenotype in all cell types, including strains deleted for this region of the chromosome [12]. The deduced protein product of this gene, MarA, is related by amino acid sequence similarity to a family of transcriptional activators, the prototype for which is the AraC regulator that controls genes involved in the metabolism of arabinose [13]. This observation suggested that the Mar phenotype resulting from a mutation at the *mar* locus was likely due to an indirect mechanism, with MarA serving to control the expression of genes located elsewhere on the chromosome. It is presumably these target genes that are the more direct effectors of antibiotic resistance.

   If overexpression of *marA* is sufficient to confer a Mar phenotype, then the *mar* locus must be capable of controlling the expression of *marA*. This proved to be the case, and the first gene in the operon, *marR*, was determined to play a critical role in this process [9]. Unlike MarA, the MarR protein, at the time of its sequencing, bore little similarity to any known genes. However, analysis of selected Mar isolates showed that the majority of these bore mutations in *marR*, and concomitantly exhibited elevated levels of the *marRAB* transcript [9,11,14]. Introduction of a wild type copy of *marR* in *trans* on a plasmid reversed the Mar phenotype, indicating that the *marR* mutations were recessive, and that this gene encoded a repressor of *marRAB* operon expression. Results of genetic experiments suggested that the target for MarR repression is the operator/promoter region of the *marRAB* operon, *marOP*, as one could titrate the repressing activity of MarR simply by introducing additional copies of *marOP* on a plasmid [9,14]. This finding was confirmed biochemically by showing that purified MarR protein bound specifically to *marOP* DNA sequences [15].

At roughly the same time as the original George and Levy experiments, it was noted that exposing *E. coli* cells to the weak aromatic acid salicylate (SAL) induced a condition of phenotypic antibiotic resistance subsequently referred to as Par [16]. Notably, SAL treatment conferred resistance to the same diverse group of antibiotics as was observed for the *mar* mutants. These findings converged mechanistically when it was found, through the use of a *mar-lacZ* fusion, that SAL treatment led to an induction of *marRAB* expression [17]. Importantly, this was the first observation that connected the *mar* regulatory locus with extracellular stimuli. Deletion of the *marRAB* operon led to a greatly reduced responsiveness to SAL as an inducer of antibiotic resistance, and to a hypersensitivity to many of the same agents that were affected by the original *mar* mutants [11,12,17]. The extent to which this hypersensitivity was observed depended on the specific *E. coli* strain background in use [8,10,11].

The crystal structure of the MarR repressor has been determined at 2.3 Å of resolution by Alekshun and co-workers [18]. The structure reveals MarR as a dimer, with each subunit composed of six helical regions that mediate a protein–protein interface in each monomer. The DNA binding domain consisting of amino acids 61 to 121 adopts a winged helix fold from amino acids 55 to 100. The formation of the MarR crystal required the presence of SAL, a strong inducer that relieves MarR-mediated repression of the Mar regulon. Based on electron density, there appear to be two SAL binding sites, both of which are positioned near the DNA binding helix. The location of these sites is consistent with the ability of SAL to alter the DNA binding properties of MarR by directly interacting with the repressor.

These studies suggested that the following hierarchy could explain inducible antibiotic resistance mediated by the *marRAB* system. The *mar* locus is normally maintained in a quiescent state due to the autorepressor activity of the *marR* gene product. Exposure to a specific inducer such as SAL leads to the binding of the inducer by MarR, antagonizing its ability to mediate transcriptional repression of the *marRAB* operon. This results in an increase in transcription of the *marRAB* genes, leading to an increase in the abundance of the products of these genes in the cell. MarA, the proximal activator of target genes involved in the antibiotic resistance response, thus becomes available in sufficient quantities to diffuse to other sites on the chromosome and activate its target genes. A more detailed discussion of the targets and inducers in the *mar* regulatory network is provided below.

THE *soxRS* SYSTEM

Exposure of *E. coli* cells to various redox cycling agents, such as paraquat, leads to the induction of a number of genes that collectively constitute the superoxide stress response [19]. Constitutive mutants have been selected in which the expression of these target genes is elevated in the absence of any inducing agent. Such regulatory mutants typically map to the *soxR* locus, located at 92 min on the *E. coli* chromosome [20]. Notably, these constitutive regulatory mutants also exhibit a concomitant antibiotic resistance phenotype, which is remarkably similar to that observed with *mar* strains. In addition, one such regulatory mutant with a very similar phenotype, known as *soxQ1*, mapped to the *marA* locus [21].

Molecular dissection of the *soxR* locus revealed two divergently transcribed regulatory genes, *soxR* and *soxS*. The constitutive *sox* mutants mapped to *soxR* and have been referred to as *soxR*(Con) alleles, to distinguish them from non-functional mutants. Gene expression studies showed that the expression of *soxR* is unaffected by either superoxide generating agents or the constitutively activating mutations [22,23]. In contrast, expression of *soxS* is induced by redox cycling agents, such as menadione or paraquat, as well as by *soxR*(Con) mutants, and an intact *soxR* gene is required for induction of *soxS* expression as well as that of superoxide stress response target genes [22,23]. Similar to findings described above for *marA*, overexpression of *soxS* was shown to be sufficient to activate the expression of superoxide stress response target genes as well as confer the antibiotic resistance phenotype [22,23]. These findings, combined with the recognition that the SoxR protein contains iron–sulfur clusters in its C-terminal region that are characteristic of those involved with redox reactions, suggested that SoxR activity (and not expression) may be modulated in response to superoxide radicals, and led to a better molecular understanding of the two-stage model for control of this regulon [24,25]. In this model, exposure to agents or conditions leading to an accumulation of superoxide radicals results in the conversion of inactive SoxR to an activated form. Activated SoxR then induces the transcription of the adjacent *soxS* gene, whose product stimulates the expression of the unlinked regulon genes, the products of which presumably engender resistance to superoxide radical-generating agents and Mar-type antibiotics. Constitutive *soxR* mutants appear to be permanently in an activated conformation, which may explain why in these strains regulon genes are expressed even in the absence of a small molecule activator.

Additional observations tied the *soxRS* regulon to the *mar* system. Along with the observations that *soxR*(Con) mutants have a Mar phenotype, and that the *soxQ1* mutant mapped near *marA*, another mutant that was initially selected based on its strong Mar phenotype was found to map to the *soxR* locus [26]. Reconciliation of these genetic observations began when it was recognized that MarA and SoxS, the proximal activators in these regulatory systems, are closely related members of the AraC family of transcription factors [13]. Thus, overexpression of either *soxS* or *marA* leads to a Mar phenotype as well as induction of the superoxide stress response target genes. However, these regulators do not behave in completely redundant ways, as there appear to be quantitative differences in the effects of these activators on the different target genes that have been studied to date. For example, *marA* overexpression tends to produce a greater level of antibiotic resistance and a smaller induction of superoxide stress response target genes, such as *nfo* (encodes endonuclease IV), than does *soxS* [21,26].

Studies of clinical isolates have verified the role of *soxRS* in resistance. In *E. coli,* fluoroquinolone-resistant clinical isolates exhibited mutations in *soxR* and *soxS* that resulted in higher levels of *soxS* expression and activation of downstream genes required for resistance [27–29]. In *Salmonella enterica* (serovar typhimurium), a quinolone-resistant isolate arose during treatment that contained a single point mutation in *soxR*. This substitution rendered SoxR constitutively active and increased expression of SoxS-dependent genes [30].

Rob—A Third Regulator?

E. coli contains another gene whose product exhibits significant amino acid sequence similarity to MarA and SoxS. This protein, known as Rob, was first identified as a factor that binds to the chromosomal origin of replication [31]. It is larger than either MarA or SoxS, and appears to contain an additional domain not found in the other two proteins. It is also different in that it is constitutively expressed at high levels, increasing in concentration as cells transition from logarithmic to stationary phase. Although higher-level induction of recombinant Rob accumulation has been shown to confer a Mar phenotype, and purified Rob protein has been shown to bind to MarA/SoxS target promoters *in vitro* [32,33], a physiological role for this protein in antibiotic resistance has yet to be demonstrated. In addition, mutants affecting intrinsic antibiotic resistance have yet to be linked to the *rob* gene. For these reasons, this interesting and mysterious protein will not be described further here.

A Single Regulon with Two Activators

As has been proposed recently, it now seems reasonable to consider the existence of a single stress response regulon that is controlled by multiple related regulators [34]. This could be called the *mar* regulon, as has been proposed, or be referred to by a more general descriptor to reflect the distinct stresses that led to its activation. Regardless, the important consequence from the perspective of this review is that intrinsic antibiotic resistance is affected. We shall now consider more distal and proximal components of this pathway.

Regulon Targets and Antibiotic Resistance

Recent work has led to a greater understanding of the target binding site in MarA and SoxS responsive promoters [34–36]. Work with MarA has suggested that this activator interacts with target promoters as a monomeric protein, and that it can bind in either of two orientations to effect transcription. However, the orientation of the binding site in a given promoter must be as it originally exists in that element; inverting it leads to a loss of MarA responsiveness. In addition, distinct spacing rules appear to exist regarding the distance between the "marbox" and the binding sites for RNA polymerase (RNAP), depending on whether the marbox is present in the + or − orientation. Marboxes that are located on the opposite strand from that of the

RNAP binding sites (-35 and -10 sequences) are positioned further upstream than are those that are found on the same strand as the RNP binding site [34,37,38]. It has been proposed that these positions and orientations allow MarA to interact productively with RNAP in either orientation.

Marboxes that have been found upstream from a number of target promoters in *E. coli* have been aligned to generate a consensus binding site [34]. Despite significant experimental work, this consensus remains quite degenerate. From the crystal structure studies of MarA, it has been proposed that MarA interacts with specific promoter elements by way of an interaction of complementary shapes that are held together by Van der Waals forces [39]. Whether the interaction of MarA with a marbox results in activation or repression of transcription appears to be related to the relative position and orientation of the marbox within a promoter element [40]. By inference, it seems reasonable to expect that many of the mechanistic observations made for MarA will also be applicable to SoxS. This is supported by biochemical studies conducted with this latter protein, and its interaction with known target genes [35,36,38,41]. Thus, several of the genes containing marbox elements in their promoters have been implicated by both genetic and biochemical methods as specific targets for MarA and/or SoxS control. Because of the focus of this volume, those key target genes implicated in antibiotic resistance are discussed in further detail here.

micF

One of the earliest physiological observations associated with the Mar phenotype was a down regulation of the major outer membrane porin OmpF [42]; this effect has also been observed following SAL treatment [43]. This porin forms a large outer membrane channel through which low-molecular-weight, water-soluble compounds can diffuse. Thus, a reduction in the abundance of this channel in *mar* mutants fits well with the reduced antibiotic accumulation phenotype observed with these strains. Studies of OmpF regulation revealed that one form of negative control involved a post-transcriptional repression mechanism mediated by the anti-sense RNA *micF* [44,45]. Experiments with *micF-lacZ* fusions as well as *micF* deletions demonstrated that *mar* mutants have elevated levels of *micF* expression, and that *mar*-mediated down regulation of OmpF requires an intact *micF* gene [12,46]. However, using strains deleted for the *ompF* gene, it was also shown that a simple loss of OmpF from the outer membrane was not sufficient to confer a Mar phenotype [12]. Thus, additional *marA* targets appeared to be required for a full Mar phenotype.

acrAB and tolC

Accumulating experimental evidence on the structure and function of efflux pumps in Gram-negative organisms [47] suggested that one of these export systems might play a role in *mar*-mediated antibiotic resistance. Subsequent genetic studies then showed that the multidrug efflux pump encoded by the *acrAB* genes is required for the Mar phenotype, as a deletion of *acrAB* completely eliminated the Mar phenotype associated with *mar* mutants [48]. Subsequently, it was noted that the promoter for the *acrAB* operon, as well as that of the *tolC* gene, whose product forms the outer membrane channel component of the AcrAB pump, contains a marbox element [34,49], which is bound by both MarA and SoxS *in vitro*. This strongly suggests that the products of *acrAB* and *tolC*, which act in concert to increase antibiotic efflux, are both controlled by MarA.

marRAB

The promoter for the *marRAB* operon also contains a marbox element and is subject to autoactivation [50]. This observation helped rationalize earlier studies, which showed that high-level expression of either *soxS* or *marA* led to increased *marRAB* operon expression. The marbox in the *marRAB* promoter region is one of the most MarA-responsive elements studied to date [34]. Moreover, *marRAB* operon expression is subject to both transcriptional and translational regulation [51].

As mentioned above, the SoxS protein is expected to bind to virtually the same set of target gene promoters as MarA. This has been largely substantiated experimentally, and in many cases a SoxS interaction was demonstrated first [41]. If this is true, then the explanation for the different effects of *marA* versus *soxS* induction on multiple antibiotic resistance, or the superoxide stress response, must lie in the quantitative ways in which these two regulators interact with their target promoters. This hypothesis is supported by recent evidence [52]. The marbox elements in different regulon promoters respond differently to MarA or SoxS induction. This difference was shown to be due to specific nucleotide sequence differences among the various marbox elements, and it was possible to vary the responsiveness of a promoter to MarA compared with SoxS by changing the sequence of a specific marbox [52]. These findings may also provide an explanation for a perplexing observation associated with certain bases in the proposed consensus sequences. Some of the invariant positions in the consensus have nonetheless been shown to be dispensable for MarA responsiveness. While one can consider it reasonable to propose that MarA and SoxS control an almost identical set of target genes (although in quantitatively different ways), it seems possible that these positions may be more important for SoxS binding than they are for MarA.

MECHANISMS OF REGULON INDUCTION AND PHYSIOLOGICAL ROLES

While much work has focused on the mechanisms by which MarA and SoxS interact with regulon target promoters, early studies were actually driven by physiological observations that gave insights into regulon induction. For the *mar* system, this work centered on the phenolic compound salicylate and its ability to stimulate *marRAB* expression [17]. As mentioned earlier, *marRAB* induction involves antagonism of the MarR repressor, apparently by a direct interaction with SAL [15,18]. The poor solubility of MarR in a purified form along with the relatively weak affinity of SAL for MarR has made biochemical characterization of this interaction difficult. In contrast, *soxRS* induction by superoxide inducing agents is somewhat better understood. Genetic and biochemical experiments demonstrated that superoxide radicals activate *soxS* transcription via their effects on SoxR [24,25]. As stated earlier, SoxR activation involves a cluster of iron–sulfur centers near the 3′ end of the protein, suggesting that a direct activation mechanism may be involved.

Studies of global regulation in *E. coli* using microarrays or macroarrays have revealed that MarA controls the expression of at least 60 genes [53] and SoxS controls at least 95 genes [54]. Array analysis of gene expression has confirmed common targets of both MarA and SoxS, such as *zwf* (encodes glucose 6-phosphate dehydrogenase), *fumC* (fumarase), *acrA*, *inaA* (pH-inducible protein involved in stress response), and *sodA* (superoxide dismutase) and also identified *acnA* (aconitase), *ribA* (GTP cyclohydrolase) and *nfsA* (nitroreductase) as new targets for both activators [53,54]. Additional genes activated by both paraquat (SoxS inducer) and sodium salicylate (MarA inducer) include: *artP* (arginine transport protein, ATP-binding), *cysK* (cysteine synthase A), *dps* (DNA protection protein induced during starvation), *deoB* (phosphopentomutase), and b1452 of unknown function [54]. Studies by Martin and Rosner indicate that the total number of genes directly activated by MarA and SoxS is less than 40 [55]. While MarA and SoxS affect the expression of what appears to be a common regulon, their impact on the expression of individual target genes is clearly not identical [56].

In a potentially intriguing connection, the *mar* and *sox* regulatory system may be linked at the sensory level, much in the same way that they share target genes. In a series of preliminary studies, a collection of naturally occurring, plant-derived phenolic compounds were tested for their ability to induce either *marA* or *soxS* expression [57]. It was noted that certain naphthoquinones that were known to induce *soxS* expression were also effective inducers of *marA* transcription [17,58]. These observations led to the proposal that compounds of this sort may be the true inducers (and substrates?) [59], or may be related to the inducers of a progenitor stress response system that has subsequently duplicated and diverged into the present-day *mar/sox* system. This proposal is supported by the finding that MarR is related at the amino acid sequence level to regulatory proteins found in other bacterial species that are known to respond to phenolic compounds [60]. For example, the HpcR repressor that controls the expression of genes involved in the catabolism of homoprotocatechuate (HPC), a plant-derived phenolic compound, is found in free-living *E. coli* C strains and is a member of the MarR family. HpcR-mediated repression of the *hpc* gene cluster is antagonized by HPC. Again, structural studies describing the specific interactions between MarR and the inducers that it binds will lead to a better understanding of the kinds of compounds that induce the *mar* system and will help shed light on this question.

The highly overlapping *mar* and *sox* systems represent intrinsic, inducible stress response networks in *E. coli* and other enteric species [1,61]. The complexity of this regulatory network suggests that more significantly diverged microbes may have also evolved their own strategies to counter these same environmental challenges, even if they lack recognizable *mar* and *sox* homologs. *mar* mutants have been identified among clinical antibiotic-resistant isolates of *E. coli* [62], and a Mar phenotype has been observed among several other Gram-negative quinolone-resistant strains, typically contributing a two- to four-fold decrease in the susceptibility of these isolates [63]. It is also possible that *mar* mutations may emerge as a consequence of agricultural usage of antibiotics whose activity is affected by the *mar* system, as this locus is also present in the notoriously pathogenic *E. coli* strain O157:H7 [64]. Observations of this sort raise the possibility that the role of *mar*-type mechanisms is substantially under-appreciated in considerations of the factors affecting both intrinsic and acquired antibiotic resistance. Thus, the identification and characterization of these systems can only help us in our efforts to predict, avoid, and counteract antibiotic resistance.

ADDITIONAL MAR-LIKE SYSTEMS INVOLVING SMALL AraC/XylS-LIKE ACTIVATORS

In *Klebsiella pneumoniae*, overexpression of the *ramA* gene encoding a transcriptional activator related to MarA/SoxS/Rob activators confers multiple antibiotic resistance [65]. Moreover, introduction of *ramA* on a multi-copy plasmid into *E. coli* also conferred multiple antibiotic resistance. In both *K. pneumoniae* and *E. coli*, RamA overexpression was associated with reduced expression of the OmpF porin. Studies have also suggested that RamA plays a significant role in clinical resistance to fluoroquinolones, where resistant isolates exhibited *ramA* overexpression [66]. In Ram overexpressing isolates, the AcrAB proteins were also overexpressed. More recent studies further support the role of RamA in AcrAB expression where a tigecycline-resistant isolate overexpressed both *ramA* and *acrAB* [67]. Moreover, an IS*903* insertion in *ramA* reversed the overexpression of *acrAB* [67]. Recent work has also shown that TetD, which is encoded by the transposon Tn*10*, can also activate the expression of a subset of genes controlled by MarA and SoxS, thereby conferring a Mar phenotype. As TetD exhibits 43% amino acid sequence identity with SoxS and MarA, it has been proposed that this protein represents an additional member of the MarA/SoxS/Rob family [68].

INTRINSIC ACETYLTRANSFERASES IN BACTERIA

While the previous section describes a mechanism by which certain Gram-negative bacteria can alter their permeability barriers to afford antibiotic resistance, a different approach involving antibiotic inactivation will now be presented. In this case, antibiotic resistance is restricted to a particular chemical class of agents, the aminoglycosides, but with apparent broad specificity among constituent components of this group. Because of the regulatory nature of many of the mutations described below, it is appropriate to consider this an additional example of intrinsic global resistance.

Aminoglycoside resistance in bacteria is primarily mediated by the presence of plasmid-encoded modifying enzymes [69]. These enzymes modify the aminoglycosides by acetylation, phosphorylation, or adenylylation [69]. In addition to these plasmid-encoded enzymes, an expanding list of chromosomally encoded aminoglycoside acetyltransferases has been identified. For each of these enzymes, the corresponding gene appears to be intrinsic to the bacterial species in which it is found [70,71].

Therefore, it is possible that these intrinsic acetyltransferases act as housekeeping enzymes involved in the acetylation of cellular substrates. Since these enzymes also acetylate aminoglycosides, there may be structural similarities between aminoglycosides with the cellular substrates for these enzymes. The AAC(2′)-Ia enzyme in *P. stuartii* has a role in peptidoglycan acetylation (see below) and the AAC(2′)-Id enzyme in *Mycobacterium smegmatis* has a role in lysozyme resistance indicating a possible function related to the cell wall [72]. Recently, the AAC(6')-Iy enzyme has been identified in *Salmonella enterica* subsp. *enterica* serotype *Enteritidis* [73]. The potential function of this enzyme may be related to sugar metabolism. Additional chromosomal acetyltransferase genes that have been identified in other bacteria include *aac(6')-Ic* in *Serratia marcescens* [74], *aac(6')-Ig* in *Acinetobacter haemolyticus* [75], *aac(6')-Ij* in *Acinetobacter* sp. 13 [76], and *aac(6')-Ik* in *Acinetobacter* sp. 6 [77]. However, the regulatory mechanisms controlling their expression have yet to be identified and these genes will not be discussed further.

AAC(2')-IA IN *PROVIDENCIA STUARTII*

PHYSIOLOGICAL FUNCTIONS

Early studies on mutants that overexpressed the AAC(2')-Ia enzyme indicated that they possessed altered cell morphology, forming small rounded cells. To further address the role of AAC(2')-Ia, a null allele was created by introducing a frameshift mutation into the *aac(2')-Ia* coding region by allelic replacement [78]. The loss of *aac(2)-Ia* resulted in cells with a slightly elongated phenotype [78]. Furthermore, the staining properties of *aac(2')-Ia* mutant cells with uranyl acetate was altered, relative to wild-type cells. The basis for this phenotype is unknown; however, it suggests changes in the surface properties of cells. These data suggested a possible role for AAC(2')-Ia that is related to the cell envelope. Work done by Payie and Clarke has revealed that AAC(2')-Ia functions as a peptidoglycan *O*-acetyltransferase [79]. The *O*-acetylation of peptidoglycan is a modification that regulates the activity of autolytic enzymes involved in peptidoglycan breakdown and turnover [80,81]. The altered cell morphology seen in cells with changes in *aac(2')-Ia* expression may be due to the changes in the activity of autolytic enzymes. The AAC(2')-Ia enzyme is capable of obtaining acetate from peptidoglycan, *N*-acetylglucosamine, and acetyl-coenzymeA [79]. Interestingly, the AAC(2')-Ia enzyme is released by osmotic shock and may be located in the periplasm. Since acetyl-CoA is not located within the periplasm, the use of this substrate as a source of acetate would require a mechanism for transfer into the periplasm. The mechanism for such a transfer is unknown in *P. stuartii*.

GENETIC REGULATION

Studies on the regulation of *aac(2')-Ia* have been conducted using *lacZ* reporter gene fusions to the *aac(2')-Ia* promoter region. Early studies demonstrated that *aac(2')-Ia* transcription was not inducible by sub-inhibitory amounts of aminoglycoside antibiotics [82]. Using these fusions, two approaches have been used to identify gene products that act in *trans* to regulate *aac(2')-Ia*. The first approach involved selecting spontaneous gentamicin resistant mutants of a *P. stuartii* strain harboring an *aac(2')-lacZ* fusion on a low-copy plasmid. One mechanism for the increased gentamicin resistance of these mutants would be increased expression of the chromosomal *aac(2')-Ia* gene. During these isolations, the predominant class of mutants were

darker blue in the presence of X-gal indicating increased transcription from the
aac(2')-Ia promoter region on the plasmid. A second approach to identify regulatory
mutants involved isolating transposon insertions (mini-Tn5Cm) that activated the
aac(2')-lacZ fusion. Insertions that resulted in aac(2')-lacZ activation were then
tested for increased expression of the chromosomal aac(2')-Ia gene. Using both of
these strategies, genes designated aar (aminoglycoside acetyltransferase regulator)
have been identified. The surprising number of regulatory genes that have been
identified suggests the importance of modifying aac(2')-Ia expression in response to
various environmental conditions. This would allow cells to fine-tune the levels of
peptidoglycan acetylation and regulate autolysis.

The aar genes are grouped into two classes. The first class of genes act phenotypically as negative regulatory genes since loss of function mutations increase
aac(2')-Ia expression. The second class of regulatory genes are those that act in a
positive manner and are required for normal levels of aac(2')-Ia expression.

NEGATIVE REGULATORS

aarA

The aarA gene encodes a very hydrophobic polypeptide of 31.1 kDa in size [83]. The
AarA protein contains at least two possible transmembrane domains, suggesting that
it is an integral membrane protein. The AarA protein has been shown to be a member
of the Rhomboid family of intramembrane serine proteases that are widely distributed in prokaryotes and eukaryotes [84,85]. The AarA protein is required for the
production or activity of an extracellular pheromone signal, AR-factor, that acts to
reduce aac(2')-Ia expression. The aarA gene was identified as a mini-Tn5Cm insertion that increased gentamicin resistance levels eight-fold above wild-type. The aarA
mutants increase aac(2')-Ia transcription 3- to 10-fold depending on the growth
phase of cells. Null mutations in aarA are highly pleiotropic and additional phenotypes include loss of production of a diffusible yellow pigment and a cell chaining
phenotype that is most prominent in cells at mid-log phase.

aarB

The aarB3 mutation originally designated aar3 [82] results in a 10- to 12-fold
increase in aac(2')-Ia transcription. In the aarB3 background, the levels of aminoglycoside resistance are increased 128-fold above wild-type, suggesting that this
mutation further increases aminoglycoside resistance in a manner independent of
aac(2')-Ia expression. The aarB3 also results in altered cell morphology and a slow
growth phenotype. The identity of the aarB gene has not been determined.

aarC

The aarC gene encodes a homolog of gcpE, a protein widely distributed in bacteria
and required for isoprenoid biosynthesis. A missense allele, aarC1, resulted in a
number of pleiotropic phenotypes including slow growth, altered cell morphology,
and increased aac(2')-Ia expression at high cell density [86]. The biochemical
function of AarC remains to be determined.

aarD

The *aarD* was identified by a mini-Tn*5Cm* insertion that resulted in a five-fold activation of an *aac(2′)-lacZ* fusion and a three-fold increase in the levels of *aac(2′)-Ia* mRNA accumulation [87]. In addition, a 32-fold increase in aminoglycoside resistance was observed in *aarD* mutants, relative to wild-type *P. stuartii*. The *aarD* locus encodes two polypeptides that are homologs of the *E. coli* CydD and CydC proteins [87–89]. The CydD and CydC proteins act in a heterodimeric ABC transporter complex required for formation of a functional cytochrome *d* oxidase complex [90–93]. *P. stuartii aarD* mutants exhibit phenotypic characteristics consistent with a defect in the cytochrome *d* oxidase including hyper-susceptibility to the respiratory inhibitors Zn^{2+} and toluidine blue [87].

The increased *aac(2′)-Ia* expression observed in the *aarD1* background contributes minimally to the overall increase in gentamicin resistance since introduction of the *aarD1* mutation into an *aac(2′)-Ia* mutant strain also results in a 32-fold increase in gentamicin resistance. Previous studies have demonstrated that uptake of aminoglycosides is dependent on the presence of a functional electron transport system [94–96]. Since electron transport is defective in the *aarD1* background [87], it is probable that a decrease in aminoglycoside uptake accounts for the high level of resistance observed in *aarD* mutants. However, the mechanism that contributes to increased *aac(2′)-Ia* transcription is unknown. A direct role for *aarD* in the regulation of *aac(2′)-Ia* is unlikely, since ABC transporters are not known to function as transcriptional regulators [97]. A regulatory protein may couple changes in the redox state of the membrane to *aac(2′)-Ia* expression (see below) [98]. Mutations in *aarD* are predicted to alter the redox state of the membrane and thus indirectly affect *aac(2′)-Ia* expression.

aarG

The *aarG* gene encodes a protein with similarity to sensor kinases of the two-component family with the strongest identity to PhoQ (57%). Immediately upstream of *aarG* is an open reading frame designated *aarR*, which encoded a protein with 75% amino acid identity to PhoP, a response regulator [99,100]. The regulatory phenotypes associated with the *aarG1* mutation may result from a failure to phosphorylate the putative response regulator AarR, which functions as a repressor of *aarP*, and possibly *aac(2′)-Ia*.

A recessive mutation (*aarG1*) results in an 18-fold increase in the expression of β-galactosidase from an *aac(2′)-lacZ* fusion [99]. Direct measurements of RNA from the chromosomal copy of *aac(2′)-Ia* have confirmed this increase occurs at the level of RNA accumulation. Taken together, these results demonstrate that loss of *aarG* results in increased *aac(2′)-Ia* transcription. The *aarG1* allele also results in enhanced expression of *aarP*, encoding a transcriptional activator of *aac(2′)-Ia* (see below) [101]. Genetic experiments have shown that in an *aarG1*, *aarP* double mutant, the expression of *aac(2′)-Ia* is significantly reduced over that seen in the *aarG1* background. However, the levels of *aac(2′)-Ia* in this double mutant are still significantly higher than in a strain with only an *aarP* mutation. Therefore, the *aarG1* mutation increases *aac(2′)-Ia* expression by both *aarP*-dependent and -independent mechanisms.

The *aarG1* allele confers a Mar phenotype to *P. stuartii*, resulting in increased resistance to tetracycline, chloramphenicol, and fluoroquinolones. This Mar phenotype in the *aarG1* background is partially due to overexpression of *aarP*, which is known to confer a Mar phenotype in both *P. stuartii* and *E. coli* (see below). However, an *aarP*-independent mechanism also accounts for increased levels of intrinsic resistance in the *aarG1* background. This mechanism could involve increased expression of a second activator with a target specificity similar to that of AarP.

POSITIVE REGULATORS OF *AAC(2')-IA*

aarE

The *aarE* gene is *ubiA*, which encodes an octaprenyltransferase required for the second step of ubiquinone biosynthesis [102]. Although the *aarE* mutations increase aminoglycoside resistance, accumulation of *aac(2')-Ia* mRNA is significantly reduced in the *aarE1* background. The loss of ubiquinone function is predicted to decrease the uptake of aminoglycosides, which accounts for the high-level aminoglycoside resistance. The decreased *aac(2')-Ia* mRNA accumulation may reflect a requirement for ubiquinone, either directly or indirectly in a regulatory process involved in *aac(2')-Ia* mRNA expression.

aarF

The *aarF* locus of *P. stuartii* acts as a positive regulator of *aac(2')-Ia* expression with the level of *aac(2')-Ia* mRNA decreased in an *aarF* null mutant [98]. Despite the lack of *aac(2')-Ia* expression, *aarF* null mutants exhibit a 256-fold increase in gentamicin resistance over the wild-type strain. *P. stuartii aarF* null mutants also exhibit severe growth defects under aerobic growth conditions and have been found to lack detectable quantities of the respiratory cofactor ubiquinone. The *aarF* gene is the *ubiB* homolog of *P. stuartii,* and heterologous complementation studies demonstrated that these genes were functionally equivalent [103].

The high-level gentamicin resistance observed in the *aarF(ubiB)* mutants is likely associated with decreased accumulation of the drug resulting from the absence of aerobic electron transport. It seems unlikely that *aarF* is directly involved in the regulation of *aac(2')-Ia*. It has been proposed that a reduced form of ubiquinone acts as an effector molecule in an uncharacterized regulatory pathway that activates the expression of *aac(2')-Ia* [98]. In ubiquinone-deficient *aarF* mutant strains, this regulatory cascade would be disrupted, resulting in decreased *aac(2')-Ia* expression (see below).

aarP

The *aarP* gene was originally isolated from a multi-copy library of *P. stuartii* chromosomal DNA based on the ability to activate *aac(2')-Ia* expression in trans [101]. The presence of *aarP* in multiple copies led to an eight-fold increase in *aac(2')-Ia* mRNA accumulation. Studies utilizing an *aac(2')-lacZ* transcriptional fusion demonstrate that this increase results from an activation of *aac(2')-Ia* transcription. Chromosomal disruption of the *aarP* locus resulted in a five-fold reduction in *aac(2')-Ia* mRNA levels and eliminated the induction of *aac(2')-Ia* expression normally observed during logarithmic growth [101]. Expression of *aarP* has been shown to be increased in the *aarB*, *aarC*, and *aarG* mutants, demonstrating that *aarP* contributes to the overexpression of *aac(2′)-Ia* in these mutant backgrounds [82,86,99].

The *aarP* gene encodes a 16 kDa protein that contains a putative DNA binding helix-turn-helix motif and belongs to the AraC/XylS family of transcriptional activators [101,104]. The AarP protein exhibits extensive homology with the *E. coli* MarA and SoxS proteins that were discussed above. AarP exhibits high homology to MarA and SoxS in the helix-turn-helix domain and was found to activate targets of both MarA and SoxS *in vivo* [101]. The purified AarP protein binds to a wild-type *aac(2′)-Ia* promoter fragment in electrophoretic mobility shift assays [105].

Expression of *aarP* appears to be governed by a mechanism that differs from those controlling MarA and SoxS expression. Unlike the MarA and SoxS proteins, which are located in operons containing a gene that regulates their expression, the *aarP* message appears to be monocistronic. Expression of *aarP* is not elevated in the presence of SAL, a potent inducer of MarA. Recent studies of *aarP* expression have revealed that the AarP message accumulates as cell density increases [106]. At least three *aar* genes (*aarB*, *aarC*, and *aarG*) are involved in *aarP* regulation [82,86,99]. In addition, we have recently identified a role for the stationary phase starvation protein SspA as an activator of *aarP* [106]. The SspA protein is a global regulator that is proposed to interact with RNA polymerase during starvation and redirect new gene expression [107,108].

ROLE OF QUORUM SENSING IN *AAC(2′)-IA* REGULATION

The regulation of *aac(2′)-Ia* expression is mediated by cell-to-cell signaling [109]. The accumulation of *aac(2′)-Ia* mRNA exhibits two levels of growth phase dependent expression. First, as cells approach mid-log phase, a significant increase is observed relative to cells at early-log phase. This increase at mid-log phase is the result of increased *aarP* expression. Second, as cells approach stationary phase, the levels of *aac(2′)-Ia* mRNA are decreased to levels that are at least 20-fold lower than those at mid-log phase. This decrease at high density is mediated by the accumulation of an extracellular factor (AR-factor) [109]. The growth of *P. stuartii* cells in spent (conditioned) media from stationary phase cultures resulted in the premature repression of *aac(2′)-Ia* in cells at mid-log phase. The ability to produce AR-factor is dependent on the AarA protein described previously.

In summary, the large number of genes that influence *aac(2′)-Ia* regulation suggest that the expression of *aac(2′)-Ia* and the subsequent *O*-acetylation of peptidoglycan must be tightly controlled in *P. stuartii*. The AAC(2′)-Ia enzyme represents a minor *O*-acetyltransferase in *P. stuartii* [78]. The physiological function of AAC(2′)-Ia may be to "fine-tune" the levels of peptidoglycan *O*-acetylation in response to different environmental conditions or phases of growth. For example, in cells at mid-log phase, there is a burst of *aac(2′)-Ia* expression that may be required for peptidoglycan turnover in rapidly growing cells. As cells increase in density and approach stationary phase, the accumulation of AR-factor leads to decreased *aac(2′)-Ia* expression at stationary phase. This may reflect a requirement for lower peptidoglycan turnover at stationary phase. The additional levels of *aac(2′)-Ia* regulation, namely, the role of ubiquinone and/or electron transport, are understood in less detail. The simplest model, proposed earlier, is that *aac(2')-Ia* expression is also coupled to electron transport via regulatory protein(s) that sense the redox status of the cell. The AarG/AarR two-component system may have a role in this process. At the present time, interplay among the *aar* genes, electron transport, and quorum sensing in controlling *aac(2')-Ia* expression is being investigated. The mechanisms identified may serve as a model for the regulation of other chromosomally encoded acetyltransferases. In addition, the identification of physiological roles for the other intrinsic acetyltransferases will allow us to better predict how the modification of intrinsic genes can lead to antibiotic resistance.

CONCLUDING REMARKS

While the *mar/sox* and *aac(2')-Ia* systems differ significantly at both the genetic and physiological levels, there are important similarities worth noting. As global, intrinsic resistance systems, they both contain regulatory components as key factors controlling the expression of resistance determinants. In the case of MarA/SoxS and AarP, the products of key regulatory genes are remarkably conserved. In addition, there is a common element of environmental sensing that is shared by these two systems. These observations support the notion that the resistance phenotypes observed for specific regulatory mutants are physiologically relevant, because they result from changes in the activity of factors that control the expression of otherwise normal effector genes. As the best-studied examples of global resistance systems, the *mar/sox* and *aac(2')-Ia* networks provide models for how subtle yet effective pathways affecting antibiotic resistance may lie buried within the complex genomes of many microorganisms. It is interesting to note that the kinds of antibiotics affected by the two systems are almost entirely complementary: aminoglycosides are among the only kinds of agents that are not impacted by the *mar/sox* pathway. Perhaps the *aac(2')-Ia* system evolved divergently to address this gap? Regardless, these pathways provide paradigms that should assist future investigators in the characterization of other global systems that affect antibiotic susceptibility.

REFERENCES

1. Alekshun MN, Levy SB. Regulation of chromosomally mediated multiple antibiotic resistance: the *mar* regulon. *Antimicrob Agents Chemother* 1997; 41:2067–2075.
2. George AM, Levy SB. Amplifiable resistance to tetracycline, chloramphenicol and other antibiotics in *Escherichia coli*: involvement of a non-plasmid-determined efflux of tetracycline. *J Bacteriol* 1983; 155:531–540.
3. Cohen SP, McMurry LM, Hooper DC, Wolfson JS, Levy SB. Cross-resistance to fluoroquinolones in multiple-antibiotic-resistant (Mar) *Escherichia coli* selected by tetracycline or chloramphenicol: decreased drug accumulation associated with membrane changes in addition to OmpF reduction. *Antimicrob Agents Chemother* 1989; 33:1318–1325.
4. McMurry LM, George AM, Levy SB. Active efflux of chloramphenicol in susceptible *Escherichia coli* strains and in multiple-antibiotic-resistant (Mar) mutants. *Antimicrob Agents Chemother* 1994; 38:542–546.
5. White DG, Goldman JD, Demple B, Levy SB. Role of the *acrAB* locus in organic solvent tolerance mediated by expression of *marA*, *soxS*, or *robA* in *Escherichia coli*. *J Bacteriol* 1997; 179:6122–6126.

6. Moken MC, McMurry LM, Levy, SB. Selection of multiple-antibiotic-resistant (Mar) mutants of *Escherichia coli* by using the disinfectant pine oil: roles of the *mar* and *acrAB* loci. *Antimicrob Agents Chemother* 1997; 41:2770–2772.

7. George AM, Levy SB. Gene in the major cotransduction gap of the *Escherichia coli* K-12 linkage map required for the expression of chromosomal resistance to tetracycline and other antibiotics. *J Bacteriol* 1983; 155:541–548.

8. Hächler H, Cohen SP, Levy SB. *marA*, regulated locus which controls expression of chromosomal multiple antibiotic resistance in *Escherichia coli*. *J Bacteriol* 1991; 173:5532–5538.

9. Cohen SP, Hächler H, Levy SB. Genetic and functional analysis of the multiple antibiotic resistance (*mar*) locus in *Escherichia coli*. *J Bacteriol* 1993; 175:1484–1492.

10. Sulavik MC, Gambino LF, Miller PF. Analysis of the genetic requirements for inducible multiple-antibiotic resistance associated with the *mar* locus in *Escherichia coli*. *J Bacteriol* 1994; 176:7754–7756.

11. Martin RG, Nyantakyi PS, Rosner JL. Regulation of the multiple antibiotic resistance (*mar*) regulon by *marORA* operator sequences in *Escherichia coli*. *J Bacteriol* 1995; 177:4176–4178.

12. Gambino L, Gracheck SJ, Miller PF. Overexpression of the MarA positive regulator is sufficient to confer multiple antibiotic resistance in *Escherichia coli*. *J Bacteriol* 1993; 175:2888–2894.

13. Gallegos MT, Schleif R, Bairoch A, Hofmann K, Ramos JL. AraC/XylS family of transcriptional regulators. *Mol Microbiol* 1997; 61:393–410.

14. Ariza RR, Cohen SP, Bachhawat N, Levy SB, Demple B. Repressor mutations in the *marRAB* operon that activate oxidative stress genes and multiple antibiotic resistance in *Escherichia coli*. *J Bacteriol* 1994; 176:143–148.

15. Martin R, Rosner JL. Binding of purified multiple antibiotic-resistance repressor protein (MarR) to *mar* operator sequences. *Proc Natl Acad Sci USA* 1995; 92:5456–5460.

16. Rosner JL. Nonheritable resistance to chloramphenicol and other antibiotics induced by salicylates and other chemotactic repellants in *Eshcerichia coli*. *Proc Natl Acad Sci USA* 1985; 82:8771–8774.

17. Cohen SP, Levy SB, Foulds J, Rosner JL. Salicylate induction of antibiotic resistance in *Escherichia coli*: activation of the *mar* operon and a *mar*-independent pathway. *J Bacteriol* 1993; 175:7856–7862.

18. Alekshun MN, Levy SB, Mealy TR, Seaton BA, Head JF. The crystal structure of MarR, a regulator of multiple antibiotic resistance, at 2.3A resolution. *Nat Struct Biol* 2001; 8:710–714.

19. Demple B. Regulation of bacterial oxidative stress genes. *Ann Rev Genet* 1991; 25:315–337.

20. Wu J, Weiss B. Two divergently transcribed genes, *soxR* and *soxS*, control a superoxide response regulon of *Escherichia coli*. *J Bacteriol* 1991; 173:2864–2871.

21. Greenberg JT, Monach P, Chou JH, Josephy PD, Demple B. Positive control of a global antioxidant defense regulon activated by superoxide-generating agents in *Escherichia coli*. *Proc Natl Acad Sci USA* 1990; 87:6181–6185.

22. Nunoshiba T, Hidalgo E, Amabile-Cuevas CF, Demple B. Two-stage control of an oxidative stress regulon: the *Escherichia coli* SoxR protein triggers redox-inducible expression of the *soxS* gene. *J Bacteriol* 1992; 174:6054–6060.

23. Wu J, Weiss B. Two-stage induction of the *soxRS* (superoxide response) regulon of *Escherichia coli*. *J Bacteriol* 1992; 174:3915–3920.

24. Hidalgo E, Demple B. An iron-sulphur center essential for transcriptional activation by the redox-sensing SoxR protein. *EMBO J* 1994; 13:138–146.

25. Wu J, Dunham WR, Weiss B. Overproduction and physical characterization of SoxR, a [2Fe-2S] protein that governs an oxidative response regulon in *Escherichia coli*. *J Biol Chem* 1995; 270:10323–10327.

26. Miller PF, Gambino LF, Sulavik MC, Gracheck SJ. Genetic relationship between *soxRS* and *mar* loci in promoting multiple antibiotic resistance in *Escherichia coli*. *Antimicrob Agents Chemother* 1994; 38:1773–1779.

27. Webber MA, Piddock LJV. Absence of mutations in *marRAB* or *soxRS* in *acrB* overexpressing fluoroquinolone resistant clinical and veterinary isolates of *Escherichia coli*. *Antimicrob Agents Chemother* 2001; 45:1550–1552.

28. Oethinger MO, Podglajen I, Kern WV, Levy SB. Overexpression of the *marA* or *soxS* rgulatory gene in clinical topoisomerase mutants of *Escherichia coli*. *Antimicrob Agents Chemother* 1998; 42:2089–2094.

29. Koutsolioutsou A, Pena-Llopis S, Demple B. Constitutive *soxR* mutations contribute to multiple antibiotic resistance in clinical *Escherichia coli* isolates. *Antimicrob Agents Chemother* 2005; 49:2746–2752.

30. Koutsolioutsou A, Martins EA, White DG, Levy SB, Demple B. A *soxRS*-constitutive mutation contributing to antibiotic resistance in a clinical isolate of *Salmonella enterica* (serovar typhimurium). *Antimicrob Agents Chemother* 2001; 45:38–43.

31. Skarstad K, Thony B, Hwang DS, Kornberg A. A novel binding protein of the origin of the *Escherichia coli* chromosome. *J Biol Chem* 1993; 268:5365–5370.

32. Ariza RR, Li Z, Ringstad N, Demple B. Activation of multiple antibiotic resistance and binding of stress-inducible promoters by *Escherichia coli* Rob protein. *J Bacteriol* 1995; 177:1655–1661.

33. Nakajima H, Kobayashi K, Kobayashi M, Asako H, Aono R. Overexpression of the *robA* gene increases organic solvent tolerance and multiple antibiotic and heavy metal ion resistance in *Escherichia coli*. *Appl Environ Microbiol* 1995; 61:2303–2307.

34. Martin RG, Gillette WK, Rhee S, Rosner JL. Structural requirements for marbox function in transcriptional activation of *mar/sox/rob* regulon promoters in *Escherichia coli*: sequence, orientation and spatial relationship to the core promoter. *Mol Microbiol* 1999; 34:431–441.

35. Li Z, Demple B. Sequence specificity for DNA binding by *Escherichia coli* SoxS and Rob proteins. *Mol Microbiol* 1996; 20:937–945.

36. Fawcett WP, Wolf RE Jr. Purification of a MalE-SoxS fusion protein and identification of the control sites of *Escherichia coli* superoxide-inducible genes. *Mol Microbiol* 1994; 14:669–679.

37. Jair K-W, Martin RG, Rosner JL, Fujita N, Ishihama A, Wolf RE Jr. Purification and regulatory properties of MarA protein, a transcriptional activator of *Escherichia coli* multiple antibiotic and superoxide resistance promoters. *J Bacteriol* 1995; 177: 7100–7104.

38. Jair K-W, Fawcett WP, Fujita N, Ishihama A, Wolf RE Jr. Ambidextrous transcriptional activation by SoxS: requirement for the C-terminal domain of the RNA polymerase alpha subunit in a subset of *Escherichia coli* superoxide-inducible genes. *Mol Microbiol* 1996; 19:307–317.

39. Rhee SR, Martin RG, Rosner JL, Davies DR. A novel DNA-binding motif in MarA: the first structure for an AraC family transcriptional activator. *Proc Natl Acad Sci USA* 1998; 95:10413–10418.

40. Schneiders T, Barbosa TM, McMurry LM, Levy SB. The *Escherichia coli* transcriptional regulator MarA directly represses transcription of *purA* and *hdeA*. *J Biol Chem* 2004; 279:9037–9042.

41. Li Z, Demple B. SoxS, an activator of superoxide stress genes in *Escherichia coli*. Purification and interaction with DNA. *J Biol Chem* 1994; 269:18371–18377.

42. Cohen SP, McMurry LM, Levy SB. *marA* locus causes decreased expression of OmpF porin in multiple-antibiotic-resistant (Mar) mutants of *Escherichia coli*. *J Bacteriol* 1988; 170:5416–5422.

43. Rosner JL, Chai T-J, Foulds J. Regulation of OmpF porin expression by salicylate in *Escherichia coli*. *J Bacteriol* 1991; 173:5631–5638.

44. Mizuno T, Chou M-Y, Inouye M. A unique mechanism regulating gene expression: translational inhibition by a complementary RNA transcript (micRNA). *Proc Natl Acad Sci USA* 1984; 81:1966–1970.
45. Anderson J, Delihas N. *micF* RNA binds to the 5′ end of *ompF* mRNA and to a protein from *Escherichia coli. Biochemistry* 1990; 29:9249–9256.
46. Chou JH, Greenberg JT, Demple B. Posttranscriptional repression of *Escherichia coli* OmpF protein in response to redox stress: positive control of the *micF* antisense RNA by the *soxRS* locus. *J Bacteriol* 1993; 175:1026–1031.
47. Nikaido H. Prevention of drug access to bacterial targets: permeability barriers and active efflux. *Science* 1994; 264:382–388.
48. Ma D, Cook DN, Alberti M, Pon NG, Nikaido H, Hearst JE. Genes of *acrA* and *acrB* encode a stress-induced efflux system of *Escherichia coli. Mol Microbiol* 1995; 16:45–55.
49. Aono R, Tsukagoshi N, Yamamoto M. Involvement of outer membrane protein TolC, a possible member of the *mar-sox* regulon, in maintenance and improvement of organic solvent tolerance of *Escherichia coli* K-12. *J Bacteriol* 1998; 180:938–944.
50. Martin RG, Jair K-W, Wolf RE Jr, Rosner JL. Autoactivation of the *marRAB* multiple antibiotic resistance operon by the MarA transcriptional activator in *Escherichia coli. J Bacteriol* 1996; 178:2216–2223.
51. Martin RG, Rosner JL. Transcriptional and translational regulation of the *marRAB* multiple antibiotic resistance operon in *Escherichia coli. Mol Microbiol* 2004; 53:183–191.
52. Martin RG, Gillette WK, Rosner JL. Promoter discrimination by the related transcriptional activators MarA and SoxS: differntial regulation by differential binding. *Mol Microbiol* 2000; 35:623–634.
53. Barbosa TM, Levy SB. Differential expression of over 60 chromosomal genes in *Escherichia coli* by constitutive expression of MarA. *J Bacteriol* 2000; 182:3467–3474.
54. Pomposiello PJ, Bennik MH, Demple B. Genome-wide transcriptional profiling of the *Escherichia coli* responses to superoxide stress and sodium salicylate. *J Bacteriol* 2001; 183:3890–3902.
55. Martin RG, Rosner JL. Genomics of the *marA/soxS/rob* regulon of *Escherichia coli:* identification of directly activated promoters by application of molecular genetics and informatics to microarray data. *Mol Microbiol* 2002; 44:1611–1624.
56. Barbosa TM, Levy SB. Activation of the *Escherichia coli nfnB* gene by MarA through a highly divergent marbox in a class II promoter. *Mol Microbiol* 2002; 45:191–202.
57. Miller PF, Sulavik M, Gambino L, Dazer M. Roles of the *marRAB* and *soxRS* regulators in protecting *Escherichia coli* from plant-derived phenolic agents. 96th General Meeting of the American Society for Microbiology, New Orleans, LA, May 19–23, 1996.
58. Seoane A, Levy SB. Characterization of MarR, the repressor of the multiple antibiotic resistance (*mar*) operon in *Escherichia coli. J Bacteriol* 1995; 177:3414–3419.
59. Miller PF, Sulavik MC. Overlaps and parallels in the regulation of intrinsic multiple-antibiotic resistance in *Escherichia coli. Mol Microbiol* 1996; 21:441–448.
60. Sulavik MC, Gambino LF, Miller PF. The MarR repressor of the multiple antibiotic resistance (*mar*) operon in *Escherichia coli*: prototypic member of a family of bacterial regulatory proteins involved in sensing phenolic compounds. *Mol Medicine* 1995; 1:436–446.
61. Cohen SP, Yan W, Levy SB. A multidrug resistance regulatory chromosomal locus is widespread among enteric bacteria. *J Infect Dis* 1993; 168:484–488.
62. Maneewannakul K, Levy SB. Identification of *mar* mutants among quinolone-resistant clinical isolates of *Escherichia coli. Antimicrob Agents Chemother* 1996; 40:1695–1698.
63. Hooper DC, Wolfson JS. Bacterial resistance to the quinolone antimicrobial agents. *Am J Med* 1989; 87:17S–23S.

64. Golding SS, Matthews KR. Intrinsic mechanism decreases susceptibility of *Escherichia coli* O157:H7 to multiple antibiotics. *J Food Protection* 2004; 67:34–39.

65. George AM, Hall RM, Stokes HW. Multidrug resistance in *Klebsiella pneumoniae:* a novel gene, *ramA,* confers a multidrug resistance phenotype in *Escherichia coli. Microbiology* 1995; 141:1909–1920.

66. Schneiders T, Amyes SGB, Levy SB. Role of AcrR and RamA in fluoroquinolone resistance in clinical *Klebsiella pneumoniae* isolates from Singapore. *Antimicrob Agents Chemother* 2003; 47:2831–2837.

67. Ruzin A, Visalli MA, Keeney D, Bradford PA. Influence of transcriptional activator RamA on expression of multidrug efflux pump AcrAB and tigecycline susceptibility in *Klebsiella pneumoniae. Antimicrob Agents Chemother* 2005; 49:1017–1022.

68. Griffith KL, Becker SM, Wolf RE Jr. Characterization of TetD as a transcriptional activator of a subset of genes of the *Escherichia coli* SoxS/MarA/Rob regulon. *Mol Microbiol* 2005; 56:1103–1117.

69. Shaw KJ, Rather PN, Hare RS, Miller GH. Molecular genetics of aminoglycoside resistance genes and familial relationships of the aminoglycoside-modifying enzymes. *Microbiol Rev* 1993; 57:138–163.

70. Rather PN. Origins of the aminoglycoside modifying enzymes. *Drug Res Updates* 1998; 5:285–291.

71. Rather PN, Macinga DR. The chromosomal 2′-N-acetyltransferase of *Providencia stuartii*: physiological functions and genetic regulation. *Front Biosci* 1999; 4:132–140.

72. Ansa JA, Pérez E, Pelicic V, Berthet F-X, Gicquel B, Martin C. Aminoglycoside 2′-N-acetyltransferase genes are universally present in mycobacteria: characterization of the aac(2′)-Ic from *Mycobacterium tuberculosis* and the aac(2′)-Id gene from *Mycobacterium smegmatis. Mol Microbiol* 1997; 24:431–441.

73. Magnet S, Courvalin P, Lambert T. Activation of the cryptic aac(6′)-Iy aminoglycoside resistance gene of *Salmonella* by a chromosomal deletion generating a transcriptional fusion. *J Bacteriol* 1999; 181:6650–6655.

74. Champion HM, Bennett PM, Lewis DA, Reeves DS. Cloning and characterization of an AAC(6′) gene from *Serratia marcescens. J Antimicrob Chemother* 1988; 22:587–596.

75. Lambert T, Gerbaud G, Galimand M, Courvalin P. Characterization of *Acinetobacter haemolyticus aac(6′)-Ig* gene encoding an aminoglycoside 6′-N-acetyltransferase which modifies amikacin. *Antimicrob Agents Chemother* 1993; 37:2093–2100.

76. Lambert T, Gerbaud G, Courvalin P. Characterization of the chromosomal *aac(6′)-Ij* gene of *Acinetobacter* sp. 13 and the *aac(6′)-Ih* plasmid gene of *Acinetobacter baumannii. Antimicrob Agents Chemother* 1998; 42:2759–2761.

77. Rudant E, Bourlioux P, Courvalin P, Lambert T. Characterization of the *aac(6′)-Ik* gene of *Acinetobacter* sp. 6. *FEMS Microbiol Letters* 1994; 124:49–54.

78. Payie KG, Rather PN, Clarke AJ. Contribution of gentamicin 2′-N-acetyltransferase to the *O*-acetylation of peptidoglycan in *Providencia stuartii. J Bacteriol* 1995; 177:4303–4310.

79. Payie KG, Clarke AJ. Characterization of gentamicin 2′-N-acetyltransferase from *Providencia stuartii*: its use of peptidoglycan metabolites for acetylation of both aminoglycosides and peptidoglycan. *J Bacteriol* 1997; 179:4106–4114.

80. Clarke AJ, Dupont C. *O*-acetylated peptidoglycan: its occurrence, pathobiological significance, and biosynthesis. *Can J Microbiol* 1992; 38:85–91.

81. Clarke AJ. Extent of peptidoglycan *O*-acetylation in the tribe Proteeae. *J Bacteriol* 1993; 175:4550–4553.

82. Rather PN, Orosz E, Shaw KJ, Hare R, Miller GH. Characterization and transcriptional regulation of the 2′-N-acetyltransferase gene from *Providencia stuartii. J Bacteriol* 1993; 175:6492–6498.

83. Rather PN, Orosz E. Characterization of *aarA*, a pleiotrophic negative regulator of the 2'-*N*-acetyltransferase in *Providencia stuartii*. *J Bacteriol* 1994; 176:5140–5144.

84. Gallio M, Sturgill G, Rather PN, Kylsten P. A common mechanism for extracellular signaling in eukaryotes and prokaryotes. *Proc Natl Acad Sci USA* 2002; 99:12208–12213.

85. Urban S, Lee JR, Freeman M. *Drosophilia* Rhomboid-1 defines a family of putative intramembrane serine protease. *Cell* 2001; 107:173–182.

86. Rather PN, Solinsky K, Paradise MR, Parojcic MM. *aarC*, an essential gene involved in density dependent regulation of the 2'-*N*-acetyltransferase in *Providencia stuartii*. *J Bacteriol* 1996; 179:2267–2273.

87. Macinga DR, Rather PN. *aarD*, a *Providencia stuartii* homologue of *cydD*: role in 2'-*N*-acetyltransferase expression, cell morphology and growth in the presence of an extracellular factor. *Mol Microbiol* 1996; 19:511–520.

88. Delaney JM, Wall D, Georgopoulos C. Molecular characterization of the *Escherichia coli htrD* gene: cloning, sequence, regulation and involvement with cytochrome d oxidase. *J Bacteriol* 1993; 175:166–175.

89. Poole RK, Hatch L, Cleeter MW, Gibson JF, Cox GB, Wu G. Cytochrome bd bio-synthesis in *Escherichia coli*: the sequences of the *cydC* and *cydD* genes suggest that they encode the components of an ABC membrane transporter. *Mol Microbiol* 1993; 10:421–430.

90. Bebbington KJ, Williams HD. Investigation of the role of the *cydD* gene product in the production of a functional cytochrome d oxidase. *FEMS Microbiol Lett* 1993; 112:19–24.

91. Poole RK, Gibson F, Wu G. The *cydD* gene product, component of a heterodimeric ABC transporter, is required for assembly of periplasmic cytochrome c and of cytochrome bd in *Escherichia coli*. *FEMS Microbiol Lett* 1994; 117:217–224.

92. Poole RK, Williams HD, Downie JA, Gibson F. Mutations affecting the cytochrome d-containing oxidase complex of *Escherichia coli* K-12: identification and mapping of a fourth locus, *cydD*. *J Gen Microbiol* 1989; 135:1865–1874.

93. Georgiou CD, Fang H, Gennis RB. Identification of the *cydC* locus required for expression of the functional form of the cytochrome d terminal oxidase complex in *Escherichia coli*. *J Bacteriol* 1987; 169:2107–2112.

94. Taber HW, Mueller JP, Miller PF, Arrow AS. Bacterial uptake of aminoglycoside antibiotics. *Microbiol Rev* 1987; 51:439–457.

95. Bryan LE, Kwan S. Roles of ribosomal binding, membrane potential, and electron transport in bacterial uptake of streptomycin and gentamicin. *Antimicrob Agents Chemother* 1983; 23:835–845.

96. Bryan LE, van den Elzen HM. Effects of membrane energy mutations and cations on streptomycin and gentamicin accumulation by bacteria: a model for entry of streptomycin and gentamicin in susceptible and resistant bacteria. *Antimicrob Agents Chemother* 1977; 12:163–177.

97. Faith MJ, Kolter R. ABC transporters: bacterial exporters. *Microbiol Rev* 1993; 57:995–1017.

98. Macinga DR, Cook GM, Poole RK, Rather PN. Identification and characterization of *aarF*, a locus required for production of ubiquinone in *Providencia stuartii* and *Escherichia coli* and for expression of 2'-*N*-acetyltransferase in *P. stuartii*. *J Bacteriol* 1998; 180:128–135.

99. Rather PN, Paradise MR, Parojcic MM, Patel S. A regulatory cascade involving *aarG*, a putative sensor kinase, controls the expression of the 2'-*N*-acetyltransferase and an intrinsic multiple antibiotic resistance (Mar) response in *Providencia stuartii*. *Mol Microbiol* 1998; 28:1345–1353.

100. Groisman EA, Heffron F, Solomon F. Molecular genetic analysis of the *Escherichia coli phoP* locus. *J Bacteriol* 1992; 174:486–491.

101. Macinga DR, Parjocic MM, Rather PN. Identification and analysis of *aarP*, a transcriptional activator of the 2′-*N*-acetyltransferase in *Providencia stuartii*. *J Bacteriol* 1995; 177:3407–3413.

102. Paradise MR, Cook G, Poole RK, Rather PN. *aarE*, a homolog of *ubiA*, is required for transcription of the *aac(2′)-Ia* gene in *Providencia stuartii*. *Antimicrob Agents Chemother* 1998; 42:959–962.

103. Poon WW, Davis DE, Huan HT, Jonassen T, Rather PN, Clarke CF. Identification of the *ubiB* gene involved in ubiquinone biosynthesis: the *aarF* gene in *P. stuartii* and *yigR* in *E. coli* correspond to *ubiB*. *J Bacteriol* 2000; 182:5139–5146.

104. Ramos JL, Rojo F, Zhou L, Timmis KN. A family of positive regulators related to the *Pseudomonas putida* TOL plasmid XylS and the *Escherichia coli* AraC activators. *Nucl Acids Res* 1990; 18:2149–2152.

105. Macinga DR, Paradise MR, Parojcic MM, Rather PN. Activation of the 2′-*N*-acetyltransferase gene (*aac(2′)-Ia*) in *Providencia stuartii* by an interaction of AarP with the promoter region. *Antimicrob Agents Chemother* 1999; 43:1769–1772.

106. Ding X, Rather PN. Unpublished data.

107. Williams MD, Ouyang TX, Flickinger MC. Starvation-induced expression of SspA and SspB: the effects of a null mutation in *sspA* on *Escherichia coli* protein synthesis and survival during growth and prolonged starvation. *Mol Microbiol* 1994; 11:1029–1043.

108. Williams MD, Ouyang TX, Flickinger MC. Glutathione S-transferase-*sspA* fusion binds to *E. coli* RNA polymerase and complements delta *sspA* mutation allowing phage P1 replication. *Biochem Biophys Res Comm* 1994; 201:123–127.

109. Rather PN, Paradise MR, Parojcic MM. An extracellular factor regulating expression of the chromosomal aminoglycoside 2′-*N*-acetyltransferase in *Providencia stuartii*. *Antimicrob Agents Chemother* 1997; 41:1749–1754.

4 Multidrug Efflux Pumps: Structure, Mechanism, and Inhibition

Olga Lomovskaya, Helen I. Zgurskaya,
Keith A. Bostian, and Kim Lewis

CONTENTS

The world of antibiotic drug discovery and development is driven by the necessity to overcome antibiotic resistance in common Gram-positive and Gram-negative pathogens. However, the lack of Gram-negative activity among both recently approved antibiotics and compounds in the developmental pipeline is a general trend. It is despite the fact that the plethora of covered drug targets is well conserved in both Gram-positive and Gram-negative bacteria. Multidrug resistance (MDR) efflux pumps play a prominent and proven role in Gram-negative intrinsic resistance. Moreover, these pumps also play a significant role in acquired clinical resistance. Together, these considerations make efflux pumps attractive targets for inhibition in that the resultant efflux pump inhibitor (EPI)/antibiotic combination drug should exhibit increased potency, enhanced spectrum of activity, and reduced propensity for acquired resistance. To date, at least one class of broad-spectrum EPI has been extensively characterized. While these efforts indicated a significant potential for developing small molecule inhibitors against efflux pumps, they did not result in a clinically useful compound. Stemming from the continued clinical pressure for novel approaches to combat drug resistant bacterial infections, second-generation programs have been initiated and show early promise to significantly improve the clinical usefulness of currently available and future antibiotics against otherwise recalcitrant Gram-negative infections. It is also apparent that some changes in regulatory decision-making regarding resistance would be very helpful in order to facilitate approval of agents aiming to reverse resistance and prevent its further development.

MDRs IN CLINICALLY RELEVANT DRUG RESISTANCE

Five families of bacterial drug efflux pumps have been identified to date [1]. However, it is mostly members of a single resistance/nodulation/division super family (RND) found in Gram-negative species that are implicated in clinically relevant resistance. In this section, we will review the structure and mechanism of RND MDRs.

Efflux is most effective when working in cooperation with other resistance mechanisms. Reduced uptake across the outer membrane of Gram-negative bacteria, which is a significant permeability barrier for both hydrophilic and hydrophobic compounds, constitutes such a mechanism [2,3]. To take advantage of the reduced uptake, some Gram-negative MDR pumps extrude their substrates across the whole cellular envelope directly into the medium, performing trans-envelope transport [4].

The RND MDRs [5] possess an astonishing breadth of substrate specificity, and in this respect surpass even the notorious ABC transporters, such as P-glycoprotein (P-gp), that are major hurdles for the effectiveness of anti-cancer therapy [6]. RND pumps can recognize and extrude positive-, negative-, or neutral-charged molecules, substances as hydrophobic as organic solvents and lipids, and compounds as hydrophilic as aminoglycoside antibiotics. They are a ubiquitous family whose members are distributed across various kingdoms. Several representatives of the RND-permease superfamily are encoded in the human genome, though the similarity to bacterial RNDs is negligible (16% identity). Examples of human RNDs include the Niemann-Pick disease, type C1 (NPC1) protein, localized in lysosomal membranes and apparently involved in intracellular cholesterol transport [7], and the homolog of *Drosophila* morphogen receptor *Patched*, thought to be crucial in the suppression of basal cell carcinoma [5,8]. Given the low similarity between the bacterial and human RNDs, the identification of highly selective inhibitors of bacterial pumps devoid of mechanism-based mammalian toxicity appears feasible.

The past few years have seen the publication of an extraordinary amount of structural information on bacterial RND transporters. Several high-resolution structures of the AcrB efflux pump from *Escherichia coli*, with and without co-crystallized substrates, as well as several mutant AcrBs, have recently emerged from several laboratories across the world [9–14]. As discussed below, RND transporters from Gram-negative bacteria function as a complex with two other types of proteins, and X-ray structures of these have also become available [15–19]. Only the structure of the tripartite complex itself is lacking. The input from this detailed structural information should dramatically facilitate discovery of inhibitors of RND transporters to improve efficacy of antibiotics against problematic bacteria that are the cause of many life-threatening infections.

In this review, we will summarize the recent advances on structure and function of RND transporters. It is important to emphasize that much research has also been devoted to understanding the complex regulation of efflux gene expression, which is usually governed by local and global regulatory mechanisms. To learn more about this topic, the reader is referred to several excellent reviews [20–23].

In addition, more and more information is becoming available regarding the significance of RND transporters in bacterial pathogenesis. Many transporters have been demonstrated to be essential for cellular invasion and resistance to natural host substances, such as bile salts and specialized host-defense molecules [24]. Some RND transporters from *Pseudomonas aeruginosa* (and other relevant pathogens) are involved in controlling the balance of quorum signal molecules important for cell-to-cell communication, and apparently for the establishment of persisting infections [25,26]. Interestingly, the Gram-positive organism *Mycobacterium tuberculosis* contains four RND transporters involved in controlling virulence [27]. A comprehensive review regarding this emerging "physiological" role of RND transporters has recently been published [24].

To perform trans-envelope efflux the inner membrane RND transporter works together with accessory proteins: a periplasmic protein belonging to the Membrane Fusion Protein (MFP) family, and the outer membrane channel, a member of the Outer Membrane Factor (OMF) family. RND pumps and accessory proteins form large multi-protein assemblies that traverse both the inner and outer membranes of Gram-negative bacteria [4,28,29]. All components are absolutely essential for transport. Working together as a well-coordinated team, they achieve the direct extrusion of substrates across the whole cell envelope and into the medium.

This is where another aspect of teamwork is evident. The integrity of the outer membrane is absolutely critical. When it is compromised no resistance is seen, even in the presence of fully functional RND-containing complexes [30]. One might infer that these MDR transporters are in fact rather sluggish machines, relying heavily for their effectiveness upon restricted diffusion of effluxed compounds back into the bacterial cell, but no reliable kinetic measurements exist for RND transporters to argue this point. An alternative explanation is that the outer membrane is simply the only barrier across, which substrates are transported, and that RND transporters, unlike all other known drug pumps, are capable of capturing their substrates in the periplasmic space rather than in the membrane or from the cytoplasm. X-ray crystallography illuminates these points.

High-resolution structures of all components of the efflux complex are now available. A steady stream of structural information, starting with the OMF TolC [18] protein from *E. coli* in 2000 and continuing to the present day, has produced a series of remarkable discoveries [9,11,12].

The best-studied AcrB transporter from *E. coli* serves as a prototype for all members of the RND family. AcrB is a protein of ca. 1100 amino acid residues that contains a transmembrane domain (TMD), consisting of 12 TM segments, and an unusually large periplasmic domain. Importantly, structural data have established that AcrB functions as a trimer (Figures 4.1 through 4.3). Inside the membrane, monomers of AcrB, which we will refer to as protomers, have very limited contact with one another [10]. In the periplasm, by contrast, they assemble into an intricate "mushroom-like" structure that protrudes about 70 Å from the membrane. The periplasmic portion of the AcrB trimer can be further sub-divided into the porter and the TolC docking domains. In the porter domain, neighboring protomers form three large vestibules that are wide open to the periplasm. These vestibules lead to a spacious central cavity.

TolC   AcrB   AcrA

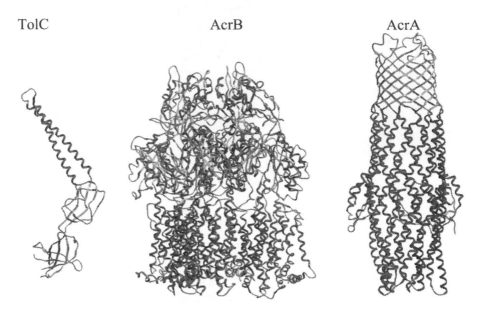

FIGURE 4.1 Structure of AcrA (PBD code 2f1m), AcrB (PBD code 1ek9), and TolC (PBD code 2dhh) proteins from *E. coli*. The figure shows a monomer of AcrA and trimers of AcrB and TolC. All proteins are shown to scale.

The structural features of the accessory proteins are consistent with their role in extending drug efflux across the outer membrane. The 3D structures of three OMFs, TolC, OprM, and VceC from *E. coli*, *P. aeruginosa*, and *Vibrio cholerae*, respectively, have been solved recently [15,18,31]. Despite very little sequence similarity, they are structurally conserved. Like AcrB, they form stable trimers organized into two-barrel structures. A 12-stranded β-barrel 40 Å long inserts into the outer membrane to form an open pore 30 Å in diameter. An unusual α-helical barrel 100 Å in length protrudes deep into the periplasm, where it reaches the TolC docking domain of AcrB. The lower half of this barrel is bounded by an equatorial domain of mixed α/β-structure. The tip of the periplasmic end of the channel is closed in an iris-like manner by interacting loops of α-helices.

Biochemical and genetic data demonstrate that MFPs interact with both the RND pump and the OM channel [32,33]. It is therefore proposed that the MFP stabilizes weak RND–OMF interactions and promotes and maintains the tripartite complex. Recently determined structures of MexA and AcrA, from *P. aeruginosa* and *E. coli*, respectively [16,17,19], are consistent with such a function. These MFPs appear to a have modular structure, with a long β-barrel domain connected to a lipoyl domain that in turn is attached to a long periplasmic α-helical hairpin. By forcing *E. coli* TolC to make a functional complex with the MexAB-translocase from *P. aeruginosa* (which otherwise is non-functional), it was established that key interactions between the MFP and OMF are located in the equatorial domain and in the vicinity of the coiled coils of the TolC entrance [34]. In addition, the MFP appears

FIGURE 4.2   Fitted model of the *E. coli* AcrAB-TolC tripartite efflux complex. Most of the AcrA trimers are removed by cross-section. There is significant evidence that the "crater" in the fully assembled pump is connected to TolC, either directly or via an adaptor composed of AcrA subunits. The tubular TolC structure serves as an "exhaust pipe" that penetrates the outer membrane, providing the exit route for the substrate efflux into the external medium. The opening indicated by the white color in the AcrB-trimer leads to the periplasmic space.

to possess significant conformational flexibility [35], which might be important to ensure the most advantageous interaction with TolC.

In modeling an "open state" of the TolC entrance, there appears to be a perfect fit with the funnel-like opening of the TolC docking domain of AcrB [36]. The possibility exists that the MFP plays an active role in the opening of the TolC channel during drug transport. It is also possible that the "open state" is the result of the AcrB–TolC interaction, and that the role of the MFP is to keep both proteins in this fixed state.

The oligomeric state of MFPs still remains controversial. Soluble forms of AcrA and MexA have been found to be monomeric *in vitro*, but cross-linking of AcrA *in vivo* suggests that the MFP works as a trimer with its two other partners [37,38].

While many details remain to be clarified, the emerging architecture of the trimeric complex provides a structural basis for understanding trans-envelope efflux. A substrate enters the tripartite transporter through the appropriate inter-envelope "substrate gate" and exits into the extracellular space through the "exhaust pipe" of TolC.

The possibility of periplasmic capture first arose based on genetic and biochemical experiments [39–42] and was subsequently reinforced with the advent of the high-resolution structures of protein–ligand co-crystals.

Interestingly, co-crystals revealed several possible drug interacting sites. In the first report, several unrelated drugs were detected in the central cavity on the membrane–periplasm interface, prompting a model in which drugs first intercalate into the phospholipid bilayer and then diffuse laterally into the central cavity of AcrB [14,43]. Drugs observed in this site are in the vicinity of a few aromatic residues that appear to interact with them through hydrophobic and stacking interactions. However, the observed ligand/protein interactions appeared weak and are insufficient to explain substrate specificity. In fact, the few amino-acid residue side-chains observed to be interacting with the ligands are highly conserved across various efflux pump proteins of broadly varying substrate recognition. Mutational analyses of AcrB and the related EmhB from *Pseudomonas fluorescens* nevertheless show that amino acid residues in the central cavity do have an impact on transporter-mediated antibiotic resistance [13,44].

In the next structure of the same protein, an additional drug binding pocket was detected in a prominent cleft on the surface of the periplasmic domain. This site might be fully exposed to the periplasm, with easy access for drug binding. The problem with this site is that it is not clear where the drug can go from there. Mutations altering several amino acids within this site were also reported to impact RND-related antibiotic resistance [13,41,45]. One intriguing possibility is that this site might play a role in regulation of the activity of the transporter by its substrates, rather than mediating the actual transport process directly.

Finally, a third structure revealed yet another, non-overlapping, multidrug binding pocket, located deep inside the periplasmic domain [9]. This voluminous pocket is extremely rich in aromatic amino acid residues capable of hydrophobic and stacking interactions. There are also a few polar residues that can form hydrogen bonds. Interestingly, the co-crystallized substrates doxorubicin and minocycline were found to be interacting with different sets of amino acid residues. This finding fits very well with the rapidly emerging paradigm of a versatile multi-specific recognition pocket, which was originally proposed based on the results obtained with soluble multidrug binding regulatory proteins, such as BmrR [46,47], QacR [48,49], and PXR [50]. These studies demonstrated (*i*) that different substrates can use different residues to bind in the same pocket; (*ii*) that the same substrates can assume multiple positions in the pocket; (*iii*) that two substrates can be bound simultaneously; and (*iv*) that this binding can give rise to negative or positive cooperativity. Numerous studies on human MDR ABC transporter P-gp, though not at the structural level, illustrate such versatility through the concept of the "induced best fit," by which a substrate can provoke rearrangements in the pocket during binding [51,52].

The first high-resolution structure of the MDR ABC transporter, Sav1866 from *Staphylococcus,* has also been determined [53]. Though the crystals lack substrate, the location of the binding pocket can be easily identified based on a variety of biochemical, mutagenesis, and cross-linking experimental data [54,55], collected over many years of research on ABC transporters. The large chamber within the membrane that is formed by two TMDs opening to the extracellular milieu may also be accessible from the lipid phase at the interfaces between the two TMDs [56], constituting a drug binding pocket for the MDR ABC transporters. The TM domains of P-gp do not contain any charged residues. Thus, ligand/protein interactions may be based solely on H-bonding, hydrophobic, and stacking interactions. The architecture of this binding pocket is very different from the one observed for AcrB. However, multidrug binding appears to follow the same rules.

In addition, this structure provides important details on the mechanism of ATP-driven ABC-mediated efflux, which has been proposed to occur using an "alternating access and release" mechanism. In essence, high affinity substrate binding induces high affinity ATP binding, which in turn induces substrate release, ATP hydrolysis, and subsequent release of ADP to re-set the system.

To return to the RND transporters: as implied earlier, a substrate can reach the binding pocket through the uptake channel. The main entrance to the channel is about 15 Å above the plane of the membrane. It is easily accessible from the periplasmic space via the "vestibules" leading to the central cavity. It seems possible that the role of the weak binding observed previously in the central cavity [43] might be to slow down the lateral diffusion of substrates in the vicinity of the entrance to the uptake channel.

Two recent studies, published side by side, provide further exciting and unexpected insights into possible transport mechanisms by RND transporters [9,11].

The first X-ray structure of the RND pump AcrB was presented as a perfectly symmetric trimer [10]. It gave rise to the so-called "elevator mechanism" [43] of transport, wherein it was proposed that substrates accumulating in the central cavity are actively transported into the upper portal space via a channel that opens along the central axis of the structure. However, in this model, a very significant conformational change associated with channel opening would have to be coupled with proton transport via the TMD in order to accommodate the passage of substrates.

The two new structures of AcrB trimers, while symmetric overall, show each protomer in a distinct conformation [9,11]. Although only one of these "new generation" structures contains co-crystallized substrates [9] the conformations of protomers in both are strikingly similar. This argues that substrate is not needed to induce asymmetry.

In the periplasmic portion, the main differences between the old and new structures are in the substrate binding pocket. First, the substrate is present in only one of the protomers, dubbed the "binding" protomer (B). The spacious drug binding pocket (described earlier) is open to the periplasm and expands far into the porter domain, almost reaching the TolC-docking funnel. The exit from the pocket into the funnel is blocked by the inclined α-helix of the central pore from the adjacent protomer. The binding pocket of this second protomer, called the "extrusion" protomer (E), is closed to the periplasm, significantly reduced in size, and opened toward the funnel. The binding cavity of the third "access" protomer (A) is largely inaccessible from either the periplasm or the exit funnel.

Based on this asymmetric structure, a new mechanism of drug transport has been proposed. This "alternate occupancy" model implies that each protomer cycles through three consecutive conformations, named after F_1F_0-ATPase as loose (L), tight (T), and open (O), corresponding to three phases of efflux [11]. This cycling is sequential, rather than synchronous, such that at any given time each protomer exists in a different phase.

Several sequences for a transport process can be envisioned. In one model [9,11], the first is the L-phase (corresponding to A-protomer), where the substrate gains limited access to the uptake channel. During the second T-phase (corresponding to B-protomer), the uptake channel expands and the substrate enters the voluminous binding pocket. In the last O-phase (corresponding to the E-protomer), the binding pocket disconnects from the periplasm and shrinks in size. The drug is pushed out of the binding pocket (presumably concomitant with the α-helical exit opening) into the funnel, where it can diffuse into the TolC channel. At the same time, one adjacent protomer receives another substrate molecule in the binding pocket, while the third protomer returns to the substrate accepting state.

It should be noted that the role of the L-phase, where the cavity is largely disconnected from either the periplasm or the funnel, is somewhat unclear. Rather than being the first, substrate accepting phase, it may actually be the second. Transport might be initiated by binding into the expanded pocket after entering through the open uptake channel (as in Figure 4.3). After a conformation change, the cavity may be closed off from the periplasm, entering the L-phase. One may hypothesize that the substrate can then be trapped, but unbound in the cavity before the transition to the funnel-open state, when it is released. The volume of the closed cavity is estimated at ~1200 Å^3, sufficient to accommodate substrates over 1000 Da MW.

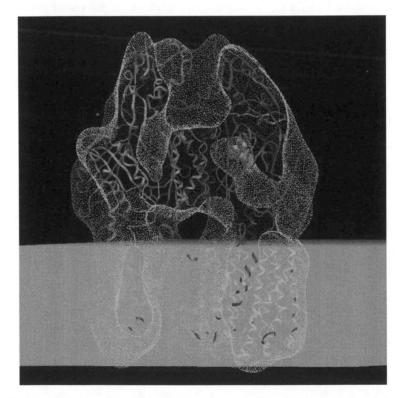

FIGURE 4.3 Structure of AcrB based on the asymmetric crystal. Lateral view of the cross-section of the AcrB/substrate complex X-ray structure. Bound substrate (aquamarine space-filling model) can be observed in the cavity near the end of the deep gorge. The gorge has an opening to the periplasm on the right. The corresponding cavity is empty and opens to the crater-like structure at the top.

Finally, another possibility is that this is in fact the third and truly ligand-free phase following drug binding and drug release.

Based on the obvious analogy, AcrB has been nicknamed a peristaltic pump [11]. Perhaps a more poetic comparison would be with the three-step rotation of the Vienna waltz. During the "dance," each protomer of AcrB functionally rotates through three different positions. This model of RND-mediated efflux is also somewhat reminiscent of an exit through "revolving doors."

Regardless of the sequence of events that produce efflux of the substrate against the concentration gradient, the affinity of the substrate to the periplasm-accessible conformation of the pump subunit cavity is expected to be higher than the affinity to the funnel-opened conformation. The transition of the ligand-bound protomer from a high-affinity to low-affinity state should require energy input, which is evidently provided by the coupled proton transport in the TMDs. While the details of the mechanism remain to be clarified, existing studies provide some initial clues.

Mutagenesis data indicate that AcrB has four electrostatically interactive residues that constitute the putative proton relay pathway, located in TM4, TM10, and TM11 [57,58]. In the binding and access protomers, Lys 940 of TM10 and Asp407 and Asp408 of TM4 are coordinated by salt bridges. However, in the extrusion protomer, Lys940 is turned toward Thr978 and the salt bridges are absent. This is turn causes twisting of TM4 and TM10. Without additional data, it is impossible to say whether or not these subtle changes in TMDs are sufficient to produce the large conformational changes in the porter domain, resulting in ultimate efflux. However, what is absolutely clear is that the residues involved in proton and substrate translocation are, as expected, far apart.

Understanding the mechanism of transport and identification of the multidrug binding pockets will help in the design of new clinically relevant approaches to inhibit drug efflux.

ANTIINFECTIVE DRUG DISCOVERY AND EFFLUX PUMP INHIBITORS

Approaches to combating drug efflux offer new opportunities to combat antibiotic resistance development across the spectrum of drugs in development and in clinical use.

The first category includes antibiotics approved by the FDA from 1998 to 2005, including rifapentine, quinupristin/dalfoprystin, moxifloxacin, gatifloxacin, linezolid, ceftidoren, ertapenem, gemifloxacin, daptomycin, telithromycin, and tigecycline [59,60]. The second category are antibiotics currently in clinical trials, such as doripenem in phase III trials, ceftobiprole, dalbavancin, telavancin, ramaplanin, sitafloxacin, and garenofloxacin, as well as more phase I and phase II fluoroquinolones and β-lactams, and the macrolide EP-420, the oxazolidinone ranbenzolid, the dihydrofolate reductase inhibitor iclaprim, the peptide deformilase inhibitor LMB-415, and the tetracycline analog PTK 0796 [61]. Finally, the third category consists of compounds at various stages of preclinical development, including more β-lactams, fluoroquinolones, oxazolidinones, and ketolides [62] as well as new analogs in the less prevalent rifamycin [63] and lincosamide [64] classes. This category also contains several compounds with novel modes of action, such as the dual GyrB/ParE inhibitor VX-692 [65] and the FabI inhibitor API-1401 [66]. Until recent termination or suspension [67,68], this category included many more compounds with novel modes of actions—compounds targeting DNA polymerase, DNA ligase, tRNA synthases, enzymes essential in cell division, as well as various essential metabolic and unexploited cell wall synthesis enzymes [69,70].

Compounds from all three categories have one thing in common. Almost all have poor activity against *P. aeruginosa* and other recalcitrant Gram-negative bacteria (such as *Acinetobacter* spp., *Stenotrophomonas maltophilia*, and *Burkholderia cepacia*). In fact, most of them lack appropriate activity against any Gram-negative bacteria at all [71]. Only tigecycline, the most recently approved antibiotic [72,73], has potent *in vitro* activity against *Acinetobacter* spp. and *S. maltophilia,* but not *P. aeruginosa* or Proteus and only the carbapenem doripenem is active against *P. aeruginosa* [74,75] (but not the strains producing metallo-β-lactamases. In this respect, it is not different from the currently approved anti-pseudomonal carbapenems imipenem and meropenem). This general trend in the lack of Gram-negative activity among both recently approved antibiotics and compounds in the developmental pipeline is rather remarkable considering the plethora of drug targets covered, both old and new, and the fact that most of the compounds have the potential to be truly broad spectrum, since they inhibit the activity of "genomically correct" targets that are well conserved in both Gram-positive and Gram-negative bacteria.

Efflux-mediated intrinsic and acquired resistance is well documented even for the limited number of antibiotics (fluoroquinolones, β-lactams, and aminoglycosides), which are available for the treatment of *P. aeruginosa* and similar recalcitrant Gram-negative bacteria [76,77].

An alternative and potentially more elegant approach is to identify and develop compounds, which avoid efflux pumps altogether rather than to resort to combination therapy with EPIs. Indeed, there are several examples where this approach has been perfectly successful for Gram-positive bacteria. In the first example, the newer fluoroquinolones, levofloxacin, moxifloxacin, gemifloxacin, gatifloxacin, and garenofloxacin are not affected by the MDR pumps NorA and PmrA, in *Staphylococcus aureus* or *Streptococcus pneumoniae,* respectively [78–83]. It is believed they are sufficiently hydrophobic, so that their rapid passive uptake overwhelms active efflux from the cell. Sparfloxacin has been shown to non-competitively inhibit the NorA transporter, which may be the reason why susceptibility to sparfloxacin is not affected by the NorA pump [84]. In the case of tetracycline derivatives tigecycline [85,86] and PTK 0796 [87], the mechanism of resistance to efflux is due to the lack of recognition by the tetracycline-specific transporters (TetA-D and TetK, M). Note, however, that the "out-foxed" efflux pumps mentioned above are either antibiotic specific or are MDR transporters from Gram-positive bacteria. The same antibiotics are still effectively extruded by MDR transporters from Gram-negative bacteria.

In the late 1990s, Microcide and Daiichi Pharmaceuticals undertook a comprehensive program to search for and develop EPIs for Gram-negative bacteria. The specific goal of their collaborative program was to potentiate the activity of levofloxacin, a substrate for multiple homologous tripartite MDR pumps belonging to the RND family of transporters. Four of these tripartite pumps in *P. aeruginosa* (MexAB-OprM, MexCD-OprJ, MexEF-OprN, and MexXY-OprM) are capable of conferring clinical resistance to this antibiotic [76]. High-throughput screening identified an inhibitor, MC-207,110, capable of reducing Mex-mediated efflux resistance to fluoroquinolones in *P. aeruginosa* [88]. This compound effectively inhibits all four clinically relevant *P. aeruginosa* pumps as well as similar RND pumps from other Gram-negative bacteria. Based on the broad spectrum of pump inhibition in various Gram-negative bacteria, compounds in this drug class are considered broad-spectrum EPIs. Interestingly, not all antibiotic substrates for a given pump are potentiated by MC-207,110. The degree of pump inhibition is dependent on the nature of the substrate. For example, MC-207,110 potentiates fluoroquinolones, macrolides/ketolides, oxazolidinones, chloramphenicol, and rifampicin, but not β-lactams or aminoglycosides. Mechanism of action studies indicated that MC-207,110 itself is a substrate of efflux pumps [89]. The assumption is that different antibiotics have non-identical binding pockets within the transporter protein and that MC-207,110 works by competing with antibiotics for binding in the substrate pocket specific to the potentiated antibiotic, but not to the binding site for the non-potentiated antibiotics, explaining the substrate-dependent inhibition. This may also explain why attempts to isolate target-based mutations conferring resistance to MC-207,110 (making the efflux pump non-susceptible to inhibition) were unsuccessful (Lomovskaya, unpublished results). Most likely, such mutations would render the pump incapable of interacting with other substrates and hence be observed as inactive. This being the case, specific targeting of the pump substrate-binding site may be a viable future strategy to design alternative or improved EPIs. It also became clear during the course of the program that in any empiric search for EPIs it is very important to identify and use specific partner antibiotics.

MC-207,110 decreased intrinsic resistance to levofloxacin about eight-fold in wild-type strains of *P. aeruginosa*, while efflux pump over-expressing strains susceptibility may be increased up to 64-fold. This same degree of potentiation is observed irrespective of the presence of target-based mutations in DNA gyrase. Recent clinical isolates of *P. aeruginosa* with a wide range of resistant phenotypes also showed increased susceptibility to levofloxacin in the presence of MC-207,110. Remarkably, both the MIC_{50} and the MIC_{90} were decreased to the same extent, 16-fold by MC-207,110, providing additional evidence that the potentiating effect is not dependent on the absolute level of resistance, but solely on the level of efflux pump expression. In the presence of the inhibitor, both MIC_{50} and MIC_{90} were below the susceptibility barrier for levofloxacin.

Of particular importance is the observation that the selection frequency for fluoroquinolone-resistant bacteria was also dramatically decreased in the presence of MC-207,110. The appearance of both efflux-mediated and target-based mutations is minimized. This is presumably because the inhibitor decreases MexAB-OprM-mediated intrinsic resistance to the level at which a single target-based mutation does not confer enough resistance to emerge under selection conditions [88]. Suppression of resistance development was also demonstrated *in vivo* using later stage compounds, in the neutropenic mouse thigh model of *P. aeruginosa* infection [90,91].

The attractiveness of MC-207,110 as a lead was based on its broad-spectrum efflux pump inhibitory activity. Such broad-spectrum activity is absolutely needed in order to have a clinically significant impact on fluoroquinolones, which are extruded by multiple efflux pumps. However, broad-spectrum EPI activity is not always essential. For example, resistance to aminoglycosides is conferred by a single pump, MexXY-OprM [92], while it is mainly MexAB-OprM, which confers resistance to β-lactams in *P. aeruginosa* [93,94].

EPIs with high selectivity toward MexAB-OprM were also identified in the Microcide-Daiichi collaboration [95–98]. One such series of compounds, the pyrido-pyrimidines, unlike MC-207,110, inhibit the efflux of all substrates of the MexAB-OprM pump, including β-lactams and fluoroquinolones, with little effect on the other Mex systems. In later studies, mutagen-induced mutations conferring resistance to potentiation by these compounds were identified in the MexB gene [Lomovskaya, unpublished results]. Of note is that these mutations were not cross resistant to MC-207,110, confirming the differences in modes of action of these two types of EPIs. Resistance studies demonstrated that while mutants with simultaneous resistance to β-lactams and fluoroquinolones due to overexpression of MexAB-OprM could be easily isolated, no such mutants were selected in the presence of MexAB-OprM-selective EPIs [Lomovskaya, unpublished results].

These studies demonstrated that multiple inhibitors of a single pump could be identified. They also validated the belief that inhibition of efflux pumps is a viable strategy to reversing antibiotic resistance and blocking its development [99–102]. Extensive efforts have been made to improve the potency and the absorption, distribution, metabolism, excretion, and toxicity (ADMET) profile of this compound series [91,90,103,104]. It has two basic moieties that were shown to be essential for activity. Unfortunately, the same moieties were found to be associated with unfavorable pharmacokinetic and toxicological profiles, and development of this lead series was suspended. While downgraded at least temporarily from drug candidate status, these compounds are widely used as a research tool, owing to broad-spectrum EPI activity, to evaluate the contribution of efflux pumps to antibiotic resistance in clinical isolates of *P. aeruginosa* and other Gram-negative bacteria [105–113].

Several other structural classes of inhibitors of the RND transporters are described in the literature, though none has been reported to have EPI activity against efflux pumps from *P. aeruginosa*. In a high-throughput screening assay for potentiators of novobiocin, scientists at Pharmacia, now Pfizer, identified 3-arylpiperidines as inhibitors of the AcrAB-TolC pump from *E. coli* [114]. In a report from University Hospital in Freiburg, Germany, selected arylpiperazines have been shown to inhibit the same transporter and to potentiate activity of its multiple antibiotic substrates [115]. Several laboratories in France undertook systematic efforts to identify and characterize various alkoxy- and alkylaminoquinolines as EPIs showing significant activity against laboratory and clinical strains of *Enterobacter aerogenes* and *Klebsiella pneumoniae* [116–119]. To the authors' knowledge, no large-scale development programs have been initiated based on these compounds.

PERSPECTIVES FOR DEVELOPING MDR INHIBITORS

It is also desirable that the antibiotic and the EPI should not engage in drug–drug interactions. In this respect, some lessons may be learned from the clinical experience with inhibitors of P-gp and other ABC-transporters as reversing agents for combination with anticancer drugs. The search for such compounds started in the mid-1970s, almost concomitant with the discovery of P-gp [120–125]. Several P-gp inhibitors have since failed in clinical trials. Perhaps the main reason for this is that P-gp and some other ABC-transporters have a distinct physiological function in the human body: they protect various cells from endogenous toxic metabolites and xenobiotics, as well as the cytotoxic anticancer drugs themselves. In addition, they participate in drug disposition. As a result, in the presence of P-pg inhibitors exposure to the co-administered cytotoxic drug in normal cells increased, resulting in toxicity. Several more potent and selective agents are undergoing clinical development at the present time.

It is expected that the introduction of inhibitors of bacterial RND transporters as antiinfective agents to the clinic might be more expeditious than for MDR-reversing agents for cancer therapy since no close human homologs exist and therefore no target-based toxicity is expected. In addition, based on the emerging significance of RND transporters in bacterial pathogenesis one can imagine EPIs as stand-alone, anti-virulence agents.

Importantly, an EPI interacting with the RND transporter in the binding site may have different modes of inhibition. They might compete directly for binding with other substrates, but they might also facilitate such binding due to positive cooperativity and thereby prevent dissociation of substrates from the pump.

In addition, based on the structure of AcrB, two of the three different conformations of the binding pocket offer the potential for inhibition without requiring that the drug reach the cytoplasm. The T-phase (or B-protomer) conformer might be accessed from the periplasm, and the O-phase (or E-protomer) from the TolC-docking funnel. These two conformations might be targeted independently.

Efflux inhibitors might also act at sites distinct from those involved in substrate binding, but whose disruption impacts overall pump activity. Such allosteric inhibitors would be expected to inhibit efflux of all substrates and therefore potentiate the activity of multiple antibiotics. A series of structurally diverse inhibitors with high selectivity toward the MexAB-OprM efflux pump from *P. aeruginosa* have been identified [95–98,126,127] and shown to negatively impact the export of all MexAB-OprM antimicrobial substrates equally [95]. It was hypothesized that these EPIs bind not to substrate-binding sites on the pump but rather to site(s) that modulate pump activity (i.e., modulation sites). Several alkoxy- and alkylaminoquinoline EPIs (Figure 4.4) showing activity against clinical strains of *E. aerogenes* have been reported [107,116,119,128] that potentiate the activities of all antimicrobials tested equally, consistent with action at a modulation site of an RND-type efflux system.

At the present time, it is unclear whether or not the RND transporters do, in fact, have a "dedicated" modulation site, but the empirical observation of a link between the ability to potentiate multiple substrates ("modulator mode") and high selectivity toward specific RND transporters is suggestive of such a feature.

Other possibilities for interfering with efflux include targeting the assembly of the pump components and blocking the TolC-like tunnel. At the present time, these are purely hypothetical; there are no reports of molecules with such activity yet.

An alternative approach is to screen libraries of known drugs. The identification of a novel mode of action in an approved drug could significantly shorten the

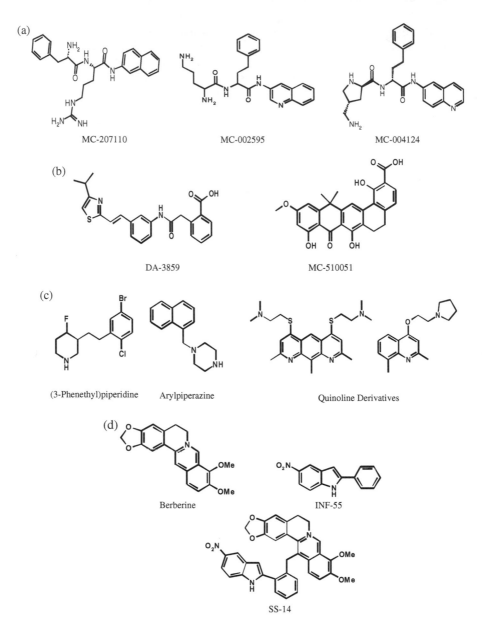

FIGURE 4.4 Various inhibitors of RND transporters. (a) Broad-spectrum efflux pump inhibitors (EPIs) with activity against multiple RND pumps, including those in *P. aeruginosa*. These inhibitors are themselves substrates of efflux pumps and most probably interact with the transporter via a substrate-binding pocket. (b) Narrow-spectrum EPIs with selective activity against the MexAB-OprM complex from *P. aeruginosa*. These inhibitors might interact with the transporter at the allosteric "modulator" site. (c) EPIs with activity against transporters from various species of Enterobacteriacea. Their mode of action is mostly uncharacterized. (d) Dual action antiinfectives with NorA pump blocker linked to anti-bacterial agent.

development pathway and mitigate the risks inherent in an new chemical entity (NCE). However, for such a proposition to be practical, the EPI activity needs to be much higher in potency compared to the original pharmacological activity. An interesting possibility would be to discover an EPI mode of action against RND transporters in compounds that themselves are antiinfective agents, but are used for a different indication. One such compound, MP-601,205 [129], was identified by scientists at Mpex Pharmaceuticals. This compound entered phase I clinical trials in cystic fibrosis patients, but the program is currently on hold due to concerns around drug tolerability.

Recently several efficient pharmaco-informatic methods have been reported for the identification of new inhibitors of P-gp [122]. Leads were discovered by ligand-based virtual screening of large, commercially available libraries of compounds. These approaches are based on machine learning algorithms and require a relatively large number of known inhibitors to be used as a training set. In the case of P-gp, many structurally unrelated compounds are available for this purpose. It is expected that as a greater number of more diverse bacterial EPIs are identified through "wet" screening, the more feasible alternative screening methods, including various virtual screening protocols, will become.

Finally, based on the characteristics of the binding sites of RND pumps, another approach might be the synthesis of flexible molecules carrying multiple aromatic moieties that could bind with high affinity into recognition cavities by inducing the best fit in the binding pocket. A complementary approach might be the synthesis of inhibitor dimers [130] and substrate-inhibitor hybrids, discussed below.

NATURAL MDR INHIBITORS

Attempts to find potent, non-toxic, broad-spectrum antibiotics from plants, and more specifically from medicinal plants, have failed, even though large-scale screens have been undertaken by both the pharmaceutical and the biotech industries [Merck, Lynn Silver, personal communication; Emergent BioSolutions Inc, Joanna Clancy, personal communication; Phytera; Shaman]. One conclusion from the failure to identify potent broad-spectrum antibacterial compounds in plants is that plants utilize a different chemical strategy for the control of microbial infections. For example, plant antibacterials may act in combinations and have little efficacy on their own. This idea has been tested in one of our laboratories (K.L.) using the plant alkaloid berberine. Berberine is widespread in nature and is present in common Barberry (*Berberis*) plants and is the principal component of the medicinal plant golden seal (*Hydrastis canadensis*). Berberine is a hydrophobic cation that increases membrane permeability and intercalates into DNA [131]. Moreover, its positive charge leads to its active accumulation in bacterial cells [132]. Nevertheless, in spite of its apparently excellent properties, berberine is ineffective as an antibacterial because it is readily extruded by pathogen-encoded MDR pumps [133]. Hydrophobic cations such as berberine are actually the preferred substrates of all classes of MDRs [133,134]. Reasoning that plants would benefit from blocking this efflux, an MDR inhibitor, 5'-methoxyhydno-carpin (MHC) (Figure 4.4) was isolated from berberis plants. MHC blocks major facilitator (MF) pumps of Gram-positive bacteria, which are drug/proton antiporters and have an especially strong bias toward cationic substrates. A combination of MHC and berberine produced a potent antibacterial [135].

The finding of MHC in berberis plants provided support for the hypothesis that in general, plant antibacterial compounds are individually relatively weak, but function in synergy. This led to a broader question—are plant antibacterials generally limited in their efficacy by MDR efflux? Testing a random collection of compounds against a panel of bacterial pathogens showed that disabling MDRs by mutation and/or addition of synthetic MDR inhibitors improved antibacterial activity in all cases tested, and for some compounds quite dramatically [136]. For example, rhein, the principal antibacterial from rhubarb, was potentiated 100- to 2000-fold (depending on the bacterial species) by disabling MDRs. Comparable potentiation was observed with plumbagin, resveratrol, gossypol, and coumestrol. The extent of potentiation was the largest in the case of Gram-negative bacteria. For example, rhein had no activity against *E. coli* at the limit of solubility (500 µg/mL), whereas disabling MDRs produced an MIC of 0.25 µg/mL. In principle, MDR efflux could be countered by simply increasing the concentration of the antibacterial above the saturation of the pumps. This is indeed how commercial antibiotics are empirically dosed. It is possible that plants may also produce antimicrobials in high concentrations for the same reason, but as the above example shows, this strategy would not work with rhein.

So far, several types of plant-derived inhibitors such as MHC mentioned above have been reported [135,137–141], but all of them are inhibitors of MF MDRs of Gram-positive bacteria. Indeed, plants seem to do very well in protecting themselves against Gram-positive bacteria (reflected, perhaps, in the fact that essentially all agriculturally significant bacterial pathogens are Gram-negatives), in part due to multiple weak antibacterials and the presence of MF MDR inhibitors. But do plants make inhibitors that are effective against RND MDRs of Gram-negative species? This remains an intriguing open question.

One of the possible approaches to EPI development is to physically combine an antibiotic with a pump inhibitor. This concept was tested by covalently linking berberine to INF_{55}, an inhibitor of MF MDRs. The resulting hybrid SS14 showed excellent penetration into cells, and was superior to berberine and INF_{55} added together [142]. INF_{55} is a fairly toxic molecule, but unlike the inhibitor, the hybrid showed no toxicity in a *Caenorhabditis elegans* model of enterococcal infection, curing worms of the pathogen [143]. This is an example of making an agent less toxic as an indirect result of improving its efficacy, by changing it so greatly as to create a new molecule. The efficacy of the hybrid provides an important proof-of-principle for this approach, and opens the way for similar compounds aimed at bypassing RND efflux in Gram-negative species.

A very different approach to bypassing MDR efflux has been examined in the design of "sterile surface" polymers. The rationale was to create a polymer of antimicrobial molecules long enough to enable penetration across the cell envelope. Polymers of hydrophobic cations, such as vinyl *N*-hexyl pyridinium and *N*-alkyl PEI attached covalently to the surface of materials, rapidly killed Gram-positive and Gram-negative bacteria [144,145]. As noted above, hydrophobic cations have the advantage of being actively accumulated in the cells of pathogens, but MDR efflux counters this ability. For example, benzalkonium chloride is a hydrophobic cation used widely as an antiseptic, and whose efficacy is limited by MDR efflux [132]. Importantly, antimicrobial activity of polymeric hydrophobic cations is not affected by MDR efflux [146]. Apparently, the pumps can extrude small molecules, but not large polymers. This enables creation of effective sterile surface materials that could prevent the spread of pathogens and serve to inhibit biofilm formation [147].

CONCLUDING REMARKS

Where does industry stand with regard to developing EPIs? There is indisputably high awareness of the contribution of efflux to both intrinsic and acquired antibiotic resistance as well as practical attention given to the implications for antibiotic development. In fact, most researchers involved in antibacterial drug discovery and development routinely evaluate the impact of efflux pumps on their "favorite" compounds, using either strains lacking efflux pumps or broad-spectrum efflux inhibitors, such as MC-207110. Despite such appreciation, to our knowledge there is no industry-wide effort to discover and develop small molecule inhibitors of RND efflux pumps from Gram-negative bacteria to overcome resistance and improve clinical outcomes of antibiotic treatment of recalcitrant infections, such as those associated with conditions such as cystic fibrosis and nosocomial respiratory diseases.

REFERENCES

1. Saier, M. H., Jr. and I. T. Paulsen. 2001. Phylogeny of multidrug transporters. *Semin Cell Dev Biol* 12:205–13.
2. Sanchez, L., W. Pan, M. Vinas, and H. Nikaido. 1997. The acrAB homolog of *Haemophilus influenzae* codes for a functional multidrug efflux pump. *J Bacteriol* 179:6855–7.
3. Sen, K., J. Hellman, and H. Nikaido. 1988. Porin channels in intact cells of *Escherichia coli* are not affected by Donnan potentials across the outer membrane. *J Biol Chem* 263:1182–7.
4. Lomovskaya, O. and M. Totrov. 2005. Vacuuming the periplasm. *J Bacteriol* 187: 1879–83.
5. Tseng, T. T., K. S. Gratwick, J. Kollman, D. Park, D. H. Nies, A. Goffeau, and M. H. Saier, Jr. 1999. The RND permease superfamily: an ancient, ubiquitous and diverse family that includes human disease and development proteins. *J Mol Microbiol Biotechnol* 1:107–25.
6. Zgurskaya, H. I. and H. Nikaido. 2002. Mechanistic parallels in bacterial and human multidrug efflux transporters. *Curr Protein Pept Sci* 3:531–40.
7. Ko, D. C., M. D. Gordon, J. Y. Jin, and M. P. Scott. 2001. Dynamic movements of organelles containing Niemann-Pick C1 protein: NPC1 involvement in late endocytic events. *Mol Biol Cell* 12:601–14.
8. Bale, A. E. and K. P. Yu. 2001. The hedgehog pathway and basal cell carcinomas. *Hum Mol Genet* 10:757–62.
9. Murakami, S., R. Nakashima, E. Yamashita, T. Matsumoto, and A. Yamaguchi. 2006. Crystal structures of a multidrug transporter reveal a functionally rotating mechanism. *Nature* 443:173–9.
10. Murakami, S., R. Nakashima, E. Yamashita, and A. Yamaguchi. 2002. Crystal structure of bacterial multidrug efflux transporter AcrB. *Nature* 419:587–93.
11. Seeger, M. A., A. Schiefner, T. Eicher, F. Verrey, K. Diederichs, and K. M. Pos. 2006. Structural asymmetry of AcrB trimer suggests a peristaltic pump mechanism. *Science* 313:1295–8.

12. Su, C. C., M. Li, R. Gu, Y. Takatsuka, G. McDermott, H. Nikaido, and E. W. Yu. 2006. Conformation of the AcrB multidrug efflux pump in mutants of the putative proton relay pathway. *J Bacteriol* 188:7290–6.

13. Yu, E. W., J. R. Aires, G. McDermott, and H. Nikaido. 2005. A periplasmic drug-binding site of the AcrB multidrug efflux pump: a crystallographic and site-directed mutagenesis study. *J Bacteriol* 187:6804–15.

14. Yu, E. W., G. McDermott, H. I. Zgurskaya, H. Nikaido, and D. E. Koshland, Jr. 2003. Structural basis of multiple drug-binding capacity of the AcrB multidrug efflux pump. *Science* 300:976–80.

15. Akama, H., M. Kanemaki, M. Yoshimura, T., et al. 2004. Crystal structure of the drug discharge outer membrane protein, OprM, of *Pseudomonas aeruginosa*: dual modes of membrane anchoring and occluded cavity end. *J Biol Chem* 279:52816–9.

16. Akama, H., T. Matsuura, S. Kashiwagi, H. Yoneyama, S. Narita, T. Tsukihara, A. Nakagawa, and T. Nakae. 2004. Crystal structure of the membrane fusion protein, MexA, of the multidrug transporter in *Pseudomonas aeruginosa*. *J Biol Chem* 279:25939–42.

17. Higgins, M. K., E. Bokma, E. Koronakis, C. Hughes, and V. Koronakis. 2004. Structure of the periplasmic component of a bacterial drug efflux pump. *Proc Natl Acad Sci USA* 101:9994–9.

18. Koronakis, V., A. Sharff, E. Koronakis, B. Luisi, and C. Hughes. 2000. Crystal structure of the bacterial membrane protein TolC central to multidrug efflux and protein export. *Nature* 405:914–9.

19. Mikolosko, J., K. Bobyk, H. I. Zgurskaya, and P. Ghosh. 2006. Conformational flexibility in the multidrug efflux system protein AcrA. *Structure* 14:577–87.

20. Grkovic, S., M. H. Brown, and R. A. Skurray. 2002. Regulation of bacterial drug export systems. *Microbiol Mol Biol Rev* 66:671–701, table of contents.

21. Grkovic, S., M. H. Brown, and R. A. Skurray. 2001. Transcriptional regulation of multidrug efflux pumps in bacteria. *Semin Cell Dev Biol* 12:225–37.

22. Miller, P. F. and M. C. Sulavik. 1996. Overlaps and parallels in the regulation of intrinsic multiple-antibiotic resistance in *Escherichia coli*. *Mol Microbiol* 21:441–8.

23. Poole, K. and R. Srikumar. 2001. Multidrug efflux in *Pseudomonas aeruginosa*: components, mechanisms and clinical significance. *Curr Top Med Chem* 1:59–71.

24. Piddock, L. J. 2006. Multidrug-resistance efflux pumps—not just for resistance. *Nat Rev Microbiol* 4:629–36.

25. Aendekerk, S., S. P. Diggle, Z. Song, N. Hoiby, P. Cornelis, P. Williams, and M. Camara. 2005. The MexGHI-OpmD multidrug efflux pump controls growth, antibiotic suscepti-bility and virulence in *Pseudomonas aeruginosa* via 4-quinolone-dependent cell-to-cell communication. *Microbiology* 151:1113–25.

26. Chan, Y. Y. and K. L. Chua. 2005. The *Burkholderia pseudomallei* BpeAB-OprB efflux pump: expression and impact on quorum sensing and virulence. *J Bacteriol* 187:4707–19.

27. Domenech, P., M. B. Reed, and C. E. Barry III. 2005. Contribution of the *Mycobacterium tuberculosis* MmpL protein family to virulence and drug resistance. *Infect Immun* 73:3492–501.

28. Poole, K. 2004. Efflux-mediated multiresistance in Gram-negative bacteria. *Clin Microbiol Infect* 10:12–26.

29. Zgurskaya, H. I. 2002. Molecular analysis of efflux pump-based antibiotic resistance. *Int J Med Microbiol* 292:95–105.

30. Nikaido, H. 2001. Preventing drug access to targets: cell surface permeability barriers and active efflux in bacteria. *Semin Cell Dev Biol* 12:215–23.

31. Federici, L., D. Du, F. Walas, H. Matsumura, J. Fernandez-Recio, K. S. McKeegan, M. I. Borges-Walmsley, B. F. Luisi, and A. R. Walmsley. 2005. The crystal structure of the outer membrane protein VceC from the bacterial pathogen *Vibrio cholerae* at 1.8 Å resolution. *J Biol Chem* 280:15307–14.

32. Tikhonova, E. B. and H. I. Zgurskaya. 2004. AcrA, AcrB, and TolC of *Escherichia coli* form a stable intermembrane multidrug efflux complex. *J Biol Chem* 279:32116–24.

33. Touze, T., J. Eswaran, E. Bokma, E. Koronakis, C. Hughes, and V. Koronakis. 2004. Interactions underlying assembly of the *Escherichia coli* AcrAB-TolC multidrug efflux system. *Mol Microbiol* 53:697–706.

34. Bokma, E., E. Koronakis, S. Lobedanz, C. Hughes, and V. Koronakis. 2006. Directed evolution of a bacterial efflux pump: Adaptation of the *E. coli* TolC exit duct to the *Pseudomonas* MexAB translocase. *FEBS Lett* 580:5339–43.

35. Ip, H., K. Stratton, H. Zgurskaya, and J. Liu. 2003. pH-induced conformational changes of AcrA, the membrane fusion protein of *Escherichia coli* multidrug efflux system. *J Biol Chem* 278:50474–82.

36. Fernandez-Recio, J., F. Walas, L. Federici, J. Venkatesh Pratap, V. N. Bavro, R. N. Miguel, K. Mizuguchi, and B. Luisi. 2004. A model of a transmembrane drug-efflux pump from Gram-negative bacteria. *FEBS Letters* 578:5–9.

37. Zgurskaya, H. I. and H. Nikaido. 1999. AcrA is a highly asymmetric protein capable of spanning the periplasm. *J Mol Biol* 285:409–420.

38. Zgurskaya, H. I. and H. Nikaido. 2000. Cross-linked complex between oligomeric periplasmic lipoprotein AcrA and the inner-membrane-associated multidrug efflux pump AcrB from *Escherichia coli*. *J Bacteriol* 182:4264–7.

39. Aires, J. R. and H. Nikaido. 2005. Aminoglycosides are captured from both periplasm and cytoplasm by the AcrD multidrug efflux transporter of *Escherichia coli*. *J Bacteriol* 187:1923–9.

40. Elkins, C. A. and H. Nikaido. 2003. Chimeric analysis of AcrA function reveals the importance of its C-terminal domain in its interaction with the AcrB multidrug efflux pump. *J Bacteriol* 185:5349–56.

41. Mao, W., M. S. Warren, D. S. Black, T. Satou, T. Murata, T. Nishino, N. Gotoh, and O. Lomovskaya. 2002. On the mechanism of substrate specificity by resistance nodulation division (RND)-type multidrug resistance pumps: the large periplasmic loops of MexD from *Pseudomonas aeruginosa* are involved in substrate recognition. *Mol Microbiol* 46:889–901.

42. Tikhonova, E. B., Q. Wang, and H. I. Zgurskaya. 2002. Chimeric analysis of the multicomponent multidrug efflux transporters from Gram-negative bacteria. *J Bacteriol* 184:6499–507.

43. Yu, E. W., J. R. Aires, and H. Nikaido. 2003. AcrB multidrug efflux pump of *Escherichia coli*: composite substrate-binding cavity of exceptional flexibility generates its extremely wide substrate specificity. *J Bacteriol* 185:5657–64.

44. Hearn, E. M., M. R. Gray, and J. M. Foght. 2006. Mutations in the central cavity and periplasmic domain affect efflux activity of the resistance-nodulation-division pump EmhB from *Pseudomonas fluorescens* cLP6a. *J Bacteriol* 188:115–23.

45. Middlemiss, J. K. and K. Poole. 2004. Differential impact of MexB mutations on substrate selectivity of the MexAB-OprM multidrug efflux pump of *Pseudomonas aeruginosa*. *J Bacteriol* 186:1258–69.

46. Heldwein, E. E. and R. G. Brennan. 2001. Crystal structure of the transcription activator BmrR bound to DNA and a drug. *Nature* 409:378–82.

47. Zheleznova, E. E., P. N. Markham, A. A. Neyfakh, and R. G. Brennan. 1999. Structural basis of multidrug recognition by BmrR, a transcription activator of a multidrug transporter. *Cell* 96:353–62.

48. Grkovic, S., K. M. Hardie, M. H. Brown, and R. A. Skurray. 2003. Interactions of the QacR multidrug-binding protein with structurally diverse ligands: implications for the evolution of the binding pocket. *Biochemistry* 42:15226–36.

49. Schumacher, M. A., M. C. Miller, and R. G. Brennan. 2004. Structural mechanism of the simultaneous binding of two drugs to a multidrug-binding protein. *EMBO J* 23:2923–30.

50. Watkins, R. E., G. B. Wisely, L. B. Moore, J. L., et al. 2001. The human nuclear xenobiotic receptor PXR: structural determinants of directed promiscuity. *Science* 292:2329–33.
51. Loo, T. W., M. C. Bartlett, and D. M. Clarke. 2003. Simultaneous binding of two different drugs in the binding pocket of the human multidrug resistance P-glycoprotein. *J Biol Chem* 278:39706–10.
52. Loo, T. W., M. C. Bartlett, and D. M. Clarke. 2003. Substrate-induced conformational changes in the transmembrane segments of human P-glycoprotein. Direct evidence for the substrate-induced fit mechanism for drug binding. *J Biol Chem* 278:13603–6.
53. Dawson, R. J. and K. P. Locher. 2006. Structure of a bacterial multidrug ABC transporter. *Nature* 443:180–5.
54. Loo, T. W. and D. M. Clarke. 2005. Recent progress in understanding the mechanism of P-glycoprotein-mediated drug efflux. *J Membr Biol* 206:173–85.
55. Pleban, K., S. Kopp, E. Csaszar, M. Peer, T. Hrebicek, A. Rizzi, G. F. Ecker, and P. Chiba. 2005. P-glycoprotein substrate binding domains are located at the transmembrane domain/transmembrane domain interfaces: a combined photoaffinity labeling-protein homology modeling approach. *Mol Pharmacol* 67:365–74.
56. Rosenberg, M. F., G. Velarde, R. C. Ford, C., et al. 2001. Repacking of the transmembrane domains of P-glycoprotein during the transport ATPase cycle. *EMBO J* 20:5615–25.
57. Goldberg, M., T. Pribyl, S. Juhnke, and D. H. Nies. 1999. Energetics and topology of CzcA, a cation/proton antiporter of the resistance-nodulation-cell division protein family. *J Biol Chem* 274:26065–70.
58. Takatsuka, Y. and H. Nikaido. 2006. Threonine-978 in the transmembrane segment of the multidrug efflux pump AcrB of *Escherichia coli* is crucial for drug transport as a probable component of the proton relay network. *J Bacteriol* 188:7284–9.
59. Bush, K. 2004. Antibacterial drug discovery in the 21st century. *Clin Microbiol Infect* 10 Suppl 4:10–7.
60. Spellberg, B., J. H. Powers, E. P. Brass, L. G. Miller, and J. E. Edwards, Jr. 2004. Trends in antimicrobial drug development: implications for the future. *Clin Infect Dis* 38:1279–86.
61. Bush, K., M. Macielag, and M. Weidner-Wells. 2004. Taking inventory: antibacterial agents currently at or beyond phase 1. *Curr Opin Microbiol* 7:466–76.
62. Barrett, J. and T. Dougherty. 2004. Antibacterial drug discovery and development summit. *Expert Opin Investig Drugs* 13:715–21.
63. Rothstein, D., S. Mullin, K. Sirokman, C. Hazlett, A. Doye, J. Gwathmey, J. Van Duzer, and C. Murphy. 2004. Presented at the 44th Interscience Conference on Antimicrobial Agents and Chemotherapy, Washington, DC.
64. Park, C., J. Blais, S. Lopez, M., et al. 2004. Presented at the 44th Interscience Conference on Antimicrobial Agents and Chemotherapy, Washington, DC.
65. Mani, N., D. Stamos, A. Grillot, J. Parsons, C. Gross, P. Charifson, and T. Grossman. 2004. Presented at the 44th Interscience Conference on Antimicrobial Agents and Chemotherapy, Washington, DC.
66. Bardouniotis, E., R. Thaladaka, N. Walsh, M. Dorsey, M. Schmid, and N. Kaplan. 2004. Presented at the 44th Interscience Conference on Antimicrobial Agents and Chemotherapy, Washington, DC.
67. Overbye, K. and J. Barrett. 2005. Antibiotics: Where did we go wrong? *Drug Discov Today* 10:45–52.
68. Silver, L. 2005. A retrospective on the failures and successes of antibacterial drug discovery. *IDrugs* 8:651–655.
69. Ali, S., A. Gill, and A. Lewendon. 2002. Novel targets for antibiotic drug design. *Curr Opin Investig Drugs* 3:1712–7.

70. Rogers, B. 2004. Bacterial targets to antimicrobial leads and development candidates. *Curr Opin Drug Discov Devel* 7:211–22.
71. Meyer, A. L. 2005. Prospects and challenges of developing new agents for tough Gram-negatives. *Curr Opin Pharmacol* 5:490–94.
72. Hoban, D. J., S. K. Bouchillon, B. M. Johnson, J. L. Johnson, and M. J. Dowzicky. 2005. *In vitro* activity of tigecycline against 6792 Gram-negative and Gram-positive clinical isolates from the global Tigecycline Evaluation and Surveillance Trial (TEST Program, 2004). *Diagn Microbiol Infect Dis* 52:215–27.
73. Noskin, G. A. 2005. Tigecycline: a new glycylcycline for treatment of serious infections. *Clin Infect Dis* 41 Suppl 5:S303–14.
74. Chen, Y., E. Garber, Q. Zhao, Y. Ge, M. A. Wikler, K. Kaniga, and L. Saiman. 2005. *In vitro* activity of doripenem (S-4661) against multidrug-resistant Gram-negative bacilli isolated from patients with cystic fibrosis. *Antimicrob Agents Chemother* 49:2510–1.
75. Mushtaq, S., Y. Ge, and D. M. Livermore. 2004. Doripenem versus *Pseudomonas aeruginosa in vitro*: activity against characterized isolates, mutants, and transconjugants and resistance selection potential. *Antimicrob Agents Chemother* 48:3086–92.
76. Poole, K. 2005. Efflux-mediated antimicrobial resistance. *J Antimicrob Chemother* 56:20–51.
77. Rossolini, G. M. and E. Mantengoli. 2005. Treatment and control of severe infections caused by multiresistant *Pseudomonas aeruginosa*. *Clin Microbiol Infect* 11 Suppl 4:17–32.
78. Harding, I. and I. Simpson. 2000. Fluoroquinolones: is there a different mechanism of action and resistance against *Streptococcus pneumoniae*? *J Chemother* 12 Suppl 4:7–15.
79. Hooper, D. C. 2000. Mechanisms of action and resistance of older and newer fluoroquinolones. *Clin Infect Dis* 31 Suppl 2:S24–8.
80. Ince, D. and D. C. Hooper. 2001. Mechanisms and frequency of resistance to gatifloxacin in comparison to AM-1121 and ciprofloxacin in *Staphylococcus aureus*. *Antimicrob Agents Chemother* 45:2755–64.
81. Ince, D., X. Zhang, L. C. Silver, and D. C. Hooper. 2002. Dual targeting of DNA gyrase and topoisomerase IV: target interactions of garenoxacin (BMS-284756, T-3811ME), a new desfluoroquinolone. *Antimicrob Agents Chemother* 46:3370–80.
82. Ince, D., X. Zhang, L. C. Silver, and D. C. Hooper. 2003. Topoisomerase targeting with and resistance to gemifloxacin in *Staphylococcus aureus*. *Antimicrob Agents Chemother* 47:274–82.
83. Jacobs, M. R., S. Bajaksouzian, A. Windau, P. C., et al. 2004. *In vitro* activity of the new quinolone WCK 771 against staphylococci. *Antimicrob Agents Chemother* 48:3338–42.
84. Yu, J. L., L. Grinius, and D. C. Hooper. 2002. NorA functions as a multidrug efflux protein in both cytoplasmic membrane vesicles and reconstituted proteoliposomes. *J Bacteriol* 184:1370–7.
85. Fritsche, T. R., P. A. Strabala, H. S. Sader, M. J. Dowzicky, and R. N. Jones. 2005. Activity of tigecycline tested against a global collection of Enterobacteriaceae, including tetracycline-resistant isolates. *Diagn Microbiol Infect Dis* 52:209–13.
86. Zhanel, G. G., K. Homenuik, K. Nichol, A., et al. 2004. The glycylcyclines: a comparative review with the tetracyclines. *Drugs* 64:63–88.
87. Weir, S., A. Macone, J. Donatelli, C. Trieber, D. Taylor, S. Tanaka, and S. Levy. 2003. Presented at the 43rd Interscience Conference on Antimicrobial Agents and Chemotherapy, Chicago, IL.
88. Lomovskaya, O., M. S. Warren, A., Lee, J., et al. 2001. Identification and characterization of inhibitors of multidrug resistance efflux pumps in *Pseudomonas aeruginosa*: novel agents for combination therapy. *Antimicrob Agents Chemother* 45:105–16.

89. Warren, M., J. C. Lee, A. Lee, K. Hoshino, H. Ishida, and O. Lomovskaya. 2000. Presented at the 40th Interscience Conference on Antimicrobial Agents and Chemotherapy, Toronto, Canada.

90. Renau, T. E., R. Leger, L. Filonova, E. M., et al. 2003. Conformationally-restricted analogues of efflux pump inhibitors that potentiate the activity of levofloxacin in *Pseudomonas aeruginosa*. *Bioorg Med Chem Lett* 13:2755–8.

91. Renau, T. E., R. Leger, R. Yen, M. W., et al. 2002. Peptidomimetics of efflux pump inhibitors potentiate the activity of levofloxacin in *Pseudomonas aeruginosa*. *Bioorg Med Chem Lett* 12:763–6.

92. Aires, J. R., T. Kohler, H. Nikaido, and P. Plesiat. 1999. Involvement of an active efflux system in the natural resistance of *Pseudomonas aeruginosa* to aminoglycosides. *Antimicrob Agents Chemother* 43:2624–8.

93. Masuda, N., E. Sakagawa, S. Ohya, N. Gotoh, H. Tsujimoto, and T. Nishino. 2000. Substrate specificities of MexAB-OprM, MexCD-OprJ, and MexXY-oprM efflux pumps in *Pseudomonas aeruginosa*. *Antimicrob Agents Chemother* 44:3322–7.

94. Okamoto, K., N. Gotoh, and T. Nishino. 2002. Extrusion of penem antibiotics by multicomponent efflux systems MexAB-OprM, MexCD-OprJ, and MexXY-OprM of *Pseudomonas aeruginosa*. *Antimicrob Agents Chemother* 46:2696–9.

95. Nakayama, K., Y. Ishida, M. Ohtsuka, H., et al. 2003. MexAB-OprM-specific efflux pump inhibitors in *Pseudomonas aeruginosa*. Part 1: discovery and early strategies for lead optimization. *Bioorg Med Chem Lett* 13:4201–4.

96. Nakayama, K., Y. Ishida, M. Ohtsuka, H., et al. 2003. MexAB-OprM specific efflux pump inhibitors in *Pseudomonas aeruginosa*. Part 2: achieving activity *in vivo* through the use of alternative scaffolds. *Bioorg Med Chem Lett* 13:4205–8.

97. Nakayama, K., H. Kawato, J. Watanabe, M., et al. 2004. MexAB-OprM specific efflux pump inhibitors in *Pseudomonas aeruginosa*. Part 3: Optimization of potency in the pyridopyrimidine series through the application of a pharmacophore model. *Bioorg Med Chem Lett* 14:475–9.

98. Nakayama, K., N. Kuru, M. Ohtsuka, Y., et al. 2004. MexAB-OprM specific efflux pump inhibitors in *Pseudomonas aeruginosa*. Part 4: Addressing the problem of poor stability due to photoisomerization of an acrylic acid moiety. *Bioorg Med Chem Lett* 14:2493–7.

99. Kaatz, G. W. 2005. Bacterial efflux pump inhibition. *Curr Opin Investig Drugs* 6:191–8.

100. Lomovskaya, O. and W. Watkins. 2001. Inhibition of efflux pumps as a novel approach to combat drug resistance in bacteria. *J Mol Microbiol Biotechnol* 3:225–36.

101. Marquez, B. 2005. Bacterial efflux systems and efflux pumps inhibitors. *Biochimie* 87:1137–47.

102. Pages, J. M., M. Masi, and J. Barbe. 2005. Inhibitors of efflux pumps in Gram-negative bacteria. *Trends Mol Med* 11:382–9.

103. Renau, T. E., R. Leger, E. M. Flamme, M. W., et al. 2001. Addressing the stability of C-capped dipeptide efflux pump inhibitors that potentiate the activity of levofloxacin in *Pseudomonas aeruginosa*. *Bioorg Med Chem Lett* 11:663–7.

104. Watkins, W. J., Y. Landaverry, R. Leger, R., et al. 2003. The relationship between physicochemical properties, *in vitro* activity and pharmacokinetic profiles of analogues of diamine-containing efflux pump inhibitors. *Bioorg Med Chem Lett* 13:4241–4.

105. Baucheron, S., C. Mouline, K. Praud, E. Chaslus-Dancla, and A. Cloeckaert. 2005. TolC but not AcrB is essential for multidrug-resistant *Salmonella enterica* serotype Typhimurium colonization of chicks. *J Antimicrob Chemother* 55:707–12.

106. Giraud, E., G. Blanc, A. Bouju-Albert, F. X. Weill, and C. Donnay-Moreno. 2004. Mechanisms of quinolone resistance and clonal relationship among *Aeromonas salmonicida* strains isolated from reared fish with furunculosis. *J Med Microbiol* 53:895–901.

107. Hasdemir, U. O., J. Chevalier, P. Nordmann, and J. M. Pages. 2004. Detection and prevalence of active drug efflux mechanism in various multidrug-resistant *Klebsiella pneumoniae* strains from Turkey. *J Clin Microbiol* 42:2701–6.

108. Kriengkauykiat, J., E. Porter, O. Lomovskaya, and A. Wong-Beringer. 2005. Use of an efflux pump inhibitor to determine the prevalence of efflux pump-mediated fluoroquinolone resistance and multidrug resistance in *Pseudomonas aeruginosa*. *Antimicrob Agents Chemother* 49:565–70.

109. Mamelli, L., J. P. Amoros, J. M. Pages, and J. M. Bolla. 2003. A phenylalanine-arginine beta-naphthylamide sensitive multidrug efflux pump involved in intrinsic and acquired resistance of *Campylobacter* to macrolides. *Int J Antimicrob Agents* 22:237–41.

110. Mamelli, L., V. Prouzet-Mauleon, J. M. Pages, F. Megraud, and J. M. Bolla. 2005. Molecular basis of macrolide resistance in *Campylobacter*: role of efflux pumps and target mutations. *J Antimicrob Chemother* 56:491–97.

111. Payot, S., L. Avrain, C. Magras, K. Praud, A. Cloeckaert, and E. Chaslus-Dancla. 2004. Relative contribution of target gene mutation and efflux to fluoroquinolone and erythromycin resistance, in French poultry and pig isolates of *Campylobacter coli*. *Int J Antimicrob Agents* 23:468–72.

112. Payot, S., A. Cloeckaert, and E. Chaslus-Dancla. 2002. Selection and characterization of fluoroquinolone-resistant mutants of *Campylobacter jejuni* using enrofloxacin. *Microb Drug Resist* 8:335–43.

113. Saenz, Y., J. Ruiz, M. Zarazaga, M. Teixido, C. Torres, and J. Vila. 2004. Effect of the efflux pump inhibitor Phe-Arg-beta-naphthylamide on the MIC values of the quinolones, tetracycline and chloramphenicol, in *Escherichia coli* isolates of different origin. *J Antimicrob Chemother* 53:544–5.

114. Thorarensen, A., A. L. Presley-Bodnar, K. R. Marotti, T. P., et al. 2001. 3-Arylpiperidines as potentiators of existing antibacterial agents. *Bioorg Med Chem Lett* 11:1903–6.

115. Bohnert, J. A. and W. V. Kern. 2005. Selected arylpiperazines are capable of reversing multidrug resistance in *Escherichia coli* overexpressing RND efflux pumps. *Antimicrob Agents Chemother* 49:849–52.

116. Chevalier, J., S. Atifi, A. Eyraud, A. Mahamoud, J. Barbe, and J. M. Pages. 2001. New pyridoquinoline derivatives as potential inhibitors of the fluoroquinolone efflux pump in resistant *Enterobacter aerogenes* strains. *J Med Chem* 44:4023–6.

117. Chevalier, J., J. Bredin, A. Mahamoud, M. Mallea, J. Barbe, and J. M. Pages. 2004. Inhibitors of antibiotic efflux in resistant *Enterobacter aerogenes* and *Klebsiella pneumoniae* strains. *Antimicrob Agents Chemother* 48:1043–6.

118. Gallo, S., J. Chevalier, A. Mahamoud, A. Eyraud, J. M. Pages, and J. Barbe. 2003. 4-alkoxy and 4-thioalkoxyquinoline derivatives as chemosensitizers for the chloramphenicol-resistant clinical *Enterobacter aerogenes* 27 strain. *Int J Antimicrob Agents* 22:270–3.

119. Mallea, M., A. Mahamoud, J. Chevalier, S. Alibert-Franco, P. Brouant, J. Barbe, and J. M. Pages. 2003. Alkylaminoquinolines inhibit the bacterial antibiotic efflux pump in multidrug-resistant clinical isolates. *Biochem J* 376:801–5.

120. Fojo, T. and S. Bates. 2003. Strategies for reversing drug resistance. *Oncogene* 22:7512–23.

121. Liang, X. J. and A. Aszalos. 2006. Multidrug transporters as drug targets. *Curr Drug Targets* 7:911–21.

122. Pleban, K., D. Kaiser, S. Kopp, M. Peer, P. Chiba, and G. F. Ecker. 2005. Targeting drug-efflux pumps—a pharmacoinformatic approach. *Acta Biochim Pol* 52:737–40.

123. Szakacs, G., J. K. Paterson, J. A. Ludwig, C. Booth-Genthe, and M. M. Gottesman. 2006. Targeting multidrug resistance in cancer. *Nat Rev Drug Discov* 5:219–34.

124. Teodori, E., S. Dei, C. Martelli, S. Scapecchi, and F. Gualtieri. 2006. The functions and structure of ABC transporters: implications for the design of new inhibitors of Pgp and MRP1 to control multidrug resistance (MDR). *Curr Drug Targets* 7:893–909.
125. Teodori, E., S. Dei, S. Scapecchi, and F. Gualtieri. 2002. The medicinal chemistry of multidrug resistance (MDR) reversing drugs. *Farmaco* 57:385–415.
126. Yoshida, K., K. Nakayama, N. Kuru, S., et al. 2006. MexAB-OprM specific efflux pump inhibitors in *Pseudomonas aeruginosa*. Part 5: Carbon-substituted analogues at the C-2 position. *Bioorg Med Chem* 14:1993–2004.
127. Yoshida, K. I., K. Nakayama, Y. Yokomizo, M., et al. 2006. MexAB-OprM specific efflux pump inhibitors in *Pseudomonas aeruginosa*. Part 6: Exploration of aromatic substituents. *Bioorg Med Chem* 14:8506–18.
128. Mallea, M., J. Chevalier, A. Eyraud, and J. M. Pages. 2002. Inhibitors of antibiotic efflux pump in resistant *Enterobacter aerogenes* strains. *Biochem Biophys Res Commun* 293:1370–3.
129. Lomovskaya, O. and K. A. Bostian. 2006. Practical applications and feasibility of efflux pump inhibitors in the clinic-A vision for applied use. *Biochem Pharmacol* 71:910–18.
130. Sauna, Z. E., M. B. Andrus, T. M. Turner, and S. V. Ambudkar. 2004. Biochemical basis of polyvalency as a strategy for enhancing the efficacy of P-glycoprotein (ABCB1) modulators: stipiamide homodimers separated with defined-length spacers reverse drug efflux with greater efficacy. *Biochemistry* 43:2262–71.
131. Amin, A. H., T. V. Subbaiah, and K. M. Abbasi. 1969. Berberine sulfate: antimicrobial activity, bioassay, and mode of action. *Can J Microbiol* 15:1067–1076.
132. Severina, I. I., M. S. Muntyan, K. Lewis, and V. P. Skulachev. 2001. Transfer of cationic antibacterial agents berberine, palmatine and benzalkonium through bimolecular planar phospholipid film and *Staphylococcus aureus* membrane. *IUBMB-Life Sciences* 52:321–324.
133. Hsieh, P. C., S. A. Siegel, B. Rogers, D. Davis, and K. Lewis. 1998. Bacteria lacking a multidrug pump: a sensitive tool for drug discovery. *Proc Natl Acad Sci USA* 95:6602–6.
134. Lewis, K. 2001. In search of natural substrates and inhibitors of MDR pumps. *J Mol Microbiol Biotechnol* 3:247–54.
135. Stermitz, F. R., P. Lorenz, J. N. Tawara, L. Zenewicz, and K. Lewis. 2000a. Synergy in a medicinal plant: antimicrobial action of berberine potentiated by 5′-methoxyhydnocarpin, a multidrug pump inhibitor. *Proc Natl Acad Sci USA* 97:1433–7.
136. Tegos, G., F. R. Stermitz, O. Lomovskaya, and K. Lewis. 2002. Multidrug pump inhibitors uncover the remarkable activity of plant antimicrobials. *Antimicrob Agents Chemother* 46:3133–41.
137. Belofsky, G., R. Carreno, K. Lewis, A. Ball, G. Casadei, and G. P. Tegos. 2006. Metabolites of the "smoke tree," *Dalea spinosa*, potentiate antibiotic activity against multidrug-resistant *Staphylococcus aureus*. *J Nat Prod* 69:261–4.
138. Belofsky, G., D. Percivill, K. Lewis, G. P. Tegos, and J. Ekart. 2004. Phenolic metabolites of *Dalea versicolor* that enhance antibiotic activity against model pathogenic bacteria. *J Nat Prod* 67:481–4.
139. Morel, C., F. R. Stermitz, G. Tegos, and K. Lewis. 2003. Isoflavone MDR efflux pump inhibitors from *Lupinus argenteus*. Synergism between some antibiotics and isoflavones. *J Agricult Food Chem* 51: 5677–9.
140. Stermitz, F. R., K. K. Cashmana, K. M. Halligana, C. Morel, G. P. Tegos, and K. Lewis. 2003. Polyacylated neohesperidosides from *Geranium caespitosum*: bacterial multidrug resistance pump inhibitors. *Bioorg Med Chem Lett* 13:1915–8.
141. Stermitz, F. R., L. N. Scriven, G. Tegos, and K. Lewis. 2002. Two flavonols from *Artemisia annua* which potentiate the activity of berberine and norfloxacin against a resistant strain of *Staphylococcus aureus*. *Planta Med* 68:1140–1.

142. Ball, A. R., G. Casadei, S. Samosorn, J. B. Bremner, F. M. Ausubel, T. I. Moy, and K. Lewis. 2006. Conjugating berberine to an MDR pump inhibitor creates an effective antimicrobial. *ACS Chem Biol* 1:594–600.

143. Moy, T. I., A. R. Ball, Z. Anklesaria, G. Casadei, K. Lewis, and F. M. Ausubel. 2006. Identification of novel antimicrobials using a live-animal infection model. *Proc Natl Acad Sci USA* 103:10414–9.

144. Milovic, N. M., J. Wang, K. Lewis, and A. M. Klibanov. 2005. Immobilized *N*-alkylated polyethylenimine avidly kills bacteria by rupturing cell membranes with no resistance developed. *Biotechnol Bioeng* 90:715–22.

145. Tiller, J. C., C. J. Liao, K. Lewis, and A. M. Klibanov. 2001. Designing surfaces that kill bacteria on contact. *Proc Natl Acad Sci USA* 98:5981–5.

146. Lin, J., J. C. Tiller, S. B. Lee, K. Lewis, and A. M. Klibanov. 2002. Insights into bactericidal action of surface-attached poly(vinyl-*N*-hexylpyridinium) chains. *Biotechnol Lett* 24:801–805.

147. Lewis, K. and A. M. Klibanov. 2005. Surpassing nature: rational design of sterile-surface materials. *Trends Biotechnol* 23:343–8.

5   Mechanisms of Aminoglycoside Antibiotic Resistance

Gerard D. Wright

CONTENTS

Aminoglycoside antibiotics are positively charged carbohydrate-containing molecules that find clinical use for the treatment of infections caused by both Gram-negative and Gram-positive bacteria. The first aminoglycosides were discovered over 60 years ago and several continue to find important clinical use including gentamicin, tobramycin, amikacin, netilmicin, and streptomycin. These antibiotics target the bacterial ribosome and interfere with protein translation. Unlike other antibiotics that block translation, most aminoglycosides are bactericidal, a highly desirable feature in an antiinfective chemotherapeutic agent. The bactericidal action of aminoglycosides is correlated with the propensity to cause misreading of the mRNA transcript resulting in the production of aberrant proteins.

Resistance to the aminoglycosides can occur through decreased uptake of the drugs, aminoglycoside efflux, mutations in the rRNA and ribosomal protein, and methylation of rRNA. However, it is the presence and action of aminoglycoside-modifying enzymes that are the most relevant in the majority of resistant clinical isolates. Three distinct classes of modifying enzyme are known: the phosphotransferases (APHs), the adenylyltransferases (ANTs), and the acetyltransferases (AACs). The APHs and ANTs are ATP-dependent enzymes, while the AACs require acetyl coenzyme A (acetylCoA). Members of each of these classes of enzyme are known and prevalent in both Gram-positive and Gram-negative clinical isolates. Several dozen distinct enzymes have been identified and these are designated by the position on the molecule where modification occurs (given by a number in parentheses), the resistance profile, represented by a Roman numeral, and the specific gene, indicated by a lowercase letter, for example, AAC(6′)-Ia is an aminoglycoside acetyltransferase modifying position 6′.

Research on aminoglycoside-modifying enzymes has greatly benefited from crystal structures of representative proteins from each class. Together with an increasing body of knowledge on the chemical mechanisms of modifying group transfer, and the molecular strategies for aminoglycoside substrate discrimination, the availability of 3D protein structural data is permitting detailed understanding of the basis for aminoglycoside antibiotic resistance and providing insight into the origins of aminoglycoside-modifying enzymes. For example, APHs have been shown to share structural similarities with protein Ser/Thr/Tyr kinases as well as the capacity to phosphorylate proteins and peptides themselves. AACs fall into a growing family of GNAT acyltransferases, which includes important protein acyltransferases, such as the histone acetyltransferases. Furthermore, ANTs are structurally similar to DNA polymerase β and share the same aspects of reaction chemistry.

Knowledge of enzyme mechanism and structure is now fueling research into specific inhibitors of these enzymes and aminoglycoside molecules that are not substrates for them and recent results in this area are promising. Furthermore, understanding of enzyme mechanism and structure can be used in the design of efficient and specific APH, ANT, and AAC inhibitors. These could find clinical application in reversing the impact of aminoglycoside resistance enzymes through the potentiation of existing aminoglycosides, and possibly the re-introduction of antibiotics no longer in use as the result of the dissemination and impact of aminoglycoside-modifying enzymes.

INTRODUCTION

AMINOGLYCOSIDE ANTIBIOTICS

The aminoglycoside antibiotics are a diverse class of clinically important antimicrobial compounds that have proven to be instrumental in the treatment of infectious diseases since their discovery in the mid-1940s. The aminoglycosides find use in the treatment of infections, caused by both Gram-positive and Gram-negative bacteria [1] and, in addition, some protozoa. In general, they are bactericidal compounds, an important trait especially for treatment of infections in immunocompromised individuals. The aminoglycosides are natural products, derived from bacterial producers, though some clinically important compounds, such as amikacin and isepamicin are semisynthetic derivatives of natural products. All aminoglycosides contain an amino-cyclitol nucleus (a six carbon ring substituted with alcohol and amino groups) and as such are more formally termed aminoglycoside-aminocyclitol antibiotics; however, the label "aminoglycosides" is generally accepted for the class.

Aminoglycoside antibiotics can be ordered into two groups depending on whether they incorporate a 2-deoxystreptamine ring or not (Table 5.1 and Figure 5.1). Within the 2-deoxystreptamine group, the most clinically relevant antibiotics are substituted on the 2-deoxystreptamine ring at positions 4 and 6 or 4 and 5. Chemical diversity in the class arises from the variety of aminohexoses and/or pentoses that decorate the aminocyclitol ring. Additional variance in these antibiotics is derived from further substitution by amino (and non-amino)-hexoses, methylation, de-oxygenation, and epimerization of various sites on the molecules (Figure 5.1). The result is a structurally rich and varied family of compounds, many of which find clinical use as antimicrobial agents. Numbering of the carbon centers, which is essential for deciphering the nomenclature of modifying enzymes, generally follows the rule that the aminocyclitol ring has no suffix while additional rings are labeled with a prime ('), double prime ("), and so on (Figure 5.1). The ubiquitous presence of amino groups confers an overall positive charge to these compounds at physiological pH, making them highly water soluble and poorly orally available.

MODE OF ACTION OF AMINOGLYCOSIDE ANTIBIOTICS: INTERACTION WITH THE BACTERIAL RIBOSOME

The cationic nature of aminoglycosides provides the electronic basis for interaction with the 16S rRNA on the small (30S) ribosomal subunit. Specifically, aminoglyco-sides bind to the region on the ribosome termed the A-site, where the aminoacyl-tRNAs dock and where cognate codon–anticodon recognition occurs. The crystal structure of the aminoglycosides paromomycin [3], spectinomycin [3], hygromycin B

TABLE 5.1
Aminoglycoside Antibiotics

2-Deoxystreptamine Aminoglycosides

4,5-Disubstituted 2-Deoxystreptamine	4,6-Disubstituted 2-Deoxystreptamine	Others
Kanamycin	Neomycin	Streptomycin
Amikacin	Butirosin	Spectinomycin
Tobramycin	Ribostamycin	Fortimicin
Gentamicins	Lividomycin	
Arbekacin		
Isepamicin		
Sisomicin		
Netlimicin		

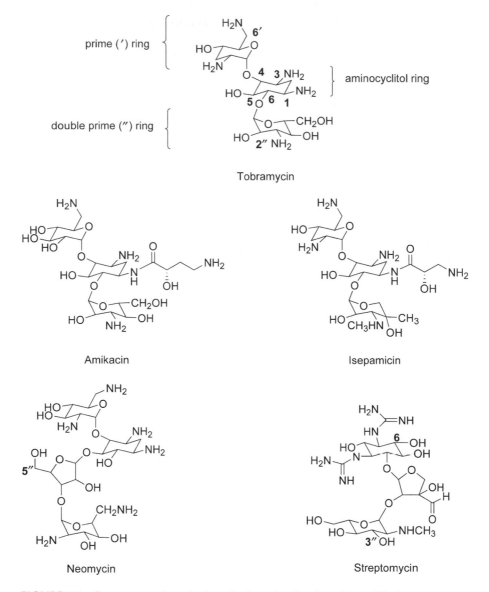

FIGURE 5.1 Structures and numbering of selected aminoglycoside antibiotics.

[4], and streptomycin [3] bound to the 30S subunit of *Thermus thermophilus* have been reported to approximately 3 Å resolution. These studies have been augmented by additional structural analysis using high-resolution NMR of complexes of amino-glycosides and A-site RNA-derived oligonucleotides [5–8]. In a parallel approach, the Westhof group has reported high-resolution (2.2 to 3.0 Å) X-ray structures of the 4,5-disubstituted 2-deoxystreptamine aminoglycoside antibiotics paromomycin, neomy-cin B, lividomycin A, and ribostamycin and the 4,6-disubstituted 2-deoxystreptamine

aminoglycoside antibiotics tobramycin, geneticin, gentamicin C1a, and kanamycin A, along with 2-ring minimal unit neamine bound to oligonucleotides containing 16S rRNA A site sequences [9–12]. This work has shown that the 2-deoxystreptamine class of aminoglycosides bind similarly to the A site 16S rRNA.

The 2-deoxystreptamine and the prime (′) ring linked to position 4 of the 2-deoxystreptamine form the active "warhead" of these antibiotics and participate in direct and indirect contacts with invariant bases and phosphates in 16S rRNA helix 44 (Figure 5.2). In particular, N1 and N3 of the 2-deoxystreptamine ring form universal contacts with A1493, G1494, and U1495. The prime (′) ring adopts a unique pucker and stacks against G1491. It also interacts with A1408 through the 6′-amino (or 6′-hydroxyl for paromomycin and lividomycin) group. Additional contacts to the rRNA are made via hydrogen bonds and van der Waals interactions between other hydroxyl and amino groups and sugars linked to the 2-deoxystreptamine, often through the intermediacy of water molecules.

The resulting tight interaction of these aminoglycosides with the A site 16S rRNA is a bulging out of adenines 1492 and 1493 into the RNA-recognition area of the A site. Recent X-ray crystallographic studies along with several decades of biochemical research have demonstrated that the ribosome participates actively in codon–anticodon discrimination to ensure translation fidelity (reviewed in ref. [13]). A key element in this recognition is a movement of G530, A1492, and A1493 from a loop region in helix 44 of the 16S rRNA to permit an interaction with the helix formed when cognate codon–anticodon pairing occurs in the A site [14]. On the other hand, binding of non-cognate anticodons to mRNA-bound ribosomes does not result in this conformational change, strongly implicating it as a key element in maintaining translational fidelity. Binding of aminoglycosides displaces A1492 and A1493 into the A site into a conformation that enables interaction of these residues with even non-cognate codon–anticodon pairs. The result is impairment of proper codon–anticodon discrimination by the ribosome in the presence of aminoglycoside antibiotics. This remarkable structural insight into the molecular mechanism of action is completely consistent with the well-established observations that aminoglycoside antibiotics cause errors in translation and that this miscoding results in formation of aberrant proteins that contribute to cell death [15,16].

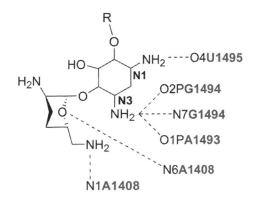

FIGURE 5.2 Conserved interactions of 2-deoxystreptamine antibiotics with the 16S rRNA.

Structural studies of aminoglycoside-RNA interactions have also provided the means to evaluate the basis for the specificity of aminoglycosides for bacterial versus eukaryotic rRNA and it has been determined that the A1408 site is critical to this selectivity. In eukaryotes, this position is generally a G, and an A1408 to G mutation in *Escherichia coli* 16S rRNA confers resistance to most aminoglycosides in the mutant bacteria [17,18]. The sensitivity of some protozoa to 6′-hydroxyl-containing aminoglycosides, such as paromomycin, may reflect A C1409-G1491 base pair, which is not found in other eukaryotes, but is shared with prokaryotes [18].

The atomic resolution structures of ribosome-aminoglycoside complexes have served to rationalize decades-old literature in the field and enlighten the link between aminoglycoside antibiotic action with subversion of the genetic code. These efforts now permit for the first time structure-based drug design approaches to the development of new aminoglycosides (see Strategies to Circumvent Aminoglycoside-Modifying Enzymes, below).

AMINOGLYCOSIDE UPTAKE

Aminoglycoside antibiotics gain entry to the cell through a multi-phase process. The initial step is passive accumulation of the positively charged aminoglycosides at the negatively charged cell surface. The antibiotics then gain entry to the bacterial cytosol apparently by diffusion through the plasma membrane. This process is dependent on the electronic potential of the membrane and is thus energy dependent. Support for this model comes from experiments in which inhibitors of membrane potential (such as CCCP and CN^-) prevent aminoglycoside entry (reviewed in ref. [19]). It is generally accepted that this mechanism of aminoglycoside uptake is ubiquitous and does not require a protein component, however, evidence for the participation of a specified protein component in aminoglycoside translocation into the cytosol has been reported (e.g., oligopeptide binding protein [20]).

MECHANISM OF BACTERICIDAL ACTION OF AMINOGLYCOSIDES

While it is well established that aminoglycosides target the bacterial ribosome, this interaction in and of itself is not sufficient to explain the bactericidal action of these compounds. Other antibiotics that target the translation machinery, such as the tetracyclines and chloramphenicol are bacteristatic rather than bactericidal [21]. As described above, once inside the cell, the aminoglycoside antibiotics bind to the decoding A site region of the ribosomes and cause mistranslation in *de novo* protein synthesis, resulting in the production of aberrant proteins [15,22–25]. There has also been a long-standing observation that aminoglycosides cause membrane damage as evidenced by the loss of ions from the cell such as K^+ [26–28]. It has since been demonstrated that the fate of some of these mistranslated proteins is interaction with the cell membrane, and that this interaction results in altered membrane permeability [29,30]. It has been proposed by Davis that aminoglycoside-mediated mistranslation followed by membrane damage caused by perturbation by the altered peptides may account for the breach of membrane integrity, which seems to be essential for the bactericidal activity of these antibiotics [16] (spectinomycin, a bacteristatic aminoglycoside, does not cause mistranslation [25] nor does it bind to the decoding A site of the ribosome [3]). Thus, aminoglycosides kill bacteria by pleotropic means involving ultimate loss of membrane integrity, however, interaction with the ribosome and mistranslation appears to be the primary and critical events.

AMINOGLYCOSIDE RESISTANCE

Bacterial resistance to the aminoglycosides can occur through four general mechanisms: (i) altered uptake, (ii) antibiotic efflux, (iii) target modification, and (iv) chemical modification.

ALTERED UPTAKE

Since uptake of aminoglycosides is an energy-requiring phenomenon, mutations that affect the membrane potential can confer aminoglycoside resistance [31,32]. Taber and Halfenger [33] isolated multiple aminoglycoside-resistant mutants of *Bacillus subtilis* that were deficient in aminoglycoside uptake and one of these was characterized as a menaquinone (a lipophilic quinone required for electron transport) auxotroph. Supplementation of the growth medium with shikimic acid (a menaquinone biosynthesis precursor) restored aminoglycoside sensitivity [34]. Similarly, quinone auxotrophs of *Staphylococcus aureus* have an aminoglycoside resistance phenotype that can be abolished by the addition of menaquinone precursors to the medium [35]. Furthermore, depletion or mutations in other electron transport components, including cytochrome *aa3* [36] and the γ-subunit of the F1F0 ATPase [37], result in aminoglycoside resistance.

While electron transport mutations can be readily isolated in the laboratory (and are not the result of exposure to aminoglycosides [33]), they appear to be infrequent sources of resistant organisms in the clinic, possibly because of the potential decreased viability of electron transport mutants in the host. Possible exceptions to this view are small colony variants of various pathogens, such as *S. aureus*. Many of these have reduced rates of aminoglycoside uptake and also have mutations in heme or menaquinone biosynthesis. More detailed discussion of this topic is found in Chapter 8 of this volume.

It has been shown that *E. coli* that harbor structural or protein expression mutations in the oligopeptide binding protein OppA, which is involved in peptide transport across the membrane, show a kanamycin-resistant phenotype [20]. These mutants failed to take up [^{14}C]-isepamicin, which suggests a possible role in aminoglycoside uptake for this protein.

AMINOGLYCOSIDE EFFLUX

Efflux-mediated resistance to aminoglycoside antibiotics is an increasing clinical problem in a select group of organisms of the genera *Acinetobacter* [38], *Pseudomonas* [39], and *Burkholderia* [40]. Most of these efflux systems are members of the tripartite resistance-nodulation-division (RND) family of efflux proteins. In *E. coli*, Nikaido's group has shown that AcrAD is a TolC-associated aminoglycoside efflux pump [41], capturing antibiotic in both the periplasm and the cytoplasm for export [42]. High-level aminoglycoside resistance in *Burkholderia pseudomallei* is mediated by the multidrug efflux systems AmrAB-OprA [40] and BpeAB-OprB [43]. It may be that high-level aminoglycoside resistance observed in other species of *Burkholderia*, such as *Burkholderia cepacia*, which is a significant pathogen in cystic fibrosis patients, will also be shown to be due to efflux. In *Pseudomonas aeruginosa*, the MexXY-OprM [44] and MexAB-OprM [39] systems are well documented to be aminoglycoside efflux systems [45]. In *Stenotrophomonas maltophilia* the SmeAB-SmeC [46] and in *Acinetobacter baumanii* the AdeAB-AdeC [38,47] systems have been associated with aminoglycoside efflux and resistance. The impact of efflux-mediated aminoglycoside resistance in important Gram-negative pathogens is therefore growing [48].

TARGET MODIFICATION

Aminoglycoside resistance through target modification can occur through two mechanisms: (*i*) point mutation of rRNA or ribosomal proteins, or (*ii*) methylation of the 16S rRNA. The latter mechanism is found in actinomycete producers of aminoglycosides where it confers high-level resistance (minimal inhibitory concentration, MIC > 500 µg/mL), for example, *Micromonospora purpurea* (gentamicin producer) [49] and *Streptomyces tenabrius* (tobramycin producer) [50]. This mechanism has now emerged in some aminoglycoside-resistant clinical strains and threatens to become a more significant problem in the future [51–56].

Ribosomal point mutation is a clinically important mechanism of resistance in the slow-growing mycobacteria (see Chapter 13 in this volume, and review in [57]). Resistance to streptomycin can occur through point mutations in the ribosomal protein S12, RpsL [58,59] through an unknown process, though conformational change at the streptomycin binding site is a likely mechanism. Resistance can also result from mutations in the aminoglycoside target 16S rRNA (*rrs* gene) [58,60]. Isolates of *Mycobacterium tuberculosis* that display resistance to kanamycin and amikacin have mutations in A1400 [61]. This base is equivalent to A1408 of the *E. coli* 16S rRNA, which has been shown by structural studies (see Mode of Action of Aminoglycoside Antibiotics, above) and mutation analysis [62], to be important to aminoglycoside recognition.

MODIFICATION OF AMINOGLYCOSIDES

Enzyme-catalyzed chemical modification of aminoglycosides remains the most relevant mechanism of resistance in the majority of clinical isolates. Chemical modification can occur through three general mechanisms: *O*-phosphorylation, *O*-adenylation, or *N*-acetylation. All three mechanisms are widespread through both Gram-negative and Gram-positive bacteria, but the latter appear to have a smaller repertoire of enzymes. Modification of key sites on the antibiotics blocks their ability to bind to the 16S rRNA in a productive fashion. It is not surprising therefore that the key important groups required for productive interaction of the antibiotics with the A site rRNA as revealed by X-ray structures (see Mode of Action of Aminoglycoside Antibiotics, above), such as N6′ and N3 are the sites of modification for some of the most effective aminoglycoside-modifying enzymes.

The various aminoglycoside-modifying enzymes are classified by the chemistry of the modifying reaction (phosphoryl, adenyl, or acetyl transfer), their site of aminoglycoside modification (regiospecificity), and by the specific isozyme sequence. Shaw et al. proposed a unifying nomenclature for all aminoglycoside-modifying enzymes where the enzyme is described by type (APH [O-phosphotransferase], AAC [N-acetyltransferase], or ANT [O-adenyltransferase]), the regiospecificity of group transfer in parentheses, for example, (3'), (2"), and so on, followed by a Roman numeral indicating a distinct phenotype (these are assigned sequentially as discovered or cloned), and finally a letter indicating the specific gene [63]. For example, APH(3')-Ia is a phosphotransferase that modifies aminoglycosides at position 3' with a distinct resistance phenotype (in this case protection against kanamycin, gentamicin B, neomycin, paromomycin, ribostamycin, and lividomycin), and is the first gene cloned with this repertoire [64]; on the other hand, APH(3')-Vc is also an aminoglycoside kinase with the same regiospecificity of phosphoryl transfer (3'-OH), but it has a different resistance phenotype (kanamycin, neomycin, paromomycin, and ribostamycin), and is the third gene cloned with these properties [65]. The list of these aminoglycoside-modifying genes continues to grow, but tables of genes, resistance phenotypes and original references can be found in several extensive reviews (e.g., [63,66]). A representative list of clinically relevant enzymes is found in Table 5.2.

While genes encoding greater than 70 aminoglycoside-modifying enzymes have already been cloned and a number are being uncovered in whole genome sequencing projects, only a subset of these genes are of significant clinical relevance today given

TABLE 5.2
Representative Aminoglycoside-Modifying Enzymes

Enzyme	Resistance Profile*	Bacterial Source
APH(3')-Ia	Kan, Neo, Rib, Livid	Enterobacteriaceae
APH(3')-IIIa	Kan, Amik, Isep, Neo, Rib, But, Livid	Enterococci, staphylococci
APH(3")-Ib	Strep	Enterobacteriaceae
APH(6)-Id	Strep	Enterobacteriaceae
AAC(6')-Ib	Kan, Tob, Amik, Neo	Enterobacteriaceae
AAC(6')-Ii	Kan, Tob, Amik, Neo	*Enterococcus faecium*
AAC(3)-Ia	Kan, Gent, Tob, Fort	Enterobacteriaceae
ANT(2")-Ia	Kan, Gent, Tob	Enterobacteriaceae
ANT(4')-Ia	Kan, Tob, Amik, Neo	*Staphylococcus aureus*
ANT(6)-Ia	Strep	*Enterococcus faecalis*
AAC(6')-(APH2")		Enterococci, staphylococci
APH activity	Kan, Gent, Amik, Isep, Neo, Rib, But, Livid	
AAC activity	Kan, Amik, Isep, Neo, Rib, But, Livid, Fort	
	(Livid is a poor substrate)	

Abbreviations: Amik, amikacin; But, butirosin; Fort, fortimicin (astromycin); Gent, gentamicin C; Isep, isepamicin; Kan, kanamycin; Livid, lividomycin A; Neo, neomycin; Rib, ribostamycin.

that usage of aminoglycosides is limited to only a few compounds (gentamicin, tobramycin, netilmicin, amikacin, and streptomycin in the United States [1]). For example, ANT(2″)-I, which confers resistance to gentamicin and tobramycin, is common in Enterobacteriaceae worldwide, but depending on aminoglycoside usage patterns, resistance to gentamicin by AAC(3)-II and AAC(3)-VI is also problematic [67–69]. Furthermore, combinations of resistance genes such as *aac(6′)*-I and *aac(3)*-II, which result in overall resistance to gentamicin, tobramycin, netilmicin, and amikacin, also are emerging in some countries [68]. In Gram-positive pathogens, such as *S. aureus*, resistance is less complex and the primary mechanism of gentamicin resistance (>90% of isolates) is a bifunctional enzyme with both aminoglycoside kinase and acetyltransferase activity, AAC(6′)-APH(2″) [69].

O-Phosphotransferases

The aminoglycoside *O*-phosphotransferases, abbreviated APH, are a common resistance mechanism. These enzymes are ATP-dependent kinases of approximately 30 kDa, which generate a phosphorylated aminoglycoside and ADP as products. The most prevalent group of aminoglycoside kinases are the APH(3′)s, which confer resistance to kanamycin and neomycin by phosphorylation of the 3′-OH (Figure 5.3). Furthermore, some of these enzymes, for example, APH(3′)-Ia, APH(3′)-IIIa, can confer resistance to the 3-deoxy-aminoglycoside lividomycin A through phosphorylation of the secondary 5″-alcohol of the pentose ring and in fact this site can be phosphorylated in other 4,5-disubstituted aminoglycosides [70]. These enzymes are common in both Gram-negative and Gram-positive bacteria and were among the first aminoglycoside resistance elements identified in bacteria [71]. The prevalence of these resistance elements motivated the search for "resistance-proof" aminoglycosides and prompted the introduction of compounds that lacked the 3′-hydroxyl such as tobramycin. Since these enzymes do not confer resistance to other important 3′-deoxy-aminoglycosides, such as gentamicin Cs or isepamicin, the clinical impact of APH(3′)s is now low, although APH(3′)-IIIa does confer resistance to amikacin in Gram-positive cocci, and is thus relevant in this context. While APH(3′)s no longer are a grave threat to modern aminoglycoside therapy, they have found use as important molecular biological tools where they are frequently used as antibiotic resistance markers; for example, APH(3′)-IIa is the common source of the "neo cassette" found in many cloning plasmids and transposons.

The APH(2″) kinases on the other hand, are important resistance elements in Gram-positive bacteria. The most relevant mechanism is the bifunctional AAC(6′)-APH(2″) that is the primary mechanism of gentamicin C resistance in staphylococci

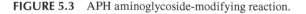

FIGURE 5.3   APH aminoglycoside-modifying reaction.

and enterococci. The APH(2″) kinase activity is located to the C-terminus of the enzyme and can efficiently use gentamicin C1, gentamicin C1a, gentamicin C2, isepamicin, netilmicin, sisomicin, and amikacin (among others), as substrates [72–74]. The site of 2″-phosphorylation has been confirmed by NMR studies [72], but is not confined to this hydroxyl, and the 3′, 5″, and 3‴ hydroxyls may also be phosphorylated on various aminoglycosides [73]. This enzyme activity is quite indiscriminant and is therefore a significant challenge for the design of new antibiotics.

APH(2″) genes have also been cloned that are not fused to a 5′-*aac(6′)* gene in *Enterococcus gallinarum* [75] and *Enterococcus casseliflavus* [76], indicating that this enzyme activity is increasing in frequency.

Other aminoglycoside kinases have been identified that modify streptomycin [APH(6), APH(3″)], spectinomycin [APH(9)], and hygromycin B [APH(4), APH(7″)]. With the exception of *strA–strB* genes found on Gram-negative R plasmids, such as RSF1010, which encode the streptomycin kinases APH(3″)-Ib and APH(6′)-Id, respectively, these kinases are not common mechanisms of clinical aminoglycoside resistance.

The 3D structures of two aminoglycoside kinases have been reported, that of APH(3′)-IIIa from Gram-positive cocci [77] and APH(3′)-IIa, which is widely distributed in many bacteria [78]. These structures are highly similar, and since all aminoglycoside kinases share a significant degree of amino acid homology, especially in the active site region, it is likely that the salient issues of enzyme mechanism will be common among these enzymes, though the specific interactions with aminoglycosides substrates, which differ widely among enzymes, will be different. The structure of APH(3′)-IIIa bound with ADP is shown in Figure 5.4. The enzyme has two distinct domains: an N-terminal region consisting largely of β-strands and a C-terminal region that is rich in α-helices. The active site lies at the junction of these domains.

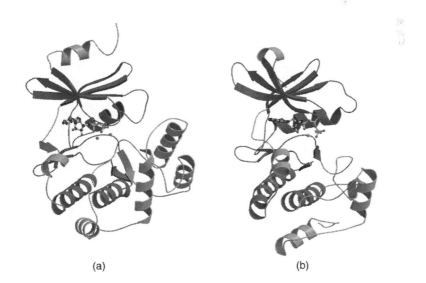

(a)   (b)

FIGURE 5.4 Structures of (a) APH(3′)-IIIa and (b) mouse protein kinase A (cAMP-dependent protein kinase, cAPK).

The structure revealed two striking features. The first was that the aminoglycoside-binding site was rich in negatively charged amino acid residues. This observation is consistent with the capacity of the enzyme to bind a broad array of positively charged aminoglycosides, which based on mutagenesis and molecular modeling studies [79], are predicted to bind to the enzyme in a number of distinct conformations.

The second important feature revealed by the 3D structure was the remarkable structural similarity between Ser/Thr/Tyr protein kinases and phosphatidylinositol kinases (Figure 5.4), despite the overall low amino acid homology (<2.5%), suggesting a possible common protein ancestor. This similarity nonetheless prompted an investigation into the potential protein kinase activities of APHs and indeed, APH(3')-IIIa and the APH activity of the bifunctional AAC(6')-APH(2''), showed the capacity to act as protein kinases [80]. A survey of several known peptide and protein substrates of protein kinases demonstrated that these two antibiotic resistance kinases could phosphorylate some peptides and proteins on Ser residues. The similarity between APHs and protein kinases was further strengthened with the demonstration that several small molecule inhibitors of protein kinases were also inhibitors of APH(3')-IIIa and APH(2'') [81] (see Strategies to Circumvent Aminoglycoside-Modifying Enzymes, below). Furthermore, extensive site-directed mutagenesis and mechanistic studies support the catalytic importance of active site Asp and Lys residues (Asp190 and Lys44 of APH(3')-IIIa), which have also been implicated as important to Ser/Thr/Tyr kinase catalysis [77,82,83].

In summary, aminoglycoside kinases and protein kinases share similarity in protein structure, enzyme mechanism, sensitivity to inhibitors, and function. These results then support a common origin for protein and aminoglycoside kinases. Furthermore, other antibiotic resistance kinases, such as the erythromycin kinases MPH (2')-I and MPH(2')-II [84,85], and viomycin kinase VPH [86], share sequence similarities within the important active site regions of APHs and protein kinases; thus these enzymes form a large superfamily of kinases.

N-Acetyltransferases

The aminoglycoside N-acetyltransferases form the largest group of aminoglycoside-modifying enzymes. They are generally 20 to 25 kDa in mass and modify positions 6', 2', and 3 of aminoglycosides in an acetylCoA-dependent fashion (Figure 5.5). AACs with the capacity to modify N-1 have also been reported [87,88]. The AAC(3)s confer resistance to gentamicin and tobramycin, and the AAC(6')s, which confer resistance to amikacin and tobramycin, are among the most abundant aminoglycoside

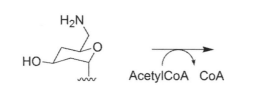
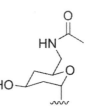

FIGURE 5.5   Reaction catalyzed by acetyltransferases.

resistance elements (over 50 isozymes). Not surprisingly then, they are very frequent causes of clinical resistance especially in Gram-negative bacteria [68]. Furthermore, AAC(6′)-Ie, which forms the N-terminal domain of the AAC(6′)-APH(2″) bifunctional enzyme noted above, is the most frequent source of aminoglycoside resistance in Gram-positive organisms.

The AAC(3) and AAC(6′) enzymes are generally encoded on mobile genetic elements, such as transposons or plasmids, although some are found in bacterial chromosomes, for example, aac(6′)-Ii in Enterococcus faecium [89]. On the other hand, the AAC(2′) enzymes are apparently universally chromosomally encoded: aac(2′)-Ia in Providencia stuartii [90] and aac(2′)-Ib-e in mycobacteria [91,92].

Unlike the case of the emergence of the APH(3′) enzymes, which prompted the replacement of 3′-hydroxyl-containing aminoglycosides, such as kanamycin, with 3′-deoxy compounds, such as tobramycin and gentamicin C, the key importance of NH_2 or OH groups at positions 6′ and 3 for binding to the target 16S rRNA and antimicrobial activity has made the presence of AAC(3) and AAC(6′) enzymes highly problematic. Furthermore, as noted above, there are a large number of these enzymes and they are frequently causes of aminoglycoside resistance. The study of AAC enzymes is therefore of key importance. Pioneering work in this area was reported by Northrop et al., who described the kinetic characterization of AAC(3)-I over 25 years ago [93–95] and AAC(6′)-Ib 15 years ago [96,97]. More recently, the kinetic mechanisms of AAC(6′)-Ii from E. faecium [98,99], AAC(2′)-Ic from M. tuberculosis [100], AAC(6′)-Iy from Salmonella enterica [101], and AAC(3)-IV from E. coli [102] have been reported. These studies demonstrated the broad aminoglycoside substrate specificity of these enzymes and established that they function through a ternary complex mechanism; that is, both acetylCoA and the aminoglycoside need to be present at the enzyme active site for acyl transfer to occur.

In addition to this research on the mechanism of AACs, mutagenesis studies have demonstrated that single amino acid substitutions can modulate the aminoglycoside substrate specificity. For example, AAC(6′)-I and AAC(6′)-II share the capacity to modify many aminoglycosides, such as kanamycin, but they differ in their propensity to acetylate amikacin and gentamicin C: AAC(6′)-I modifies amikacin but not gentamicin, while AAC(6′)-II is incapable of amikacin acetylation, but does modify gentamicin C. Shaw and colleagues prepared a series of hybrid AAC(6′) enzymes consisting of various portions of AAC(6′)-Ib and AAC(6′)-IIa and demonstrated that the key elements that conferred amikacin versus gentamicin recognition were in the C-terminus [103]. Spontaneous and site-directed mutagenesis studies indicated that modification of amino acid 119 from Ser to Leu could toggle between gentamicin resistance and amikacin sensitivity. Similar Ser→Leu mutants resulting from a single C-to-T transition characterized by amikacin sensitivity and gentamicin resistance, have been isolated in aac(6′)-Ib from a clinical isolate of P. aeruginosa, demonstrating the exquisite balance between antibiotic resistance and sensitivity [104]. Point mutants cannot only expand the aminoglycoside substrate repertoire of AAC, but can also result in catalysts capable of modification of new classes of antibiotics as demonstrated by the characterization of a variant of AAC(6′)-Ib with ciprofloxacin (a fluoroquinolone antibiotic) modification and resistance activity [105]. This demonstrates the important substrate possibility of the family.

The 3D structures of four AACs are known: AAC(3)-Ia, encoded on plasmids in *Serratia marcescens* and other Enterobacteriaceae, determined to 2.3 Å resolution bound to CoA [106]; AAC(2′)-Ic, chromosomally encoded by *M. tuberculosis*, determined to 1.8 to 1.5 Å in apo form and in ternary complexes with aminoglycosides and CoA [107]; AAC(6′)-Ia, chromosomally encoded by *E. faecium*, to 1.8 Å in acetylCoA and CoA-bound forms [108,109]; and AAC(6′)-Iy, chromosomally encoded by *Salmonella enterica*, determined to 2.0 to 3.2 Å in apo form and in ternary complexes with ribostamycin and CoA [110]. Despite low amino acid sequence identity (<11%), there is remarkable conservation in 3D structures (Figure 5.6). Furthermore, reminiscent of the relationship between the structures of APH and protein kinases, there is significant 3D protein structure similarity between the structures of these AACs and other acyltransferases including histone acetyltransferases. These proteins are all members of the GCN5 superfamily of acyltransferases [111].

(a)   (b)

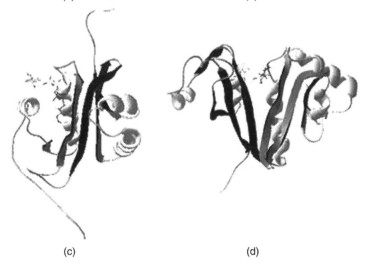

(c)   (d)

FIGURE 5.6   Structures of (a) AAC(3)-Ia, (b) AAC(2′)-Ic, (c) AAC(6′)-Iy, and (d) AAC(6′)-Ii.

The structural similarity between aminoglycoside resistance and protein acyltransferases has been extended to include function, as both AAC(6′)-Ii [108] and AAC(6′)-Iy [110] have been shown to have protein acetyltransferase activity in addition to aminoglycoside modification capacity. Both of these dual function acetyltransferases are chromosomally encoded in bacteria, and this may reflect the evolution of an antibiotic resistance element from an acyltransferase with other function or antibiotic modification may simply be a fortuitous additional activity.

Bacterial genome sequences have revealed a plethora of predicted GCN5-like acyltransferases of unknown function, and many have been annotated as putative aminoglycoside-modifying enzymes. In *M. tuberculosis*, protein Rv1347c was annotated as a predicted aminoglycoside acetyltransferase; however, upon biochemical characterization of purified recombinant enzyme, no antibiotic inactivation was observed [112]. The 3D structure of Rv1347c has been determined by X-ray crystallography, demonstrating unequivocally that it has the GCN5 fold, and additional study suggests it may be involved in siderophore biosynthesis [113]. These results and the observation that a variant AAC(6′)-Ib modifies and confers resistance to the fluoroquinolone antibiotic ciprofloxacin [105] help explain the dominance of the AAC family of aminoglycoside resistance enzymes. The number of genes and apparent malleability of substrate recognition sites in these enzymes make these formidable agents in the evolution of antibiotic resistance.

O-Nucleotidyltransferases

The aminoglycoside *O*-nucleotidyltransferases (ANTs) (Figure 5.7) represent the smallest group of aminoglycoside-modifying enzymes in terms of numbers of reported isozymes (<10), but they have significant impact on clinical aminoglycoside resistance. In particular, ANT(2″)-I is a major source of gentamicin and tobramycin resistance in Enterobacteriaceae [67]. Unlike the APH family of enzymes, the ANTs are quite diverse at the amino acid level with similarities around 20%, and also differ in predicted molecular mass from ~28 to 38 kDa. The most conserved sequence motif, GlySer(Xaa)10-12(Asp or Glu)Xaa(Asp or Glu), where X is any amino acid, is found in the *N*-terminal region of ANTs. Northrop's group has purified ANT(2″)-Ia enzyme from *E. coli* extracts [114], determined the substrate specificity [115], established the kinetic mechanism [116], and identified the rate-limiting step (AMP-aminoglycoside release) [117]. Furthermore, the stereochemistry of AMP-transfer has been shown to occur with inversion of configuration at the α-phosphorus, implicating a mechanism of direct nucleotidyl transfer to the aminoglycoside hydroxyl, that is, no AMP-enzyme intermediate [118]. The interactions of substrates and

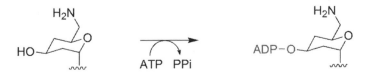

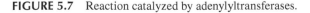

FIGURE 5.7 Reaction catalyzed by adenylyltransferases.

enzyme have been probed by calorimetric approaches that show that binding of MgATP improves aminoglycoside affinity up to three-fold [119].

These mechanistic results can be evaluated in light of the 3D structure of ANT(4′)-Ia in both the apo and ternary complex forms (Figure 5.8) [120,121]. This enzyme was originally obtained from *S. aureus*, where it confers resistance to tobramycin and amikacin [122] and shows 27% amino acid homology (10% identity) to the more predominant ANT(2″)-Ia.

ANT(4′)-Ia is a dimer consisting of two identical subunits and reveals two active sites. Each active site is located at the interface of the dimer and each monomer contributes residues that interact with Mg-ATP and the aminoglycoside (kanamycin in the crystal structure) [120]. Not surprisingly, the signature motif GlySer(Xaa) 10-12(Asp or Glu)Xaa(Asp or Glu) is involved in nucleotide binding, where the conserved Ser interacts with the γ-phosphate of ATP and the conserved Asp/Glu residues are Mg^{2+} ligands. The aminoglycoside-binding pocket is lined with negatively charged residues. This general strategy is conserved in all the aminoglycoside resistance enzyme structures determined to date, and is consistent with the requirements for binding a diverse array of positively charged aminoglycoside substrates.

The 3D structure of ANT(4′) is similar to the fold of mammalian DNA polymerase β [123]. It has been proposed that the ANTs form part of a large polymerase β-like superfamily of nucleotidyltransferases and points to the divergence of a minimal nucleotidyltransferase into a variety of important protein groups with diverse function, but similar chemical cleavage of NTPs [124].

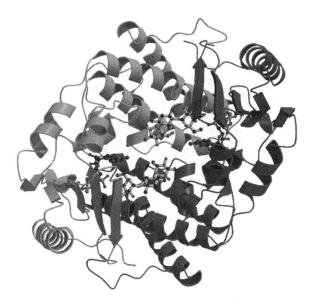

FIGURE 5.8   Structure of ANT(4′)-Ia. The dimer bound with two molecules of kanamycin A and adenosine 5′-(β,γ-methylene)triphosphate (AMPCPP) is shown. (From Pedersen, L.C., Benning, M.M., and Holden, H.M., Structural investigation of the antibiotic and ATP-binding sites in kanamycin nucleotidyltransferase, *Biochemistry* 34, 13305–13311, 1995.)

STRATEGIES TO CIRCUMVENT AMINOGLYCOSIDE-MODIFYING ENZYMES

The challenge presented by the dissemination of aminoglycoside resistance elements can be met with two strategies: (*i*) the discovery of new aminoglycoside antibiotics that are not susceptible to modifying enzymes, and (*ii*) the use of inhibitors of modifying enzymes to potentiate the activity of existing aminoglycosides. The first strategy has been the mainstay of the response to resistance over the past several decades. Thus tobramycin, a 3'-deoxyaminoglycoside, was introduced in the years following the characterization of aminoglycoside modification by APH(3') (Figure 5.1). These enzymes are incapable of tobramycin modification and in fact this compound is a good competitive inhibitor of APH(3') [125]. Similarly, dibekacin (3',4'-dideoxykanamycin B) is effective against some resistant bacteria as well [126]. Comparison of the crystal structure of tobramycin [10] and kanamycin A [12], which differ only by the substitutions of a hydrogen for a hydroxyl at position 3', and amino for hydroxyl at position 2', reveals a predictable loss of an interaction between the O_2 phosphate of A1492 with the 3'-hydroxyl of kanamycin.

The early observation that butirosin, which is derivatized on *N*-1 of the 2-deoxystreptamine ring by an (S)-4-amino-2-hydroxybutyryl (AHB) group, is poorly modified by APH(3')-I (reviewed in reference [127]), prompted the synthesis of other AHB aminoglycosides including amikacin, 1-*N*-AHB kanamycin A [128], which has proven to be an effective and clinically important aminoglycoside antibiotic. Similarly, isepamicin, 1-*N*-(S-3-amino-2-hydroxypropionyl)-gentamicin B, also has found important clinical application [129]. Other *N*1-alkylated aminoglycosides, such as netilmicin (1*N*-ethylsisomicin) [130], have been clinically used.

N-alkylated aminoglycosides including *N*-6' derivatives have been prepared [131,132] and these generally evade modification by the abundant AAC(6')s; however, they sacrifice antimicrobial activity by sterically interfering with a key interaction between the antibiotic and a key 16S rRNA element.

Arbekacin, (1-*N*-(S-3-amino-2-hydroxybutyryl)-3',4'-kanamycin B) [133], has found clinical use in Japan against aminoglycoside-resistant MRSA. Nonetheless, this compound is a substrate for the bifunctional AAC(6')-APH(2") [134]. Novel acetylation of the primary amino group of the AHB moiety of arbekacin in cell-free extracts of arbekacin-resistant MRSA has been reported, though not yet associated with a specific resistance enzyme [135]. 2"-Amino derivatives of arbekacin have been synthesized and show improved antimicrobial activity against *S. aureus* harboring the bifunctional enzyme [136].

The challenge in these synthetic and semisynthetic approaches to circumvent aminoglycoside resistance by alteration of the sites of enzymatic modification is to preserve antibacterial activity, as it is these sites on the molecules that are frequently important in 16S rRNA interaction. Alkylation of *N*-6', for example, results in a parallel decrease of affinity for resistance enzymes and antibacterial activity. Mobashery and colleagues have probed the importance of the *N*-2', *N*-6', *N*-1, and *N*-3 sites through the synthesis of deaminated neamine and kanamycin derivatives [137]. They have shown that loss of these strategic amines can result in dramatic reduction in enzymatic modification by APH(3')-Ia and APH(3')-IIa while in many cases still retaining significant antibacterial activity. While encouraging, these results may not be generally applicable as these same compounds are good substrates for APH(3′)-IIIa [138].

Research in this area continues in several laboratories. For example, Wong's group has prepared several novel aminoglycosides in recent years based on the neamine nucleus [139–141]. Some of these compounds show good antibacterial activity and *in vitro* inhibition of translation, and overcome some, but not all, aminoglycoside-modifying enzymes (e.g., compound 1, Figure 5.9) [141]. The pyranmycins are semisynthetic derivatives of aminoglycosides developed by the Chang group with promising activity against aminoglycoside-resistant Gram-negative bacteria (e.g., compound 2, Figure 5.9) [142–144]. A series of conformationally constrained neomycin analogs designed to stabilize the interactions with the ribosome over resistance enzymes have been reported with promising activity (e.g., compound 3, Figure 5.9) [145]. Finally, Mobashery has designed and synthesized derivatives of neamine with improved binding to the ribosome and antibiotic activity (e.g., compound 4, Figure 5.9) [146–148]. This is an exciting area of research that can leverage the strides made in understanding aminoglycoside action and resistance over recent years.

The other route to evade aminoglycoside resistance by modifying enzymes is through the use of specific inhibitors of these activities. This approach would rescue the antibacterial properties of pharmacologically well-understood aminoglycosides, such as gentamicin C, amikacin, or tobramycin, through the co-administration of inhibitors of common resistance enzymes. There is in fact excellent precedent for this approach in the β-lactam area, where co-administration of β-lactam antibiotics and β-lactamase inhibitors is now well established in clinical practice; however, there are several challenges to this approach in the aminoglycoside field. First is the

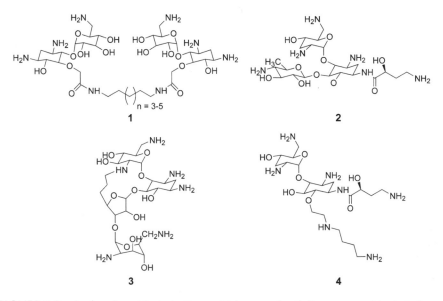

FIGURE 5.9 Aminoglycoside derivatives with improved activity versus resistant strains.

fact that there are dozens of known aminoglycoside inactivating enzymes and these use three chemically distinct routes of modification: phosphorylation, adenylylation, and acetylation. Since many of the best enzyme inhibitors are based on enzyme mechanism or structure of the predicted transition state, it would be unrealistic to envision an inhibitor that would show activity against all of these mechanisms and enzymes. Nonetheless, all aminoglycoside resistance enzymes share the capacity to bind these structurally diverse molecules, and the available 3D structures of all three classes of modifying enzymes have shown that they all have a highly negatively charged substrate binding site. Therefore, compounds that mimic the structure and charge of aminoglycosides, without the capacity to be modified by these enzymes, could act as "universal" inhibitors.

However, a search for such broad-spectrum compounds may not be necessary. It is known that in fact there are predominant resistance elements in the clinic [67–69], and thus only a few mechanisms need be targeted to achieve significant rescue of aminoglycoside activity. Furthermore, in many cases, resistance is genus or even species specific and one could envision cases where targeted molecules would be of great benefit, for example, versus AAC(6′)-APH(2″) in staphylococci and enterococci.

The second challenge to the inhibitor approach is a general requirement for thorough understanding of enzyme mechanism and structure. Modern drug design approaches demand superior knowledge of mechanism, structure, and inhibition to optimize the likelihood of selecting lead compounds with chemotherapeutic potential. Even in high throughput random chemical library screens, downstream optimization of leads by traditional medicinal chemistry or combinatorial methods is greatly facilitated by comprehensive knowledge of enzyme mechanism. As indicated above, this information is now becoming available for all three classes of aminoglycoside-modifying enzyme and examples of new inhibitors have been reported.

The similarity between aminoglycoside kinases and protein kinases has been exploited in a survey of known Ser/Thr/Tyr kinase inhibitors against APH(3′)-IIIa and the kinase activity of AAC(6′)-APH(2″) [81]. For example, this screen demonstrated that the flavonoid quercetin was an APH(3′)-IIIa inhibitor. The isoquinoline sulfonamide inhibitors, such as H-7, H-9, CKI-7, and CKI-9 (Figure 5.10), are known to inhibit protein kinases by binding to the ATP binding site. These compounds behave similarly in APHs, showing competitive inhibition of ATP and non-competitive inhibition of aminoglycosides, consistent with binding in the ATP site. These inhibitors also exhibit different affinities for APH(3′)-IIIa versus APH(2″), and point to the capacity to engineering APH-specific inhibitors based on the isoquinoline nucleus. Although none of these inhibitors could reverse aminoglycoside resistance in bacterial cultures, these studies provide the proof of principle for screening libraries of protein kinase inhibitors as potentiators of aminoglycoside antibiotics against resistant isolates.

A similar screen of inositide kinase inhibitors revealed that APH(2″)-Ia, but not APH(3′)-IIIa, is susceptible to inhibition by inhibitors of inositide kinases, such as LY294002 and wortmanin (Figure 5.10). The latter is in fact an inactivator of APH(2″), covalently modifying Lys52 [149].

Mobashery's group has designed two novel approaches based on the synthesis of aminoglycoside analogs that have the potential to evade the resistance caused by

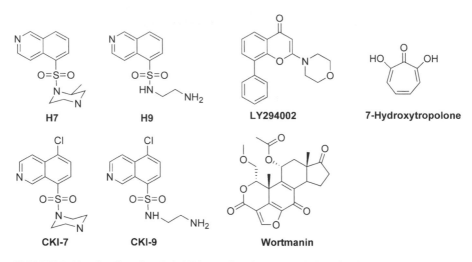

FIGURE 5.10　Small molecule inhibitors of aminoglycoside inactivating enzymes.

APH(3′)s. The first approach required the preparation of aminoglycosides with a nitro group in position 2′ [150]. These compounds were found to be mechanism-based inactivators of APH(3′)-Ia and APH(3′)-IIa, and a mode of action has been proposed that suggests that phosphorylation of the aminoglycoside at position 3′ is followed by elimination of phosphate and the generation of electrophilic nitroalkene in the enzyme active site (Figure 5.11). Such compounds readily undergo nucleophilic attack and thus have the potential to alkylate the enzyme through reaction with amino acid side chains, for example, SH of Cys, NH_2 of Lys.

The second approach involved the synthesis of a kanamycin analog with a ketone at position 3′, rather than a hydroxyl group [151]. In aqueous solution, the ketone is hydrated to form the *gem*-diol. This can act as a substrate for APH(3′), but the phosphate group is unstable in this configuration and the ketone readily regenerated with loss of inorganic phosphate. This compound showed poor biological activity, but the strategy of reversible phosphorylation has been demonstrated and future analogs may prove useful.

ANT(2″)-Ia has been shown to be inhibited by 7-hydroxytropolone (Figure 5.10), a natural product produced by *Streptomyces neyagawaensis* [152]. This compound

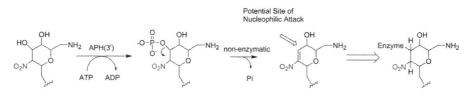

FIGURE 5.11 Proposed mechanism of APH inactivation by 2′-NO_2-containing aminoglycosides.

was competitive inhibitor of ATP and was identified by its capacity to potentiate tobramycin in ANT(2″)-Ia expressing *E. coli*, but not cells expressing AAC(6′), AAC(3), or ANT(3″). These studies indicate that the concept of reversing aminoglycoside resistance through co-administration of resistance enzyme inhibitors is valid.

The observation that the 3D structures of aminoglycoside inactivating enzymes all share a common highly negatively charged aminoglycoside binding site and that AACs and APHs can bind and modify peptides and proteins, inspired a screen of cationic peptides as possible inhibitors of aminoglycoside-modifying enzymes. Several peptides with inhibitory activity were identified including some that inhibited more than one enzyme and mechanism [153].

There are excellent reasons to be optimistic for the discovery of inhibitory compounds that could find clinical use for the reversal of aminoglycoside resistance. The growing understanding of enzyme mechanism in addition to protein structural data provides the requisite foundation for a concerted effort in this area. The fact that well-established enzyme assays are in place for all three classes of aminoglycoside-modifying enzymes and that these are amenable for high-throughput screening methods is also of great benefit. Screens against small molecule libraries may uncover new, non-aminoglycoside leads that may prove to be starting points for inhibitors that can potentiate the activity of aminoglycoside antibiotics in resistant organisms.

SUMMARY AND CONCLUSIONS

Aminoglycosides are clinically important antibiotics that interact with the bacterial ribosome and disrupt proper translation. Aminoglycoside resistance in clinical isolates is largely the result of enzymes that phosphorylate, adenylate, or acetylate the antibiotics. Recent efforts in understanding the mechanisms and structures of these enzymes now open the possibility for the design of high-affinity inhibitors that could reverse resistance and potentiate existing aminoglycoside antibiotics. Several challenges remain to be addressed, however, including issues of the number and diversity of resistance enzymes, transport of inhibitors across cell membranes, and that aminoglycoside usage patterns select for different resistance mechanisms. At the same time, genome sequencing efforts have shown that a number of potential or cryptic aminoglycoside resistance genes are located within the genomes of many bacteria including *M. tuberculosis*, *B. subtilis*, and *P. aeruginosa*. The impact of the presence of these elements remains to be assessed, but it speaks to the prevalence and diversity of aminoglycoside resistance within bacterial populations. Nonetheless, the efficacy of these antibiotics and decades of experience in managing the toxicity associated with the class suggest that a combination of a potent resistance enzyme inhibitor that covers the important clinically relevant enzymes along with a well-tolerated aminoglycoside, such as gentamicin or streptomycin, could have significant clinical utility for the treatment of infection in environments, where the resistance burden is significant, for example, intensive care units and battlefields. Such efforts, coupled with programs to dampen or eliminate the renal and oto-toxicity associated with aminoglycosides, could have important impact on reviving pharmaceutical sector interest in the class. The key elements in this endeavor are continued research into all aspects of aminoglycoside resistance as well as interest in pursuing the molecular details of aminoglycoside toxicity in mammals.

ACKNOWLEDGMENTS

The author thanks Dr. Albert Berghuis and Tushar Shakya for assistance in preparing figures and the Canadian Institutes of Health Research and the Canada Research Chair program for funding aminoglycoside research in the author's laboratory.

REFERENCES

1. Edson, R. S. and Terrell, C. L., The aminoglycosides, *Mayo Clin. Proc.* 74 (5), 519–528, 1999.
2. Hoepelman, A. I., Current therapeutic approaches to cryptosporidiosis in immunocompromised patients, *J. Antimicrob. Chemother.* 37 (5), 871–880, 1996.
3. Carter, A. P., Clemons, W. M., Brodersen, D. E., Morgan-Warren, R. J., Wimberly, B. T., and Ramakrishnan, V., Functional insights from the structure of the 30S ribosomal subunit and its interactions with antibiotics, *Nature* 407 (6802), 340–348, 2000.
4. Brodersen, D. E., Clemons, W. M., Jr., Carter, A. P., Morgan-Warren, R. J., Wimberly, B. T., and Ramakrishnan, V., The structural basis for the action of the antibiotics tetracycline, pactamycin, and hygromycin B on the 30S ribosomal subunit, *Cell* 103 (7), 1143–1154, 2000.
5. Fourmy, D., Recht, M. I., Blanchard, S. C., and Puglisi, J. D., Structure of the A site of *Escherichia coli* 16S ribosomal RNA complexed with an aminoglycoside antibiotic, *Science* 274, 1367–1371, 1996.
6. Fourmy, D., Recht, M. I., and Puglisi, J. D., Binding of neomycin-class aminoglycoside antibiotics to the A-site of 16 S rRNA, *J. Mol. Biol.* 277, 347–362, 1998.
7. Yoshizawa, S., Fourmy, D., and Puglisi, J. D., Structural origins of gentamicin antibiotic action, *EMBO J.* 17 (22), 6437–6448, 1998.
8. Lynch, S. R., Gonzalez, R. L., and Puglisi, J. D., Comparison of X-ray crystal structure of the 30S subunit-antibiotic complex with NMR structure of decoding site oligonucleotide-paromomycin complex, *Structure* 11 (1), 43–53, 2003.
9. Vicens, Q. and Westhof, E., Crystal structure of paromomycin docked into the eubacterial ribosomal decoding A site, *Structure* 9 (8), 647–658, 2001.
10. Vicens, Q. and Westhof, E., Crystal structure of a complex between the aminoglycoside tobramycin and an oligonucleotide containing the ribosomal decoding A site, *Chem. Biol.* 9 (6), 747–755, 2002.
11. Vicens, Q. and Westhof, E., Crystal structure of geneticin bound to a bacterial 16S ribosomal RNA A site oligonucleotide, *J. Mol. Biol.* 326 (4), 1175–1188, 2003.
12. Francois, B., Russell, R. J., Murray, J. B., Aboul-ela, F., Masquida, B., Vicens, Q., and Westhof, E., Crystal structures of complexes between aminoglycosides and decoding A site oligonucleotides: role of the number of rings and positive charges in the specific binding leading to miscoding, *Nucleic Acids Res.* 33 (17), 5677–5690, 2005.
13. Ogle, J. M. and Ramakrishnan, V., Structural insights into translational fidelity, *Annu. Rev. Biochem.* 74, 129–177, 2005.
14. Ogle, J. M., Brodersen, D. E., Clemons, W. M., Jr., Tarry, M. J., Carter, A. P., and Ramakrishnan, V., Recognition of cognate transfer RNA by the 30S ribosomal subunit, *Science* 292 (5518), 897–902, 2001.
15. Davies, J., Gorini, L., and Davis, B. D., Misreading of RNA codewords induced by aminoglycoside antibiotics, *Mol. Pharmacol.* 1, 93–106, 1965.

16. Davis, B. D., Mechanism of action of aminoglycosides, *Microbiol. Rev.* 51, 341–350, 1987.

17. Recht, M. I., Douthwaite, S., Dahlquist, K. D., and Puglisi, J. D., Effect of mutations in the A site of 16 S rRNA on aminoglycoside antibiotic-ribosome interaction, *J. Mol. Biol.* 286 (1), 33–43, 1999.

18. Recht, M. I., Douthwaite, S., and Puglisi, J. D., Basis for prokaryotic specificity of action of aminoglycoside antibiotics, *EMBO J.* 18 (11), 3133–3138, 1999.

19. Taber, H. W., Mueller, J. P., Miller, P. F., and Arrow, A. S., Bacterial uptake of amino-glycoside antibiotics, *Microbiol. Rev.* 51, 439–457, 1987.

20. Kashiwagi, K., Tsuhako, M. H., Sakata, K., Saisho, T, Igarashi, A., da Costa, S. O., and Igarashi, K., Relationship between spontaneous aminoglycoside resistance in *Escherichia coli* and a decrease in oligopeptide binding protein, *J. Bacteriol.* 180 (20), 5484–5488, 1998.

21. Walsh, C. T., *Antibiotics, Actions, Origins, Resistance.* ASM Press, Washington, D.C., 2003.

22. Davies, J. and Davis, B. D., Misreading of ribonucleic acid code words induced by aminoglycoside antibiotics. The effect of drug concentration, *J. Biol. Chem.* 243 (12), 3312–3316, 1968.

23. Davies, J., Jones, D. S., and Khorana, H. G., A further study of misreading of codons induced by streptomycin and neomycin using ribopolynucleotides containing two nucleotides in alternating sequence as templates, *J. Mol. Biol.* 18 (1), 48–57, 1966.

24. Gorini, L., Streptomycin and misreading of the genetic code, in *Ribosomes*, Nomura, M., Tissères, A., and Lengyel, P. Cold Spring Harbor Laboratory, Cold Spring Harbor, NY, 1974, pp. 791–803.

25. Lando, D., Cousin, M. A., and Privat de Garilhe, M., Misreading, a fundamental aspect of the mechanism of action of several aminoglycosides, *Biochemistry* 12, 4528–4533, 1973.

26. Anand, N. and Davis, B. D., Damage by streptomycin to the cell membrane of *Escherichia coli*, *Nature* 185, 22–23, 1960.

27. Dubin, D. T. and Davis, B. D., The effect of streptomycin on potassium flux in *Escherichia coli*, *Biochim. Biophys. Acta* 52, 400–402, 1961.

28. Hancock, R., Early effects of streptomycin on *Bacillus megaterium*, *J. Bacteriol.* 88, 633–639, 1961.

29. Busse, H.-J., Wöstmann, C., and Bakker, E. P., The bactericidal action of streptomycin: membrane permeabilization caused by the insertion of mistranslated proteins into the cytoplasmic membrane of *Escherichia coli* and subsequent caging of the antibiotic inside the cells due to degradation of these proteins, *J. Gen. Microbiol.* 138, 551–561, 1992.

30. Davis, B. D., Chen, L. L., and Tai, P. C., Misread protein creates membrane channels: an essential step in the bactericidal action of aminoglycosides, *Proc. Natl. Acad. Sci. USA* 83, 6164–6168, 1986.

31. Bryan, L. E. and Van Den Elzen, H. M., Effects of membrane-energy mutations and cations on streptomycin and gentamicin accumulation by bacteria: a model for entry of streptomycin and gentamicin in susceptible and resistant bacteria, *Antimicrob. Agents Chemother.* 12, 163–177, 1977.

32. Muir, M. E. and Wallace, B. J., Isolation of mutants of *Escherichia coli* uncoupled in oxidative phosphorylation using hypersensitivity to streptomycin, *Biochim. Biophys. Acta* 547 (2), 218–229, 1979.

33. Taber, H. and Halfenger, G. M., Multiple-aminoglycoside-resistant mutants of *Bacillus subtilis* deficient in accumulation of kanamycin, *Antimicrob. Agents Chemother.* 9 (2), 251–259, 1976.

34. Taber, H. W., Sugarman, B. J. U., and Halfenger, G. M., Involvement of menaquinone in the active accumulation of aminoglycosides by *Bacillus subtilis*, *J. Gen. Microbiol.* 123, 143–149, 1981.

35. Miller, M. H., Edberg, S. C., Mandel, L. J., Behar, C. F., and Steigbigel, N. H., Gentamicin uptake in wild-type and aminoglycoside-resistant small-colony mutants of *Staphylococcus aureus*, *Antimicrob. Agents Chemother.* 18, 722–729, 1980.

36. McEnroe, A. S. and Taber, H. W., Correlation between cytochrome *aa3* concentrations and streptomycin accumulation in *Bacillus subtilis*, *Antimicrob. Agents Chemother.* 26 (4), 507–512, 1984.

37. Humbert, R. and Altendorf, K., Defective gamma subunit of ATP synthase (F1F0) from *Escherichia coli* leads to resistance to aminoglycoside antibiotics, *J. Bacteriol.* 171, 1435–1444, 1989.

38. Magnet, S., Courvalin, P., and Lambert, T., Resistance-nodulation-cell division-type efflux pump involved in aminoglycoside resistance in *Acinetobacter baumannii* strain BM4454, *Antimicrob. Agents Chemother.* 45 (12), 3375–3380, 2001.

39. Aires, J. R., Kohler, T., Nikaido, H., and Plesiat, P., Involvement of an active efflux system in the natural resistance of *Pseudomonas aeruginosa* to aminoglycosides, *Antimicrob. Agents Chemother.* 43, 2624–2628, 1999.

40. Moore, R. A., DeShazer, D., Reckseidler, S., Weissman, A., and Woods, D. E., Efflux-mediated aminoglycoside and macrolide resistance in *Burkholderia pseudomallei*, *Antimicrob. Agents Chemother.* 43 (3), 465–470, 1999.

41. Rosenberg, E. Y., Ma, D., and Nikaido, H., AcrD of *Escherichia coli* is an aminoglycoside efflux pump, *J. Bacteriol.* 182 (6), 1754–1756, 2000.

42. Aires, J. R. and Nikaido, H., Aminoglycosides are captured from both periplasm and cytoplasm by the AcrD multidrug efflux transporter of *Escherichia coli*, *J. Bacteriol.* 187 (6), 1923–1929, 2005.

43. Chan, Y. Y., Tan, T. M., Ong, Y. M., and Chua, K. L., BpeAB-OprB, a multidrug efflux pump in *Burkholderia pseudomallei*, *Antimicrob. Agents Chemother.* 48 (4), 1128–1135, 2004.

44. Mine, T., Morita, Y., Kataoka, A., Mizushima, T., and Tsuchiya, T., Expression in *Escherichia coli* of a new multidrug efflux pump, MexXY, from *Pseudomonas aeruginosa*, *Antimicrob. Agents Chemother.* 43 (2), 415–417, 1999.

45. Poole, K., Aminoglycoside resistance in *Pseudomonas aeruginosa*, *Antimicrob. Agents Chemother.* 49 (2), 479–487, 2005.

46. Li, X. Z., Zhang, L., and Poole, K., SmeC, an outer membrane multidrug efflux protein of *Stenotrophomonas maltophilia*, *Antimicrob. Agents Chemother.* 46 (2), 333–343, 2002.

47. Marchand, I., Damier-Piolle, L., Courvalin, P., and Lambert, T., Expression of the RND-type efflux pump AdeABC in *Acinetobacter baumannii* is regulated by the AdeRS two-component system, *Antimicrob. Agents Chemother.* 48 (9), 3298–3304, 2004.

48. Poole, K., Efflux-mediated multiresistance in Gram-negative bacteria, *Clin. Microbiol. Infect.* 10 (1), 12–26, 2004.

49. Kelemen, G. H., Cundliffe, E., and Financsek, I., Cloning and characterization of gentamicin-resistance genes from *Micromonospora purpurea* and *Micromonospora rosea*, *Gene* 98, 53–60, 1991.

50. Skeggs, P. A., Holmes, D. J., and Cundliffe, E., Cloning of aminoglycoside-resistance determinants from *Streptomyces tenebrarius* and comparison with related genes from other actinomycetes, *J. Gen. Microbiol.* 133, 915–923, 1987.

51. Galimand, M., Courvalin, P., and Lambert, T., Plasmid-mediated high-level resistance to aminoglycosides in *Enterobacteriaceae* due to 16S rRNA methylation, *Antimicrob. Agents Chemother.* 47 (8), 2565–2571, 2003.

52. Yokoyama, K., Doi, Y., Yamane, K., Kurokawa, H., Shibata, N., Shibayama, K., Yagi, T., Kato, H., and Arakawa, Y., Acquisition of 16S rRNA methylase gene in *Pseudomonas aeruginosa*, *Lancet* 362 (9399), 1888–1893, 2003.

53. Yamane, K., Doi, Y., Yokoyama, K., Yagi, T., Kurokawa, H., Shibata, N., Shibayama, K., Kato, H., and Arakawa, Y., Genetic environments of the *rmtA* gene in *Pseudomonas aeruginosa* clinical isolates, *Antimicrob. Agents Chemother.* 48 (6), 2069–2074, 2004.

54. Doi, Y., Yokoyama, K., Yamane, K., Wachino, J., Shibata, N., Yagi, T., Shibayama, K., Kato, H., and Arakawa, Y., Plasmid-mediated 16S rRNA methylase in *Serratia marcescens* conferring high-level resistance to aminoglycosides, *Antimicrob. Agents Chemother.* 48 (2), 491–496, 2004.

55. Gonzalez-Zorn, B., Catalan, A., Escudero, J. A., Dominguez, L., Teshager, T., Porrero, C., and Moreno, M. A., Genetic basis for dissemination of *armA*, *J. Antimicrob. Chemother.* 56 (3), 583–585, 2005.

56. Galimand, M., Sabtcheva, S., Courvalin, P., and Lambert, T., Worldwide disseminated *armA* aminoglycoside resistance methylase gene is borne by composite transposon Tn*1548*, *Antimicrob. Agents Chemother.* 49 (7), 2949–2953, 2005.

57. Musser, J. M., Antimicrobial agent resistance in mycobacteria: molecular genetic insights, *Clin. Microbiol. Rev.* 8, 496–514, 1995.

58. Finken, M., Kirschner, P., Meier, A., Wrede, A., and Bottger, E. C., Molecular basis of streptomycin resistance in *Mycobacterium tuberculosis*: alterations of the ribosomal protein S12 gene and point mutations within a functional 16S ribosomal RNA pseudoknot, *Mol. Microbiol.* 9 (6), 1239–1246, 1993.

59. Nair, J., Rouse, D. A., Bai, G. H., and Morris, S. L., The *rpsL* gene and streptomycin resistance in single and multiple drug-resistant strains of *Mycobacterium tuberculosis*, *Mol. Microbiol.* 10 (3), 521–527, 1993.

60. Douglass, J. and Steyn, L. M., A ribosomal gene mutation in streptomycin-resistant *Mycobacterium tuberculosis* isolates, *J. Infect. Dis.* 167 (6), 1505–1506, 1993.

61. Alangaden, G. J., Kreiswirth, B. N., Aouad, A., Khetarpal, M., Igno, F. R., Moghazeh, S. L., Manavathu, E. K., and Lerner, S. A., Mechanism of resistance to amikacin and kanamycin in *Mycobacterium tuberculosis*, *Antimicrob. Agents Chemother.* 42 (5), 1295–1297, 1998.

62. De Stasio, E. A., Moazed, D., Noller, H. F., and Dahlberg, A. E., Mutations in 16S ribosomal RNA disrupt antibiotic-RNA interactions, *EMBO J.* 8 (4), 1213–1216, 1989.

63. Shaw, K. J., Rather, P. N., Hare, R. S., and Miller, G. H., Molecular genetics of aminoglycoside resistance genes and familial relationships of the aminoglycoside-modifying enzymes, *Microbiol. Rev.* 57, 138–163, 1993.

64. Oka, A., Sugisaki, H., and Takanami, M., Nucleotide sequence of the kanamycin resistance transposon Tn*903*, *J. Mol. Biol.* 147, 217–226, 1981.

65. Hoshiko, S., Nojiri, C., Matsunaga, K., Katsumata, K., Satoh, E., and Nagaoka, K., Nucleotide sequence of the ribostamycin phosphotransferasegene and of its control region in *Streptomyces ribosidificus*, *Gene* 68, 285–296, 1988.

66. Wright, G. D., Berghuis, A. M., and Mobashery, S., Aminoglycoside antibiotics: Structures, function and resistance, in *Resolving the Antibiotic paradox: Progress in Drug Design and Resistance*, Rosen, B. P. and Mobashery, S. Plenum Press, New York, 1998, pp. 27–69.

67. Miller, G. H., Sabatelli, F. J., Hare, R. S., Glupczynski, Y., Mackey, P., Shlaes, D., Shimizu, K., Shaw, K. J., and Groups, a. A. R. S., The most frequent aminoglycoside resistance mechanisms-changes with time and geographic area: a reflection of aminoglycoside usage patterns? *Clin. Infect. Dis.* 24, S46–S62, 1997.

68. Miller, G. H., Sabatelli, F. J., Naples, L., Hare, R. S., Shaw, K. J., and Groups, a. t. A. R. S., The most frequently occuring aminoglycoside resistance mechanisms—combined results of surveys in eight regions of the world, *J. Chemother.* 7 suppl. 2, 17–30, 1995.

69. Miller, G. H., Sabatelli, F. J., Naples, L., Hare, R. S., Shaw, K. J., and Groups, a. t. A. R. S., The changing nature of aminoglycoside resistance mechanisms and the role of isepamicin— a new broad-spectrum aminoglycoside, *J. Chemother.* 7 suppl. 2, 31–44, 1995.

70. Thompson, P. R., Hughes, D. W., and Wright, G. D., Regiospecificity of aminoglycoside phosphotransferase from *Enterococci* and *Staphylococci* (APH(3')-IIIa), *Biochemistry* 35, 8686–8695, 1996.

71. Umezawa, H., Okanishi, M., Kondo, S., Hamana, K., Utahara, R., Maeda, K., and Mitsuhashi, S., Phosphorylative inactivation of aminoglycopside antibioics by *Escherichia coli* carrying R factor, *Science* 157, 1559–1561, 1967.

72. Azucena, E., Grapsas, I., and Mobashery, S., Properties of a bifunctional bacterial antibiotic resistance enzyme that catalyzes ATP-dependent 2″-phosphorylation and acetyl-CoA-dependent 6′-acetylation of aminoglycosides, *J. Am. Chem. Soc.* 119, 2317–2318, 1997.

73. Daigle, D. M., Hughes, D. W., and Wright, G. D., Prodigious substrate specificity of AAC(6′)-APH(2″), an aminoglycoside antibiotic resistance determinant in enterococci and staphylococci, *Chem. Biol.* 6, 99–110, 1999.

74. Ferretti, J. J., Gilmore, K. S., and Courvalin, P., Nucleotide sequence analysis of the gene specifying the bifunctional 6′-aminoglycoside acetyltransferase 2″-aminoglycoside phosphotransferase enzyme in *Streptococcus faecalis* and identification and cloning of gene regions specifying the two activities, *J. Bacteriol.* 167, 631–638, 1986.

75. Chow, J. W., Zervos, M. J., Lerner, S. A., Thal, L. A., Donabedian, S. M., Jaworski, D. D., Tsai, S., Shaw, K. J., and Clewell, D. B., A novel gentamicin resistance gene in *Enterococcus*, *Antimicrob. Agents Chemother.* 41, 511–514, 1997.

76. Tsai, S. F., Zervos, M. J., Clewell, D. B., Donabedian, S. M., Sahm, D. F., and Chow, J. W., A new high-level gentamicin resistance gene, *aph(2″)-Id*, in *Enterococcus* spp., *Antimicrob. Agents Chemother.* 42, 1229–1232, 1998.

77. Hon, W. C., McKay, G. A., Thompson, P. R., Sweet, R. M., Yang, D. S. C., Wright, G. D., and Berghuis, A. M., Structure of an enzyme required for aminoglycoside resistance reveals homology to eukariotic protein kinases, *Cell* 89, 887–895, 1997.

78. Nurizzo, D., Shewry, S. C., Perlin, M. H., Brown, S. A., Dholakia, J. N., Fuchs, R. L., Deva, T., Baker, E. N., and Smith, C. A., The crystal structure of aminoglycoside-3′-phosphotransferase-IIa, an enzyme responsible for antibiotic resistance, *J. Mol. Biol.* 327 (2), 491–506, 2003.

79. Thompson, P. R., Schwartzenhauer, J., Hughes, D. W., Berghuis, A. M., and Wright, G. D., The COOH terminus of aminoglycoside phosphotransferase (3′)-IIIa is critical for antibiotic recognition and resistance, *J. Biol. Chem.* 274 (43), 30697–30706, 1999.

80. Daigle, D. M., McKay, G. A., Thompson, P. R., and Wright, G. D., Aminoglycoside phosphotransferases required for antibiotic resistance are also Serine protein kinases, *Chem. Biol.* 6, 11–18, 1998.

81. Daigle, D. M., McKay, G. A., and Wright, G. D., Inhibition of aminoglycoside antibiotic resistance enzymes by protein kinase inhibitors, *J. Biol. Chem.* 272, 24755–24758, 1997.

82. Madhusudan, Trafny, E. A., Xuong, N.-H., Adams, J. A., Ten Eyck, L. F., Taylor, S. S., and Sowadski, J. M., cAMP-dependent protein kinase: Crystallographic insights into substrate recognition and phosphotransfer, *Prot. Sci.* 3, 176–187, 1994.

83. Thompson, P. R., Boehr, D. D., Berghuis, A. M., and Wright, G. D., Mechanism of aminoglycoside antibiotic kinase APH(3′)-IIIa: role of the nucleotide positioning loop, *Biochemistry* 41 (22), 7001–7007, 2002.

84. Noguchi, N., Emura, A., Matsuyama, H., O'Hara, K., Sasatsu, M., and Kono, M., Nucleotide sequence and characterization of erythromycin resistance determinant that encodes macrolide 2′-phosphotransferase-I in *Escherichia coli*, *Antimicrob. Agents Chemother.* 39, 2359–2363, 1995.

85. Noguchi, N., Katayama, J., and O'Hara, K., Cloning and nucleotide sequence of the *mphB* gene for macrolide 2′-phosphotransferase-II in *Escherichia coli*, *FEMS Microbiol. Lett.* 144, 197–202, 1996.

86. Bibb, M. J., Bibb, M. J., Ward, J. M., and Cohen, S. N., Nucleotide sequences encoding and promoting expression of three antibiotic resistance genes indigenous to *Streptomyces*, *Mol. Gen. Genet.* 199, 26–36, 1985.

87. Lovering, A. M., White, L. O., and Reeves, D. S., AAC(1): a new aminoglycoside-acetylating enzyme modifying the Cl amino group of apramycin, *J. Antimicrob. Chemother.* 20, 803–813, 1987.

88. Sunada, A., Nakajima, M., Ikeda, Y., Kondo, S., and Hotta, K., Enzymatic 1-*N*-acetylation of paromomycin by an actinomycete strain #8 with multiple aminoglycoside resistance and paromomycin sensitivity, *J. Antibiot.* 52, 809–814, 1999.

89. Costa, Y., Galimand, M., Leclercq, R., Duval, J., and Courvalin, P., Characterization of the chromosomal *aac(6′)-Ii* gene specific for *Enterococcus faecium*, *Antimicrob. Agents Chemother.* 37, 1896–1903, 1993.

90. Rather, P. N., Orosz, E., Shaw, K. J., Hare, R., and Miller, G., Characterization and transcriptional regulation of the 2′-*N*-acetyltransferase gene from *Providencia stuartii*, *J. Bacteriol.* 175, 6492–6498, 1993.

91. Aínsa, J. A., Martin, C., Gicquel, B., and Gomez-Lus, R., Characterization of the chromosomal aminoglycoside 2′-*N*-acetyltransferase gene from *Mycobacterium fortuitum*, *Antimicrob. Agents Chemother.* 40, 2350–2355, 1996.

92. Aínsa, J. A., Pérez, E., Pelicic, V., Berthet, F. X., Gicquel, B., and Martín, C., Aminoglycoside 2′-*N*-acetyltransferase genes are universally present in mycobacteria: characterization of the *aac(2′)-Ic* gene from *Mycobacterium tuberculosis* and the *aac(2′)-Id* gene from *Mycobacterium smegmatis*, *Mol. Microbiol.* 24, 431–441, 1997.

93. Williams, J. W. and Northrop, D. B., Kinetic mechanism of gentamicin acetyltransferase I, *J. Biol. Chem.* 253, 5902–5907, 1978.

94. Williams, J. W. and Northrop, D. B., Substrate specificity and structure-activity relationships of gentamicin acetyltransferase I, *J. Biol. Chem.* 253, 5908–5914, 1978.

95. Williams, J. W. and Northrop, D. B., Synthesis of a tight-binding, multisubstrate analog inhibitor of gentamicin acetyltransferase, *J. Antibiot.* 32, 1147–1154, 1979.

96. Radika, K. and Northop, D. B., Substrate specificities and structure-activity relationships for acylation of antibiotics catalyzed by kanamycin acetyltransferase, *Biochemistry* 23, 5118–5122, 1984.

97. Radika, K. and Northrop, D., The kinetic mechanism of kanamycin acetyltransferase derived from the use of alternative antibiotics and coenzymes, *J. Biol. Chem.* 259, 12543–12546, 1984.

98. Draker, K. A., Northrop, D. B., and Wright, G. D., Kinetic mechanism of the GCN5-related chromosomal aminoglycoside acetyltransferase AAC(6′)-Ii from *Enterococcus faecium*: evidence of dimer subunit cooperativity, *Biochemistry* 42 (21), 6565–6574, 2003.

99. Draker, K. A. and Wright, G. D., Molecular mechanism of the enterococcal aminoglycoside 6′-*N*-acetyltransferase: role of GNAT-conserved residues in the chemistry of antibiotic inactivation, *Biochemistry* 43 (2), 446–454, 2004.

100. Hegde, S. S., Javid-Majd, F., and Blanchard, J. S., Overexpression and mechanistic analysis of chromosomally encoded aminoglycoside 2′-*N*-acetyltransferase (AAC(2′)-Ic) from *Mycobacterium tuberculosis*, *J. Biol. Chem.* 276 (49), 45876–45881, 2001.

101. Magnet, S., Lambert, T., Courvalin, P., and Blanchard, J. S., Kinetic and mutagenic characterization of the chromosomally encoded *Salmonella enterica* AAC(6′)-Iy aminoglycoside *N*-acetyltransferase, *Biochemistry* 40 (12), 3700–3709, 2001.

102. Magalhaes, M. L. and Blanchard, J. S., The kinetic mechanism of AAC3-IV aminoglycoside acetyltransferase from *Escherichia coli*, *Biochemistry* 44 (49), 16275–16283, 2005.

103. Rather, P. N., Munayyer, H., Mann, P. A., Hare, R. S., Miller, G. H., and Shaw, K. J., Genetic analysis of bacterial acetyltransferases: Identification of amino acids determining the specificities of the aminoglycoside 6′-*N*-acetyltransferase Ib and IIa proteins, *J. Bacteriol.* 175, 3196–3203, 1992.

104. Lambert, T., Ploy, M.-C., and Courvalin, P., A spontaneous point mutation in the *aac(6')-Ib'* gene results in altered substrate specificity of aminoglycoside 6'-*N*-acetyltransferase of a *Pseudomonas fluorescens* strain, *FEMS Microbiol. Lett.* 115, 297–304, 1994.

105. Robicsek, A., Strahilevitz, J., Jacoby, G. A., Macielag, M., Abbanat, D., Hye Park, C., Bush, K., and Hooper, D. C., Fluoroquinolone-modifying enzyme: a new adaptation of a common aminoglycoside acetyltransferase, *Nat. Med.* 12 (1), 83–88, 2006.

106. Wolf, E., Vassilev, A., Makino, Y., Sali, A., Nakatani, Y., and Burley, S. K., Crystal structure of a GCN5-related *N*-acetyltransferase: *Serratia marcescens* aminoglycoside 3-*N*-acetyltransferase, *Cell* 94, 439–449, 1998.

107. Vetting, M. W., Hegde, S. S., Javid-Majd, F., Blanchard, J. S., and Roderick, S. L., Aminoglycoside 2'-*N*-acetyltransferase from *Mycobacterium tuberculosis* in complex with coenzyme A and aminoglycoside substrates, *Nat. Struct. Biol.* 9 (9), 653–658, 2002.

108. Wybenga-Groot, L., Draker, K. A., Wright, G. D., and Berghuis, A. M., Crystal structure of an aminoglycoside 6'-*N*-acetyltransferase: defining the GCN5-related *N*-acetyltransferase superfamily fold, *Structure* 7, 497–507, 1999.

109. Burk, D. L., Ghuman, N., Wybenga-Groot, L. E., and Berghuis, A. M., X-ray structure of the AAC(6')-Ii antibiotic resistance enzyme at 1.8 Å resolution; examination of oligomeric arrangements in GNAT superfamily members, *Protein Sci.* 12 (3), 426–437, 2003.

110. Vetting, M. W., Magnet, S., Nieves, E., Roderick, S. L., and Blanchard, J. S., A bacterial acetyltransferase capable of regioselective *N*-acetylation of antibiotics and histones, *Chem. Biol.* 11 (4), 565–573, 2004.

111. Vetting, M. W., LP, S. d. C., Yu, M., Hegde, S. S., Magnet, S., Roderick, S. L., and Blanchard, J. S., Structure and functions of the GNAT superfamily of acetyltransferases, *Arch. Biochem. Biophys.* 433 (1), 212–226, 2005.

112. Draker, K. A., Boehr, D. D., Elowe, N. H., Noga, T. J., and Wright, G. D., Functional annotation of putative aminoglycoside antibiotic modifying proteins in *Mycobacterium tuberculosis* H37Rv, *J. Antibiot. (Tokyo)* 56 (2), 135–142, 2003.

113. Card, G. L., Peterson, N. A., Smith, C. A., Rupp, B., Schick, B. M., and Baker, E. N., The crystal structure of Rv1347c, a putative antibiotic resistance protein from *Mycobacterium tuberculosis*, reveals a GCN5-related fold and suggests an alternative function in siderophore biosynthesis, *J. Biol. Chem.* 280 (14), 13978–13986, 2005.

114. Van Pelt, J. E. and Northrop, D. B., Purification and properties of gentamicin nucleotidyltransferase from *Escherichia coli*: Nucleotide specificity, pH optimum, and the separation of two electrophoretic variants, *Arch. Biochem. Biophys.* 230, 250–263, 1984.

115. Gates, C. A. and Northrop, D. B., Substrate specificities and structure-activity relationships for the nucleotidylation of antibiotics catalyzed by aminoglycoside nucleotidyltransferase 2''-I, *Biochemistry* 27, 3820–3825, 1988.

116. Gates, C. A. and Northrop, D. B., Alternative substrate and inhibition kinetics of aminoglycoside nucleotidyltransferase 2''-I in support of a Theorell-Chance kinetic mechanism, *Biochemistry* 27, 3826–3833, 1988.

117. Gates, C. A. and Northrop, D. B., Determination of the rate-limiting segment of aminoglycoside nucleotidyltransferase 2''-I by pH- and viscosity-dependent kinetics, *Biochemistry* 27, 3834–3842, 1988.

118. Van Pelt, J. E., Iyengar, R., and Frey, P. A., Gentamicin nucleotidyltansferase. Stereochemical inversion at phosphorus in enzymatic 2'-deoxyadenylyl transfer to tobramycin, *J. Biol. Chem.* 261, 15995–15999, 1986.

119. Wright, E. and Serpersu, E. H., Enzyme-substrate interactions with an antibiotic resistance enzyme: aminoglycoside nucleotidyltransferase(2'')-Ia characterized by kinetic and thermodynamic methods, *Biochemistry* 44 (34), 11581–11591, 2005.

120. Pedersen, L. C., Benning, M. M., and Holden, H. M., Structural investigation of the antibiotic and ATP-binding sites in kanamycin nucleotidyltransferase, *Biochemistry* 34, 13305–13311, 1995.
121. Sakon, J., Liao, H. H., Kanikula, A. M., Benning, M. M., Rayment, I., and Holden, H. M., Molecular structure of kanamycin nucleotidyl transferase determined to 3 Å resolution, *Biochemistry* 32, 11977–11984, 1993.
122. Matsumura, M., Katakura, Y., Imanaka, T., and Aiba, S., Enzymatic and nucleotide sequence studies of a kanamycin-inactivating enzyme encoded by a plasmid from thermophilic bacilli in comparison with that encoded by plasmid pUB110, *J. Bacteriol.* 160, 413–420, 1984.
123. Holm, L. and Sander, C., DNA polymerase β belongs to an ancient nucleotidyltransferase superfamily, *Trends Biol. Chem.* 20, 345–347, 1995.
124. Aravind, L. and Koonin, E. V., DNA polymerase β-like nucleotidyltransferase superfamily: identification of three new families, classification and evolutionary history, *Nucleic Acids Res.* 27, 1609–1618, 1999.
125. McKay, G. A. and Wright, G. D., Kinetic mechanism of aminoglycoside phosphotransferase type IIIa: Evidence for a Theorell-Chance mechanism, *J. Biol. Chem.* 270, 24686–24692, 1995.
126. Umezawa, H., Umezawa, S., Tsuchiya, T., and Okazaki, Y., 3′,4′-Dideoxykanamycin B active against kanamycin-resistant *Escherichia coli* and *Pseudomonas aeruginosa*, *J. Antibiot.* 24, 485–487, 1971.
127. Umezawa, H. and Kondo, S., Mechanisms of resistance to aminoglycoside antibiotics, in *Aminoglycoside antibiotics*, Umezawa, H. and Hooper, I. R. Springer-Verlag, Berlin, 1982, pp. 267–292.
128. Kawaguchi, H., Naito, T., Nakagowa, S., and Fuijawa, K., BBK8, a new semisynthetic aminoglycoside antibiotic, *J. Antibiot.* 25, 695, 1972.
129. Nagabhushan, T. L., Cooper, A. B., Tsai, H., Daniels, P. J., and Miller, G. H., The syntheses and biological properties of 1-N-(S-4-amino-2-hydroxybutyryl)-gentamicin B and 1-N-(S-3-amino-2-hydroxypropionyl)-gentamicin B, *J. Antibiot.* 31, 681–687, 1978.
130. Wright, J. J., Synthesis of 1-N-ethylsisomicin: a broad spectrum semithinthetic aminoglycoside antibiotic, *J. Chem. Soc. Chem. Commun.* 206–208, 1976.
131. Umezawa, H., Iinuma, K., Kondo, S., and Maeda, K., Synthesis and antibacterial activity of 6′-N-alkyl derivatives of 1-N- [(S)-4-amino-2-hydroxybutyryl]-kanamycin, *J. Antibiot. (Tokyo)* 28 (6), 483–485, 1975.
132. Umezawa, H., Nishimura, Y., Tsuchiya, T., and Umezawa, S., Syntheses of 6′-N-methyl-kanamycin and 3′,4′-dideoxy-6′-N-methylkanamycin B active against resistant strains having 6′-N-acetylating enzymes, *J. Antibiot.* 25, 743–745, 1972.
133. Kondo, S., Iinuma, K., Yamamoto, H., Maeda, K., and Umezawa, H., Syntheses of 1-N-(S)-4-amino-2-hydroxybutyryl)-kanamycin B and 3′-, 4′-dideoxykanamycin B active against kanamycin-resistant bacteria, *J. Antibiot.* 26 (7), 412–415, 1973.
134. Kondo, S., Tamura, A., Gomi, S., Ikeda, Y., Takeuchi, T., and Mitsuhashi, S., Structures of enzymatically modified products of arbekacin by methicillin-resistant *Staphylococcus aureus*, *J. Antibiot.* 46, 310–315, 1993.
135. Fujimura, S., Tokue, Y., Takahashi, H., Nukiwa, T., Hisamichi, K., Mikami, T., and Watanabe, A., A newly recognized acetylated metabolite of arbekacin in arbekacin-resistant strains of methicillin-resistant *Staphylococcus aureus*, *J. Antimicrob. Chemother.* 41, 495–497, 1998.
136. Kondo, S., Ikeda, Y., Ikeda, D., Takeuchi, T., Usui, T., Ishii, M., Kudo, T., Gomi, S., and Shibahara, S., Synthesis of 2″-amino-2″-deoxyarbekacin and its analogs having potent activity against methicillin-resistant *Staphylococcus aureus*, *J. Antibiot.* 47 (7), 821–832, 1994.

137. Roestamadji, J., Grapsas, I., and Mobashery, S., Loss of individual electrostatic interactions between aminoglycoside antibiotics and resistance enzymes as an effective means to overcoming bacterial drug resistance, *J. Am. Chem. Soc.* 117, 11060–11069, 1995.

138. McKay, G. A., Roestamadji, J., Mobashery, S., and Wright, G. D., Recognition of aminoglycoside antibiotics by enterococcal-staphylococcal aminoglycoside 3′-phosphotransferase type IIIa: Role of substrate amino groups, *Antimicrob. Agents Chemother.* 40, 2648–2650, 1996.

139. Alper, P. B., Hendrix, M., Sears, P., and Wong, C.-H., Probing the specificity of aminoglycoside-ribosomal RNA interactions with designed synthetic analogs, *J. Am. Chem. Soc.* 120, 1965–1978, 1998.

140. Greenberg, W. A., Priestley, E. S., Sears, P., Alper, P. B., Rosenbohm, C., Hendrix, M., Hung, S.-C., and Wong, C.-H., Design and synthesis of new aminoglycoside antibiotics containing neamine as an optimal core strucutre: Correlation of antibiotic activity with in vitro inhibition of translation, *J. Am. Chem. Soc.* 121, 6527–6541, 1999.

141. Sucheck, S. J., Wong, A. L., Koeller, K. M., Boehr, D. D., Draker, K.-A., Sears, P., Wright, G. D., and Wong, C.-H., Design of bifunctional antibiotics that target bacterial rRNA and inhibit resistance-causing enzymes, *J. Am. Chem. Soc.* 122, 5230–5231, 2000.

142. Rai, R., Chen, H. N., Czyryca, P. G., Li, J., and Chang, C. W., Design and synthesis of pyrankacin: A pyranmycin class of broad-spectrum aminoglycoside antibiotic, *Org. Lett.* 8, 887–889, 2006.

143. Elchert, B., Li, J., Wang, J., Hui, Y., Rai, R., Ptak, R., Ward, P., Takemoto, J. Y., Bensaci, M., and Chang, C. W., Application of the synthetic aminosugars for glycodiversification: synthesis and antimicrobial studies of pyranmycin, *J. Org. Chem.* 69 (5), 1513–1523, 2004.

144. Chang, C. W., Hui, Y., Elchert, B., Wang, J., Li, J., and Rai, R., Pyranmycins, a novel class of aminoglycosides with improved acid stability: the SAR of D-pyranoses on ring III of pyranmycin, *Org. Lett.* 4 (26), 4603–4606, 2002.

145. Bastida, A., Hidalgo, A., Chiara, J. L., Torrado, M., Corzana, F., Perez-Canadillas, J. M., Groves, P., Garcia-Junceda, E., Gonzalez, C., Jimenez-Barbero, J., and Asensio, J. L., Exploring the use of conformationally locked aminoglycosides as a new strategy to overcome bacterial resistance, *J. Am. Chem. Soc.* 128 (1), 100–116, 2006.

146. Haddad, J., Kotra, L. P., Llano-Sotelo, B., Kim, C., Azucena, E. F., Jr., Liu, M., Vakulenko, S. B., Chow, C. S., and Mobashery, S., Design of novel antibiotics that bind to the ribosomal acyltransfer site, *J. Am. Chem. Soc.* 124 (13), 3229–3237, 2002.

147. Russell, R. J., Murray, J. B., Lentzen, G., Haddad, J., and Mobashery, S., The complex of a designer antibiotic with a model aminoacyl site of the 30S ribosomal subunit revealed by X-ray crystallography, *J. Am. Chem. Soc.* 125 (12), 3410–3411, 2003.

148. Murray, J. B., Meroueh, S. O., Russell, R. J., Lentzen, G., Haddad, J., and Mobashery, S., Interactions of designer antibiotics and the bacterial ribosomal aminoacyl-tRNA site, *Chem. Biol.* 13 (2), 129–138, 2006.

149. Boehr, D. D., Lane, W. S., and Wright, G. D., Active site labeling of the gentamicin resistance enzyme AAC(6′)-APH(2″) by the lipid kinase inhibitor wortmannin, *Chem. Biol.* 8 (8), 791–800, 2001.

150. Roestamadji, J., Grapsas, I., and Mobashery, S., Mechanism-based inactivation of bacterial aminoglycoside 3′-phosphotransferases, *J. Am. Chem. Soc.* 117, 80–84, 1995.

151. Haddad, J., Vakulenko, S., and Mobashery, S., An antibiotic cloaked by its own resistance enzyme, *J. Am. Chem. Soc.* 121, 11922–11923, 1999.

152. Allen, N. E., Alborn, W. E., Jr., Hobbs, J. N., Jr., and Kirst, H. A., 7-Hydroxytropolone: An inhibitor of aminoglycoside-2″-*O*-adenyltransferase, *Antimicrob. Agents Chemother.* 22, 824–831, 1982.
153. Boehr, D. D., Draker, K., Koteva, K., Bains, M., Hancock, R. E., and Wright, G. D., Broad-spectrum peptide inhibitors of aminoglycoside antibiotic resistance enzymes, *Chem. Biol.* 10 (2), 189–196, 2003.

6 Resistance to β-Lactam Antibiotics Mediated by β-Lactamases: Structure, Mechanism, and Evolution

Jooyoung Cha, Lakshmi P. Kotra, and Shahriar Mobashery

CONTENTS

The catalytic function of β-lactamases is the primary cause of resistance to β-lactam antibiotics. These enzymes hydrolyze the β-lactam ring of these versatile antibiotics, a process that inactivates the drugs. Over 470 β-lactamases are known that are grouped into four distinct functional classes (classes A, B, C, and D). The members of each class operate by distinct catalytic mechanisms. A series of recently discovered β-lactamases exhibit wide breadth for their substrate preferences, which often include penicillins, cephalosporins, and carbapenems among other substrates. These so-called extended-spectrum β-lactamases (ESBLs) are being identified among all classes of β-lactamases. The breadth of phenotypic traits for each of these enzymes collectively includes all known types of β-lactam antibiotics. The global distribution of the pathogens that harbor these various enzymes is sufficiently different at the present that obsolescence of β-lactam antibiotics has not happened to date. This chapter discusses the various properties of these microbial enzymes.

INTRODUCTION

Resistance to antimicrobials is a serious clinical problem, with more than 70% of the bacteria that cause hospital-acquired infections resistant to at least one of the drugs that are currently used for treatment of infections [1]. Indeed, resistance to one class of antibiotic in any given organism is rare and resistance to multiple classes is common. Infections are the leading cause of death on a global scale, and drug resistance is expected to aggravate the already serious situation in the immediate future [2]. β-Lactam antibiotics remain the most commonly used antibacterial agents in the present chemotherapeutic armamentarium, and β-lactamases, the enzymes that hydrolyze β-lactam antibiotics are the major cause of resistance to these compounds [3]. This is in general a bigger problem in Gram-negative organisms, as Gram-positive organisms have evolved additional strategies in countering β-lactam antibiotics, such as modification of penicillin-binding proteins (PBPs). The fact that we rely so heavily on β-lactams to the present day is remarkable in light of the fact that β-lactamases were discovered before their widespread clinical use [4], and to date, over 470 novel β-lactamases (www.lahey.org/Studies) have been identified to complicate their therapeutic use. The genes for β-lactamases may be chromosomal, plasmid-borne, or found on transposable elements. Furthermore, their existence on integrons has also been documented [5,6]. Hence, there is ample opportunity for bacteria to share these drug resistance genes, and indeed this has happened extensively [7–10]. It would seem that the diversity in structures of β-lactamases, and in mechanisms of genetic dissemination should have put an end to viability of β-lactam antibacterials. The difficulties in treatment of resistant organisms harboring these enzymes are becoming acute, but the demise of these versatile antibacterial agents has not yet happened. In fact, we will remain dependent on β-lactam antibiotics for the foreseeable future. Nonetheless, the search for non-β-lactam antibiotics as potential replacements for β-lactams continues unabated [11–14]. Meanwhile, it is clear that we need to develop a detailed knowledge of the properties of these antibiotics and of their mechanisms of resistance. In this chapter, we discuss β-lactamases from the perspective of their mechanisms and structures. We also explore the means by which random mutation and selection have provided opportunities for these enzymes to extend their substrate specificities such that resistance to virtually any β-lactam antibiotic has been observed.

CLASSIFICATION OF β-LACTAMASES

Various classification schemes have been proposed for β-lactamases based on the characteristics of the enzymes and/or their substrate profiles [15,16]. Bush proposed a comprehensive functional classification of β-lactamases in 1989, which was expanded in 1995 to include just under 190 β-lactamases [15]. This classification system utilized an extensive set of kinetic data on various enzymes and categorized β-lactamases according to substrate preferences and inhibition characteristics. Four major groups are recognized in this classification. Group 1 consists of cephalosporinases which are not inhibited by clavulanic acid. Group 2 consists of penicillinases, including broad-spectrum penicillinases that are generally inhibited by the active-site-directed β-lactamase inhibitors. Subgroups of enzymes, namely 2a, 2b, 2be, 2br, 2c, 2d, 2e, and 2f, were defined based on the rates of hydrolysis of carbenicillin, cloxacillin, extended-spectrum β-lactams ceftazidime, cefotaxime, or aztreonam, and of inhibition profile by clavulanate, respectively. Enzymes that are inhibited by the metal-chelating agent EDTA are classified as group 3. Group 4 consists of other β-lactamases that are not inhibited by clavulanic acid.

However, a functional classification scheme for β-lactamases proposed by Ambler has found common usage [16,17]. Ambler classifies these enzymes into four classes, A, B, C, and D [17–19]. Whereas classes A, C, and D have evolved dependence on an active-site serine as their key mechanistic feature, class B enzymes are zinc dependent and hence different. The catalytic process for turnover of the members of the former group involves acylation at the active-site serine by the β-lactam antibiotic, followed by deacylation of the acyl-enzyme species. It is noteworthy that these enzymes do not share any sequence homologies, structural similarities, or mechanistic features with serine or zinc-dependent proteases. Class A β-lactamases generally prefer penicillins as substrates, whereas class C enzymes turn over cephalosporins better (Scheme 1). Class B enzymes can hydrolyze a broad range of substrates including carbapenems, which resist hydrolysis by most of the other classes of enzymes. Class D β-lactamases, on the other hand, hydrolyze oxacillin-type β-lactams efficiently. Classes A and C of β-lactamase are the most common and the second most common enzymes, respectively [20–22]. We also reported a unique β-lactamase activity for a certain *T. pallidum* PBP (referred to as Tp47). This catalytic activity is not metal dependent and does not involve acylation of the protein in the course of the turnover process. Whereas other examples of this type of enzyme are not known presently, potential variants could serve as a fifth class of β-lactamases [23,24]. The general properties of all these enzymes, which operate by distinct mechanisms, will be discussed in the following sections. We add that several other reviews on β-lactamases have appeared in the literature that complement this chapter in various ways [3,22,25–31].

ORIGIN OF β-LACTAMASES

It is now accepted that β-lactamases evolved from PBPs, which experience covalent modification by penicillins and other β-lactams. These biosynthetic enzymes assemble the bacterial cell wall and regulate its function. Certain PBPs carry out the cross-linking reaction, which imparts rigidity to the bacterial cell wall. Penicillins bind to PBPs and acylate an active-site serine. The resultant acyl-enzyme species is sufficiently stable to provide effective inhibition of the biological function of the PBP, an inhibition event that leads to bacterial cell death.

The kinship of PBPs and β-lactamases has been established based on extensive multiple-sequence alignment and structural data [3,25–28]. In essence, nature discovered that the same structural motif that binds penicillin (that of PBP) could be used to destroy the drug. Insofar as the resultant acyl-enzyme species between a β-lactam antibiotic and a PBP was relatively stable, evolution of the drug-resistant phenotype had to render it unstable. For such an evolutionary scheme to be successful, the nascent β-lactamase should have been able to experience active-site acylation (inherited from the parental PBP; Scheme 2, species 3), followed by the deacylation of the acyl-enzyme species. This process, of course, would take place as a consequence of random mutation and selection. It must have taken place in incremental steps to result in liberation of the PBP from inhibition by the β-lactam. As such, the example of the β-lactamase activity of the PBP Tp47 of *T. pallidum* PBP is significant [23,24]. Here we have a PBP that serves a function in the biosynthesis of cell wall, yet has acquired an activity in hydrolytic turnover of penicillins. It is intuitive that the driving force for liberation from inhibition must have been to make the active PBP available again for it to function in cell wall assembly. It is interesting that once acylated by a β-lactam, the modern PBPs undergo slow deacylation with a wide range of deacylation rate constants [23,32–35] indicative of the diversification among PBPs. Such diversification over an evolutionary time scale has introduced substantial sequence divergence among these proteins [27,36].

The evolutionary advent of a PBP that would undergo acylation by β-lactam antibiotics followed by a reasonably rapid deacylation step would have had a clear advantage for the bacterium. Ultimately, the strategy must have been so successful that the PBP that underwent the process of acylation and deacylation fairly effectively started to become more specialized in hydrolysis of the β-lactam antibiotics, so they served the role of *bona fide* resistance enzymes. Along the way, this *bona fide* resistance enzyme would detach itself from the surface of the bacterial plasma membrane—the vast majority of PBPs are membrane-bound proteins—so it could serve as a vanguard in fighting the in-coming antibiotics in solution.

Clearly PBPs are ancient proteins as bacteria came into existence over 3.8 billion years ago [37], and the evolution of cell wall must have followed suit at about the same time. But the development of β-lactamases is a relatively recent event, which must have taken place after the evolution of the first biosynthetic pathways for the natural β-lactam antibiotics [22,38–40]. The diversification of function has been impressive, in light of various catalytic functions for PBPs and the breadth of the substrate profile for β-lactamases [3,41–43]. The process of evolution for β-lactamases has been accelerated by the extensive use of β-lactams in clinic over the past 60 years [3,36,44–46]. As a result, although the degree of sequence homology among these proteins is low, it would appear from the emerging structural information that the three-dimensional fold of these proteins is preserved (*vide infra*) [27]. Multiple X-ray structures for PBPs and β-lactamases are available, giving a wealth of structural information for understanding the functions of these bacterial proteins. These also include a handful of ultra-high resolution structures that shed light on the protonation states of certain active site residues.

CLASS A β-LACTAMASES

Enzymes that belong to this class are the most common among pathogens. These β-lactamases are the best understood in many aspects of their chemistry, biochemistry, and molecular biology. Furthermore, these enzymes have reached catalytic perfection by performing their reaction at the diffusion limit [47]. Evolution of class A β-lactamases from Gram-negative bacteria has proceeded by selection of mutations that broaden the phenotypic properties, such as substrate profile or resistance to inhibitors. Currently 144 variants of the TEM and 72 of the SHV β-lactamases are known. It is interesting that such diversification by selection of point mutants has been absent in the evolution of class A enzymes from Gram-positive bacteria.

We previously analyzed an extensive amino acid sequence alignment of over 140 members of all classes of β-lactamases and PBPs [27]. The sequences are so divergent that there are no significant homologies in general, except a Ser-X-X-Lys sequence motif is seen in the vast majority of these enzymes. The serine corresponds to the active-site residue of both β-lactamases and PBPs that experiences acylation by the substrate. A lysine, three residues to the carboxyl-terminal side of the serine, is clearly important for the functions of both types of enzymes as that too is absolutely conserved. Since the only mechanistic feature that all these proteins share is the active-site serine acylation, the Ser-X-X-Lys sequence is a minimal essential motif for this event.

The previous comparison of two X-ray structures, one of the *Staphylococcus aureus* PC1 β-lactamase modified by a phosphonate [48], and another of the TEM-1 β-lactamase modified by the same phosphonate [49], indicated a pathway for the formation of the acyl-enzyme intermediate. The phosphonate modified the active-site serine in both enzymes, and the structures mimicked the transition state for the acylation process. However, there are intriguing differences between the two structures. In the *S. aureus* PC1 β-lactamase structure the side chains of Ser70 and Lys73 interact closely, giving the appearance of an interaction of a base (Lys73) abstracting the proton from serine. On the other hand, the structure for the TEM-1 enzyme shows strong interactions between Ser130 and the phosphonate oxygen corresponding to the leaving group, indicating that the complex mimics the collapse of the tetrahedral species en route to the formation of the acyl-enzyme intermediate. The process would take place by the transfer of a proton from Ser130 to the departing amine in the β-lactam substrate. Then, a proton would be transferred to Ser130 from the now-protonated Lys73 [49]. These analyses collectively argue for the existence of Lys73 in its unprotonated form. Yet, others have argued for a protonated Lys73, which would be incapable of serving as a base in these proton-transfer events [50–53]. As such, a protonated Lys73 would mandate a different residue serving the role of a base in serine activation. The only alternative candidate is Glu166.

The catalytic mechanism actually appears to be more complicated. Recent analyses of the kinetics with a γ-thialysine mutant at position 73 of the TEM-1 β-lactamase, the ^{15}N-NMR characterization for titrational analyses of all lysines in the same protein, and free-energy calculation using the thermodynamic integration method all support a pK_a in the range of 8.0 to 8.5 for Lys73 [54]. These disparate methods all point to the fact that Lys73 of class A β-lactamases is unique in that its pK_a is attenuated by 2 to 3 pK units. The attenuated pK_a for Lys73, of course, indicates that it is a weaker base compared to a typical lysine. However, it also indicates that this lysine is poised for the proton transfer events that are critical for the enzyme to perform its catalytic function.

It is significant to note that mutagenesis of either Glu166 or Lys73 does not abolish acylation of class A β-lactamases by their substrates [51,53]. Furthermore, crystal structures reveal that acylation clearly takes place in the presence of mutant variants at position 166, which would not be able to participate in abstraction of protons from the active site serine [52,55]. Our recent *ab initio* quantum mechanical/molecular mechanical (QM/MM) calculation reveals for the first time that a duality of mechanistic possibilities is in hand [56]. Two distinct pathways exist for enzyme acylation, a full discussion of which is beyond the scope of this chapter. The essence of these findings is that an energetically downhill path toward a protonated Glu166 and a free-base Lys73 exists that would predispose Lys73 in serving as the base for the active site serine activation. In the absence of Lys73 (incidentally, not seen in any of the clinical variants), Glu166 remains as the viable residue to serve the function. Energetically, the two pathways are comparable, which explains the *in vitro* findings that enzyme acylation could take place for the mutant variants at each position.

A consensus has emerged indicating that residue Glu166 promotes a water molecule for the deacylation event [55,57]. This assertion is supported by the results from studies with site-directed mutagenesis [51,53,58–61], and by studies of β-lactam molecules that acylate the enzymes but resist deacylation [62–66]. It is worthy of comment that the water molecule approaches the ester of the acyl-enzyme intermediate from the α-face (Figure 6.1).

In general, various members of class A β-lactamases enjoy considerable conservation of the three-dimensional fold, regardless of whether they are from

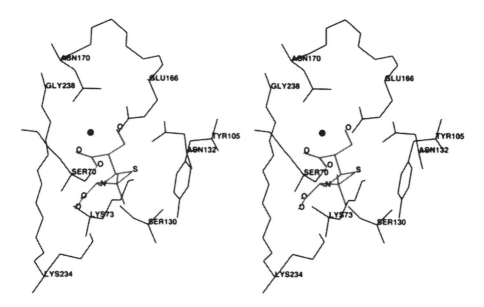

FIGURE 6.1   Stereo view of the active site of TEM-1 β-lactamase (class A enzyme) complexed with 6α-hydroxymethyl penicillanic acid (pdb code 1TEM). The hydrolytic water molecule (shown as a sphere) is approaching the ester moiety from the α-face, which is also seen interacting with the hydroxymethyl moiety of the bound inhibitor.

Gram-negative or Gram-positive bacteria. This observation is demonstrated below for the structures of three Gram-negative class A β-lactamases from *Escherichia coli* (TEM-1), from *Klebsiella pneumoniae* (SHV-1), and from *Enterobacter cloacae* (NMC-A), and for the Gram-positive enzyme from *S. aureus* (Figure 6.2). Figure 6.3 shows details of the active site structure for the TEM-1 β-lactamases.

Two broadly defined phenotypic properties for class A β-lactamases are emerging from the clinical isolates. One is the inhibitor-resistant phenotype that was first limited to the TEM subfamily, for which the term "inhibitor-resistant TEM" (IRT) was coined. This type of phenotype has now been discovered in the SHV family as well. Some other variants have broadened their substrate profiles to recent cephalosporins (ceftazidime, cefotaxime, and monobactam aztreonam) and carbapenems. Hence, the term extended-spectrum β-lactamases (ESBLs) was introduced. These ESBLs are exemplified by the Imi-1, Per-1, Sme-1, Toho-1, and NMC-A β-lactamases, among others [67–70]. Both these phenotypes are cause for concern in the clinic.

The IRT β-lactamases were first discovered in 1992 in France and England [71,72]. These variants of the TEM-1 β-lactamase emerged as a consequence of

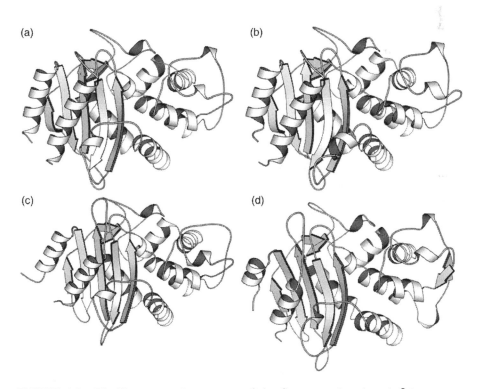

FIGURE 6.2 The X-ray crystal structures of the Gram-negative class A β-lactamases from *Escherichia coli* (TEM-1; a), *Klebsiella pneumoniae* (SHV-1; b) and from *Enterobacter cloacae* (NMC-A; c), and for the Gram-positive enzyme from *Staphylococcus aureus* (d) (pdb codes 1TEM, 1SHV, 1BUL, and 1BLC, respectively). (These figures were prepared using the program MOLSCRIPT.)

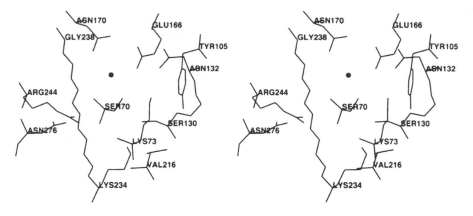

FIGURE 6.3 Stereo view of the active site of the TEM-1 β-lactamase with important residues labeled. The hydrolytic water, coordinated to Glu-166 and Asn-170, is shown as a black sphere.

substantial use of mechanism-based inhibitors (clavulanate and sulbactam) for class A β-lactamases, the first of which, clavulanic acid, was introduced to the clinic in 1984 in combination with amoxicillin [38,73]. The inhibitors impair the function of the class A β-lactamases, so that the co-administered penicillin would have the opportunity to inhibit the PBP. These combinations of β-lactamase inhibitors and penicillins have led to selection of the phenotype that resists inhibition of β-lactamase. IRT β-lactamases often are the result of single mutations at key amino acid positions [74–76]. Additional mutations may also be present, but these are in conjunction with mutations at a few key positions. Error-prone PCR mutagenesis has identified four mutations of consequence for the IRT β-lactamases [77]. These are mutations at positions 69, 130, 244, and 276. Biochemical analyses have shed light on the functions of these amino acids in the mechanism of inhibition of β-lactamases. At position 69, methionine, clearly not a conserved TEM residue, is located in the structurally constrained region [78,79]. According to the studies by Meroueh et al. [78], the Met69Lys mutant variant does not cause any structural perturbation compared to the wild-type, and the effect of the mutation is subtle. Molecular dynamic simulations revealed that the wild-type and the Met69Leu mutant forms experience differences in their dynamical behavior that result in elevation of the dissociation constant for the inhibitor from the enzyme prior to the onset of covalent chemistry. This mutant variant (designated TEM-33) exhibits comparable catalytic ability compared to the wild-type enzyme [78]. Another mechanism has also been offered to explain the effect of subtle changes at position 69 [79,80]. Wang et al. argued that the related variant Met69Ile/Met182Thr (TEM-32) or Met69Val (TEM-34) exhibit a disruption of the active site Ser130 interactions, which involve effects during the covalent phase of inhibition of the enzyme by clavulanate [79].

Properties of the Ser130Gly mutant variants of TEM and SHV have been studied by both crystallography and enzymology [81–83]. The mutation of conserved Ser130 is very rare (3 out of 25 IRTs) and only a single change to glycine has been documented (www.lahey.org/Studies/temtable). Inhibition of the TEM β-lactamases by clavulanic acid commences by acylation of Ser70, the active-site serine, and interaction with Ser130 as a step in the proton transfer events [74,79,84,85]. The involvement of Ser130 hydroxyl in the proton transfer events was thought to be critical for the catalytic process. However, since Ser130Gly mutant would not be able to engage this residue in these processes, the observation of this mutant was rather enigmatic. Recent crystallographic findings reveal that substitution of serine to glycine has created a cavity suitable for sequestering a water molecule within the active site [81,82]. This water molecule serves in the proton transfer events and is an effective surrogate for serine in the turnover events with typical substrates as well as in events that lead to covalent chemistry by clavulanate [74,75]. However, Ser130 hydroxyl is trapped in a crosslinking event with the inhibitor, which would not take place in the glycine mutant. This gives rise to the IRT phenotype.

Arg244 is another important residue for the IRT phenotype. The side chain of Arg244 participates in substrate recognition as a counter ion to the carboxylate of substrates and inhibitors [74,79,86]. This arginine coordinates to a structurally conserved water molecule important for interactions with clavulanate [74]. Mutations at position 244, such as in Arg244Ser, would not be able to fix the water molecule in the requisite position, and hence would impair the ability of clavulanic acid to inhibit the enzyme.

The effects of mutations at positions 244 and 276 are related to each other because their side chains are in communication with each other; however, the mechanistic reasons for the IRT phenotype associated with mutations at each site are distinct. The side chain of Arg244 is hydrogen bonded to the side chain of Asn276. The mutation Asn276Asp, seen in IRTs, enhances the strength of interactions between residues at 244 and 276 [87]. The interaction in the mutant protein also influences the water molecule coordinated to Arg244, such that the rate constant for inhibition of the enzyme is also affected negatively. The structural effects with these kinetic consequences are indeed quite subtle, as perceived from the X-ray structure of the Asn276Asp variant of TEM β-lactamase (Figure 6.4) [87].

ESBLs class A β-lactamase have also appeared among resistance strains, falling into three groups. The TEM and SHV ESBL variants were found initially among *Enterobacteriaceae,* but increasingly also among *Pseudomonas* strains [3,88] (http://www.lahey.org/Studies/webt.htm). The VEB and PER variants are found among *Pseudomonas.* The last group is the CTX-M family (Toho-1, and -2) [89,90]. These enzymes are of serious concern because their genes originated from nosocomial strains that now have spread to community-derived strains [3,89].

The structural factors that result in the extended-spectrum phenotypes are quite diverse, and we will not be able to discuss them fully in this review, although this subject has been discussed previously [25,91]. In general, ESBLs do not hydrolyze oxyiminocephalosporins (ceftazidime and cefotaxime), aztreonam, and carbapenems [90]. ESBLs with ability to turn over imipenem (a carbapenem) [4] and cefoxitin (a cephamycin) are of special concern. Imipenem is an exceedingly poor substrate for class A β-lactamases and the mechanistic and structural bases for such poor turnover of imipenem by the TEM-1 enzyme have been elucidated [64,92]. The structural reason behind the poor turnover is the interactions of the 6α-hydroxyethyl

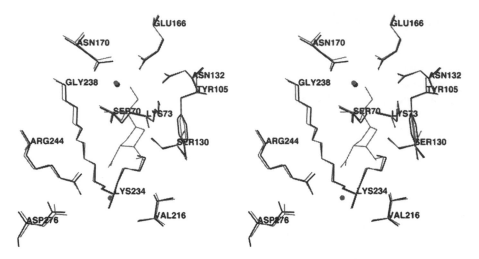

FIGURE 6.4 Stereo view of the active sites of wild-type TEM-1 β-lactamase (shown in gray) superimposed with the Asn-276-Asp mutant of TEM-1 β-lactamase (shown in black). Inhibitor of wild-type enzyme, 6α-hydroxymethylpenicillanate covalently bound to Ser-70 of the wild-type TEM-1 enzyme is shown to illustrate the interactions. The side chain of Asp-276 (at seven o'clock, in black) is seen shifted toward Arg-244, as is Arg-244 toward Asp-276, resulting in a stronger interaction between these two residues. The coordinated water molecule to Arg-244 (black sphere at six o'clock position) is not seen in the crystal structure of the mutant enzyme. The hydrolytic water, however, is seen in both crystal structures (shown here as spheres at twelve o'clock).

group of imipenem and 7α-methoxy group of cefoxitin with the enzyme. It is evident that these substituents interfere with the approach of the hydrolytic water molecule to the ester carbonyl at the acyl-enzyme stage. Also, these interactions force the acyl-enzyme complex to assume a new conformation, which is no longer predisposed for the deacylation process [64]. On the hand, the explanation for poor turnover with cefoxitin is reasonably straightforward. Cefoxitin possesses the 7α-methoxy group on the cephalosporin nucleus. This substitution not only displaces the catalytic water but also dislocates Asn132 located on the Ω-loop, giving rise to a stable acyl-enzyme complex [93,94]. For a class A enzyme to become adept at turning over imipenem, it has to eliminate the unfavorable interactions of the enzyme with the 6α-hydroxyethyl group of imipenem. This is indeed what has been seen for the NMC-A (non-metallo carbapenemase of class A) β-lactamase from *E. cloacae* [95]. This is also an example of an extremely subtle change in the structure of the enzyme to give a profound phenotypic consequence [95,96]. The NMC-A β-lactamase is, in every respect of the catalytic machinery, similar to other prototypic class A enzymes, such as the TEM-1 β-lactamase (Figure 6.5). But the nearly 100-fold improvement of ability to hydrolyze imipenem [95] in this enzyme has been attributed to the new position of Asn132, which has moved away from the active site by a mere 1 Å. This repositioning has enlarged the cavity in which the 6α-hydroxyethyl group of imipenem would fit, and has eliminated the unfavorable steric interactions that were seen in the related TEM-1 enzyme.

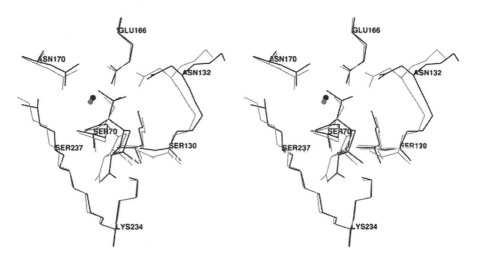

FIGURE 6.5 Stereo view of the superimposition of the TEM-1 (shown in gray) and NMC-A (shown in black) β-lactamases complexed with 6α-hydroxymethylpenicillanic acid and 6α-hydroxypropylpenicillanic acid, respectively. The active site of the NMC-A enzyme near the hydrolytic water is expanded by the relocation of the Asn-132 away from the active site, such that this enzyme could accommodate substituents, such as the hydroxypropyl moiety in the inhibitor's structure.

The three-dimensional structure of the SHV-1 β-lactamase that possesses a somewhat broader substrate profile than the TEM-1 enzyme, shows a similar overall fold to that of the TEM-1 enzyme (Figure 6.2). There are, however, few subtle differences that differentiate the active sites of the two enzymes. The Ser130 to Asn132 loop and the neighboring Asp104/Tyr105 loop in the SHV-1 enzyme have shifted away from the active site by about 0.7 to 1.2 Å, in comparison to the TEM-1 enzyme, thus widening the active site [97]. A similar shift of Asn132, seen in the NMC-A β-lactamase, was observed for the SHV-1 enzyme as well (see above).

A set of important amino acid substitutions giving rise to the ESBLs of the TEM or SHV groups are Arg164Ser/His, Glu104Lys, Gly238Ser/Ala, and Ala237Thr. The Gly238Ser SHV-2 increases the MIC of *E. coli* from 2 to 8 μg/mL for ceftazidime and from 0.125 to 16 μg/mL for cefotaxime [98]. Crystal structure revealed that the Gly238Ser mutation opens the active site sufficiently to accommodate the larger side chains of the more recent generations of β-lactam antibiotics, while preserving the role of catalytic residues [99,100].

The X-ray crystal structure of another ESBL, CTX-M Toho-1, also was solved [101–103]. The studies with Toho-1 acyl-enzyme species with cefotaxime, cephalothin, and benzylpenicillin reveal distinct features, such as the displacement of the Ω-loop to avoid steric hindrance with the bulky side chain of cefotaxime and also favorable interactions with residues Asn104 and Asp240 for improved binding.

The PER-1 β-lactamase is expressed from a chromosome-encoded gene of *Pseudomonas aeruginosa*. It was first detected and isolated from a Turkish patient in 1993 [104] before it was found in nosocomial strains of *S. typhimurium* and *Acinetobacter baumanii* [105,106]. The X-ray crystal structure of the PER-1 β-lactamase shows a number of differences when compared to other class A β-lactamases, such as the TEM-1 enzyme (Figure 6.6) [107]. Two of most conserved features in class A β-lactamases were not seen in PER-1. These include the fold of the Ω-loop and the *cis* conformation of peptide bond between residues 166 and 167. In class A β-lactamases, the *cis* conformation is required to proper positioning of Glu166, a catalytically critical residue.

CLASS B β-LACTAMASES

Metallo-β-lactamases (MBLs) were first isolated in 1966 and were subsequently categorized as class B β-lactamases in 1980 [17]. These enzymes are dependent of zinc ions for their activity. The initial studies of the chromosomal MBLs were performed on the *Bacillus cereus* and *Stenotrophomonas maltophilia* enzymes [108,109]. Since the first discoveries, many more MBLs from various bacterial pathogens have emerged [110,111]. This class of β-lactamases shows an unprecedented breadth of substrate preference, which includes many of recent generations of cephalosporins, carbapenems, and β-lactam inhibitors of β-lactamases, such as clavulanate and penam sulfones [43,112]. This is a cause for concern because carbapenem antibiotics

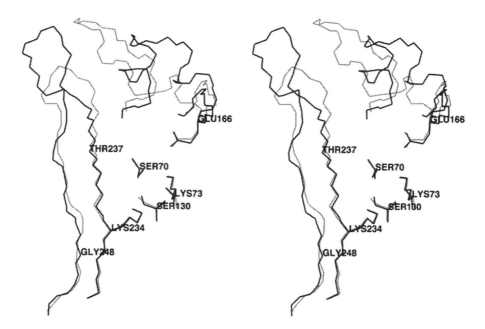

FIGURE 6.6 Superimposition of the TEM-1 (shown in gray) and PER-1 (shown in black) β-lactamases depicting the differences in the Ω-loop and the β3-sheet regions (top and left regions of the figure, respectively). The region near the Ω-loop is substantially larger in the PER-1 β-lactamase, compared to the TEM-1 enzyme, a factor that is linked to the extended-spectrum activity of the former.

are generally resistant to the action of the other classes of β-lactamases, and have enjoyed longevity in the clinic since their introduction in the mid-1980s. MBLs possess less than 25% amino acid identity with each other, and they do not show any apparent evolutionary relationship to other classes of β-lactamases [110].

The report that the genes for class B β-lactamases could be plasmid-borne has been disconcerting [113,114]. Dissemination of MBL genes would seem to be inevitable by future use of extended-spectrum β-lactams. These genes appear on gene cassettes that include other resistance genes for disparate antibiotics, such as kanamycin, neomycin, streptomycin, among others [110,115].

The crystal structures for MBLs show the presence of two zinc ions [116]. It has been asserted that among known B1 subclass MBLs, despite the remarkable degree of similarity of their active sites, the relative affinities of each zinc ion to the enzyme are not equal [117,118]. CcrA and IMP-1 enzymes possess two high-affinity metals [119,120]. BcII is active with one or two zinc ions but fully active when both metals are bound [121].

MBLs from several organisms have been crystallized, but only two were in complex with β-lactams [122]. The structure of IMP-1 from *P. aeruginosa* was solved by Concha et al. [111]. Furthermore, the MBLs from *B. fragilis*, *B. cereus*, and *S. maltophilia* have also been crystallized, and their structures have been useful in understanding this class of enzymes and their diversities (Figures 6.7 and 6.8) [123–126]. MBLs largely exhibit an αββα motif and their respective active sites are superimposable. The active site resides at the end of the β-sandwich and it has either one or two divalent zinc ions separated by ~3.5 Å. The first zinc ion is coordinated to three histidines and one hydroxide/water molecule, which bridges to the other zinc ion. The second zinc ion is coordinated by a cysteine, a histidine, an aspartate, and an apical water molecule. Crystal structures of different MBLs reveal that the bond

(a)   (b)

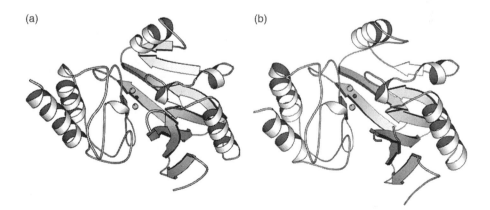

FIGURE 6.7 Structures of the zinc-dependent β-lactamases from *B. fragilis* (a) and *B. cereus* 549/H/9 (b) (pdb codes 1A7T and 1BC2, respectively). Two zinc ions in the active sites of the enzymes are shown as gray spheres and a water molecule (identified by an arrow) is shown as a black sphere coordinated to the two zinc ions. (The figures were prepared using the program MOLSCRIPT.)

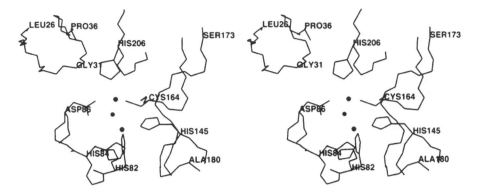

FIGURE 6.8 Stereo view of the active site of a class B β-lactamase (from *B. fragilis*; pdb code 1A7T). Side chains of the residues that coordinate to the zinc ions are shown. A water molecule (identified by an arrow)—in between the two zinc ions—is shown as a sphere that closely interacts with the zinc ions.

distance between the first zinc ion and O is as short as 1.9 Å, which also supports the hypothesis of a hydroxide ion bound at neutral pH [111,123,127]. This hydroxide is reported to be bound as a bridge between two zinc ions at pH 5 and higher [118]. The bridging hydroxide (pK_a = 4.9–5.6) [128,129] is proposed to transfer an electron to the open β-lactam ring. According to the proposed mechanism by Wang et al., the enzyme detaches the bound hydroxide from the second zinc ion prior to its functioning as a nucleophile [130].

There are no clinically useful inhibitors for this class of β-lactamases at present, although work is making progress in that direction. In light of the fact that the various members of this family of enzymes may operate by somewhat different mechanisms, general inhibition of class B β-lactamases by one type of inhibitor—such as achieved by clavulanic acid for class A enzymes—may prove difficult. However, from the MBL substrate spectrum, this class of enzyme is not able to hydrolyze aztreonam, a monocyclic *N*-sulfonyl β-lactam. A computational study (density functional calculations and docking analyses) suggests that aztreonam binds to the enzyme but in a non-productive orientation [131].

CLASS C β-LACTAMASES

Although class C β-lactamases were believed to be exclusively of chromosomal origin, plasmid-borne variants were soon identified [25,132]. This in part explains why they are only found in Gram-negative bacteria, and why there are not many mutant variants of the various members of this family of enzymes. Indeed, class C β-lactamases are largely homogeneous as far as their kinetic properties are concerned [28]. These enzymes are somewhat larger—approximately 39 kDa—than their class A counterparts.

In contrast to class A β-lactamases, class C enzymes have evolved an entirely distinct mechanism for their deacylation of the acyl-enzyme intermediate. They lack any residue that could correspond to Glu166 of the class A β-lactamases. It has been generally accepted that Tyr150 could be a player in both the acylation and deacylation processes [27,133,134]. Although any direct evidence has been lacking, many have assumed that Tyr150 should be in its deprotonated basic form. Hence, Tyr150 has been proposed as the base for activation of Ser64 in the acylation event [135–138]. After the activation of Ser64, Tyr150 transfers a proton to β-lactam ring nitrogen to allow collapse of the tetrahedral intermediates. At the same time, the very same deprotonated Tyr150 was also proposed as the base for activation of a water molecule in hydrolysis of the acyl-enzyme species [135]. However, the possibility of a protonated Tyr150 was also considered [139,140]. A recent NMR titration study by Kato-Toma et al. provided experimental evidence indicating that the pK_a of Tyr150 is 8.3, suggesting that the native enzyme has the residue protonated [141]. In the absence of Tyr150 as a viable base for activation of serine, the only other alternative for this function is Lys67. It is important to note that side chains of both Lys67 and Tyr150 are in hydrogen bonding contact with Ser64. A recent molecular dynamics study with the class C β-lactamase from *Citrobacter freundii* and its Michaelis complex with aztreonam supports the direct involvement of Lys67 as an activator of Ser64 for acylation [142]. From their simulation, neutral Lys67 is much more stable than anionic Tyr150 (20 kcal/mol). A QM/MM calculation on the deacylation process supports the involvement of the protonated Tyr150 [143]. In the absence of a residue equivalent to Glu166 of class A β-lactamase, we had argued that if the approach of the hydrolytic water from the α-face was not possible for lack of a general base on that side, then it is likely that the hydrolytic water would approach the ester of the acyl-enzyme intermediate from the opposite β-direction. Such a route for the hydrolytic water would bring it into the coordination sphere of the amine of the acyl-enzyme intermediate—formerly the β-lactam nitrogen—and in contact with the side chain hydroxyl of Tyr150 (Figure 6.9) [144]. It is noteworthy that the direction of the approach of the hydrolytic water is the opposite of that for the class A enzymes, hence the evolution of the deacylation steps in the two classes of enzymes is entirely distinct [27].

Class C β-lactamases belong to a group of a handful of enzymes that operate by "substrate-assisted catalysis" (Figure 6.10). When water attacks the acyl-enzyme complex, water approaches to the position where it is stabilized by both Tyr150 and β-lactam ring nitrogen. A crystal structure of the acyl-enzyme species of AmpC β-lactamase and moxalactam supports this proposal further [136]. According to the recent ultrahigh resolution crystallographic structure of AmpC β-lactamase, Tyr150 is protonated throughout the reaction. Moreover, the structure is consistent with the role of Tyr150 coordinating for activation of water in cooperation with substrate β-lactam nitrogen [145].

We have previously shown that class C β-lactamases operate at the diffusion limit for turnover of their preferred cephalosporin substrates [146]. Diffusion-controlled catalysis was also shown for turnover of the preferred penicillin substrates by class A β-lactamases as well [47]. Therefore, it would appear that evolution of classes A and C of β-lactamases took entirely different courses, obviously due to different selection pressures: in one case penicillins, in the other cephalosporins. Furthermore, an extensive sequence alignment of β-lactamases and PBPs indicated

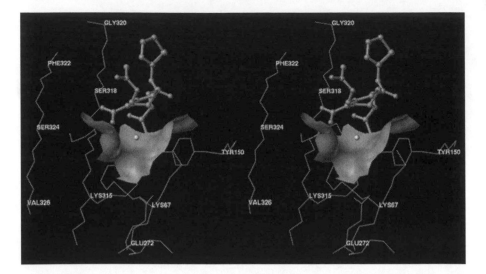

FIGURE 6.9 Stereo view of the energy-minimized model for the acyl-enzyme complex of the *Enterobacter cloacae* P99 β-lactamase (a class C enzyme; pdb code 1BLS) with cephalothin. The hydrolytic water is seen coordinated between Tyr-150 and amine of the dihydrothiazine ring, positioned to approach the ester moiety from the β-face. A Connolly water-accessible surface (in gray) is shown for the binding site of the water molecule.

that classes A and C of β-lactamases evolved from two different groups of PBPs [27]. All these observations collectively and conclusively indicate that the two classes of β-lactamases had different evolutionary experiences, but each reached its full potential by becoming "perfect" (i.e., diffusion-controlled) in its catalytic competence.

Several class C β-lactamases have been crystallized to date (Figure 6.11) [100,135–137,145,147–151]. These enzymes are very similar in their structures, and the similarity is even more pronounced within the active sites. The active sites of the class C β-lactamases are in general wider than those of class A enzymes, which explains why they readily bind the relatively bulkier cephalosporin substrates. The active site of *E. cloacae* P99 is shown in Figure 6.12.

One of the important features of class C β-lactamases is that they can accommodate cephalosporins in their active site and are not inactivated by class A β-lactamase inhibitors, such as clavulanic acid. Lobkovsky et al. suggested that the machinery present in class A β-lactamases to process the clinical inhibitor clavulanic acid in the course of the inactivation chemistry, namely a residue, such as the Arg244 of the class A enzymes and the water molecule coordinated to it, does not have a structural counterpart in class C enzymes [137]. An *E. coli* strain was recently described that produces an inhibitor-sensitive AmpC class C β-lactamase from Japan [152]. The amino acid sequence of the enzyme contains a tripeptide deletion (Gly286-Ser287-Asp288) and several substitutions. The level of inhibition was greater with sulbactam and tazobactam than with clavulanic acid. Molecular modeling indicated structural changes around a helix, which caused alteration in the substrate binding site.

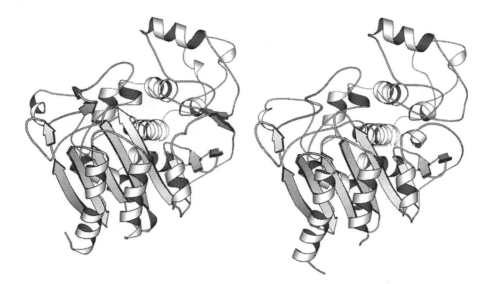

FIGURE 6.10 Stereo view of the active site of the *Enterobacter cloacae* P99 β-lactamase (in gray, pdb code: 1BLS) superimposed onto the active site of the D-Ala-D-Ala transpeptidase—a penicillin-binding protein from Streptomyces R61 (in black; pdb code 1CEG)—modified covalently by cephalothin (also in black). The modeled hydrolytic water (gray sphere) is shown in the active site of the *E. cloacae* P99 β-lactamase as per Figure 6.9. The crystallographic water molecule in the structure of D-Ala-D-Ala transpeptidase (shown as a black sphere) is seen coordinated to Arg-285 and Tyr-159; this interaction does not activate the water molecule for hydrolysis of the ester moiety. Arg-285 is replaced by Glu-272 in the β-lactamase structure. There is no opportunity for direct contact between the side chain of Glu-272 and the hydrolytic water.

The catalytic processes of β-lactamases may entail conformational changes. Such has been shown for class A enzymes, the TEM-1 and NMC-A β-lactamases, by X-ray crystallography and molecular dynamics simulations [63,64]. Similarly, the class C β-lactamase from *C. freundii* studied by infrared spectroscopy was shown to have multiple carbonyl stretches in the course of turnover of a penicillin. These observations were interpreted to be the result of different conformations for the acyl-enzyme intermediate [153].

CLASS D β-LACTAMASES

To date, more than 50 class D β-lactamases have been identified [154]. These enzymes are grouped together because they prefer oxacillin or cloxacillin as their substrate (hence, the "OXA" designation). These enzymes are becoming important clinically with the discovery of new variants such as OXA-14, OXA-15, OXA-16, and OXA-18, which show the extended-spectrum property against carbapenems and third-generation cephalosporins [155]. The substrate profile for OXA-18, for example, includes penicillins (oxacillin, amoxicillin, and ticarcillin), cephalosporins

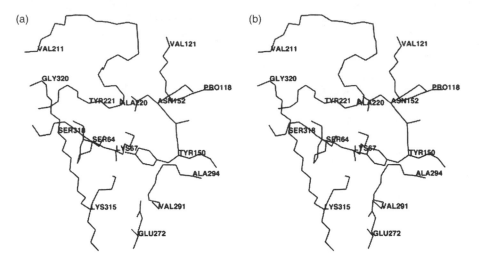

FIGURE 6.11 X-ray crystal structures of the class C β-lactamases from *Escherichia coli* K12 (a) and *Enterobacter cloacae* P99 (b) (pdb codes 2BLS and 2BLT, respectively). (Structures were drawn using the program MOLSCRIPT.)

(cephalothin, ceftazidime, and cefotaxime), and the monobactam aztreonam [156]. Ceftazidime, cefotaxime, and aztreonam are three important clinical antibiotics.

Class D β-lactamases are the smallest among active-site-serine β-lactamases. For example, the entire sequence of the OXA-1 class D β-lactamase (including the signal peptide) is 246 amino acids, compared to 286 amino acids for the *E. coli* TEM-1 (class A) and 381 amino acids for the *E. cloacae* P99 (class C) β-lactamases. There does not appear to be any striking similarities in their sequences when compared to those of the classes A and C of β-lactamases [27]. Detailed analyses of these enzymes had not taken place until recently, in large part due to the historical difficulties of *in vitro* assay [157–159]. Complicating matters farther, these enzymes were known to have often non-reproducible biphasic kinetics.

An important discovery was that these enzymes are *N*-carboxylated on their active site lysine three residues to the carboxyl end of the serine that experiences acylation [160,161]. This is a post-translational modification that takes place by the addition of carbon dioxide to the free-base form of the lysine. The crystal structure of the OXA-10 β-lactamase revealed that the *N*-carboxylated lysine is stabilized by the active site environment by three specific hydrogen bonding interactions [161,162]. The carboxylated lysine was shown to be the base that promoted the active site serine for acylation and activated a water molecule for the deacylation event [160]. Hence, substrate turnover by class D β-lactamases enjoys symmetry in catalysis.

The theory of how the active site lysine experiences *N*-carboxylation in this class of enzymes was investigated recently [163]. Furthermore, the theory revealed that in the course of catalysis, the *N*-carboxylated lysine could shuttle protons via either the carbamate oxygen or nitrogen. When the oxygen is protonated transiently, the resulting carbamic acid is stable. However, when the nitrogen of the carbamate

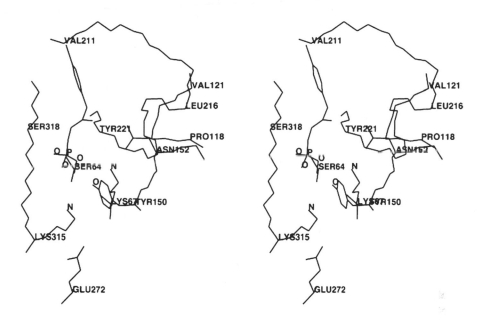

FIGURE 6.12   Stereo view of the active site of a class C β-lactamase from *Enterobacter cloacae* P99 (pdb code 2BLT). Important residues, Ser-64, Lys-67, Tyr-150, Lys-315 and Ser-318 are labeled, and their side chains are shown.

is protonated, it experiences spontaneous and barrierless decarboxylation. The *N*-carboxylation/*N*-decarboxylation is reversible, and the equilibrium constant for this process lies in the micromolar range. In addition, since concentration of carbon dioxide in biological systems is in the millimolar range, it is expected that class D β-lactamases remain fully carboxylated *in vivo* [160].

The loss of lysine *N*-carboxylation results in an inactive enzyme [160]. In light of the reversible *N*-carboxylation of lysine the basis for biphasic turnover for some substrates became clear. In mid-catalysis, if the carbamate nitrogen were to abstract a proton, the barrierless loss of carbon dioxide would result in an inactive enzyme. Catalysis could not resume until the lysine residue is *N*-recarboxylated, hence the biphasic kinetics [160]. Previously the biphasic kinetics were explained by invocation of monomer and dimer formation for the enzyme [157,158,164,165].

PERSPECTIVES

Introduction of β-lactam antibiotics for use in treatment of infections was one of the most important medical contributions of the 20th century. Penicillin G was the first antibiotic that received widespread clinical application, and its success paved the way for the discovery of other β-lactams and antibiotics of different classes. Despite the presence of the ubiquitous β-lactamases, the clinical importance of β-lactams remains very high. They remain the most commonly used antibiotics in the clinic, and we will have to rely on them for the foreseeable future [166]. The entire genomes of many bacteria have been sequenced, and others are being actively investigated [167–171; http://www.tigr.org]. Knowledge of these microbial enzymes and the proteins they encode has the potential for scientific impact in the future [172–175]. We hasten to say that the developments in this field in the past decade have not yet met the expectation. However, if history is any indication, resistance to any drug will develop in due time. Indeed, the literature indicates that for the eight major classes of known antibacterials, resistance has developed within one to four years from the time of the clinical introduction [176,177]. This is the vindication of Paul Ehrlich's prophetic statement that "drug resistance follows the drug like a faithful shadow."

We have described in this chapter the different classes of β-lactamases that have arisen in response to the challenge of β-lactam antibiotics. It is important to note that it would appear that the processes of random mutation and selection have resulted in at least four known classes of β-lactamases, all operating by distinct mechanisms. The many variants of the members of these classes of drug resistance determinants, which essentially cover the full spectrum of phenotypic traits needed to give resistance to all β-lactam antibiotics, document further the power of the evolutionary processes at work in microorganisms. We do not expect to have any obvious replacements for β-lactam antibiotics in the near future. Only time will tell if the renewed interest in the pharmaceutical industry in development of antiinfectives will meet the clinical challenges before the arrival of what has been presaged as a "post-antimicrobial era" [178].

ACKNOWLEDGMENTS

This work was supported by the U.S. National Institutes of Health grants AI33170 and GM61629.

REFERENCES

1. The problem of antibiotic resistance, National Institute of Allergy and Infectious Disease, National Institutes of Health, Bethesda, 2004.
2. F. A. Waldvogel. Infectious diseases in the 21st century: old challenges and new opportunities. *Int. J. Infect. Dis.* 8, 5–12, 2004.
3. J. F. Fisher, S. O. Meroueh, and S. Mobashery. Bacterial resistance to β-lactam antibiotics: Compelling opportunism, compelling opportunity. *Chem. Rev.* 105, 395–424, 2005.
4. H. Welch and C. N. Lewis, Eds. *Antibiotic Therapy*, Arunde Press, Washington, DC, 1951.
5. C. Mabilat, J. Lourencaovital, S. Goussard, and P. Courvalin. A new example of physical linkage between Tn*1* and Tn*21*—the antibiotic multiple-resistance region of plasmid Pcff04 encoding extended-spectrum β-lactamase TEM-3. *Mol. Gen. Genet.* 235, 113–121, 1992.
6. H. W. Stokes and R. M. Hall. The integron In1 in plasmid R46 includes 2 copies of the Oxa2 gene cassette. *Plasmid* 28, 225–234, 1992.
7. A. Y. Peleg, C. Franklin, J. M. Bell, and D. W. Spelman. Dissemination of the metallo-β-lactamase gene *bla*(IMP-4) among Gram-negative pathogens in a clinical setting in Australia. *Clin. Infect. Dis.* 41, 1549–1556, 2005.
8. D. M. Livermore. Bacterial resistance: Origins, epidemiology, and impact. *Clin. Infect. Dis.* 36, S11–S23, 2003.

9. T. L. Wu, J. H. Chia, L. H. Su, A. J. Kuo, C. S. Chu, and C. H. Chiu. Dissemination of extended-spectrum β-lactamase-producing Enterobacteriaceae in pediatric intensive care units. *J. Clin. Microbiol.* 41, 4836–4838, 2003.

10. R. Canton, T. M. Coque, and F. Baquero. Multi-resistant Gram-negative bacilli: from epidemics to endemics. *Curr. Opin. Infect. Dis.* 16, 315–325, 2003.

11. J. P. Stahl. Tigecycline: a new antibiotic in ongoing clinical development. *Medecine et Maladies Infectieuses* 35, 62–67, 2005.

12. K. L. LaPlante and M. J. Rybak. Daptomycin—a novel antibiotic against Gram-positive pathogens. *Exp. Opin. Pharmacother.* 5, 2321–2331, 2004.

13. C. J. Thomson, E. Power, H. Ruebsamen-Waigmann, and H. Labischinski. Antibacterial research and development in the 21st century—an industry perspective of the challenges. *Curr. Opin. Microbiol.* 7, 445–450, 2004.

14. T. E. Long. Recent progress toward the clinical development of new anti-MRSA antibiotics. *IDrugs* 6, 351, 2003.

15. K. Bush, G. A. Jacoby, and A. A. Medeiros. A functional classification scheme for β-lactamases and its correlation with molecular structure. *Antimicrob. Agents Chemother.* 39, 1211–1233, 1995.

16. R. P. Ambler, A. F. W. Coulson, J. M. Frere, J. M. Ghuysen, B. Joris, M. Forsman, R. C. Levesque, G. Tiraby, and S. G. Waley. A standard numbering scheme for the class-A β-lactamases. *Biochem. J.* 276, 269–270, 1991.

17. R. P. Ambler. The structure of β-lactamases. *Phil. Trans. Royal Soc. London B* 289, 321–331, 1980.

18. B. Jaurin and T. Grundstrom. Ampc cephalosporinase of *Escherichia-coli* K-12 has a different evolutionary origin from that of β-lactamases of the penicillinase type. *Proc. Natl. Acad. Sci. U.S.A.* 78, 4897–4901, 1981.

19. M. Ouellette, L. Bissonnette, and P. H. Roy. Precise insertion of antibiotic-resistance determinants into Tn*21*-like transposons—nucleotide-sequence of the Oxa-1 β-lactamase gene. *Proc. Natl. Acad. Sci. U.S.A.* 84, 7378–7382, 1987.

20. A. Beceiro and G. Bou. Class C β-lactamases: an increasing problem worldwide. *Rev. Med. Microbiol.* 15, 141–152, 2004.

21. Y. J. Yang, B. A. Rasmussen, and D. M. Shlaes. Class A β-lactamases—enzyme-inhibitor interactions and resistance *Pharmacol. Ther.* 83, 141–151, 1999.

22. K. Bush and S. Mobashery. How β-lactamases have driven pharmaceutical drug discovery. *Resolving the Antibiotic Paradox*, B. Rosen and S. Mobashery, Eds., Springer-Verlag, New York, 1998.

23. J. Y. Cha, A. Ishiwata, and S. Mobashery. A novel β-lactamase activity from a penicillin-binding protein of *Treponema pallidum* and why syphilis is still treatable with penicillin. *J. Biol. Chem.* 279, 14917–14921, 2004.

24. R. K. Deka, M. Machius, M. V. Norgard, and D. R. Tomchick. Crystal structure of the 47-kDa lipoprotein of *Treponema pallidum* reveals a novel penicillin-binding protein. *J. Biol. Chem.* 277, 41857–41864, 2002.

25. B. G. Hall and M. Barlow. Evolution of the serine β-lactamases: past, present, and future. *Drug Resist. Updates* 7, 111–123, 2004.

26. K. S. Thomson and E. S. Moland. Version 2000: the new β-lactamases of Gram-negative bacteria at the dawn of the new millenium. *Microbes Infect.* 2, 1225–1235, 2000.

27. I. Massova and S. Mobashery. Kinship and diversification of bacterial penicillin binding proteins and β-lactamases. *Antimicrob. Agents Chemother.* 42, 1–17, 1998.

28. A. Matagne, A. Dubus, M. Galleni, and J. M. Frere. The β-lactamase cycle: a tale of selective pressure and bacterial ingenuity. *Natural Product Reports* 16, 1–19, 1999.

29. A. J. Wright. The penicillins. *Mayo Clin. Proc.* 74, 290–307, 1999.

30. C. Goffin and J. M. Ghuysen. The multimodular penicillin-binding proteins: an enigmatic family of orthologs and paralogs. *Microbiol. Mol. Biol. Rev.* 62, 1079–1093, 1998.

31. A. Philippon, J. Dusart, B. Joris, and J. M. Frere. The diversity, structure and regulation of β-lactamases. *Cell. Mol. Life Sci.* 54, 341–346, 1998.
32. C. Fuda, M. Suvorov, S. B. Vakulenko, and S. Mobashery. The basis for resistance to β-lactam antibiotics by penicillin-binding protein 2a of methicillin-resistant *S. aureus*. *J. Biol. Chem.* 279, 40802–40806, 2004.
33. L. Chesnel, A. Zapun, N. Mouz, O. Dideberg, and T. Vernet. Increase of the deacylation rate of PBP2x from *Streptococcus pneumoniae* by single point mutations mimicking the class A β-lactamases. *Eur. J. Biochem.* 269, 1678–1683, 2002.
34. A. M. Di Guilmi, A. Dessen, O. Dideberg, and T. Vernet. Functional characterization of penicillin-binding protein 1b from *Streptococcus pneumoniae*. *J. Bacteriol.* 185, 1650–1658, 2003.
35. S. Lepage, M. Galleni, B. Lakaye, B. Joris, I. Thamm, and J. M. Frere. Kinetic-properties of the bacillus-licheniformis penicillin-binding proteins. *Biochem. J.* 309, 49–53, 1995.
36. A. L. Koch. Penicillin binding proteins, β-lactams, and lactamases: Offensives, attacks, and defensive countermeasures. *Crit. Rev. Microbiol.* 26, 205–220, 2000.
37. H. D. Holland. Geochemistry—evidence for life on Earth more than 3850 million years ago. *Science* 275, 38–39, 1997.
38. A. A. Medeiros. Evolution and dissemination of β-lactamases accelerated by generations of β-lactam antibiotics. *Clin. Infect. Dis.* 24, S19–S45, 1997.
39. J. F. Martin and S. Gutierrez. Genes for β-lactam antibiotic biosynthesis. *Antonie Van Leeuwenhoek J. Microbiol.* 67, 181–200, 1995.
40. J. J. R. Coque, P. Liras, and J. F. Martin. Genes for a β-lactamase, a penicillin-binding protein and a transmembrane protein are clustered with the cephamycin biosynthetic genes in *Nocardia lactamdurans*. *EMBO J.* 12, 631–639, 1993.
41. H. Lode. Management of serious nosocomial bacterial infections: do current therapeutic options meet the need? *Clin. Microbiol. Infect.* 11, 778–787, 2005.
42. P. M. Hawkey and C. J. Munday. Multiple resistance in Gram-negative bacteria. *Rev. Med. Microbiol.* 15, 51–61, 2004.
43. P. Nordmann and L. Poirel. Emerging carbapenemases in Gram-negative aerobes. *Clin. Microbiol. Infect.* 8, 321–331, 2002.
44. B. H. Normark and S. Normark. Evolution and spread of antibiotic resistance. *J. Int. Med.* 252, 91–106, 2002.
45. D. A. Rowe-Magnus, J. Davies, and D. Mazel. Impact of integrons and transposons on the evolution of resistance and virulence. *Pathogenicity Islands and the Evolution of Pathogenic Microbes, Vol. 2*, J. Hacker and J. B. Kaper, Eds., Springer-Verlag, Berlin, 2002.
46. R. F. Pratt. Functional evolution of the serine β-lactamase active site. *J. Chem. Soc. Perkin Transact. 2* 851–861, 2002.
47. L. W. Hardy and J. F. Kirsch. Diffusion-limited component of reactions catalyzed by *Bacillus cereus* β-lactamase I. *Biochemistry* 23, 1275–1282, 1984.
48. C. C. H. Chen, J. Rahil, R. F. Pratt, and O. Herzberg. Structure of a phosphonate-inhibited β-lactamase—an analog of the tetrahedral transition-state intermediate of β-lactam hydrolysis. *J. Mol. Biol.* 234, 165–178, 1993.
49. L. Maveyraud, R. F. Pratt, and J. P. Samama. Crystal structure of an acylation transition-state analog of the TEM-1 β-lactamase. Mechanistic implications for class A β-lactamases. *Biochemistry* 37, 2622–2628, 1998.
50. C. Damblon, X. Raquet, L. Y. Lian, J. Lamotte-Brasseur, E. Fonze, P. Charlier, G. C. K. Roberts, and J. M. Frere. The catalytic mechanism of β-lactamases: NMR titration of an active-site lysine residue of the TEM-1 enzyme. *Proc. Natl. Acad. Sci. U.S.A.* 93, 1747–1752, 1996.

51. G. Guillaume, M. Vanhove, J. Lamotte-Brasseur, P. Ledent, M. Jamin, B. Joris, and J. M. Frere. Site-directed mutagenesis of glutamate 166 in two β-lactamases. Kinetic and molecular modeling studies. *J. Biol. Chem.* 272, 5438–5444, 1997.

52. J. R. Knox, P. C. Moews, W. A. Escobar, and A. L. Fink. A catalytically-impaired class-a β-lactamase—2 Å crystal structure and kinetics of the *Bacillus licheniformis* E166a mutant. *Protein Eng.* 6, 11–18, 1993.

53. R. M. Gibson, H. Christensen, and S. G. Waley. Site-directed mutagenesis of β-lactamase-I—single and double mutants of Glu-166 and Lys-73. *Biochem. J.* 272, 613–619, 1990.

54. D. Golemi-Kotra, S. O. Meroueh, C. Kim, S. B. Vakulenko, A. Bulychev, A. J. Stemmler, T. L. Stemmler, and S. Mobashery. The importance of a critical protonation state and the fate of the catalytic steps in class A β-lactamases and penicillin-binding proteins. *J. Biol. Chem.* 279, 34665–34673, 2004.

55. N. C. Strynadka, H. Adachi, S. E. Jensen, K. Johns, A. Sielecki, C. Betzel, K. Sutoh, and M. N. James. Molecular structure of the acyl-enzyme intermediate in β-lactam hydrolysis at 1.7 Å resolution. *Nature* 359, 700–705, 1992.

56. S. O. Meroueh, J. F. Fisher, H. B. Schlegel, and S. Mobashery. The *ab initio* QM/MM study of class A β-lactamase acylation: dual participation of glu166 andlys73 in a concerted base promotion of ser70. *J. Am. Chem. Soc.* 127, 15397–15407, 2005.

57. M. Hata, Y. Fujii, M. Ishii, T. Hoshino, and M. Tsuda. Catalytic mechanism of class A β-lactamase. I. The role of Glu166 and Ser130 in the deacylation reaction. *Chem. Pharm. Bull.* 48, 447–453, 2000.

58. M. W. Makinen, A. Sosa-Peinado, V. W. Huang, and D. Mustafi. Active site structure of TEM1 and CcrA β-lactamases in action. *Biophys. J.* 80, 279A–279A, 2001.

59. D. Mustafi, A. Sosa-Peinado, and M. W. Makinen. ENDOR structural characterization of a catalytically competent acylenzyme reaction intermediate of wild-type TEM-1 β-lactamase confirms glutamate-166 as the base catalyst. *Biochemistry* 40, 2397–2409, 2001.

60. H. Adachi, T. Ohta, and H. Matsuzawa. Site-directed mutants, at position 166, of RTEM-1 β-lactamase that form a stable acyl-enzyme intermediate with penicillin. *J. Biol. Chem.* 266, 3186–3191, 1991.

61. W. A. Escobar, A. K. Tan, and A. L. Fink. Site-Directed mutagenesis of β-lactamase leading to accumulation of a catalytic intermediate. *Biochemistry* 30, 10783–10787, 1991.

62. L. Maveyraud, I. Massova, C. Birck, K. Miyashita, J. P. Samama, and S. Mobashery. Crystal structure of 6 α-(hydroxymethyl)penicillanate complexed to the TEM-1 β-lactamase from *Escherichia coli*: evidence on the mechanism of action of a novel inhibitor designed by a computer-aided process. *J. Am. Chem. Soc.* 118, 7435–7440, 1996.

63. L. Mourey, L. P. Kotra, J. Bellettini, A. Bulychev, M. O'Brien, M. J. Miller, S. Mobashery, and J. P. Samama. Inhibition of the broad spectrum non-metallo carbapenamase of class A (NMC-A) β-lactamase from *Enterobacter cloacae* by monocyclic β-lactams. *J. Biol. Chem.* 274, 25260–25265, 1999.

64. L. Maveyraud, L. Mourey, L. P. Kotra, J. D. Pedelacq, V. Guillet, S. Mobashery, and J. P. Samama. Structural basis for clinical longevity of carbapenem antibiotics in the face of challenge by the common class A β-lactamase from the antibiotic-resistant bacteria. *J. Am. Chem. Soc.* 120, 9748–9752, 1998.

65. I. Massova and S. Mobashery. Molecular bases for interactions between β-lactam antibiotics and β-lactamases. *Acc. Chem. Res.* 30, 162–168, 1997.

66. K. Miyashita, I. Massova, P. Taibi, and S. Mobashery. Design, synthesis, and evaluation of a potent mechanism-based inhibitor for the TEM β-lactamase with implications for the enzyme mechanism. *J. Am. Chem. Soc.* 117, 11055–11059, 1995.

67. P. Nordmann, S. Mariotte, T. Naas, R. Labia, and M. H. Nicolas. Biochemical properties of a carbapenem-hydrolyzing β-lactamase from *Enterobacter cloacae* and cloning of the gene into *Escherichia coli*. *Antimicrob. Agents Chemother.* 37, 939–946, 1993.

68. T. Naas, L. Vandel, W. Sougakoff, D. M. Livermore, and P. Nordmann. Cloning and sequence-analysis of the gene for a carbapenem-hydrolyzing class-A β-lactamase, Sme-1, from serratia-marcescens S6. *Antimicrob. Agents Chemother.* 38, 1262–1270, 1994.

69. B. A. Rasmussen, K. Bush, D. Keeney, Y. J. Yang, R. Hare, C. Ogara, and A. A. Medeiros. Characterization of IMI-1 β-lactamase, a class A carbapenem-hydrolyzing enzyme from *Enterobacter cloacae*. *Antimicrob. Agents Chemother.* 40, 2080–2086, 1996.

70. B. A. Rasmussen and K. Bush. Carbapenem-hydrolyzing β-lactamases. *Antimicrob. Agents Chemother.* 41, 223–232, 1997.

71. G. Vedel, A. Belaaouaj, L. Gilly, R. Labia, A. Philippon, P. Nevot, and G. Paul. Clinical isolates of *Escherichia coli* producing tri β-lactamases—novel TEM-enzymes conferring resistance to β-lactamase inhibitors. *J. Antimicrob. Chemother.* 30, 449–462, 1992.

72. C. J. Thomson and S. G. B. Amyes. Trc-1—emergence of a clavulanic acid-resistant TEM β-lactamase in a clinical strain. *FEMS Microbiol. Lett.* 91, 113–117, 1992.

73. M. M. M. Canica, C. Y. Lu, R. Krishnamoorthy, and G. C. Paul. Molecular diversity and evolution of *bla* (TEM) genes encoding β-lactamases resistant to clavulanic acid in clinical *E. coli*. *J. Mol. Evol.* 44, 57–65, 1997.

74. U. Imtiaz, E. Billings, J. R. Knox, E. K. Manavathu, S. A. Lerner, and S. Mobashery. Inactivation of class A β-lactamases by clavulanic acid: the role of arginine-244 in a proposed nonconcerted sequence of events. *J. Am. Chem. Soc.* 115, 4435–4442, 1993.

75. Y. Yang, K. Janota, K. Tabei, N. Huang, M. M. Siegel, Y. I. Lin, B. A. Rasmussen, and D. M. Shlaes. Mechanism of inhibition of the class A β-lactamases PC1 and TEM-1 by tazobactam: observation of reaction products by ESI MS. *J. Biol. Chem.* 275, 26674–26682, 2000.

76. A. P. Kuzin, M. Nukaga, Y. Nukaga, A. Hujer, R. A. Bonomo, and J. R. Knox. Inhibition of the SHV-1 β-lactamase by sulfones: crystallographic observation of two reaction intermediates with tazobactam. *Biochemistry* 40, 1861–1866, 2001.

77. S. B. Vakulenko, B. Geryk, L. P. Kotra, S. Mobashery, and S. A. Lerner. Selection and characterization of β-lactam–β-lactamase inactivator-resistant mutants following PCR mutagenesis of the TEM-1 β-lactamase gene. *Antimicrob. Agents Chemother.* 42, 1542–1548, 1998.

78. S. O. Meroueh, P. Roblin, D. Golemi, L. Maveyraud, S. B. Vakulenko, Y. Zhang, J. P. Samama, and S. Mobashery. Molecular dynamics at the root of expansion of function in the M69L inhibitor-resistant β-lactamase from *Escherichia coli*. *J. Am. Chem. Soc.* 124, 9422–9430, 2002.

79. X. Wang, G. Minasov, and B. K. Shoichet. The structural bases of antibiotic resistance in the clinically derived mutant β-lactamases TEM-30, TEM-32, and TEM-34. *J. Biol. Chem.* 277, 32149–32156, 2002.

80. M. S. Helfand, A. M. Hujer, F. D. Sonnichsen, and R. A. Bonomo. Unexpected advanced generation cephalosporinase activity of the M69F variant of SHV β-lactamase. *J. Biol. Chem.* 277, 47719–47723, 2002.

81. T. Sun, C. R. Bethel, R. A. Bonomo, and J. R. Knox. Inhibitor-resistant class A β-lactamases: consequences of the Ser130-to-Gly mutation seen in Apo and tazobactam structures of the SHV-1 variant. *Biochemistry* 43, 14111–14117, 2004.

82. V. L. Thomas, D. Golemi-Kotra, C. Kim, S. B. Vakulenko, S. Mobashery, and B. K. Shoichet. Structural consequences of the inhibitor-resistant Ser130Gly substitution in TEM-lactamase. *Biochemistry* 44, 9330–9338, 2005.

83. D. Sulton, D. Pagan-Rodriguez, X. Zhou, Y. D. Liu, A. M. Hujer, C. R. Bethel, M. S. Helfand, J. M. Thomson, V. E. Anderson, J. D. Buynak, L. M. Ng, and R. A. Bonomo. Clavulanic acid inactivation of SHV-1 and the inhibitor-resistant S130G SHV-1 β-lactamase—insights into the mechanism of inhibition. *J. Biol. Chem.* 280, 35528–35536, 2005.

84. R. P. A. Brown, R. T. Aplin, and C. J. Schofield. Inhibition of TEM-2 β-lactamase from *Escherichia coli* by clavulanic acid: observation of intermediates by electrospray ionization mass spectrometry. *Biochemistry* 35, 12421–12432, 1996.

85. R. P. A. Brown, R. T. Aplin, C. J. Schofield, and C. H. Frydrych. Mass spectrometric studies on the inhibition of TEM-2 β-lactamase by clavulanic acid derivatives. *J. Antibiotics* 50, 184–185, 1997.

86. G. Zafaralla, E. K. Manavathu, S. A. Lerner, and S. Mobashery. Elucidation of the role of arginine-244 in the turnover processes of class A β-lactamases. *Biochemistry* 31, 3847–3852, 1992.

87. P. Swaren, D. Golemi, S. Cabantous, A. Bulychev, L. Maveyraud, S. Mobashery, and J. P. Samama. X-ray structure of the Asn276Asp variant of the *Escherichia coli* TEM-1 β-lactamase: direct observation of electrostatic modulation in resistance to inactivation by clavulanic acid. *Biochemistry* 38, 9570–9576, 1999.

88. G. F. Weldhagen, L. Poirel, and P. Nordmann. Ambler class A extended-spectrum β-lactamases in *Pseudomonas aeruginosa*: novel developments and clinical impact. *Antimicrob. Agents Chemother.* 47, 2385–2392, 2003.

89. R. Bonnet. Growing group of extended spectrum β-lactamases: the CTX-M enzymes. *Antimicrob. Agents Chemother.* 48, 1–14, 2004.

90. M. S. Helfand and R. A. Bonomo. Current challenges in antimicrobial chemotherapy: the impact of extended-spectrum β-lactamases and metallo-β-lactamases on the treatment of resistant Gram-negative pathogens. *Curr. Opin. Pharmacol.* 5, 452–458, 2005.

91. J. R. Knox. Extended-spectrum and inhibitor-resistant TEM-type β-lactamases—mutations, specificity, and 3-dimensional structure. *Antimicrob. Agents Chemother.* 39, 2593–2601, 1995.

92. T. Taibi and S. Mobashery. Mechanism of turnover of imipenem by the TEM β-lactamase revisited. *J. Am. Chem. Soc.* 117, 7600–7605, 1995.

93. B. Vilanova, J. Donoso, J. Frau, and F. Munoz. Kinetic and molecular modelling studies of reactions of a class A β-lactamase with compounds bearing a methoxy group on the β-lactam ring. *Helv. Chim. Acta* 82, 1274–1288, 1999.

94. E. Fonzé, M. Vanhove, G. Dive, E. Sauvage, J. M. Frère, and P. Charlier. Crystal structures of the *Bacillus licheniformis* BS3 class A β-lactamase and of the acyl-enzyme adduct formed with cefoxitin. *Biochemistry* 41, 1877–1885, 2002.

95. L. Mourey, K. Miyashita, P. Swaren, A. Bulychev, J. P. Samama, and S. Mobashery. Inhibition of the NMC-A β-lactamase by a penicillanic acid derivative and the structural bases for the increase in substrate profile of this antibiotic resistance enzyme. *J. Am. Chem. Soc.* 120, 9382–9383, 1998.

96. P. Swaren, L. Maveyraud, X. Raquet, S. Cabantous, C. Duez, J. D. Pedelacq, S. Mariotte-Boyer, L. Mourey, R. Labia, M. H. Nicolas-Chanoine, P. Nordmann, J. M. Frere, and J. P. Samama. X-ray analysis of the NMC-A β-lactamase at 1.64-Å resolution, a class A carbapenemase with broad substrate specificity. *J. Biol. Chem.* 273, 26714–26721, 1998.

97. A. P. Kuzin, M. Nukaga, Y. Nukaga, A. M. Hujer, R. A. Bonomo, and J. R. Knox. Structure of the SHV-1 β-lactamase. *Biochemistry* 38, 5720–5727, 1999.

98. A. M. Hujer, K. M. Hujer, and R. A. Bonomo. Mutagenesis of amino acid residues in the SHV-1 β-lactamase: the premier role of Gly238Ser in penicillin and cephalosproin resistance. *Biochim. Biophys. Acta* 1547, 37–50, 2001.

99. M. C. Orencia, J. S. Yoon, J. E. Ness, W. P. C. Stemmer, and R. C. Stevens. Predicting the emergence of antibiotic resistance by directed evolution and structural analysis. *Nature Struct. Biol.* 8, 238–242, 2001.

100. M. Nukaga, T. Abe, A. M. Venkatesan, T. S. Mansour, R. A. Bonomo, and J. R. Knox. Inhibition of class A and class C β-lactamases by penems: crystallographic structures of a novel 1, 4-thiazepine intermediate. *Biochemistry* 42, 13152–13159, 2003.

101. A. Ibuka, A. Taguchi, M. Ishiguro, S. Fushinobu, Y. Ishii, S. Kamitori, K. Okuyama, K. Yamaguchi, M. Konno, and H. Matsuzawa. Crystal structure of the E166A mutant of extended-spectrum β-lactamase Toho-1 at 1.8 Å resolution. *J. Mol. Biol.* 285, 2079–2087, 1999.

102. A. S. Ibuka, Y. Ishii, M. Galleni, M. Ishiguro, K. Yamaguchi, J. M. Frère, H. Matsuzawa, and H. Sakai. Crystal structure of extended-spectrum β-lactamase Toho-1: insights into the molecular mechanism for catalytic reaction and substrate specificity expansion. *Biochemistry* 42, 10634–10643, 2003.

103. T. Shimamura, A. S. Ibuka, S. Fushinobu, T. Wakagi, M. Ishiguro, Y. Ishii, and H. Matsuzawa. Acyl-intermediate structures of the extended-spectrum class of β-lactamase Toho-1 in complex with cefotaxime, cephalothin, and benzylpenicillin. *J. Biol. Chem.* 277, 46601–46608, 2002.

104. P. Nordmann, E. Ronco, T. Naas, C. Duport, Y. Michelbriand, and R. Labia. Characterization of a novel extended-spectrum β-lactamase from *Pseudomonas aeruginosa*. *Antimicrob. Agents Chemother.* 37, 962–969, 1993.

105. L. Poirel, A. Karim, A. Mercat, I. Le Thomas, H. Vahaboglu, C. Richard, and P. Nordmann. Extended-spectrum β-lactamase-producing strain of *Acinetobacter baumannii* isolated from a patient in France. *J. Antimicrob. Chemother.* 43, 157–158, 1999.

106. H. Vahaboglu, R. Ozturk, G. Aygun, F. Coskunkan, A. Yaman, A. Kaygusuz, H. Leblebicioglu, I. Balik, K. Aydin, and M. Otkun. Widespread detection of PER-1-type extended-spectrum β-lactamases among nosocomial *Acinetobacter* and *Pseudomonas aeruginosa* isolates in Turkey: a nationwide multicenter study. *Antimicrob. Agents Chemother.* 41, 2265–2269, 1997.

107. S. Tranier, A. T. Bouthors, L. Maveyraud, V. Guillet, W. Sougakoff, and J. P. Samama. The high resolution crystal structure for class A β-lactamase PER-1 reveals the bases for its increase in breadth of activity. *J. Biol. Chem.* 275, 28075–28082, 2000.

108. H. M. Lim, J. J. Pene, and R. W. Shaw. Cloning, nucleotide-sequence, and expression of the *Bacillus cereus* 5/B/6 β-lactamase-Ii structural gene. *J. Bacteriol.* 170, 2873–2878, 1988.

109. T. R. Walsh, L. Hall, S. J. Assinder, W. W. Nichols, S. J. Cartwright, A. P. Macgowan, and P. M. Bennett. Sequence-analysis of the L1 metallo-β-lactamase from *Xanthomonas maltophilia*. *Biochim. Biophys. Acta* 1218, 199–201, 1994.

110. T. R. Walsh, M. A. Toleman, L. Poirel, and P. Nordmann. Metallo-β-lactamases: the quiet before the storm? *Clin. Microbiol. Rev.* 18, 306–325, 2005.

111. N. O. Concha, C. A. Janson, P. Rowling, S. Pearson, C. A. Cheever, B. P. Clarke, C. Lewis, M. Galleni, J. M. Frère, D. J. Payne, J. H. Bateson, and S. S. Abdel-Meguid. Crystal structure of the IMP-1 metallo-β-lactamase from *Pseudomonas aeruginosa* and its complex with a mercaptocarboxylate inhibitor: binding determinants of a potent, broad-spectrum inhibitor. *Biochemistry* 39, 4288–4298, 2000.

112. D. Livermore and N. Woodford. Carbapenemases: a problem in waiting? *Curr. Opin. Microbiol.* 3, 489–495, 2000.

113. G. F. Weldhagen. Integrons and β-lactamases: A novel perspective on resistance. *Int. J. Antimicrob. Agents* 23, 556–562, 2004.

114. F. Luzzaro, J. D. Docquier, C. Colinon, A. Endimiani, G. Lombardi, G. Amicosante, G. Rossolini, and A. Toniolol. Emergence in *Klebsiella pneumoniae* and *Enterobacter cloacae* clinical isolates of the VIM-4 metallo-β-lactamase encoded by a conjugative plasmid. *Antimicrob. Agents Chemother.* 48, 648–650, 2004.

115. T. R. Walsh, M. A. Toleman, W. Hryniewicz, P. M. Bennett, and R. N. Jones. Evolution of an integron carrying bla(VIM-2) in Eastern Europe: report from the SENTRY Antimicrobial Surveillance Program. *J. Antimicrob. Chemother.* 52, 116–119, 2003.

116. M. Galleni, J. Lamotte-Brasseur, G. M. Rossolini, J. Spencer, O. Dideberg, and J. M. Frère. Standard numbering scheme for the class B β-lactamases. *Antimicrob. Agents Chemother.* 45, 660–663, 2001.

117. R. M. Rasia and A. J. Vila. Exploring the role and the binding affinity of a second zinc equivalent in *B. cereus* metallo-β-lactamase. *Biochemistry* 41, 1853–1860, 2002.

118. A. M. Davies, R. M. Rasia, A. J. Vila, B. J. Sutton, and S. M. Fabiane. Effect of pH on the active site of an Arg121Cys mutant of the metallo-β-lactamase from *Bacillus cereus*: implications for the enzyme mechanism. *Biochemistry* 44, 4841–4849, 2005.

119. M. W. Crowder, Z. G. Wang, S. L. Franklin, E. P. Zovinka, and S. J. Benkovic. Characterization of the metal-binding sites of the β-lactamase from *Bacteroides fragilis*. *Biochemistry* 35, 12126–12132, 1996.

120. N. Laraki, N. Franceschini, G. M. Rossolini, P. Santucci, C. Meunier, E. de Pauw, G. Amicosante, J. M. Frere, and M. Galleni. Biochemical characterization of the *Pseudomonas aeruginosa* 101/1477 metallo-β-lactamase IMP-1 produced by *Escherichia coli*. *Antimicrob. Agents Chemother.* 43, 902–906, 1999.

121. E. G. Orellano, J. E. Girardini, J. A. Cricco, E. A. Ceccarelli, and A. J. Vila. Spectroscopic characterization of a binuclear metal site in *Bacillus cereus* β-lactamase II. *Biochemistry* 37, 10173–10180, 1998.

122. G. Garau, C. Bebrone, C. Anne, M. Galleni, J. M. Frere, and O. Dideberg. A metallo-β-lactamase enzyme in action: crystal structures of the monozinc carbapenemase CphA and its complex with biapenem. *J. Mol. Biol.* 345, 785–795, 2005.

123. N. O. Concha, B. A. Rasmussen, K. Bush, and O. Herzberg. Crystal structure of the wide-spectrum binuclear zinc β-lactamase from *Bacteroides fragilis*. *Structure* 4, 823–836, 1996.

124. A. Carfi, E. Duee, R. Paul-Soto, M. Galleni, J. M. Frere, and O. Dideberg. X-ray structure of the Zn-II β-lactamase from *Bacteroides fragilis* in an orthorhombic crystal form. *Acta Crystal. D* 54, 47–57, 1998.

125. A. Carfi, S. Pares, E. Duee, M. Galleni, C. Duez, J. M. Frere, and O. Dideberg. The 3-D structure of a zinc metallo-β-lactamase from *Bacillus cereus* reveals a new type of protein fold. *EMBO J.* 14, 4914–4921, 1995.

126. J. H. Ullah, T. R. Walsh, I. A. Taylor, D. C. Emery, C. S. Verma, S. J. Gamblin, and J. Spencer. The crystal structure of the L1 metallo-β-lactamase from *Stenotrophomonas maltophilia* at 1.7 Å resolution. *J. Mol. Biol.* 284, 125–136, 1998.

127. S. M. Fabiane, M. K. Sohi, T. Wan, D. J. Payne, J. H. Bateson, M. T., and B. J. Sutton. Crystal structure of the zinc dependent β-lactamase from *B. cereus* at 1.9 Å resolution: binuclear active site with features of a mononuclear enzyme. *Biochemistry* 37, 12404–12411, 1998.

128. W. Fast, Z. Wang, and S. J. Benkovic. Familial mutations and zinc stoichiometry determine the rate-limiting step of nitrocefin hydrolysis by metallo-β-lactamase from *Bacteroides fragilis*. *Biochemistry* 40, 1640–1650, 2001.

129. Z. G. Wang and S. J. Benkovic. Purification, characterization, and kinetic studies of a soluble *Bacteroides fragilis* metallo-β-lactamase that provides multiple antibiotic resistance. *J. Biol. Chem.* 273, 22402–22408, 1998.

130. Z. G. Wang, W. Fast, and S. J. Benkovic. On the mechanism of the metallo-β-lactamase from *Bacteroides fragilis*. *Biochemistry* 38, 10013–10023, 1999.

131. N. Díaz, T. L. Sordo, D. Suárez, R. Mendez, and J. Martín-Villacorta. Zn²⁺ catalysed hydrolysis of β-lactams: experimental and theoretical studies on the influence of the β-lactam structure. *New J. Chem.* 28, 15–25, 2004.

132. L. B. Rice. Do we really need new anti-infective drugs? *Curr. Opin. Pharmacol.* 3, 459–464, 2003.

133. L. P. Kotra and S. Mobashery. β-Lactam antibiotics, β-lactamases and bacterial resistance. *Bulletin de L' Institut Pasteur* 96, 139–150, 1998.
134. K. Bush. The evolution of β-lactamases. *Antibiotic Resistance: Origins, Evolution, Selection and Spread*, Ciba Foundation, Ed., John Wiley & Sons, New York, 1997.
135. C. Oefner, A. D'Arcy, J. J. Daly, K. Gubernator, R. L. Charnas, I. Heinze, C. Hubschwerlen, and F. K. Winkler. Refined crystal structure of β-lactamase from *Citrobacter freundii* indicates a mechanism for β-lactam hydrolysis. *Nature* 343, 284–288, 1990.
136. A. Patera, L. C. Blaszczak, and B. K. Shoichet. Crystal structures of substrate and inhibitor complexes with AmpC β-lactamase: possible implications for substrate-assisted catalysis. *J. Am. Chem. Soc.* 122, 10504–10512, 2000.
137. E. Lobkovsky, P. C. Moews, H. Liu, H. Zhao, J. M. Frere, and J. R. Knox. Evolution of an enzyme activity: crystallographic structure at 2 Å resolution of cephalosporinase from the *ampC* gene of *Enterobacter cloacae* P99 and comparison with a class A penicillinase. *Proc. Natl. Acad. Sci. U.S.A.* 90, 11257–11261, 1993.
138. M. I. Page, B. Vilanova, and N. J. Layland. pH-Dependence of and kinetic solvent isotope effects on the methanolysis and hydrolysis of β-lactams catalyzed by class-C β-lactamase. *J. Am. Chem. Soc.* 117, 12092–12095, 1995.
139. J. Lamotte-Brasseur, A. Dubus, and R. C. Wade. pKa Calculations for class C β-lactamases: the role of Tyr-150. *Proteins* 40, 23–28, 2000.
140. B. M. Beadle, I. Trehan, P. J. Focia, and B. K. Shoichet. Structural milestones in the reaction pathway of an amide hydrolase: substrate, acyl, and product complexes of cephalothin with AmpC β-lactamase. *Structure* 10, 413–424, 2002.
141. Y. Kato-Toma, T. Iwashita, K. Masuda, Y. Oyama, and M. Ishiguro. pKa Measurements from NMR of tyrosine-150 in class C β-lactamase. *Biochem. J.* 371, 175–181, 2003.
142. N. Diaz, D. Suarez, and T. L. Sordo. Molecular dynamics simulations of class C β-lactamase from *Citrobacter freundii*: insights into the base catalyst for acylation. *Biochemistry* 45, 439–451, 2006.
143. B. F. Gherman, S. D. Goldberg, V. W. Cornish, and R. A. Friesner. Mixed quantum mechanical/molecular mechanical (QM/MM) study of the deacylation reaction in a penicillin binding protein (PBP) versus in a class C β-lactamase. *J. Am. Chem. Soc.* 126, 7652–7664, 2004.
144. A. Bulychev, I. Massova, K. Miyashita, and S. Mobashery. Nuances of mechanisms and their implications for evolution of the versatile β-lactamase activity: from biosynthetic enzymes to drug resistance factors. *J. Am. Chem. Soc.* 119, 7619–7625, 1997.
145. Y. Chen, G. Minasov, T. A. Roth, F. Prati, and B. K. Shoichet. The deacylation mechanism of AmpC β-lactamase at ultrahigh resolution. *J. Am. Chem. Soc.* In press.
146. A. Bulychev and S. Mobashery. Class C β-lactamases operate at the diffusion limit for turnover of their preferred cephalosporin substrates. *Antimicrob. Agents Chemother.* 43, 1743–1746, 1999.
147. E. Lobkovsky, E. M. Billings, P. C. Moews, J. Rahil, R. F. Pratt, and J. R. Knox. Crystallographic structure of a phosphonate derivative of the *Enterobacter cloacae* P99 cephalosporinase—mechanistic interpretation of a β-lactamase transition-state analog. *Biochemistry* 33, 6762–6772, 1994.
148. K. C. Usher, L. C. Blaszczak, G. S. Weston, B. K. Shoichet, and S. J. Remington. Three-dimensional structure of AmpC β-lactamase from *Escherichia coli* bound to a transition-state analogue: possible implications for the oxyanion hypothesis and for inhibitor design. *Biochemistry* 37, 16082–16092, 1998.
149. G. V. Crichlow, A. P. Kuzin, M. Nukaga, K. Mayama, T. Sawai, and J. R. Knox. Structure of the extended-spectrum class C β-lactamase of *Enterobacter cloacae* GC1, a natural mutant with a tandem tripeptide insertion. *Biochemistry* 38, 10256–10261, 1999.
150. E. Caselli, R. A. Powers, B. K. Blaszczak, C. Y. E. Wu, F. Prati, and B. K. Shoichet. Energetic, structural, and antimicrobial analyses of β-lactam side chain recognition by β-lactamases. *Chem. Biol.* 8, 17–31, 2001.

151. B. M. Beadle and B. K. Shoichet. Structural basis for imipenem inhibition of class C β-lactamases. *Antimicrob. Agents Chemother.* 46, 3978–3980, 2002.

152. Y. Doi, J. Wachino, M. Ishiguro, H. Kurokawa, K. Yamane, N. Shibata, K. Shibayama, K. Yokoyama, H. Kato, T. Yagi, and Y. Arakawa. Inhibitor-sensitive AmpC β-lactamase variant produced by an *Escherichia coli* clinical isolate resistant to oxyiminocephalosporins and cephamycins. *Antimicrob. Agents Chemother.* 48, 2652–2658, 2004.

153. A. S. Wilkinson, P. K. Bryant, S. O. Meroueh, M. G. P. Page, S. Mobashery, and C. W. Wharton. A dynamic structure for the acyl-enzyme species of the antibiotic aztreonam with the *Citrobacter freundii* β-lactamase revealed by infrared spectroscopy and molecular dynamics simulations. *Biochemistry* 42, 1950–1957, 2003.

154. C. Heritier, L. Poirel, and P. Nordmann. Genetic and biochemical characterization of a chromosome-encoded carbapenem-hydrolyzing ambler class D β-lactamase from *Shewanella algae. Antimicrob. Agents Chemother.* 48, 1670–1675, 2004.

155. G. Bou and J. Mart. Cloning, nucleotide sequencing, and analysis of the gene encoding an AmpC β-lactamase in *Acinetobacter baumannii. Antimicrob. Agents Chemother.* 44, 428–432, 2000.

156. L. N. Philippon, T. Naas, A. T. Bouthors, V. Barakett, and P. Nordmann. OXA-18, a class D clavulanic acid-inhibited extended-spectrum β-lactamase from *Pseudomonas aeruginosa. Antimicrob. Agents Chemother.* 41, 2188–2195, 1997.

157. F. Danel, J. M. Frere, and D. M. Livemore. Evidence of dimerization among class D β-lactamases: kinetics of OXA-14 β-lactamase. *Biochim. Biophys. Acta* 1546, 132–142, 2001.

158. N. Franceschini, B. Segatore, M. Perilli, S. Vessilier, L. Franchino, and G. Amicosante. Meropenam stability to β-lactamase hydrolysis and comparative *in vitro* activity against several β-lactamase-producing Gram-negative strains. *J. Antimicrob. Chemother.* 49, 395–398, 2002.

159. L. Pernot, F. Frenois, T. Rybkine, G. L'Hermite, S. Petrella, J. Delettre, V. Jarlier, E. Collatz, and W. Sougakoff. Crystal structures of the class D β-lactamase OXA-13 in the native form and in complex with meropenem. *J. Mol. Biol.* 310, 859–874, 2001.

160. D. Golemi, L. Maveyraud, S. Vakulenko, J. P. Samama, and S. Mobashery. Critical involvement of a carbamylated lysine in catalytic function of class D β-lactamases. *Proc. Natl. Acad. Sci. U.S.A.* 98, 14280–14285, 2001.

161. L. Maveyraud, D. Golemi, L. P. Kotra, S. Tranier, S. Vakulenko, S. Mobashery, and J. P. Samama. Insights into class D β-lactamases are revealed by the crystal structure of the OXA-10 enzyme from *Pseudomonas aeruginosa. Structure Fold. Des.* 8, 1289–1298, 2000.

162. L. Maveyraud, D. Golemi-Kotra, A. Ishiwata, O. Meroueh, S. Mobashery, and J. P. Samama. High resolution X-ray structure of an acyl-enzyme species for the class D OXA-10 β-lactamase. *J. Am. Chem. Soc.* 124, 2461–2465, 2002.

163. Z. Li, J. B. Cross, T. Vreven, S. O. Meroueh, S. Mobashery, and H. B. Schlegel. Lysine carboxylation in proteins: OXA-10 β-lactamase. *Proteins Struct. Funct. Bioinform.* 61, 246–257, 2005.

164. N. Franceschini, L. Boschi, S. Pollini, R. Herman, M. Perilli, M. Galleni, J. M. Frere, G. Amicosante, and G. M. Rossolini. Characterization of OXA-29 from *Legionella (Fluoribacter) gormanii*: molecular class D β-lactamase with unusual properties. *Antimicrob. Agents Chemother.* 45, 3509–3516, 2001.

165. F. Danel, M. Paetzel, N. C. J. Strynadka, and M. G. P. Page. Effect of divalent metal cations on the dimerization of OXA-10 and-14 class D β-lactamases from *Pseudomonas aeruginosa. Biochemistry* 40, 9412–9420, 2001.

166. E. D. Modugno and A. Felici. The renewed challenge of β-lactams to overcome bacterial resistance. *Curr. Opin. Anti-Infect. Invest. Drugs* 1, 26–39, 1999.

167. T. A. White and D. B. Kell. Comparative genomic assessment of novel broad-spectrum targets for antibacterial drugs. *Comp. Funct. Genom.* 5, 304–327, 2004.

168. B. L. Wang, Z. M. Li, and H. J. Zang. Influence of genomics on structure-based drug design. *Prog. Chem.* 15, 505–511, 2003.

169. B. Fritz and G. A. Raczniak. Bacterial genomics—potential for antimicrobial drug discovery. *Biodrugs* 16, 331–337, 2002.

170. T. J. Dougherty, J. F. Barrett, and M. J. Pucci. Microbial genomics and novel antibiotic discovery: new technology to search for new drugs. *Curr. Pharm. Des.* 8, 1119–1135, 2002.

171. C. M. Tang and E. R. Moxon. The impact of microbial genomics on antimicrobial drug development. *Ann. Rev. Genomics Hum. Genet.* 2, 259–269, 2001.

172. A. Coates, Y. Hu, R. Bax, and C. Page. The future challenges facing the development of new antimicrobial drugs. *Nature Rev. Drug Disc.* 1, 895–910, 2002.

173. S. D. Mills. The role of genomics in antimicrobial discovery. *J. Antimicrob. Chemother.* 51, 749–752, 2003.

174. N. Kaldalu, R. Mei, and K. Lewis. Killing by ampicillin and ofloxacin induces overlapping changes in *Escherichia coli* transcription profile. *Antimicrob. Agents Chemother.* 48, 890–896, 2004.

175. B. Hutter, C. Schaab, S. Albrecht, M. Borgmann, N. A. Brunner, C. Freiberg, K. Ziegelbauer, C. O. Rock, I. Ivanov, and H. Loferer. Prediction of mechanisms of action of antibacterial compounds by gene expression profiling. *Antimicrob. Agents Chemother.* 48, 2838–2844, 2004.

176. K. K. Wong and D. L. Pompliano. Peptidoglycan biosynthesis—unexploited antibacterial targets within a familiar pathway. *Resolving the Antibiotic Paradox*, B. Rosen and S. Mobashery, Eds., Springer-Verlag, New York, 1998.

177. D. Golemi-Kotra, S. Vakulenko, and S. Mobashery. Evolution of multiple mechanisms of resistance to β-lactam antibiotics; Forum on Emerging Infections "The Resistance Phenomenon in Microbes and Infectious Disease Vectors: Implications for Human Health and Strategies for Containment," Institute of Medicine and S. Knobler, Eds., National Academy Press, Washington, D.C., 2003.

178. M. L. Cohen. Epidemiology of drug-resistance: implications for a postantimicrobial era. *Science* 257, 1050–1055, 1992.

7 Target Modification as a Mechanism of Antimicrobial Resistance

David C. Hooper

CONTENTS

Alteration in the target of an antimicrobial drug is a widely used bacterial mechanism of drug resistance and, in addition to reduced drug permeation to its target and drug modification, is one of the three major mechanisms. Resistance by the general mechanism of target modification can be brought about, however, by a remarkable variety of specific means, which have been exploited by different clinically important bacteria. The modification mechanism often results in an altered structure of the original drug target structure that binds the drug poorly or not at all. This alteration in structure can be brought about by naturally occurring spontaneous mutations in the gene(s) encoding the drug target that modify single or limited numbers of amino acids in the target protein, often in the region of a known or putative drug binding site. Quinolone resistance due to alterations in the target enzymes DNA gyrase and topoisomerase IV involved in DNA synthesis, rifampin resistance due to alterations in the β subunit of the target RNA polymerase involved in RNA synthesis, and low-level penicillin resistance in *Streptococcus pneumoniae* due to alterations in the target transpeptidases (penicillin-binding proteins [PBPs]) involved in cell wall synthesis are examples of this category.

More extensive modifications of a drug target often require other genetic mechanisms. In the case of high-level penicillin resistance in *S. pneumoniae*, more extensive modifications of the target transpeptidases involved in cell wall synthesis are possible because of the ability of this organism to exchange DNA segments with related bacterial species, some of which have transpeptidases that bind penicillin poorly, allowing the generation of mosaic transpeptidases with extensively modified regions of these target enzymes in *S. pneumoniae*. In other cases, such as glycopeptide resistance in enterococci and macrolide resistance in many bacteria, the target structures to which these drugs bind (the cell wall for the glycopeptides and the bacterial ribosome for the macrolides) are exogenously modified by enzymes encoded by DNA acquired on mobile genetic elements, such as plasmids or transposons, which can be transferred between bacteria. In other cases, such as tetracycline resistance in many bacteria and plasmid-encoded quinolone resistance due to Qnr proteins in enteric Gram-negative bacteria, the drug targets are protected from drug action, but not modified by the resistance-determining proteins.

Another novel variation of the altered target mechanism is the overexpression of unmodified drug target binding sites in such a way that binding of drug to these extra sites limits access of drug to a subset of critical target binding sites, as is thought to be the cause of low-level glycopeptide resistance in staphylococci. Finally, in a number of cases, such as resistance to methicillin and other β-lactams in staphylococci, resistance to mupirocin in staphylococci, and resistance to trimethoprim in many species, bacteria have acquired genes, sometimes on mobile genetic elements, that encode an alternative or bypass drug-resistant target enzyme. This enzyme then provides the functions that would have otherwise been inhibited, allowing growth in the presence of drug. Thus, the creativity of nature in developing resistance mechanisms under selective pressure has as yet been fully capable of meeting the challenge of new drug development.

INTRODUCTION

Modification of the targets of antimicrobial agents is one of the three principal mechanisms by which bacteria effect resistance to antimicrobial agents, in addition to alteration in drug permeation to its target, and drug modification. Within this general mechanism category, however, there is remarkable bacterial diversity in the means by which target modification is accomplished. In this chapter, six different bacterial strategies (Table 7.1) will be discussed for bringing about target modifications that cause antimicrobial resistance, with a focus on specific examples of each strategy that are of general clinical importance.

In many cases, target modification by various means produces alteration of an existing natural target such that it has reduced drug binding affinity, but in other notable cases the resistance determinant may block drug access by interaction with the drug target, and resistance may occur from target overproduction that sequesters drug, thereby limiting its access to a subset of target molecules at a critical cellular location, or the bacterium may acquire a resistant drug target that can carry out the functions of the natural sensitive target molecule in the presence of drug. Modification of a natural drug target may result from spontaneous chromosomal mutation resulting in single or multiple amino acid substitutions, or from homologous recombination with exogenous DNA containing gene segments that encode portions of proteins with reduced drug binding properties. Analysis of modifications of this type is particularly useful in understanding the structural basis of drug binding to its target. Genes acquired on mobile genetic elements, such as plasmids and transposons, may also contribute to resistance by target modification by encoding proteins that themselves modify the drug target, or block drug access to the target, or act as drug-resistant (bypass) targets. Thus, multiple variations on the general mechanism of drug target modification have evolved in nature.

SPONTANEOUS MUTATIONAL CHANGES IN TARGET PROTEINS

QUINOLONE RESISTANCE

Quinolones are widely used antimicrobial agents with a generally broad spectrum of activity. There are two intracellular targets of the quinolone class of antimicrobials, DNA gyrase and topoisomerase IV, two related enzymes both of which are essential for bacterial DNA replication. Each enzyme is a tetramer composed of two A and two B subunits. In the case of DNA gyrase the subunits are GyrA and GyrB, and in the case of topoisomerase IV the subunits are ParC (or GrlA) and ParE (or GrlB). GyrA is homologous to ParC, and GyrB is homologous to ParE [1]. Quinolones interact with DNA gyrase-DNA complexes and with topoisomerase IV-DNA complexes to trap the enzymes as stabilized reaction intermediates in which broken DNA strands are covalently linked to a tyrosine in the enzyme's active site. These stabilized complexes form a barrier to DNA replication and are necessary but not sufficient for bacterial cell death [2,3]. Under some conditions, such as treatment of cells with the quinolone nalidixic acid together with an inhibitor of protein or RNA synthesis, inhibition of DNA synthesis by the quinolone is unaffected but cell death does not occur [4]. This dissociation has suggested that other factors involving new protein or RNA synthesis are necessary for cell death to occur after interaction of the quinolone with its target enzyme-DNA complex. The specific nature of this other factor(s), however, has remained elusive, and for many newer fluoroquinolones inhibition of protein and RNA synthesis has little or no effect on their bactericidal activity. Release

TABLE 7.1
Target Modifications Resulting in Resistance to Antibiotics in Clinical Use

Mechanism	Drug	Targets	Bacteria Involved	Location of Resistance Determinants	Alteration Affecting Drug Interaction with its Target
Spontaneous mutational changes in a target protein	Quinolones	DNA gyrase and topoisomerase IV	Multiple Gram-negative and Gram-positive species	Chromosomal	Reduced target affinity for drug
	Rifamycins	RNA polymerase	Many Gram-positive bacteria	Chromosomal	Reduced target affinity for drug
	β-Lactams	Transpeptidases (penicillin-binding proteins)	*Streptococcus pneumoniae*	Chromosomal	Reduced target affinity for drug
Generation of mosaic target proteins by homologous recombination	β-Lactams	Transpeptidases	*Streptococcus pneumoniae*	Chromosomal	Reduced target affinity for drug
Target remodeling by acquired metabolic pathways	Glycopeptides	Cell wall peptide side chains with D-Ala-D-Ala	Enterococci and *Staphylococcus aureus*	Plasmids and transposons	Reduced target affinity for drug

Mechanism	Drug	Target	Organism	Genetic basis	Effect
Target overproduction as a means to block access to critical target sites	Glycopeptides	Cell wall peptide side chains with D-Ala-D-Ala	Staphylococcus aureus	Chromosomal	Sequestration of drug and blocking of diffusion to critical target sites
Target modification or protection by acquired proteins	Macrolides, lincomycins, streptogramin B	23S RNA of 50S ribosomal subunit	Multiple Gram-positive species	Plasmids and transposons	Reduced target affinity for drug
	Tetracyclines	30S ribosomal subunit	Multiple Gram-positive and Gram-negative species	Plasmids	Reduced target affinity for drug by target protection
	Quinolones	DNA gyrase and topoisomerase IV	Multiple Gram-negative species	Plasmids and integrons	Reduced target affinity by target protection
Acquired bypass targets	β-Lactams	Transpeptidases	Staphylococci	Chromosomal	Alternative target with reduced drug affinity
	Mupirocin	Isoleucyl tRNA synthetase	Staphylococci	Plasmids	Alternative target with reduced drug affinity
	Trimethoprim	Dihydrofolate reductase	Multiple Gram-positive and Gram-negative bacteria	Plasmids, transposons, and gene cassettes within integrons	Alternative target with reduced drug affinity

of double-strand DNA breaks from the quinolone-trapped enzyme-DNA complex, which might occur with different efficiencies for different quinolones, has been postulated to be the ultimate lethal cellular lesion from which cellular DNA repair mechanisms may be only poorly able to recover [2]. Certain mutants that are bacteriostatically inhibited by otherwise bactericidal antibiotics, such as β-lactams and vancomycin, are also bacteriostatically inhibited by quinolones, suggesting that common cellular pathways mediate the final events that result in cell death after interactions of diverse drugs with diverse targets [5,6]. In the case of *S. pneumoniae*, these pathways have been linked to autolytic activity [6].

Point mutations encoding single amino acid changes in either DNA gyrase or topoisomerase IV can cause quinolone resistance. These resistance mutations have most commonly been localized to the amino terminal domain of GyrA or ParC and are in proximity to the active site tyrosine [7]. This domain has been termed the "quinolone resistance determining region (QRDR)" of GyrA and ParC [8]. The most common site of mutation in GyrA of *Escherichia coli* is at serine (Ser) 83 (or a Ser at equivalent positions of GyrA of other species or equivalent positions of ParC), which may be changed to tryptophan (Trp), leucine, alanine, or other amino acids [9]. Ser83Trp and Ser83Leu mutations of *E. coli* GyrA have been associated with reduced binding of the quinolones norfloxacin and enoxacin to gyrase-DNA complexes [10–12]. Many of the common mutations appear to have little effect on the enzyme's catalytic efficiency [13].

Mutations in specific domains of GyrB and ParE have also been shown to cause quinolone resistance [14,15], although these mutations appear to be substantially less common in resistant clinical bacterial isolates than mutations in GyrA or ParC. GyrB resistance mutations have also been shown to have reduced binding of enoxacin to enzyme-DNA complexes [10]. The QRDR of GyrB (or ParE) appears to be distant from the QRDR of GyrA (or ParC) based on the X-ray crystallographic structure of the homologous enzyme, topoisomerase II of yeast [16]. More recent crystal structures of yeast topoisomerase II, however, have identified other enzyme conformations in which the regions homologous to the QRDRs of GyrA and GyrB are in proximity, suggesting such a conformation may be important for forming the site of quinolone binding [17]. Thus, it appears that mutations in the QRDRs of both GyrA and GyrB act by reducing the affinity of quinolones for the enzyme-DNA complex. Although there are no direct data on quinolone binding to complexes of wildtype and mutant topoisomerase IV, it is presumed that similar mutations in ParC and ParE also reduce quinolone binding affinity for topoisomerase IV-DNA complexes because of the similarity in overall structure, and the strong conservation of amino acid sequence in QRDRs of DNA gyrase and topoisomerase IV.

The magnitude of resistance conferred by a single amino acid change in the subunits of DNA gyrase or topoisomerase IV varies both by bacterial species and by quinolone. The variation in the phenotype of a given resistance genotype relates at least in part to the relative sensitivities of DNA gyrase and topoisomerase IV to a given quinolone. Because quinolone interaction with either enzyme target is sufficient to block cell growth and trigger cell death [2], the level of susceptibility of a wildtype bacterium is determined by the more sensitive of the two target enzymes. Interestingly for many quinolones in clinical use, topoisomerase IV is the more sensitive enzyme in Gram-positive bacteria, and DNA gyrase is the more sensitive enzyme in Gram-negative bacteria [18]. Thus, mutations in the subunits of the more sensitive target enzyme generally occur in first-step mutants, providing a genetic definition of the primary target enzyme [15,19–22]. The magnitude of the resistance increment from such a first-step mutation may be determined by either the magnitude of the effect of the mutation on enzyme sensitivity or the intrinsic level of sensitivity of the secondary target enzyme, whichever of the two is less. Thus, quinolones that have highly similar activities against both DNA gyrase and topoisomerase IV of a given species may require mutations in a subunit of both enzymes before the mutant bacterium exhibits a substantial resistance phenotype [23,24]. Sequential mutations in subunits of both target enzymes have been shown to provide increasing levels of quinolone resistance. In some species in which high-level quinolone resistance is common, such as clinical isolates of methicillin-resistant *Staphylococcus aureus* (MRSA), mutations in subunits of both enzymes are common [25]. There are also several species, *Mycobacterium tuberculosis*, *Helicobacter pylori*, and *Treponema pallidum*, for which genome sequencing has revealed the absence of genes for topoisomerase IV subunits [9], indicating that in these organisms gyrase is likely the only quinolone target. Thus, selection of mutations with substantial resistance phenotypes would be predicted to occur readily in these pathogens, an inference that is supported by clinical data indicating the frequent occurrence of resistance with clinical use of quinolones without use of additional active agents to treat patients with *M. tuberculosis* and *H. pylori* infections.

RIFAMPIN RESISTANCE

Rifampin and related rifamycins, rifabutin, rifapentene, and rifaximin, are inhibitors of the essential bacterial enzyme DNA-dependent RNA polymerase and have been used for treatment of mycobacterial and other bacterial infections or colonizations. RNA polymerase appears to be the sole target of rifampin action. Core RNA polymerase is composed of three subunits, β, β', and α. Core polymerase combines with one of several σ subunits to enable specific binding to promoters and initiation of transcription. Rifampin forms a 1:1 complex with RNA polymerase and blocks the initiation of transcription [26,27].

Resistance to rifampin occurs by mutations in the *rpoB* gene that encode amino acid changes in the β subunit of *E. coli* RNA polymerase [28]. These mutations are clustered in three highly conserved regions of *rpoB* in the midportion of the gene (cluster I—codons 507–511 and 513–533, cluster II—codons 563–564 and 572, and cluster III—codon 687) [28]. These regions appear to be involved in the polymerase antitermination process, because most resistance mutations affect the polymerase readthrough of termination signals, although the importance of this occurrence for the fitness of rifampin-resistant mutants *in vivo* is uncertain [29]. It is presumed that these changes reduce the affinity of RNA polymerase for rifampin, although direct binding studies have not been reported.

Resistance to rifampin has been associated with mutations in similar regions of the *rpoB* genes of *M. tuberculosis* [30], *Mycobacterium leprae* [31], *S. aureus* [32,33], *S. pneumoniae* [34], and *Neisseria meningitidis* [35]. In the case of *M. tuberculosis*, single mutations were associated with high-level resistance to rifampin [30], and single mutations in *S. aureus* selected *in vitro* and *in vivo* have been associated with both high and low levels of resistance depending on the nature of the amino acid change [32,33]. Some clinical isolates of both *S. aureus* and *S. pneumoniae* have been found to have multiple mutations in the cluster regions [34]. Isolates of *Rickettsia typhi* and *R. prowazekii* from patients failing treatment with rifampin have also been found to have homologous mutations, and similar mutations have been found in naturally resistant species of *Rickettsia* [36].

Thus, the ability of single spontaneous mutations to confer high-level rifampin resistance correlates with clinical observations that resistance develops rapidly in clinical settings when rifampin is used alone for therapy of established infections.

LOW-LEVEL PENICILLIN RESISTANCE IN *STREPTOCOCCUS PNEUMONIAE*

β-Lactams target a set of enzymes involved in cell wall biosynthesis and thus, like quinolones, have multiple targets within the bacterial cell. These target enzymes are transpeptidases that crosslink the peptidoglycan lattice providing osmotic and structural stability to the cell [37]. Because these enzymes bind penicillins, they are commonly referred to as PBPs. In *S. pneumoniae*, high-molecular-weight PBPs 1 (1a, 1b) and 2 (2a, 2b, and 2x) are essential for cell viability.

Single amino acid changes in individual PBPs cause only low-level resistance to penicillins and cephalosporins, and perhaps for this reason have been found only in laboratory mutants. Higher levels of resistance, which have occurred in clinical isolates of *S. pneumoniae*, involve another target modification mechanism that will be discussed in the next section. Amino acid substitutions in PBP 2x have been found in laboratory mutants in domains near the penicillin-binding motifs, and often several amino acid changes are required for substantial reductions in the affinity of PBP 2x for penicillin. Increments in the MIC of penicillin, however, were limited to a change from 0.02 to 0.32 μg/mL even in the presence of as many as four amino acid changes in PBP 2x [38]. The need for multiple mutations to reduce drug binding affinity suggests that there are multiple contact points between penicillins and PBP 2x. In addition, the limited resistance phenotype of PBP 2x mutants even when penicillin's affinity for this PBP is reduced suggests that more than one PBP must be changed in order to effect high-level resistance to penicillin by target modification [39]. This circumstance is similar in principle to that for fluoroquinolones interacting with dual targets (as discussed above), in which the most sensitive of multiple essential drug targets (be they mutant or wild type) in a given bacterial cell determines the level of drug susceptibility.

OXAZOLIDINONE RESISTANCE IN ENTEROCOCCI AND STAPHYLOCOCCI

Oxazolidinones inhibit bacterial protein synthesis by interacting with the ribosome and interrupting the formation of the initiation complex of mRNA and the 30S and 50S ribosomal subunits. Linezolid is currently the only oxazolidinone antimicrobial in clinical use, with indications for treatment of infections due to vancomycin-resistant enterococci (VRE) and MRSA. Most human pathogens, including enterococci and staphylococci, have multiple genomic copies of ribosomal RNA (rRNA) genes, and thus resistance caused by mutations in rRNA genes requires that multiple genes be altered, thereby reducing the likelihood of mutational resistance. Nonetheless, linezolid-resistant strains of VRE have been reported to emerge in patients treated with extended courses of vancomycin [40,41] or uncommonly without such exposure [42]; nosocomial transmission of resistant strains has been reported [43]. Similarly, but in lesser numbers, clinical strains of linezolid-resistant MRSA have also been identified [44]. In both enterococci and *S. aureus*, mutations changing G to U at position 2576 of 23S RNA have been found in the resistant clinical isolates. Laboratory-selected resistant strains have also revealed additional resistance mutations in the central region of domain V of 23S RNA, suggesting the specific site of linezolid interaction with the ribosome [45–47]. It is not yet clear to what extent sequential mutations in rRNA genes occurring under selective pressure and/or duplication of the initial mutation in other gene copies by gene conversion contribute to multiple mutated gene copies. In one report, resistance and mutation emerged with linezolid treatment, reverted when linezolid was stopped, and re-emerged with resumption of linezolid therapy [48]. As with any acquired resistance mechanism, the prevalence of resistant strains can be amplified by transmission of resistant strains from patient to patient along with continued selective pressure from antimicrobial use [43].

MACROLIDE RESISTANCE IN MYCOBACTERIA AND *HELICOBACTER PYLORI*

In at least two clinically important bacterial pathogens, mycobacteria and *H. pylori*, there are single genomic copies of rRNA genes. Macrolides are inhibitors of bacterial protein synthesis that interact with the 50S ribosomal subunit. Thus, as noted above for oxazolidinones, resistance to these classes of antimicrobials rarely occurs by mutation (see "Target Modification by Acquired Proteins" below for discussion of the more common mechanism of resistance to macrolides), except in the case of mycobacteria and *H. pylori*. Strains of *Mycobacterium avium* resistant to the macrolide azithromycin emerge when this antimicrobial is used without other active agents for treatment of established infection in patients with HIV/AIDS. Azithromycin has, however, been used for prophylaxis of *M. avium* infections in at-risk patients with HIV/AIDS without emergence of resistance. This is likely due to the low numbers of *M. avium* present, and thus the low likelihood that a spontaneous resistance mutation is present in the bacterial population that is exposed to drug for this application. Similarly, use of clarithromycin for treatment of *H. pylori* infection can result in selection of resistant strains unless additional agents are also used for treatment. In addition, *Mycoplasma* spp. and *Propionobacterium* spp. also have small numbers of copies of rRNA genes. Resistance in all of these species has been associated with mutations in rRNA genes that substitute guanine, cytosine, or uracil in place of adenine at position 2058 [49,50] or alter other bases in the peptidyltransferase circle [51].

GENERATION OF MOSAIC TARGET PROTEINS BY HOMOLOGOUS RECOMBINATION

HIGH-LEVEL PENICILLIN RESISTANCE IN *STREPTOCOCCUS PNEUMONIAE*

As discussed above, individual amino acid changes in the transpeptidase (PBP) enzyme targets of penicillins and cephalosporins cause only limited levels of resistance to these antimicrobials. In addition, β-lactamases, which degrade penicillin, have not been described as a mechanism of β-lactam resistance in *S. pneumoniae*. This organism, however, has been able to develop substantial resistance to penicillins by target altera- tion using a mechanism of transformation and homologous recombination made possi- ble by several distinctive factors. First, *S. pneumoniae* is naturally competent to take up exogenous DNA by the process called transformation. If this DNA has sufficient sequence similarity to DNA on the pneumococcal chromosome, then *S. pneumoniae* can recombine the imported DNA into its chromosomal DNA, creating mosaic genes consisting of segments of both original host and imported DNA. Second, viridans streptococci, many strains of which now contain PBPs with low affinity for penicillins, are genetically related to pneumococci and thus have highly similar genes encoding PBPs. Third, viridans streptococci are normal inhabitants of the upper respiratory tract, which may also contain pneumococci during periods of colonization or infection, providing a natural opportunity for exchange of genetic information between these organisms.

Clinical isolates of *S. pneumoniae* with reduced susceptibility to penicillin and other β-lactams have been found to have such mosaic genes encoding modified PBP 2x, PBP 2b, and PBP 1a [39]. DNA encoding mosaic PBP 2b from penicillin- resistant strains of *S. pneumoniae* is capable of transforming a recipient pneumococcus that has preexisting changes (causing low affinity for penicillin) in PBPs 1a and 2x to higher levels of resistance to penicillin [52]. That *S. pneumoniae* without altera- tions in PBPs 1a and 2x cannot be similarly transformed illustrates the requirements for multiple changes in PBPs necessary to effect high-level resistance. A set of spe- cific amino acid changes at positions 371 and 575–577 found in the mosaic segments of PBP 1a from all resistant clinical isolates from South Africa have been shown genetically to contribute to resistance [53].

In the case of the mosaic gene encoding PBP 2b from a penicillin-resistant pneumococcus, the non-pneumococcal segments of DNA most closely resemble similar segments of DNA from the gene for PBP 2b from penicillin-susceptible strains of *Streptococcus mitis*. The mosaic pneumococcal gene, however, also contains other segments of DNA from an unidentified third species [52]. Clinical isolates of penicillin-resistant and penicillin-susceptible strains of other viridans streptococci, such as *Streptococcus sanguis* and *Streptococcus oralis*, as well as *S. pneumoniae* have been found to have mosaic genes for PBPs, highlighting the extent of genetic exchange, but leaving uncertain the directionality of transfer and the original source of the gene segments causing resistance.

The extensive modification of multiple PBPs necessary to cause high-level penicillin resistance in clinical isolates of pneumococci may come at a price. In this regard, the catalytic activity of PBP 2x purified from a resistant clinical isolate has been shown to be substantially lower than that of PBP 2x purified from a susceptible isolate [54]. Since penicillin-resistant pneumococci remain capable of colonizing and infecting humans, it is presumed that they have acquired compensatory mecha- nisms, as yet poorly defined, to ensure their fitness *in vivo*. It has also been suggested that the stringency of these compensatory requirements may be responsible for the limited number of serotypes of penicillin-resistant pneumococci, which have never- theless been quite successful in spreading throughout the world [37].

TARGET REMODELING BY ACQUIRED METABOLIC PATHWAYS

GLYCOPEPTIDE RESISTANCE IN ENTEROCOCCI AND STAPHYLOCOCCI

Vancomycin, teicoplanin, and other glycopeptides bind to components of the bacterial cell wall, which is a peptidoglycan lattice composed of polymers of alternating *N*-acetyl glucosamine and *N*-acetyl muramic acid residues that are cross-linked via attached short peptide chains. The peptide chains attached to the polymer backbone are the substrates for the crosslinking reaction, which is catalyzed by transpeptidases (PBPs). Vancomycin binds specifically to the terminal two D-alanine (D-Ala) residues of the peptide side chain. This binding inhibits several reactions in cell wall biosynthesis, including transfer of precursors from a membrane lipid carrier to the peptidoglycan backbone, D,D-carboxypeptidase activity, and transpeptidase activity. To alter the target of vancomycin thus requires remodeling of peptidoglycan structure.

Acquired vancomycin resistance results from substitution of D-lactate (D-Lac) for the terminal D-Ala of the peptide side chain, a change that decreases vancomycin affinity for the cell wall by 1000-fold [55]. Cellular transpeptidases in most cases remain able to catalyze crosslinking using peptide side chains terminated in D-Ala-D-Lac. The production of a cell wall with peptide side chains terminated in D-Ala-D-Lac is engineered in both *Enterococcus faecium* and *Enterococcus faecalis* by acquired clusters of genes located on mobile genetic elements, the best studied of which is transposon Tn*1546* [56]. Three of the eight genes identified on Tn*1546* are necessary for vancomycin resistance [57]. *vanA* encodes a ligase enzyme that catalyzes attachment of D-Lac to D-Ala. *vanH* encodes a dehydrogenase that catalyzes production of D-Lac from pyruvate. *vanX* encodes an enzyme that hydrolyzes D-Ala-D-Ala precursors, thereby blocking parallel production of normal peptide side chains with D-Ala-D-Ala termini, which could serve as residual targets for binding of vancomycin [58]. Two other genes encode accessory proteins not required for vancomycin resistance. *vanY* appears to encode a D,D-carboxypeptidase that hydrolyzes the terminal D-Ala on the cytoplasmic precursor peptide side chain [59]. When VanX blocks synthesis of D-Ala-D-Ala terminated peptides, VanY may have little effect on resistance. In contrast, under conditions in which overall activity of VanX is limited, VanY may additionally remove residual D-Ala-D-Ala peptides as vancomycin targets. The *vanZ* gene appears to contribute to resistance to teicoplanin, but not vancomycin when VanH, VanA, and VanX are produced at low levels [60].

In Tn*1546*, *vanH*, *vanA*, and *vanX* are cotranscribed from a promoter located between two upstream genes, *vanR* and *vanS*, which encode, respectively, response and sensor proteins of a two-component regulatory system [57,61–63]. Expression of the *van* gene cluster is thought to be regulated by this sensor-response system, but the signal for induction of expression is not yet known. Induction of expression of high-level glycopeptide resistance by exposure to either vancomycin or teicoplanin is characteristic of the VanA phenotype of the Tn*1546* resistance element, but induction can also occur with exposure to structurally unrelated, non-glycopeptide antibiotics [64]. Thus, induction is not solely due to the structural features of vancomycin or teicoplanin. A related resistance phenotype, VanB, is distinguished by a lower level of resistance and a lack of inducibility by teicoplanin and is encoded on other mobile elements by a cluster of genes (designated as *vanB* instead of *vanA* for the ligase genes and for the other homologous genes with a subscript B) similar to that found on Tn1546 [65–69]. Vancomycin but not teicoplanin is an inducer of expression of $vanR_B$-$vanS_B$, thereby providing an explanation for the lack of induction of resistance by teicoplanin in VanB strains [65]. VanC, a third phenotype of intrinsic constitutive low-level resistance to glycopeptides found in *Enterococcus gallinarum,* is due to peptide side chains terminated in D-Ala-D-serine [70], and naturally highly resistant species, such as *Lactobacillus casei, Leuconostoc mesenteroides,* and *Pediococcus pentosaceus* have been shown to have D-Ala-D-Lac termini on their peptide side chains [70–72].

The genetic and mechanistic complexity of vancomycin resistance represents a remarkable feat of natural genetic engineering that suggests that bacterial plasticity given sufficient time and selective advantage is likely to be capable of circumventing the activity of any antimicrobial agent developed for clinical use, that is, resistance may ultimately be inevitable, and it is only a matter of how rapidly or slowly it emerges, as determined by mechanistic and epidemiologic factors. The original source of the vancomycin resistance gene cluster remains uncertain. Homologs of *vanH, vanA,* and *vanX* have been found in *Streptomyces toyocaensis, Amycolatopsis orientalis* (in the same orientation as in enterococci), and other glycopeptide-producing species of bacteria, suggesting the possibility that the original evolution of such gene clusters might have occurred in these or related species as a means of protection from their own antimicrobial products [73]. The G+C content of these genes, however, differs in *S. toyocaensis* and *A. orientalis* (65%) in comparison to enterococci (44% to 49%), and in enterococci the G+C content of *vanH, vanA,* and *vanX* is 5% to 10% higher than that of the flanking genes *vanR, vanS, vanY,* and *vanZ.* Thus, the *vanHAX* gene cluster could have been mobilized *en bloc* from some donor species, but the exact nature of the donor remains uncertain, as does the origin of a possible intermediate recipient organism containing *vanR, vanS, vanY,* and *vanZ.* There also may be heterogeneity in such a putative intermediate recipient, since some resistant enterococci lack *vanZ* and the location of *vanY* varies between VanA and VanB resistance elements [65]. If the source of *vanHAX* is from one of the glycopeptide-producing species, then considerable evolution of these genes must have occurred before they appeared in enterococci, and the organism(s) in which such evolution occurred remains to be defined.

In some early laboratory experiments, it was possible to transfer vancomycin-resistance determinants from *E. faecalis* to a methicillin-susceptible strain of *S. aureus* [74]. Subsequently, transfer appears to have occurred in clinical settings in only a few instances in which vancomycin-resistant *E. faecalis* or *E. faecium* transferred resistance to MRSA [75,76]. An analysis of the vancomycin-resistant MRSA isolates indicated that additional genetic events occurred in these isolates, in particular the mobilization of the vancomycin resistance gene cassette onto an endogenous staphylococcal plasmid, possibly reflecting poor stability of the enterococcal plasmid in *S. aureus* [77]. In the two vancomycin-resistant *S. aureus* isolates studied in most detail, one contained the intact Tn1546 transposon of *E. faecalis* and the other had a modifed Tn1546 with a partial deletion of one region and inversion of another, establishing that the two isolates likely represented separate transfer events [78]. Once established, the staphylococcal plasmid carrying vancomycin-resistance elements was capable of transfer to other staphylococci in the laboratory [77]. Transmission of vancomycin-resistant clinical MRSA isolates from person to person, however, has not yet been documented [76].

TARGET OVERPRODUCTION AS A MEANS OF BLOCKING ACCESS TO CRITICAL TARGET SITES

INTERMEDIATE GLYCOPEPTIDE RESISTANCE IN STAPHYLOCOCCI

In addition to high-level glycopeptide resistance in the few isolates of S. *aureus* that acquired the enterococcal vancomycin resistance gene cassette, other clinical isolates of staphylococci have been found to have reduced susceptibility to glycopeptides by different mechanisms. The mechanisms of intermediate resistance to vancomycin in staphylococci may differ between coagulase-negative and -positive species, and complete details have not been elucidated in any species.

In S. *aureus*, some clinical isolates from patients in Japan and the United States who have failed vancomycin therapy have been found to have either (*i*) heterogeneously expressed low-level resistance to vancomycin in subpopulations of cells that cannot be detected by standard MIC testing methods in clinical laboratories (MIC of vancomycin = 2 μg/mL) [79], or (*ii*) a higher level of resistance (MIC = 8 μg/mL), which is also difficult to detect by disk-diffusion, but not MIC methods [80,81]. In a case-control study, patients with MRSA infections caused by isolates with MICs of vancomycin of 4 or 8 μg/mL were compared with those with MRSA infections caused by isolates with vancomycin MICs of ≤2 μg/mL. Attributable mortality in the two groups was 63% and 12%, respectively, and higher vancomycin MICs were associated with prior vancomycin use [82]. Recently, the clinical laboratory breakpoint criteria for identification of vancomycin-intermediate S. *aureus* have been modified such that only isolates with MICs of vancomycin of ≤2 μg/mL are classified as susceptible.

Common among these strains has been preexisting methicillin resistance and prolonged exposure to vancomycin *in vivo*. The prototypic Japanese vancomycin-heteroresistant and -resistant strains, Mu3 and Mu50, respectively, have been shown to differ from fully vancomycin-susceptible strains in several features, including enhanced incorporation of N-acetylglucosamine into the cell wall, an increased pool of the cytoplasmic cell wall precursor monomer UDP-N-acetylmuramyl-pentapeptide, enhanced autolysis, and increased production of PBP 2 [83]. Mu50, the more resistant strain, exhibited, in addition, a two-fold increased thickness of the cell wall, a substantially higher proportion of peptide side chains in which the glutamine residue is non-amidated (side chains composed of L-Ala-D-Glu-L-Lys-D-Ala-D-Ala instead of L-Ala-D-Gln-L-Lys-D-Ala-D-Ala), and a slight increase in the number of uncrosslinked peptide chains, the latter possibly due to the preference of PBPs for crosslinking the normal rather than the non-amidated peptide side chains [84]. The observed increased numbers of uncrosslinked pentapeptide chains, which retain their terminal D-Ala-D-Ala binding site for vancomycin (crosslinking removes these residues) and an increased thickness of the cell wall has led to the proposal of the false target model [85]. In this model, vancomycin binding to increased numbers of non-critical D-Ala-D-Ala targets present in the thickened and poorly cross-linked cell wall protects the critical D-Ala-D-Ala target sites at the point of action of the transglycosyl-ase enzymes near to the cell membrane. It has been estimated that Mu50 has a three-fold increase in free D-Ala-D-Ala termini. Increased binding of vancomycin to this strain has been reported [85], and more recently anomalous vancomycin diffusion was demonstrated, suggesting a phenomenon in which binding of vancomycin molecules to distal termini reduces or clogs vancomycin diffusion to key drug target sites [86]. Thus, vancomycin is sequestered by binding to extra target D-Ala-D-Ala sites far from the site of new cell wall synthesis adjacent to the cell membrane, and this binding, in addition, reduces drug diffusion across the modified peptidoglycan [86].

The genetic determinants of this mechanism of resistance are multiple and have not yet been fully defined, but 17 genes were found to be overexpressed in a compari-son of six pairs of vancomycin-resistant and -susceptible strains and were shown to contribute to a low-level resistance phenotype when overexpressed in strain N315 [87]. Of these genes, three, *graF*, *msrA*, and *mgrA*, were also shown to contribute to increased resistance to oxacillin, and *graF* and *msrA2* overexpression were shown to produce increased cell wall thickness in association with vancomycin resistance. Overall, many of these 17 genes appeared to be involved in cell wall biosynthetic pathways (*murA*, *murZ*, *sgtB*, and *gcaD*), nutrient uptake (*lysC*, *asd*, *dapA*, *dapD*, *proP*, *opuD*, and *opp-2B*), and regulatory functions. Since it is possible to select resis-tance at the level of Mu50 from Mu3 by growth in the presence of vancomycin in the laboratory, acquired genes do not appear to be necessary for these steps. Laboratory strains of *S. aureus* selected for resistance by repetitive exposure to vancomycin (MIC = 5 µg/mL) have exhibited some of the properties of Mu3 and Mu50, including increased production of PBP2, an increased muropeptide monomer pool, and thicken-ing of the cell wall [88]. These strains, however, lacked non-amidated pentapeptide chains or alterations in crosslinking. *femC* mutants of *S. aureus*, which have reduced glutamine synthetase activity, have been found to have increased numbers of non-amidated peptide side chains [84], so that *femC*-type mutations might be contributory. The *mecA* gene, which is necessary for resistance to methicillin and encodes PBP2A (see below), appears not to be necessary for vancomycin intermediate resistance, since this resistance persists after excision of *mecA* [89]. The association of vancomycin resistance with MRSA strains is likely due to the importance of exposure to vanco-mycin in selecting resistance, since vancomycin would be used commonly for treat-ment of patients with MRSA infections and less often for treatment of patients with methicillin-susceptible strains of *S. aureus*.

There is additional complexity and probably other mechanisms of resistance to glycopeptides in coagulase-negative staphylococci as well as additional properties of teicoplanin resistance that differ from those of vancomycin resistance in *S. aureus* that are beyond the scope of this chapter. Thus, multiple and complex mechanisms may occur in different settings, but the model of resistance suggested by the results in Mu3 and Mu50 strains of *S. aureus* represents another potential variation on the altered target mechanism. In this case, in contrast to high-level vancomycin resistance in enterococci and staphylococci in which the target is remodeled, vancomycin resis-tance in *S. aureus* appears to involve target overproduction that titrates vancomycin away from critical target sites, and also blocks its further diffusion to sites near the cell membrane at which its binding blocks peptidoglycan synthetic enzymes.

TARGET MODIFICATION AND PROTECTION DUE TO ACQUIRED GENES AND PROTEINS

MACROLIDE, STREPTOGRAMIN, AND LINCOSAMIDE RESISTANCE

Macrolides, streptogramins, and lincosamides inhibit bacterial protein synthesis by binding to the 23S rRNA component of the bacterial 50S ribosomal subunit [90]. One of the most common mechanisms of resistance to these three classes of antimicrobials involves posttranscriptional alteration of a specific base of rRNA that results in ribosomes with reduced drug affinity [51]. Specifically, methylation of an adenine at position 2058 of 23S rRNA is accomplished by members of the Erm family of N^6 methyltransferases, the family name being an acronym for erythromycin resistance methylase. Dimethylated A2058 causes high-level resistance to all generations of macrolides, to lincosamides such as clindamycin, and to the group B streptogramins represented by pristinamycin and quinupristin, which is a derivative of pristinamycin A_1 and a component of quinupristin-dalfopristin, which is in clinical use [91]. This resistance phenotype is referred to as MLS_B. The methylated base is located in the peptidyltransferase loop of domain V of the 23S RNA, which is thought to contain at least part of the site of macrolide binding to the ribosome [51].

The new ketolide class of antimicrobials is related to the macrolides and appears to interact with the ribosome in domain V, which contains A2058, and at domain II, particularly position A752 [92]. Monomethylation of A2058 of the ribosomes of streptococci does not have a major effect on the activity of telithromycin, the only currently marketed ketolide [93]. Dimethylation of A2058, however, produces resistance to telithromycin despite the drug's interaction with domain II [94]. Macrolide-resistant strains of streptococci with *erm* genes that are susceptible to telithromycin appear to have either inefficient ribosomal methylation, or macrolide-inducible but not ketolide-inducible expression of the *erm* genes [95]. As with streptococci, staphylococci with constitutively expressed, but not inducible *erm*-mediated resistance to macrolides are also resistant to telithromycin [96].

erm genes are most often acquired on plasmids and may have arisen from macrolide-producing actinomycetes [97]. Enzymatic methylation rather than mutational modification of rRNA at A2058 is important in many bacteria, because they contain multiple copies of rRNA genes, placing a requirement for multiple mutations were resistance to occur by chromosomal mutation alone. *erm* genes may also be present on the chromosome of macrolide-resistant streptococci.

The expression of Erm methylases is often inducible by low concentrations of erythromycin, but not by the ketolide, telithromycin [95]. Expression of *erm* is negatively regulated by transcriptional and translational attenuator mechanisms related to the secondary structure of the *erm* mRNA leader sequence. In the absence of erythromycin, stem-loop structures of the mRNA mask the first two codons and the ribosome binding site of *ermC*, thereby reducing the efficiency of translation of ErmC methylase [90]. Upstream of *ermC* is an open reading frame encoding a 19-amino acid leader peptide, translation of which stabilizes the stem-loop conformation that masks critical sites for *ermC* translation downstream. In the presence of erythromycin bound to a ribosome, translation of this leader peptide is stalled, destabilizing the blocking stem-loop structures and facilitating a change in the conformation

of the attached downstream mRNA that unmasks the sites critical for *ermC* translation. The unmasked *ermC* transcript must then dissociate from the erythromycin-
bound ribosome in order for translation to occur on other ribosomes to which
erythromycin has not bound. Thus, the low concentrations of erythromycin associated with induction may allow the presence of both erythromycin-bound and unbound
ribosomes within the cell. Once sufficient ErmC methylase is translated, then many
ribosomes can be modified and high-level resistance to erythromycin ensues.

AMINOGLYCOSIDE RESISTANCE

Resistance to aminoglycosides among clinical bacterial isolates has generally been
due to acquisition of genes encoding enzymes that modify the structure of specific
aminoglycosides, but recently methylation of 16S ribosomal RNA has been found in
Gram-negative clinical isolates as a mechanism of high-level resistance to a broad
range of aminoglycosides [98,99]. Methyltransferases conferring aminoglycoside
resistance in various species of actinomycetes that produce these antibiotics have
been previously recognized as protection mechanisms in these producer organisms,
but it was not until 2003 that genes encoding other methyltransferases were found to
cause aminoglycoside resistance in *Pseudomonas aeruginosa* [99] and a range of
enteric bacteria [98,100,101]. *armA*, found initially in a resistant strain of *Klebsiella
pneumoniae*, encodes a methyltransferase thought to methylate G1405 of 16S
ribosomal RNA, based on the concordance of its phenotype of resistance to 4,6-
di-substituted deoxystreptamines (gentamicin, tobramycin, netilmicin, amikacin,
and isepamicin) and fortimicin and that conferred by methylation of G1405 by other
means [98]. The *armA* gene has also been found in a broad range of enteric bacteria
and to be present on a plasmid within a composite transposon together with other
genes encoding resistance to streptomycin-spectinomycin, sulfonamides, and trimethoprim, likely accounting for its dissemination in Europe and the Far East [101–103].
The origin of *armA* is uncertain since its guanine + cytosine content differs from
that of actinomycetes.

Several other 16S RNA methyltransferases have also been identified in Gram-
negative bacteria. RmtA was found in broadly aminoglycoside-resistant isolates of
P. aeruginosa from Japanese hospitals and showed 30% to 35% amino acid identity
to methyltransferases in producer actinomycetes [99]. Extracts of cells containing
cloned *rmtA* were able to methylate 16S rRNA. Related methyltransferases encoded
by *rmtB* and *rmtC* have now been identified on conjugative plasmids in several species
of enteric bacteria in Japan and Taiwan [100,101,104]. Thus, ribosomal protection as an
additional mechanism of aminoglycoside resistance has emerged in human and veterinary Gram-negative pathogens in recent years. This mechanism has the potential of
substantial clinical importance because of the range of aminoglycosides affected,
including amikacin, which has heretofore often remained active in the face of resistance to gentamicin and tobramycin conferred by specific drug-modifying enzymes.

RIBOSOMAL PROTECTION FROM TETRACYCLINES

In addition to target modification, target protection without modification by acquired
proteins can occur and is one of the two principal mechanisms of resistance to tetracyclines. Tetracyclines reversibly inhibit bacterial protein synthesis by disrupting the

interaction of aminoacyl-tRNA with the ribosome [105]. Tetracyclines bind to a high affinity site on the 30S ribosomal subunit, although low affinity sites have been identified on both 30S and 50S subunits. Binding appears to involve the S7 ribosomal protein, but other ribosomal proteins such as S3, S8, S14, and S19 appear to contribute to an optimal drug-binding conformation [106–108]. Drug binding is also thought to be in proximity to the 16S rRNA component of the ribosome.

Tetracycline resistance determinants TetM, TetO, and OtrA are proteins that interact with ribosomes to protect them from the action of tetracycline, and structurally similar TetS, TetT, TetQ, TetB(P), TetW, Tet(32), and Tet(36) are thought to function in the same way. These resistance determinants are found on plasmids and transposons and have been identified in Gram-positive and Gram-negative bacteria [109–113]. Best studied of these determinants is *tetM* encoding the TetM protein in streptococci. Ribosomes isolated from *tetM*-containing cells are resistant to tetracycline in *in vitro* translation systems if extracted under low-salt but not high-salt conditions, and purified TetM protein confers tetracycline resistance on ribosomes isolated from tetracycline-susceptible cells [114]. Binding of tetracycline to the ribosome, however, is not altered by TetM, and resistance occurs even when ribosomes are in substantial excess of TetM *in vitro*, suggesting that TetM acts catalytically [115]. TetM and TetO have structural similarities to elongation factors EF-Tu and EF-G, which bind ribosomes, and also have GTPase activity. TetM mediates the release of tetracycline from the ribosome in a GTP-dependent manner [116], and TetM and EF-G compete for binding to the ribosome [117].

Host factors also appear to be important in that chromosomal *miaA* mutations, which cause defects in Δ^2-isopentenylpyrophosphate transferase, an enzyme that modifies adenosine at position 37 of tRNA, had partial loss of tetracycline resistance in *tetM* cells [115], and *rpsL* mutations causing streptomycin resistance by alteration in the S12 protein also reduce *tetM*- and *tetO*-mediated tetracycline resistance [118]. Modification of base A37 of tRNA is important for accuracy of translation, but it is uncertain if the effects of *miaA* mutations on tetracycline resistance reflect a direct or indirect effect on tetracycline interaction with the ribosome [119]. Current hypotheses about the exact mechanism of ribosome protection include an interaction between TetM and the ribosome that eliminates the binding of tetracycline specifically to ribosomes that are actively engaged in protein synthesis or that allows aminoacyl-tRNA to enter the A site on the ribosome in the presence of tetracycline [120,121].

Expression of tetracycline resistance by *tetM*, like expression of *erm*-mediated resistance to macrolides, is inducible and regulated by transcriptional and translational attenuation [122]. Upstream of *tetM* is an open reading frame (ORF) that encodes two regions of GC-rich RNA inverted repeats flanked by a series of uracil (U) residues, promoting formation of hairpin secondary structures, which cause RNA polymerase to pause. The series of U residues downstream also produces an unstable RNA/DNA hybrid that facilitates destabilization of binding of RNA polymerase. In the absence of tetracycline, translation of the ORF is thought to be retarded because five of the first eight codons require rare aminoacyl-tRNAs. Read-through transcription of *tetM* is more likely to occur if the translating ribosome is in proximity to the transcription complex, thereby destabilizing the hairpin structures that would otherwise retard transcription [123]. In this model, in the presence of tetracycline as an inducer, the A and P sites on the ribosome are occupied by drug. Transcription and translation are thus delayed and availability of aminoacyl-tRNAs is increased, allowing proximity of the ribosome to the transcribing RNA polymerase and subsequent transcription and translation of *tetM*. Other tetracycline resistance determinants, such as *tetO* of *Campylobacter* spp., however, are not inducible by tetracycline and appear to be expressed constitutively [124].

Most recently tigecycline, a glycylcycline derivative of minocycline, has been released for clinical use. Tigecycline remains active against strains containing *tetM* and other established mechanisms of plasmid-mediated tetracycline resistance [125,126].

GYRASE PROTECTION FROM QUINOLONES

As noted above, quinolone resistance can occur by mutations in the subunits of DNA gyrase and DNA topoisomerase IV that reduce drug binding to enzyme-DNA complexes. In addition, more recently described has been low-level quinolone resistance conferred by the plasmid-encoded Qnr proteins, which interact with DNA gyrase and topoisomerase IV and reduce quinolone action [127–129]. Thus far three major Qnr proteins have been identified, QnrA [128], QnrB [130], and QnrS [131], with additional minor variants (e.g., QnrA1, QnrA2, etc.) within each group that differ by a few amino acids. All Qnr proteins are members of the pentapeptide repeat family of proteins, as are McbG, a plasmid-encoded protein that protects gyrase from microcin B17 in strains that produce this microcin, and MfpA, a chromosomally encoded protein involved in quinolone resistance in *Mycobacterium smegmatis* and *M. tuberculosis* [132]. The crystal structure of MfpA was found to resemble the structure of DNA, and QnrA has been shown to reduce DNA binding to gyrase, suggesting that MfpA and QnrA effect quinolone resistance by competing with DNA for formation of gyrase-DNA complexes, which are the target of quinolones. Reduction of inhibitory gyrase-DNA-quinolone complex formation thus has been proposed to confer resistance to quinolones, which function as poisons of bacterial DNA synthesis [133]. Although MfpA has been shown to inhibit gyrase function [133], QnrA and QnrB protect from quinolone action at concentrations at which there is no detectable gyrase inhibition [127,130], suggesting that additional functions of these proteins may be important for effecting resistance. The effects of Qnr on binding of quinolones to gyrase-DNA complexes has not yet been studied. Since quinolones are synthetic antimicrobial agents and are unlikely until recently to have provided selective advantage to bacteria bearing Qnr proteins, it has been suggested that the normal function of Qnr and related pentapeptide repeat proteins is in regulation of the function of cellular DNA-binding proteins [133].

Genes encoding QnrA variants have been found on the chromosome of the environmental water organism *Shewanella algae* [134], and other related genes have been found in the genome sequences of *Photobacterium profundum*, *Vibrio parahemolyticus*, and *Vibrio vulnificus*, which also live in water habitats [135]. Plasmid-encoded *qnrA* and *qnrB* both now circulate on multidrug resistance plasmids found in a variety of species of enteric bacteria, including *Klebsiella pneumoniae*, *Enterobacter* spp., *E. coli*, and *Citrobacter freundii* throughout the world [136–139]. Although these genes confer only low-level resistance alone, they facilitate selection of higher levels of quinolone resistance due to chromosomal mutations without affecting strain mutability [140]. They have been found both in strains classified as susceptible as well as those classified as resistant to quinolones by clinical microbio-logic criteria, and in the former instance could be insidiously promoting selection of quinolone resistance [136]. Both *qnrA* and *qnrB* have been found on type 1 integrons and are linked within these structures to genes encoding a variety of resistances to other antimicrobial agents, particularly β-lactams and aminoglycosides [130,141]. *qnrS* has been found in strains of *Shigella* spp. and *Salmonella* spp. and like *qnrA* and *qnrB* appears to be transferable to other bacteria. *qnrS* has thus far been found on transposon-like structures, but not on integrons [131]. Thus, plasmid-encoded quino-lone resistance due to target protection now commonly augments established chromo-somal target mutations in clinical isolates of Gram-negative bacteria. No such quinolone resistance proteins have yet been identified in Gram-positive bacteria.

BYPASS TARGETS

METHICILLIN RESISTANCE IN STAPHYLOCOCCI

Methicillin, other semisynthetic penicillins, such as oxacillin and nafcillin, and most cephalosporins are not degraded by the common staphylococcal penicillinase enzyme, and thus many of these antibiotics are often used as antistaphylococcal agents. Some strains of staphylococci, including both *S. aureus* and coagulase-negative staphylococci, however, are resistant to methicillin by a mechanism that renders them also resistant to all current β-lactam antibiotics. Methicillin-resistant strains of staphylococci contain an insertion of DNA containing the *mecA* gene that occurs at a specific site on the chromosome, *orfX*, near the origin of replication. *mecA* encodes a transpeptidase, PBP 2a, which has low affinity for β-lactams and appears to be capable of serving the functions of the essential native transpeptidases of staphylococci. Thus, PBP 2a serves as a single, resistant bypass enzyme for the several normal targets of β-lactams in staphylococci [142]. The insertion elements that contain the *mecA* gene range in size from 21 to 67 kb and are referred to as SCC*mec* (staphylococcal chromosome cassette *mec*), which also contains type-specific *ccr* genes, which encode for recombinases that mediate integration and excision of the element at *orfX* [143]. Insertion elements also within SCC*mec*, such as IS*431*, function as the site of integration of plasmids and transposons carrying other antibiotic resistance elements. SCC*mec* types I, II, and III are found in health-care-associated strains and typically have additional resistance determinants in addi-tion to *mecA*. SCC*mec* types IV and V have been found in community-associated strains and typically have *mecA* as their sole resistance determinant [144].

Expression of methicillin resistance varies among strains of staphylococci [145]. Two regulatory genes, *mecI* and *mecR*, immediately upstream of the *mecA* promoter are transcribed divergently. *mecI* is homologous to *blaI*, which encodes a repressor of β-lactamase expression, and *mecR* is homologous to *blaR*, which encodes a pro-tein that binds penicillins and leads to transcription of *blaZ*, the structural gene for β-lactamase, thereby leading to expression of β-lactamase. Thus, MecI and MecR proteins are thought to perform analogous functions to those of the BlaI and BlaR proteins, respectively, in regulating expression of PBP 2a. Mutation and disruption of *mecI* and *mecR* or mutations in the *mecA* promoter are now commonly found in clinical methicillin-resistant isolates of *S. aureus* such that production of PBP 2a is usually sufficient to provide a resistance phenotype [146]. This phenotype, however, can vary due to other as yet incompletely defined factors that do not correlate simply with the level of PBP 2a production [147,148]. BlaI and BlaR may also be involved in regulation of expression of PBP 2a when MecI and MecR are altered [149].

The methicillin resistance phenotype may be either homogeneous or heterogeneous. Heterogeneous resistance results in a varying level of resistance depending on the culture conditions and the β-lactam antibiotic used, and often only one in 10^6 cells in a population may express high-level resistance [150,151]. This proportion is higher (1 in 10^2) if cells are grown at 30°C or in medium supplemented with NaCl [152]. Stably homogeneously resistant strains can be selected from heterogeneous strains by passage on β-lactam antibiotics (see below) [153].

Several chromosomal loci have been identified that are important to allow for expression of methicillin resistance [154]. Among these, the *fem* (factors essential for methicillin resistance) genes encode proteins involved in the synthesis of the pentaglycine crosslinking chains of the muramyl peptide component of the cell wall [145]. The chains are involved in the transpeptidase reactions catalyzed by PBP 2a and other PBPs to generate crosslinking of the peptidoglycan. For example, the *femA* gene encodes an enzyme necessary to add the second and third glycines of the pentaglycine side chain [155]. Other mutants, such as *fmtA* have in common with *fem* mutants alterations in cell wall structure [156]. That *fem* mutants express methicillin resistance poorly or not at all implies that PBP 2a functions poorly (or at least less well than native PBPs) in the setting of such modified cell wall precursor structures. The specific molecular interactions of PBP 2a with its substrates, however, remain to be defined.

The genetic basis for the change from heterogeneous to homogeneous methicillin resistance is beginning to be defined. Stable mutants of *S. aureus* with homogeneous resistance can be selected from heterogeneously resistant strains after single exposure to β-lactams [157]. These mutants, termed *chr**, have mutation(s) at several non-*fem* and non-*mec* loci. Extragenic revertants of *fem* mutants can also express high-level homogeneous resistance and differ from the *chr** mutants. Thus, multiple types of mutations may allow homogeneous expression of methicillin resistance. In some cases, these mutants have had alterations in genes encoding enzymes with cell wall lytic activity thought to be involved in normal cell wall remodeling [158], including the *lytH* gene [154]. and *hmrA*, which encodes a putative aminohydrolase [159]. The means by which reduced lytic enzymes or other changes contribute to homogeneous methicillin resistance remains to be determined, but it is reasonable to speculate that, as with the *fem* mutants, some may involve cell wall structural changes that improve rather than impair the function of PBP 2a.

The site-specific insertion of *mec* DNA in the chromosome of *S. aureus* appears to be an infrequent event, since most clinical MRSA strains appear to have a clonal lineage [160]. In staphylococci, *mec* is located within the larger genetic element SCC-*mec* [161] of which three major types exist among classical strains that have circulated in healthcare settings. The original source of *mec* DNA is uncertain, but has been suggested to be coagulase-negative staphylococci [162] with the most closely related *mec* homolog found in *Staphylococcus sciuri* [163,164]. As with high-level resistance to penicillin in pneumococci, the apparent infrequency of the genetic events generating new resistant clones has not been a barrier to the persistence and dissemination of methicillin resistance in staphylococci in hospital settings worldwide. More recently, in addition there has been a rapid emergence of MRSA strains with the novel type IV SCC*mec* element IV that have spread in community settings throughout the United States and elsewhere [165].

MUPIROCIN RESISTANCE IN STAPHYLOCOCCI

Mupirocin (pseudomonic acid) is an antibiotic derived from cultures of *Pseudomonas fluorescens* that inhibits isoleucyl tRNA synthetase and thus indirectly inhibits bacterial protein synthesis by depriving the cell of the ability to incorporate a common amino acid into protein. Mupirocin is used in topical preparations to eradicate nasal colonization with *S. aureus* and for treatment of certain staphylococcal skin infections or colonizations. High-level resistance to mupirocin in *S. aureus* has been found to be encoded on a variety of transferable plasmids [166–168]. Two isoleucyl tRNA synthetase activities have been isolated from high-level resistant strains. One was also found in susceptible strains and those with low-level resistance. The other was found only in high-level resistant strains, and it had substantially reduced sensitivity to mupirocin [169,170]. A gene, *ileS2* (originally termed *mupA*), encoding a mupirocin-resistant enzyme was identified on resistance plasmids and found to encode a protein with 57% identity and 30% similarity in amino acid sequence relative to the native IleS protein [171]. The origin of the *ileS2* gene is uncertain, but it has been found on plasmids in both *S. aureus* and coagulase-negative staphylococci and has also been found in a chromosomal location in some strains of *S. aureus* [172,173]. High-level mupirocin resistance has been transferred between coagulase-negative staphylococci and *S. aureus* on plasmids [174,175]. On some plasmids, *ileS2* is flanked by direct repeats of the insertion sequence IS257 [176], suggesting that it is located on a transposable element.

Low-level mupirocin resistance can occur on exposure of *S. aureus* to mupirocin by selection of mutations in the native *ileS* gene. A mutation changing valine at position 588 to phenylalanine is common and appears to confer little fitness cost [177]. Further selections for higher level resistance are associated with additional mutations in *ileS*, some of which have apparent fitness costs, but compensatory mutations can mitigate these costs [178]. In addition, some strains of *S. aureus* with low levels of mupirocin resistance also appear to contain the usually plasmid-encoded *ile2* gene on the chromosome [168,173,179,180].

Thus, an acquired resistant isoleucyl tRNA synthetase that bypasses the sensitive native enzyme allows protein synthesis to proceed in the presence of mupirocin. That the level of resistance to mupirocin is higher when the gene encoding this enzyme is located on plasmids relative to the chromosome suggests that differences in gene expression due either to plasmid copy number or to differences in promoter strength or regulation of expression in the two locations may be responsible for the different levels of resistance.

TRIMETHOPRIM RESISTANCE

Trimethoprim is a synthetic structural analog of folic acid and a competitive inhibitor of the bacterial enzyme dihydrofolate reductase (DHFR), which in the presence of

NADPH converts dihydrofolate to tetrahydrofolate. N^5,N^{10}-methylenetetrahydrofolate is a cofactor for thymidylate synthase, donating a methyl group to convert deoxyuridylate to thymidylate, which is required for DNA synthesis. Although sometimes used alone, trimethoprim has most commonly been combined with a sulfonamide, such as sulfamethoxazole, which is an inhibitor of dihydropteroate synthase, the enzyme preceding DHFR in the tetrahydrofolate synthesis pathway that converts p-aminobenzoic acid to tetrahydrofolate. DHFRs are essential enzymes found in all living cells, but the human enzyme is intrinsically resistant to trimethoprim, accounting for the trimethoprim's selective antibacterial activity [181].

In clinical settings, acquired bacterial resistance to trimethoprim results most frequently from exogenous acquisition of drug-resistant DHFRs on plasmids or transposons [182]. These resistant DHFRs are widely distributed and have been studied extensively [183]. The resistant enzyme exists in the cell in addition to the native sensitive enzyme and is able to provide the necessary bypass enzymatic function to generate tetrahydrofolic acid in the presence of trimethoprim. Although there are numerous types of trimethoprim-resistant DHFRs, these types with some exceptions appear to fall into two principal families. The DHFRs of family 1, comprising types I, V, VI, VII, and Ib, have in common a polypeptide length of 157 amino acids, with the type I enzyme being dimeric. Family 1 enzymes all have increased half-inhibitory concentration (IC_{50}) values for trimethoprim (1 to 100 µM) [182] relative to those for the chromosomally encoded native enzymes of *E. coli* (0.01 µM) [184] and *Haemophilus influenzae* (0.001 µM) [185]. All DHFRs of family 1 mediate resistance to trimethoprim producing MICs of host cells of substantially >1 mg/mL. Levels of resistance conferred by a particular resistant enzyme may vary due to different levels of expression of that enzyme from its plasmid- or transposon-encoded gene. *dfrA1* is the most common acquired *dfr* gene, perhaps, because of its robust resistance phenotype [183].

DHFRs of family 2 comprise types IIa, IIb, and IIc and have in common a tetrameric structure of four 78-amino acid subunits [182]. All members of this family have exceedingly high levels of resistance to trimethoprim with IC_{50} values of >100 µM [185,186]. The three-dimensional structure of the type IIa resistant enzyme [187] and the native *E. coli* enzyme [188] have been solved and differ substantially from one another, including in their active sites. Comparisons of these structures with that of the intrinsically resistant avian DHFR [181,188] have led to models in which resistance may now be explained at the molecular level by loss of trimethoprim affinity for the enzyme. This loss is due to steric alterations in contact points between trimethoprim and key glutamate and threonine side chains in the DHFR active site as well as by the absence of a key hydrogen bond between the DHFR valine 115 and the 4-amino group of trimethoprim. The structural differences between the *E. coli* and type IIa enzymes are of sufficient magnitude to make implausable any hypothesis that the type IIa resistant enzyme is evolutionarily derived from the *E. coli* enzyme.

Other DHFRs that are not members of families 1 or 2 (types III, IIIb, IIIc, IV, VIII, IX, X, XII, and S1) are heterogeneous, and tend to confer only low levels of resistance to trimethoprim [182]. The origins of resistant DHFRs of any of the types are not certain, but type III is the enzyme most closely related to the native *E. coli* enzyme (~50% identity) [189], and the type S1 is the enzyme most closely related to the *S. aureus* native enzyme (80% identity) [190,191].

Resistant DHFRs are generally encoded by genes on mobile genetic elements, including transposons (e.g., *dfrI* on Tn*7*, *dfrVII* on Tn*5086*, *dfrIIc* on Tn*5090*) and plasmids (e.g., *dfrI* on pLMO150, *dfrIIa* on R67, *dfrIIb* on R388) most commonly, and thus may be able to spread readily among bacteria [182]. Trimethoprim resistance was in fact the earliest identified example of plasmid-mediated antibiotic resistance by an altered or bypass target mechanism. Transposon Tn*7* commonly inserts into a specific site on the chromosome of *E. coli* and other bacteria [192]. In addition, many (if not most) *dfr* genes occur on transferable gene cassettes as part of integrons, which are genetic structures consisting of a series of one or more gene cassettes in tandem and a gene encoding an integrase enzyme that catalyzes the recombination of gene cassettes into the integron structure [193]. Thus, resistance to trimethoprim is often associated with resistance to other antibiotics, such as sulfonamides and aminoglycosides, the resistance determinants of which are also often in gene cassettes and often found in integrons.

Less frequently in clinical bacterial isolates, acquired bacterial resistance to trimethoprim can occur by chromosomal mutation (*i*) leading to thymine auxotrophy, (*ii*) causing an altered native DHFR with reduced affinity for trimethoprim, (*iii*) causing overexpression of the native DHFR, or (*iv*) a combination of mechanisms 2 and 3 [182,194–196]. In the case of thymine auxotrophs, in the presence of exogenous thymine, which is required for growth, the need for the product of DHFR action, tetrahydrofolate, in thymidylate synthesis is bypassed, and thus enzyme inhibition has a minimal effect on cell growth. A combination of both structural changes in the chromosomally encoded enzyme and its overproduction appear to be necessary for high-level trimethoprim resistance in the absence of acquired resistant DHFRs [185]. These constraints and the potential fitness disadvantage of thymine auxotrophy may explain why chromosomal trimethoprim resistance is less common than acquired resistant bypass DHFRs in clinical settings.

SUMMARY AND CONCLUSIONS

Although in most cases target modification ultimately results in reduced affinity for drug binding to the target as the ultimate cause of drug resistance (Table 7.1), overexpression of natural drug targets does produce resistance on a more limited basis. This limited occurrence of gene overexpression is in contrast to the usual circumstances with resistance mechanisms involving drug modification and active efflux in which overexpression of resistance determinants is a common and mechanistically important occurrence. Current models suggest that intermediate vancomycin resistance in *S. aureus* occurs by target overexpression [85,86], and the infrequent occurrence of resistance to trimethoprim mediated by mutant or wildtype chromosomally encoded DHFRs appears to require their overexpression as well [182]. In the former case, target overexpression acts to sequester drug from and block its diffusion to critical target sites. In the latter example, DHFR overexpression may act because of the competitive nature of inhibition by trimethoprim. It is noteworthy that for some drugs, such as quinolones, in which the drug-target complex itself forms the toxic lesion, target enzyme overexpression is predicted to cause increased drug susceptibility rather than resistance. There are a few other examples of target overexpression mediating resistance, such as resistance to isoniazid, ethambutol, and other drugs in mycobacteria [197]. It is argued, however, that this mechanism is used by mycobacteria because in these organisms resistance is almost completely dependent on chromosomal mutation due to the only limited occurrence of plasmids and transposons in these organisms.

Although target modification mechanisms can in most cases be simplified to reductions in affinity of the target for the drug, the means to that end is impressively diverse among different bacterial pathogens and different drugs used against them. The range of possible resistance mechanisms affects which mechanisms may become dominant in the long run as well as the rapidity with which resistance develops after initial introduction of a drug into clinical settings. For chromosomal mutations, which occur spontaneously in large bacterial populations due to low-frequency errors in DNA replication, there is the potential to select for resistance by drug exposure in any bacterium. The likelihood that such spontaneous mutational changes will emerge as the dominant resistance mechanism is in part determined by (*i*) the magnitude of the increase in resistance possible with single mutations thereby affecting the ability of the mutant bacterium to survive in the presence of drug and (*ii*) the consequences of these mutations on the fitness of the mutant bacterium to compete in Nature in both the absence and presence of antibiotic selection pressures, and (*iii*) the presence of more than a single drug target within the cell [198]. Thus, the potential level of such mutational resistance is low for penicillin resistance in *S. pneumoniae*, for example, because of the multiple PBPs to which penicillin binds. In this circumstance, other mechanisms may be drawn on by a bacterial population in order to generate successful survivors of drug exposure. The importance of genetic exchange is thus highlighted by the exploitation of transformation and recombination by *S. pneumoniae* to generate alteration in multiple targets and levels of resistance to penicillin not readily possible by chromosomal mutational mechanisms.

The importance of genetic exchange in expanding the range of possibilities for resistance is perhaps best highlighted by the multiplicity of mobile genetic elements that have become important vectors for altered target resistance. Both exogenous target modification and protection mechanisms and bypass targets mediating resistance to macrolides, tetracyclines, mupirocin, quinolones, and trimethoprim are widely dispersed on plasmids, transposons, and integrons, which have become the dominant source of resistance to some of these drugs. The rapidity with which resistance may emerge with initial introduction of antibiotics in this circumstance is likely determined by the presence or absence of complete or partial cross-resistance mechanisms existing in natural bacterial populations due to naturally produced antibiotics or in non-human reservoirs of bacteria under antibiotic selection pressure that may be introduced into human populations (e.g., antibiotic use in food animals). Risk of such cross-resistance may be more likely in antibiotics that are natural products, the presence of which in Nature may have provided pressures for prior evolution of resistance mechanisms that may be incorporated into mobile genetic elements. The mechanism of resistance to vancomycin in enterococci involving multiple genes from apparently diverse sources assembled into a large mobile genetic element is perhaps the most impressive example of exploitation of genetic exchange mechanisms in the service of antibiotic resistance and bodes poorly for any hope of ever developing an antibiotic for which resistance will not emerge ultimately in the presence of persisting opportunities for selection.

REFERENCES

1. Wang, J.C. DNA topoisomerases, *Annual Review of Biochemistry*, 65, 635, 1996.
2. Drlica, K.; Zhao, X.L. DNA gyrase, topoisomerase IV, and the 4-quinolones, *Microbiological Reviews*, 61, 377, 1997.
3. Hiasa, H.; Yousef, D.O.; Marians, K.J. DNA strand cleavage is required for replication fork arrest by a frozen topoisomerase-quinolone-DNA ternary complex, *Journal of Biological Chemistry*, 271, 26424, 1996.
4. Dietz, W.H.; Cook, T.M.; Goss, W.A. Mechanism of action of nalidixic acid on *Escherichia coli*. III. Conditions required for lethality, *Journal of Bacteriology*, 91, 768, 1966.
5. Wolfson, J.S. et al. Isolation and characterization of an *Escherichia coli* strain exhibiting partial tolerance to quinolones, *Antimicrobial Agents and Chemotherapy*, 33, 705, 1989.
6. Novak, R. et al. Emergence of vancomycin tolerance in *Streptococcus pneumoniae*, *Nature*, 399, 590, 1999.
7. Morais Cabral, J.H. et al. Crystal structure of the breakage-reunion domain of DNA gyrase, *Nature*, 388, 903, 1997.
8. Yoshida, H. et al. Quinolone resistance-determining region in the DNA gyrase *gyrA* gene of *Escherichia coli*, *Antimicrobial Agents and Chemotherapy*, 34, 1271, 1990.
9. Hooper, D.C. Mechanisms of quinolone resistance, in *Quinolone antimicrobial agents*, D.C. Hooper; E. Rubinstein, Eds., ASM Press: Washington, D.C., 2003; Chapter 3, pp. 41–67.
10. Yoshida, H. et al. Mechanism of action of quinolones against *Escherichia coli* DNA gyrase, *Antimicrobial Agents and Chemotherapy*, 37, 839, 1993.
11. Willmott, C.J.; Maxwell, A.A single point mutation in the DNA gyrase A protein greatly reduces binding of fluoroquinolones to the gyrase-DNA complex, *Antimicrobial Agents and Chemotherapy*, 37, 126, 1993.
12. Willmott, C.J. et al. The complex of DNA gyrase and quinolone drugs with DNA forms a barrier to transcription by RNA polymerase, *Journal of Molecular Biology*, 242, 351, 1994.
13. Aleixandre, V. et al. New *Escherichia coli gyrA* and *gyrB* mutations which have a graded effect on DNA supercoiling, *Molecular and General Genetics*, 219, 306, 1989.
14. Yoshida, H. et al. Quinolone resistance-determining region in the DNA gyrase *gyrB* gene of *Escherichia coli*, *Antimicrobial Agents and Chemotherapy*, 35, 1647, 1991.
15. Breines, D.M. et al. Quinolone resistance locus *nfxD* of *Escherichia coli* is a mutant allele of *parE* gene encoding a subunit of topoisomerase IV, *Antimicrobial Agents and Chemotherapy*, 41, 175, 1997.
16. Berger, J.M. et al. Structure and mechanism of DNA topoisomerase II, *Nature*, 379, 225, 1996.
17. Fass, D.; Bogden, C.E.; Berger, J.M. Quaternary changes in topoisomerase II may direct orthogonal movement of two DNA strands, *Nature Structural Biology*, 6, 322, 1999.
18. Blanche, F. et al. Differential behaviors of *Staphylococcus aureus* and *Escherichia coli* type II DNA topoisomerases, *Antimicrobial Agents and Chemotherapy*, 40, 2714, 1996.
19. Trucksis, M.; Wolfson, J.S.; Hooper, D.C. A novel locus conferring fluoroquinolone resistance in *Staphylococcus aureus*, *Journal of Bacteriology*, 173, 5854, 1991.

20. Ng, E.Y.; Trucksis, M.; Hooper, D.C. Quinolone resistance mutations in topoisomerase IV: relationship of the *flqA* locus and genetic evidence that topoisomerase IV is the primary target and DNA gyrase the secondary target of fluoroquinolones in *Staphylococcus aureus*, *Antimicrobial Agents and Chemotherapy*, 40, 1881, 1996.

21. Panel designs model strategic plan for tackling in-hospital antimicrobial resistance, *American Journal of Health-System Pharmacy*, 53, 600, 1996.

22. Breines, D.M.; Burnham, J.C. Modulation of *Escherichia coli* type 1 fimbrial expression and adherence to uroepithelial cells following exposure of logarithmic phase cells to quinolones at subinhibitory concentrations, *Journal of Antimicrobial Chemotherapy*, 34, 205, 1994.

23. Pan, X.S.; Fisher, L.M. *Streptococcus pneumoniae* DNA gyrase and topoisomerase IV: overexpression, purification, and differential inhibition by fluoroquinolones, *Antimicrobial Agents and Chemotherapy*, 43, 1129, 1999.

24. Pan, X.S.; Fisher, L.M. DNA gyrase and topoisomerase IV are dual targets of clinafloxacin action in *Streptococcus pneumoniae*, *Antimicrobial Agents and Chemotherapy*, 42, 2810, 1998.

25. Schmitz, F.J. et al. Characterization of *grlA*, *grlB*, *gyrA*, and *gyrB* mutations in 116 unrelated isolates of *Staphylococcus aureus* and effects of mutations on ciprofloxacin MIC, *Antimicrobial Agents and Chemotherapy*, 42, 1249, 1998.

26. Kumar, K.P.; Chatterji, D. Differential inhibition of abortive transcription initiation at different promoters catalysed by *E. coli* RNA polymerase. Effect of rifampicin on purine or pyramidine-initiated phosphodiester synthesis, *FEBS Letters*, 306, 46, 1992.

27. Kumar, K.P.; Reddy, P.S.; Chatterji, D. Proximity relationship between the active site of *Escherichia coli* RNA polymerase and rifampicin binding domain: a resonance energy-transfer study, *Biochemistry*, 31, 7519, 1992.

28. Jin, D.J.; Gross, C.A. Mapping and sequencing of mutations in the *Escherichia coli* *rpoB* gene that lead to rifampicin resistance, *Journal of Molecular Biology*, 202, 45, 1988.

29. Jin, D.J. et al. Effects of rifampicin resistant *rpoB* mutations on antitermination and interaction with *nusA* in *Escherichia coli*, *Journal of Molecular Biology*, 204, 247, 1988.

30. Telenti, A. et al. Direct, automated detection of rifampin-resistant *Mycobacterium tuberculosis* by polymerase chain reaction and single-strand conformation polymorphism analysis, *Antimicrobial Agents and Chemotherapy*, 37, 2054, 1993.

31. Honore, N.; Cole, S.T. Molecular basis of rifampin resistance in *Mycobacterium leprae*, *Antimicrobial Agents and Chemotherapy*, 37, 414, 1993.

32. Wichelhaus, T.A. et al. Molecular characterization of *rpoB* mutations conferring cross-resistance to rifamycins on methicillin-resistant *Staphylococcus aureus*, *Antimicrobial Agents and Chemotherapy*, 43, 2813, 1999.

33. Aubry-Damon, H.; Soussy, C.J.; Courvalin, P. Characterization of mutations in the *rpoB* gene that confer rifampin resistance in *Staphylococcus aureus*, *Antimicrobial Agents and Chemotherapy*, 42, 2590, 1998.

34. Padayachee, T.; Klugman, K.P. Molecular basis of rifampin resistance in *Streptococcus pneumoniae*, *Antimicrobial Agents and Chemotherapy*, 43, 2361, 1999.

35. Carter, P.E. et al. Molecular characterization of rifampin-resistant *Neisseria meningitidis*, *Antimicrobial Agents and Chemotherapy*, 38, 1256, 1994.

36. Drancourt, M.; Raoult, D. Characterization of mutations in the *rpoB* gene in naturally rifampin-resistant *Rickettsia* species, *Antimicrobial Agents and Chemotherapy*, 43, 2400, 1999.

37. Chambers, H.F. Penicillin-binding protein-mediated resistance in pneumococci and staphylococci, *Journal of Infectious Diseases*, 179, S353, 1999.

38. Laible, G.; Hakenbeck, R. Five independent combinations of mutations can result in low-affinity penicillin binding protein 2x of *Streptococcus pneumoniae*, *Journal of Bacteriology*, 173, 6986, 1991.

39. Hakenbeck, R. et al. β-lactam resistance in *Streptococcus pneumoniae*: penicillin-binding proteins and non-penicillin-binding proteins, *Molecular Microbiology*, 33, 673, 1999.

40. Gonzales, R.D. et al. Infections due to vancomycin-resistant *Enterococcus faecium* resistant to linezolid, *Lancet*, 357, 1179, 2001.

41. Burleson, B.S. et al. *Enterococcus faecalis* resistant to linezolid: case series and review of the literature, *Pharmacotherapy*, 24, 1225, 2004.

42. Jones, R N. et al. Linezolid-resistant *Enterococcus faecium* isolated from a patient without prior exposure to an oxazolidinone: Report from the SENTRY Antimicrobial Surveillance Program, *Diagnostic Microbiology and Infectious Disease*, 42, 137, 2002.

43. Herrero, I.A.; Issa, N.C.; Patel, R. Nosocomial spread of linezolid-resistant, vancomycin-resistant *Enterococcus faecium*, *New England Journal of Medicine*, 346, 867, 2002.

44. Tsiodras, S. et al. Linezolid resistance in a clinical isolate of *Staphylococcus aureus*, *Lancet*, 358, 207, 2001.

45. Xiong, L.Q. et al. Oxazolidinone resistance mutations in 23S rRNA of *Escherichia coli* reveal the central region of domain V as the primary site of drug action, *Journal of Bacteriology*, 182, 5325, 2000.

46. Kloss, P. et al. Resistance mutations in 23S rRNA identify the site of action of the protein synthesis inhibitor linezolid in the ribosomal peptidyl transferase center, *Journal of Molecular Biology*, 294, 93, 1999.

47. Prystowsky, J. et al. Resistance to linezolid: characterization of mutations in rRNA and comparison of their occurrences in vancomycin-resistant enterococci, *Antimicrobial Agents and Chemotherapy*, 45, 2154, 2001.

48. Swoboda, S. et al. Varying linezolid susceptibility of vancomycin-resistant *Enterococcus faecium* isolates during therapy: a case report, *Journal of Antimicrobial Chemotherapy*, 56, 787, 2005.

49. Meier, A. et al. Identification of mutations in the 23S ribosomal RNA gene of clarithromycin resistant *Mycobacterium intracellulare*, *Antimicrobial Agents and Chemotherapy*, 38, 381, 1994.

50. Versalovic, J. et al. Point mutations in the 23S rRNA gene of *Helicobacter pylori* associated with different levels of clarithromycin resistance, *Journal of Antimicrobial Chemotherapy*, 40, 283, 1997.

51. Weisblum, B. Erythromycin resistance by ribosome modification, *Antimicrobial Agents and Chemotherapy*, 39, 577, 1995.

52. Dowson, C.G. et al. Evolution of penicillin resistance in *Streptococcus pneumoniae*; the role of *Streptococcus mitis* in the formation of a low affinity PBP2B in *S. pneumoniae*, *Molecular Microbiology*, 9, 635, 1993.

53. Smith, A.M.; Klugman, K.P. Alterations in PBP 1A essential for high-level penicillin resistance in *Streptococcus pneumoniae*, *Antimicrobial Agents and Chemotherapy*, 42, 1329, 1998.

54. Zhao, G.S. et al. Biochemical characterization of penicillin-resistant and -sensitive penicillin-binding protein 2x transpeptidase activities of *Streptococcus pneumoniae* and mechanistic implications in bacterial resistance to β-lactam antibiotics, *Journal of Bacteriology*, 179, 4901, 1997.

55. Bugg, T.D.H. et al. Molecular basis for vancomycin resistance in *Enterococcus faecium* BM4147: biosynthesis of a depsipeptide peptidoglycan precursor by vancomycin resistance proteins VanH and VanA, *Biochemistry*, 30, 10408, 1991.

56. Arthur, M. et al. Characterization of Tn*1546*, a Tn*3*-related transposon conferring glycopeptide resistance by synthesis of depsipeptide peptidoglycan precursors in *Enterococcus faecium* BM4147, *Journal of Bacteriology*, 175, 117, 1993.

57. Arthur, M.; Courvalin, P. Genetics and mechanisms of glycopeptide resistance in enterococci, *Antimicrobial Agents and Chemotherapy*, 37, 1563, 1993.

58. Reynolds, P.E. et al. Glycopeptide resistance mediated by enterococcal transposon Tn*1546* requires production of VanX for hydrolysis of D-alanyl-D-alanine, *Molecular Microbiology*, 13, 1065, 1994.

59. Arthur, M. et al. Contribution of VanY D, D-carboxypeptidase to glycopeptide resistance in *Enterococcus faecalis* by hydrolysis of peptidoglycan precursors, *Antimicrobial Agents and Chemotherapy*, 38, 1899, 1994.

60. Arthur, M. et al. The *vanZ* gene of Tn*1546* from *Enterococcus faecium* BM4147 confers resistance to teicoplanin, *Gene*, 154, 87, 1995.

61. Wright, G.D.; Holman, T.R.; Walsh, C.T. Purification and characterization of VanR and the cytosolic domain of VanS: a two-component regulatory system required for vancomycin resistance in *Enterococcus faecium* BM4147, *Biochemistry*, 32, 5057, 1993.

62. Holman, T.R. et al. Identification of the DNA-binding site for the phosphorylated VanR protein required for vancomycin resistance in *Enterococcus faecium*, *Biochemistry*, 33, 4625, 1994.

63. Arkwright, P.D. et al. Age-related prevalence and antibiotic resistance of pathogenic staphylococci and streptococci in children with infected atopic dermatitis at a single-specialty center, *Archives of Dermatology*, 138, 939, 2002.

64. Lai, M.H.; Kirsch, D.R. Induction signals for vancomycin resistance encoded by the *vanA* gene cluster in *Enterococcus faecium*, *Antimicrobial Agents and Chemotherapy*, 40, 1645, 1996.

65. Evers, S.; Courvalin, P. Regulation of VanB-type vancomycin resistance gene expression by the VanS$_B$-VanR$_B$ two-component regulatory system in *Enterococcus faecalis* V583, *Journal of Bacteriology*, 178, 1302, 1996.

66. Quintiliani, R, Jr.; Courvalin, P. Conjugal transfer of the vancomycin resistance determinant *vanB* between enterococci involves the movement of large genetic elements from chromosome to chromosome, *FEMS Microbiology Letters*, 119, 359, 1994.

67. Carias, L.L. et al. Genetic linkage and cotransfer of a novel, *vanB*-containing transposon (Tn*5382*) and a low-affinity penicillin-binding protein 5 gene in a clinical vancomycin-resistant *Enterococcus faecium* isolate, *Journal of Bacteriology*, 180, 4426, 1998.

68. Arthur, M.; Quintiliani, R, Jr. Regulation of VanA- and VanB-type glycopeptide resistance in enterococci, *Antimicrobial Agents and Chemotherapy*, 45, 375, 2001.

69. Bonafede, M.E.; Carias, L.L.; Rice, L.B. Enterococcal transposon Tn*5384*: Evolution of a composite transposon through cointegration of enterococcal and staphylococcal plasmids, *Antimicrobial Agents and Chemotherapy*, 41, 1854, 1997.

70. Billot-Klein, D. et al. Modification of peptidoglycan precursors is a common feature of the low-level vancomycin-resistant VANB-type *Enterococcus* D366 and of the naturally glycopeptide-resistant species *Lactobacillus casei*, *Pediococcus pentosaceus*, *Leuconostoc mesenteroides*, and *Enterococcus gallinarum*, *Journal of Bacteriology*, 176, 2398, 1994.

71. Handwerger, S. et al. Vancomycin-resistant *Leuconostoc mesenteroides* and *Lactobacillus casei* synthesize cytoplasmic peptidoglycan precursors that terminate in lactate, *Journal of Bacteriology*, 176, 260, 1994.

72. Al-Obeid, S. et al. Replacement of the essential penicillin-binding protein 5 by high-molecular mass PBPs may explain vancomycin-β-lactam synergy in low-level vancomycin-resistant *Enterococcus faecium* D366, *FEMS Microbiology Letters*, 70, 79, 1992.

73. Poyart, C. et al. Emergence of vancomycin resistance in the genus *Streptococcus*: Characterization of a *vanB* transferable determinant in *Streptococcus bovis*, *Antimicrobial Agents and Chemotherapy*, 41, 24, 1997.
74. Noble, W.C.; Virani, Z.; Cree, R.G.A. Co-transfer of vancomycin and other resistance genes from *Enterococcus faecalis* NCTC 12201 to *Staphylococcus aureus*, *FEMS Microbiology Letters*, 93, 195, 1992.
75. Tenover, F.C. et al. Vancomycin-resistant *Staphylococcus aureus* isolate from a patient in Pennsylvania, *Antimicrobial Agents and Chemotherapy*, 48, 275, 2004.
76. Chang, S. et al. Infection with vancomycin-resistant *Staphylococcus aureus* containing the *vanA* resistance gene, *New England Journal of Medicine*, 348, 1342, 2003.
77. Weigel, L.M. et al. Genetic analysis of a high-level vancomycin-resistant isolate of *Staphylococcus aureus*, *Science*, 302, 1569, 2003.
78. Clark, N.C. et al. Comparison of Tn*1546*-like elements in vancomycin-resistant *Staphylococcus aureus* isolates from Michigan and Pennsylvania, *Antimicrobial Agents and Chemotherapy*, 49, 470, 2005.
79. Hiramatsu, K. et al. Dissemination in Japanese hospitals of strains of *Staphylococcus aureus* heterogeneously resistant to vancomycin, *Lancet*, 350, 1670, 1997.
80. Tenover, F.C. et al. Characterization of staphylococci with reduced susceptibilities to vancomycin and other glycopeptides, *Journal of Clinical Microbiology*, 36, 1020, 1998.
81. Smith, T.L. et al. Emergence of vancomycin resistance in *Staphylococcus aureus*, *New England Journal of Medicine*, 340, 493, 1999.
82. Fridkin, S.K. et al. Epidemiological and microbiological characterization of infections caused by *Staphylococcus aureus* with reduced susceptibility to vancomycin, United States, 1997–2001, *Clinical Infectious Diseases*, 36, 429, 2003.
83. Hanaki, H. et al. Activated cell-wall synthesis is associated with vancomycin resistance in methicillin-resistant *Staphylococcus aureus* clinical strains Mu3 and Mu50, *Journal of Antimicrobial Chemotherapy*, 42, 199, 1998.
84. Strandén, A.M.; Roos, M.; Berger-Bachi, B. Glutamine synthetase and heteroresistance in methicillin-resistant *Staphylococcus aureus*, *Microbial Drug Resistance*, 2, 201, 1996.
85. Hiramatsu, K. Vancomycin resistance in staphylococci, *Drug Resistance Updates*, 1, 135, 1998.
86. Cui, L. et al. Novel mechanism of antibiotic resistance originating in vancomycin-intermediate *Staphylococcus aureus*, *Antimicrobial Agents and Chemotherapy*, 50, 428, 2006.
87. Cui, L. et al. DNA microarray-based identification of genes associated with glycopeptide resistance in *Staphylococcus aureus*, *Antimicrobial Agents and Chemotherapy*, 49, 3404, 2005.
88. Moreira, B. et al. Increased production of penicillin-binding protein 2, increased detection of other penicillin-binding proteins, and decreased coagulase activity associated with glycopeptide resistance in *Staphylococcus aureus*, *Antimicrobial Agents and Chemotherapy*, 41, 1788, 1997.
89. Hiramatsu, K.; Hanaki, H. Glycopeptide resistance in staphylococci, *Current Opinion in Infectious Diseases*, 11, 653, 1998.
90. Weisblum, B. Macrolide resistance, *Drug Resistance Updates*, 1, 29, 1998.
91. el Solh, N.; Allignet, J. Staphylococcal resistance to streptogramins and related antibiotics, *Drug Resistance Updates*, 1, 169, 1998.
92. Douthwaite, S.; Champney, W.S. Structures of ketolides and macrolides determine their mode of interaction with the ribosomal target site, *Journal of Antimicrobial Chemotherapy*, 48, 1, 2001.
93. Liu, M.F.; Douthwaite, S. Activity of the ketolide telithromycin is refractory to erm monomethylation of bacterial rRNA, *Antimicrobial Agents and Chemotherapy*, 46, 1629, 2002.

94. Douthwaite, S.; Jalava, J.; Jakobsen, L. Ketolide resistance in *Streptococcus pyogenes* correlates with the degree of rRNA dimethylation by Erm, *Molecular Microbiology*, 58, 613, 2005.
95. Bonnefoy, A. et al. Ketolides lack inducibility properties of MLS$_B$ resistance phenotype, *Journal of Antimicrobial Chemotherapy*, 40, 85, 1997.
96. Shortridge, V.D. et al. Comparison of in vitro activities of ABT-773 and telithromycin against macrolide-susceptible and -resistant streptococci and staphylococci, *Antimicrobial Agents and Chemotherapy*, 46, 783, 2002.
97. Graham, M.Y.; Weisblum, B. 23S ribosomal ribonucleic acid of macrolide-producing streptomycetes contains methylated adenine, *Journal of Bacteriology*, 137, 1464, 1979.
98. Galimand, M.; Courvalin, P.; Lambert, T. Plasmid-mediated high-level resistance to aminoglycosides in *Enterobacteriaceae* due to 16S rRNA methylation, *Antimicrobial Agents and Chemotherapy*, 47, 2565, 2003.
99. Yokoyama, K. et al. Acquisition of 16S rRNA methylase gene in *Pseudomonas aeruginosa*, *Lancet*, 362, 1888, 2003.
100. Doi, Y. et al. Plasmid-mediated 16S rRNA methylase in *Serratia marcescens* conferring high-level resistance to aminoglycosides, *Antimicrobial Agents and Chemotherapy*, 48, 491, 2004.
101. Yan, J.-J. et al. Plasmid-mediated 16S rRNA methylases conferring high-level aminoglycoside resistance in *Escherichia coli* and *Klebsiella pneumoniae* isolates from two Taiwanese hospitals, *Journal of Antimicrobial Chemotherapy*, 54, 1007, 2004.
102. Galimand, M. et al. Worldwide disseminated *armA* aminoglycoside resistance methylase gene is borne by composite transposon Tn*1548*, *Antimicrobial Agents and Chemotherapy*, 49, 2949, 2005.
103. González-Zorn, B. et al. Genetic basis for dissemination of *armA*, *Journal of Antimicrobial Chemotherapy*, 56, 583, 2005.
104. Wachino, J.-I. et al. Novel plasmid-mediated 16S rRNA methylase, RmtC, found in a *Proteus mirabilis* isolate demonstrating extraordinary high-level resistance against various aminoglycosides, *Antimicrobial Agents and Chemotherapy*, 50, 178, 2006.
105. Epe, B.; Woolcy, P.; Hornig, H. Competition between tetracycline and tRNA at both P and A sites of the ribosome of *Escherichia coli*, *FEBS Letters*, 213, 443, 1987.
106. Goldman, R.A. et al. Photoincorporation of tetracycline into *Escherichia coli* ribosomes. Identification of the major proteins photolabeled by native tetracycline and tetracycline photoproducts and implications for the inhibitory action of tetracycline on protein synthesis, *Biochemistry*, 22, 359, 1983.
107. Buck, M.A.; Cooperman, B.S. Single protein omission reconstitution studies of tetracycline binding to the 30S subunit of *Escherichia coli* ribosomes, *Biochemistry*, 29, 5374, 1990.
108. Arbuck, S.G.; Takimoto, C.H. An overview of topoisomerase I—Targeting agents, *Seminars in Hematology*, 35, 3, 1998.
109. Martin, P.; Tieu-Cuot, P.; Courvalin, P. Nucleotide sequence of the *tetM* tetracycline resistance determinant of the streptococcal conjugative shuttle transposon Tn*1545*, *Nucleic Acids Research*, 14, 7047, 1986.
110. LeBlanc, D.J. et al. Nucleotide sequence analysis of tetracycline resistance gene *tetO* from *Streptococcus mutans* DL5, *Journal of Bacteriology*, 170, 3618, 1988.
111. Taylor, D.E. Plasmid-mediated tetracycline resistance in *Campylobacter jejuni*: expression in *Escherichia coli* and identification of homology with streptococcal class M determinant, *Journal of Bacteriology*, 165, 1037, 1986.
112. Atkinson, B.A.; Abu-Al-Jaibat, A.; LeBlanc, D.J. Antibiotic resistance among enterococci isolated from clinical specimens between 1953 and 1954, *Antimicrobial Agents and Chemotherapy*, 41, 1598, 1997.

113. Ahmed, M. et al. A protein that activates expression of a multidrug efflux transporter upon binding the transporter substrates, *Journal of Biological Chemistry*, 269, 28506, 1994.

114. Burdett, V. Purification and characterization of Tet(M), a protein that renders ribosomes resistant to tetracycline, *Journal of Biological Chemistry*, 266, 2872, 1991.

115. Burdett, V. tRNA modification activity is necessary for Tet(M)-mediated tetracycline resistance, *Journal of Bacteriology*, 175, 7209, 1993.

116. Burdett, V. Tet(M)-promoted release of tetracycline from ribosomes is GTP-dependent, *Journal of Bacteriology*, 178, 3246, 1996.

117. Dantley, K.A.; Dannelly, H.K.; Burdett, V. Binding interaction between Tet(M) and the ribosome: requirements for binding, *Journal of Bacteriology*, 180, 4089, 1998.

118. Taylor, D.E. et al. Host mutations (*miaA* and *rpsL*) reduce tetracycline resistance mediated by Tet(O) and Tet(M), *Antimicrobial Agents and Chemotherapy*, 42, 59, 1998.

119. Schnappinger, D.; Hillen, W. Tetracyclines: Antibiotic action, uptake, and resistance mechanisms, *Archives of Microbiology*, 165, 359, 1996.

120. Taylor, D.E.; Chau, A. Tetracycline resistance mediated by ribosomal protection, *Antimicrobial Agents and Chemotherapy*, 40, 1, 1996.

121. Connell, S.R. et al. Ribosomal protection proteins and their mechanism of tetracycline resistance, *Antimicrobial Agents and Chemotherapy*, 47, 3675, 2003.

122. Widdowson, C.A.; Klugman, K.P. The molecular mechanisms of tetracycline resistance in the pneumococcus, *Microbial Drug Resistance*, 4, 79, 1998.

123. Su, Y.A.; He, P.; Clewell, D.B. Characterization of the *tet*(M) determinant of Tn*916*: evidence for regulation by transcriptional attenuation, *Antimicrobial Agents and Chemotherapy*, 36, 769, 1992.

124. Taylor, D.E. et al. Characterization and expression of a cloned tetracycline resistance determinant from *Campylobacter jejuni* plasmid pUA466, *Journal of Bacteriology*, 169, 2984, 1987.

125. Rasmussen, B.A.; Gluzman, Y.; Tally, F.P. Inhibition of protein synthesis occurring on tetracycline-resistant TetM-protected ribosomes by a novel class of tetracyclines, the glycylcyclines, *Antimicrobial Agents and Chemotherapy*, 38, 1658, 1994.

126. Bergeron, J. et al. Glycylcyclines bind to the high-affinity tetracycline ribosomal binding site and evade Tet(M)- and Tet(O)-mediated ribosomal protection, *Antimicrobial Agents and Chemotherapy*, 40, 2226, 1996.

127. Tran, J.H.; Jacoby, G.A.; Hooper, D.C. Interaction of the plasmid-encoded quinolone resistance protein Qnr with *Escherichia coli* DNA gyrase, *Antimicrobial Agents and Chemotherapy*, 49, 118, 2005.

128. Tran, J.H.; Jacoby, G.A. Mechanism of plasmid-mediated quinolone resistance, *Proceedings of the National Academy of Sciences of the United States of America*, 99, 5638, 2002.

129. Tran, J.H.; Jacoby, G.A.; Hooper, D.C. Interaction of the plasmid-encoded quinolone resistance protein QnrA with *Escherichia coli* topoisomerase IV, *Antimicrobial Agents and Chemotherapy*, 49, 3050, 2005.

130. Jacoby, G.A. et al. Another plasmid-mediated gene for quinolone resistance, *qnrB*, *Antimicrobial Agents and Chemotherapy*, 50, 1178, 2006.

131. Hata, M. et al. Cloning of a novel gene for quinolone resistance from a transferable plasmid in *Shigella flexneri* 2b, *Antimicrobial Agents and Chemotherapy*, 49, 801, 2005.

132. Vetting, M.W. et al. Pentapeptide repeat proteins, *Biochemistry*, 45, 1, 2006.

133. Hegde, S.S. et al. A fluoroquinolone resistance protein from *Mycobacterium tuberculosis* that mimics DNA, *Science*, 308, 1480, 2005.

134. Poirel, L. et al. Origin of plasmid-mediated quinolone resistance determinant QnrA, *Antimicrob Agents Chemother,* 49, 3523, 2005.

135. Saga, T. et al. *Vibrio parahaemolyticus* chromosomal qnr homologue VPA0095: demonstration by transformation with a mutated gene of its potential to reduce quinolone susceptibility in *Escherichia coli, Antimicrobial Agents and Chemotherapy*, 49, 2144, 2005.

136. Robicsek, A. et al. Broader distribution of plasmid-mediated quinolone resistance in the United States, *Antimicrobial Agents and Chemotherapy*, 49, 3001, 2005.

137. Wang, M. et al. Emerging plasmid-mediated quinolone resistance associated with the qnr gene in *Klebsiella pneumoniae* clinical isolates in the United States, *Antimicrobial Agents and Chemotherapy*, 48, 1295, 2004.

138. Wang, M. et al. Plasmid-mediated quinolone resistance in clinical isolates of *Escherichia coli* from Shanghai, China, *Antimicrobial Agents and Chemotherapy*, 47, 2242, 2003.

139. Nordmann, P.; Poirel, L. Emergence of plasmid-mediated resistance to quinolones in *Enterobacteriaceae, The Journal of Antimicrobial Chemotherapy*, 56, 463, 2005.

140. Martínez-Martínez, L. et al. Interaction of plasmid and host quinolone resistance, *Journal of Antimicrobial Chemotherapy*, 51, 1037, 2003.

141. Wang, M. et al. Plasmid-mediated quinolone resistance in clinical isolates of *Escherichia coli* from Shanghai, China, *Antimicrobial Agents and Chemotherapy*, 47, 2242, 2003.

142. Chambers, H.F. Methicillin resistance in staphylococci: Molecular and biochemical basis and clinical implications, *Clinical Microbiology Reviews*, 10, 781, 1997.

143. Katayama, Y.; Ito, T.; Hiramatsu, K. A new class of genetic element, staphylococcus cassette chromosome *mec*, encodes methicillin resistance in *Staphylococcus aureus, Antimicrobial Agents and Chemotherapy*, 44, 1549, 2000.

144. Hiramatsu, K. et al. The emergence and evolution of methicillin-resistant *Staphylococcus aureus, Trends in Microbiology*, 9, 486, 2001.

145. Berger-Bachi, B.; Rohrer, S. Factors influencing methicillin resistance in staphylococci, *Archives of Microbiology*, 178, 165, 2002.

146. Hiramatsu, K.; Konodo, N.; Ito, T. Genetic basis for molecular epidemiology of MRSA, *Journal of Infection and Chemotherapy*, 2, 117, 1996.

147. Kuwahara-Arai, K. et al. Suppression of methicillin resistance in a *mecA*-containing pre-methicillin-resistant *Staphylococcus aureus* strain is caused by the *mecI*-mediated repression of PBP 2' production, *Antimicrobial Agents and Chemotherapy*, 40, 2680, 1996.

148. Niemeyer, D.M. et al. Role of *mecA* transcriptional regulation in the phenotypic expression of methicillin resistance in *Staphylococcus aureus, Journal of Bacteriology*, 178, 5464, 1996.

149. Hackbarth, C.J.; Chambers, H.F. *blaI* and *blaR1* regulate beta-lactamase and PBP 2a production in methicillin-resistant *Staphylococcus aureus, Antimicrobial Agents and Chemotherapy*, 37, 1144, 1993.

150. Hartman, B.J.; Tomasz, A. Expression of methicillin resistance in heterogeneous strains of *Staphylococcus aureus, Antimicrobial Agents and Chemotherapy*, 29, 85, 1986.

151. Matthews, P.R.; Stewart, P.R. Resistance heterogeneity in methicillin-resistant *Staphylococcus aureus, FEMS Microbiology Letters*, 22, 161, 1984.

152. Sabath, L.D.; Wallace, S.J. Factors influencing methicillin resistance in staphylococci, *Annals of the New York Academy of Sciences*, 236, 258, 1974.

153. Chambers, H.F.; Hackbarth, C.J. Effect of NaCl and nafcillin on penicillin-binding protein 2a and heterogeneous expression of methicillin resistance in *Staphylococcus aureus, Antimicrobial Agents and Chemotherapy*, 31, 1982, 1987.

154. Berger-Bachi, B.; Tschierske, M. Role of Fem factors in methicillin resistance, *Drug Resistance Updates*, 1, 325, 1998.

155. Berger-Bachi, B. et al. FemA, a host-mediated factor essential for methicillin resistance in *Staphylococcus aureus*: molecular cloning and characterization, *Molecular and General Genetics*, 219, 263, 1989.
156. Komatsuzawa, H. et al. Characterization of *fmtA*, a gene that modulates the expression of methicillin resistance in *Staphylococcus aureus*, *Antimicrobial Agents and Chemotherapy*, 43, 2121, 1999.
157. Ryffel, C. et al. Mechanisms of heteroresistance in methicillin-resistant *Staphylococcus aureus*, *Antimicrobial Agents and Chemotherapy*, 38, 724, 1994.
158. Fujimura, T.; Murakami, K. Increase of methicillin resistance in *Staphylococcus aureus* caused by deletion of a gene whose product is homologous to lytic enzymes, *Journal of Bacteriology*, 179, 6294, 1997.
159. Kondo, N. et al. Eagle-type methicillin resistance: New phenotype of high methicillin resistance under *mec* regulator gene control, *Antimicrobial Agents and Chemotherapy*, 45, 815, 2001.
160. Kreiswirth, B. et al. Evidence for a clonal origin of methicillin resistance in *Staphylococcus aureus*, *Science*, 259, 227, 1993.
161. Hiramatsu, K. et al. Molecular genetics of methicillin-resistant *Staphylococcus aureus*, *International Journal of Medical Microbiology*, 292, 67, 2002.
162. Archer, G.L. et al. Dissemination among staphylococci of DNA sequences associated with methicillin resistance, *Antimicrobial Agents and Chemotherapy*, 38, 447, 1994.
163. Couto, I. et al. Development of methicillin resistance in clinical isolates of *Staphylococcus sciuri* by transcriptional activation of the *mecA* homologue native to the species, *Journal of Bacteriology*, 185, 645, 2003.
164. Wu, S.W.; De Lancastre, H.; Tomasz, A. Recruitment of the *mecA* gene homologue of *Staphylococcus sciuri* into a resistance determinant and expression of the resistant phenotype in *Staphylococcus aureus*, *Journal of Bacteriology*, 183, 2417, 2001.
165. Boyce, J.M. et al. Meticillin-resistant *Staphylococcus aureus*, *The Lancet Infectious Diseases*, 5, 653, 2005.
166. Cookson, B.D. The emergence of mupirocin resistance: a challenge to infection control and antibiotic prescribing practice, *Journal of Antimicrobial Chemotherapy*, 41, 11, 1998.
167. Morton, T.M. et al. Characterization of a conjugative staphylococcal mupirocin resistance plasmid, *Antimicrobial Agents and Chemotherapy*, 39, 1272, 1995.
168. Bradley, S.F. et al. Mupirocin resistance: clinical and molecular epidemiology, *Infection Control and Hospital Epidemiology*, 16, 354, 1995.
169. Farmer, T.H.; Gilbart, J.; Elson, S.W. Biochemical basis of mupirocin resistance in strains of *Staphylococcus aureus*, *Journal of Antimicrobial Chemotherapy*, 30, 587, 1992.
170. Gilbart, J.; Perry, C.R.; Slocombe, B. High-level mupirocin resistance in *Staphylococcus aureus*: evidence for two distinct isoleucyl-tRNA synthetases, *Antimicrobial Agents and Chemotherapy*, 37, 32, 1993.
171. Hodgson, J.E. et al. Molecular characterization of the gene encoding high-level mupirocin resistance in *Staphylococcus aureus* J2870, *Antimicrobial Agents and Chemotherapy*, 38, 1205, 1994.
172. Yun, H.J. et al. Prevalence and mechanisms of low- and high-level mupirocin resistance in staphylococci isolated from a Korean hospital, *The Journal of Antimicrobial Chemotherapy*, 51, 619, 2003.
173. Udo, E.E.; Al-Sweih, N.; Noronha, B.C. A chromosomal location of the *mupA* gene in *Staphylococcus aureus* expressing high-level mupirocin resistance, *The Journal of Antimicrobial Chemotherapy*, 51, 1283, 2003.

174. Janssen, D.A. et al. Detection and characterization of mupirocin resistance in *Staphylococcus aureus*, *Antimicrobial Agents and Chemotherapy*, 37, 2003, 1993.
175. Hurdle, J.G. et al. *In vivo* transfer of high-level mupirocin resistance from *Staphylococcus epidermidis* to methicillin-resistant *Staphylococcus aureus* associated with failure of mupirocin prophylaxis, *The Journal of Antimicrobial Chemotherapy*, 56, 1166, 2005.
176. Needham, C. et al. An investigation of plasmids from *Staphylococcus aureus* that mediate resistance to mupirocin and tetracycline, *Microbiology*, 140, 2577, 1994.
177. Hurdle, J.G.; O'Neill, A.J.; Chopra, I. The isoleucyl-tRNA synthetase mutation V588F conferring mupirocin resistance in glycopeptide-intermediate *Staphylococcus aureus* is not associated with a significant fitness burden, *The Journal of Antimicrobial Chemotherapy*, 53, 102, 2004.
178. Hurdle, J.G. et al. Analysis of mupirocin resistance and fitness in *Staphylococcus aureus* by molecular genetic and structural modeling techniques, *Antimicrobial Agents and Chemotherapy*, 48, 4366, 2004.
179. Ramsey, M.A. et al. Identification of chromosomal location of *mupA* gene, encoding low-level mupirocin resistance in staphylococcal isolates, *Antimicrobial Agents and Chemotherapy*, 40, 2820, 1996.
180. Bonilla, H.F. et al. Susceptibility of ciprofloxacin-resistant staphylococci and enterococci to trovafloxacin, *Diagnostic Microbiology and Infectious Disease*, 26, 17, 1996.
181. Matthews, D.A. et al. Dihydrofolate reductase. The stereochemistry of inhibitor selectivity, *Journal of Biological Chemistry*, 260, 392, 1985.
182. Huovinen, P. et al. Trimethoprim and sulfonamide resistance, *Antimicrobial Agents and Chemotherapy*, 39, 279, 1995.
183. Blahna, M.T. et al. The role of horizontal gene transfer in the spread of trimethoprim-sulfamethoxazole resistance among uropathogenic *Escherichia coli* in Europe and Canada, *The Journal of Antimicrobial Chemotherapy*, 57, 666, 2006.
184. Amyes, S.G.B.; Smith, J.T. The purification and properties of trimethoprim-resistant dihydrofolate reductase mediated by the R-factor, R 388, *European Journal of Biochemistry*, 61, 597, 1976.
185. Flensburg, J.; Sköld, O. Massive overproduction of dihydrofolate reductase in bacteria as a response to the use of trimethoprim, *European Journal of Biochemistry*, 162, 473, 1987.
186. Smith, S.L. et al. R plasmid dihydrofolate reductase with subunit structure, *Journal of Biological Chemistry*, 254, 6222, 1979.
187. Matthews, D.A. et al. Crystal structure of a novel trimethoprim-resistant dihydrofolate reductase specified in *Escherichia coli* by R-plasmid R67, *Biochemistry*, 25, 4194, 1986.
188. Matthews, D.A. et al. Refined crystal structures of *Escherichia coli* and chicken liver dihydrofolate reductase containing bound trimethoprim, *Journal of Biological Chemistry*, 260, 381, 1985.
189. Joyner, S.S. et al. Characterization of an R-plasmid dihydrofolate reductase with a monomeric structure, *Journal of Biological Chemistry*, 259, 5851, 1984.
190. Dale, G.E.; Then, R.L.; Stuber, D. Characterization of the gene for chromosomal trimethoprim-sensitive dihydrofolate reductase of *Staphylococcus aureus* ATCC 25923, *Antimicrobial Agents and Chemotherapy*, 37, 1400, 1993.
191. Rouch, D.A. et al. Trimethoprim resistance transposon Tn*4003* from *Staphylococcus aureus* encodes genes for a dihydrofolate reductase and thymidylate synthetase flanked by three copies of IS257, *Molecular Microbiology*, 3, 161, 1989.
192. Lichtenstein, C.; Brenner, S. Site-specific properties of Tn*7* transposition into the *E. coli* chromosome, *Molecular and General Genetics*, 183, 380, 1981.

193. Barlow, R.S. et al. Isolation and characterization of integron-containing bacteria without antibiotic selection, *Antimicrobial Agents and Chemotherapy*, 48, 838, 2004.
194. Then, R.L. Mechanisms of resistance to trimethoprim, the sulfonamides, and trimethoprim-sulfamethoxazole, *Rev Infect Dis*, 4, 261, 1982.
195. Huovinen, P. Resistance to trimethoprim-sulfamethoxazole, *Clinical Infectious Diseases*, 32, 1608, 2001.
196. Maskell, J.P.; Sefton, A.M.; Hall, L.M. Multiple mutations modulate the function of dihydrofolate reductase in trimethoprim-resistant *Streptococcus pneumoniae*, *Antimicrobial Agents and Chemotherapy*, 45, 1104, 2001.
197. Chopra, I. Over-expression of target genes as a mechanism of antibiotic resistance in bacteria, *Journal of Antimicrobial Chemotherapy*, 41, 584, 1998.
198. Spratt, B.G. Resistance to antibiotics mediated by target alterations, *Science*, 264, 388, 1994.

8 Antibiotic Permeability

Harry Taber

CONTENTS

The ability of antimicrobial compounds to enter bacterial cells generally is a prerequisite to their antibacterial action. In order to penetrate the outer layers of the cell—the cell envelope—semi-permeable membranes and polymeric cell wall structures must be negotiated. Depending on the chemical nature of the antibiotic (hydrophilic or hydrophobic), penetration may occur by use of transmembrane pores, by localized disorganization of the membrane, by diffusion through the lipid bilayer, or by transport processes involving the co-opting of nutrient transport systems. Bacterial species differ widely in their envelope structures, and hence in their intrinsic resistance to antibiotics.

The energetic requirements for antibiotic entry appear to vary widely by antibiotic class. Cationic compounds such as the aminoglycosides appear to respond to a threshold level of membrane potential in order to cross the cytoplasmic membrane, while others, such as albomycin, depend on transporters of the ATP-binding cassette (ABC) type.

Permeability of the cell envelope can be modified by exposure of bacterial cells to non-antibiotic (e.g., cationic) compounds, a maneuver that will often significantly reduce the MIC for a particular antibiotic-bacterium combination. However, since intrinsic permeability is a balance between influx (often a passive process) and efflux (usually an active process), mechanistic interpretations of "permeabilizing" treatments often are flawed.

Bacterial cells contained in biofilms appear to occur as unique, physiologically heterogeneous populations with a variety of mechanisms for preventing antibiotic action on individuals within the biofilm. Many of these mechanisms are not yet understood, but in large part do not depend on genetic modifications. It seems likely that decreased permeability accounts for at least some of this resistance.

INTRODUCTION

As with other portions of this volume, this chapter does not presume to be comprehensive about its subject, even if attention is confined to the most recent literature. Rather, citations are made to reviews and to publications in the primary literature that illustrate advances in the field or problems that have not yielded adequate solutions. Bibliographies contained within the references that are cited will lead the reader to the very considerable depth that the permeability literature enjoys. Except for those concerned with antimicrobial peptides and with efflux of antimicrobial agents, general reviews on bacterial permeability to antibiotics have not appeared in recent years. Thorough coverage of the earlier literature can be found in the article by Chopra and Ball [1]. For permeability of Gram-negative bacteria, the reader is referred to the publication by Hancock and Bell [2].

Because of the intimate association between uptake of antibiotics into bacterial cells and the drug efflux pathways that have become the focus of so much study during the past several years, discussions of antibiotic permeability are often found embedded within publications, the principal intent of which is to explore contributions of efflux mechanisms to antibiotic resistance. The reader should be alert to this, and carefully probe these publications for relationships between drug influx and efflux phenomena. In the text that follows, several illustrations of these relationships are found.

WHAT DO WE MEAN BY PERMEABILITY?

Permeability in its broadest sense means the properties of a cell (in this case, a bacterial cell) that allow an ion or a molecule to traverse one or more of its boundary structures (its *envelope*, see below) and enter the cell. As applied to antimicrobial agents, this might involve penetration only of the outermost layer of the cell envelope, because the particular agent in question would not have to enter the bacterial cell cytoplasm proper in order to exert its inhibitory action. Alternatively, many antibiotics act only after reaching the cytoplasm (for example, those that affect DNA, RNA, or protein synthesis), and thus penetration of the cytoplasmic membrane is required.

The mechanisms for entry of antimicrobial agents vary widely, depending upon the chemical nature of the agent, and the characteristics of the envelope structure being penetrated. Some of these mechanisms will be discussed below, but in many cases, they are not known. A general term to describe entry is *antibiotic uptake*,

TABLE 8.1
PATHS FOR ANTIBIOTIC UPTAKE BY BACTERIA

Uptake by passive diffusion

a. Saturable, apparently dependent on a facilitator(s) contained within the cell envelope; this facilitator could be either in the outer or the inner membrane, and would be rate-limiting for uptake.

b. Nonsaturable, not dependent on interaction with a facilitator. Transit of the outer membrane could occur either through a porin-based structure, or—for lipophilic compounds—through the membrane bilayer. Lipophilic compounds could also readily transit the cytoplasmic membrane bilayer, but charged antimicrobials would encounter a barrier unless some mechanism such as self-promoted uptake (see below) were operating.

Energy-requiring uptake

Accumulation of antimicrobial drugs inside the bacterial cell occurs against a concentration gradient. Operationally, inhibitors of energy metabolism will prevent uptake. Uptake could depend either on ATP or on the proton-motive force as a source of energy.

Self-promoted uptake

The antimicrobial agent interacts with a cell envelope structure in such a way as to increase the permeability of that structure to the agent; this seems to occur for certain positively charged antibiotics as they encounter the outer membrane of Gram-negative bacteria.

Illicit uptake

The antimicrobial agent utilizes an uptake system normally used by the bacterial cell for uptake of a metabolite or nutrient.

---

which simply states that the molecule is moving inward from the environment. A related term is *antibiotic accumulation*, which carries the implication that concentrations of the molecule inside the bacterial cell are higher than outside, that is, that inward movement is occurring against a concentration gradient, and that energy is being expended in this accumulation process. Finally, use of the term *transport* indicates that the uptake process is both specific and saturable and involves some type of carrier component in one or another of the boundary layers. All of these terms are brought to bear in describing antibiotic permeability in the literature. Table 8.1 summarizes modes of uptake of antimicrobial compounds by bacterial cells.

PERMEABILITY BARRIERS OF THE BACTERIAL CELL

STRUCTURE OF BACTERIAL CELL ENVELOPES

The envelope of the bacterial cell consists of all structures external to the cytoplasm, including the cytoplasmic membrane (CM), which immediately overlies the cytoplasm; the cell wall, composed largely of the polymer peptidoglycan (cross linked to various degrees in different species); and, in Gram-negative bacteria, the outer membrane (OM). Gram-positive bacteria lack an OM, but have in general a much thicker cell wall layer than do Gram-negatives. In addition, some bacteria, particularly pathogens, possess a capsule lying outside the OM. As a consequence of their two-membrane envelope, Gram-negative bacteria have a compartment, called the *periplasmic space*, located between the CM and OM, containing many proteins and certain other macromolecules. Gram-positive bacteria have a somewhat comparable compartment within their cell walls, called appropriately, the *wall space*. Depending upon the particular Gram-positive species, the wall space also contains proteins, some anchored in the CM, others retained within this space by mechanisms not yet elucidated. Any antimicrobial compound that requires access to the cell in order to act as an inhibitory or bactericidal agent must traverse the bacterial cell envelope, since it completely surrounds the cell. It is not surprising to find, then, that antibiotic inactivating enzymes are found in the envelope: β-lactamases and aminoglycoside-modifying enzymes are examples.

Envelope Components Relevant to Antibiotic Permeability

Lipopolysaccharide (LPS) is a complex molecule characteristic of the Gram-negative OM, and maintenance of its integrity is essential to proper functioning of the outer membrane as a permeability barrier. It has been known for some time that this structure provides an effective impediment to free movement of antibiotics into the bacterial cell [3–5]. Detailed analysis of precisely which constituents are essential for LPS integrity has revealed that they are contained within the oligosaccharide *core region* of the molecule, which is strongly negatively charged. The presence of LPS only in the outer leaflet of the OM, with negative charges partially neutralized by divalent cations, such as Mg^{++} and Ca^{++}, provides a "carpet" of surface negative charge, contributing to the role of the OM as a selective permeability barrier. The core region is being dissected genetically [6] in order to find the molecular requirements for LPS integrity; so-called "deep-rough" mutations convey hypersensitivity to certain antibiotics (e.g., novobiocin) and to surface-active agents, such as sodium dodecyl sulfate.

A second class of constituents of the Gram-negative OM that produces its permeability characteristics is the porins. These are proteins capable of forming trans-OM channels ("pores") that have an aqueous interior and permit influx of hydrophilic small molecules, such as nutrients, and efflux of waste products [5,7,8]. In many cases, the pores exclude antibiotics because of their lipophilicity, and in addition restrict entry by size. The so-called "classical" porins occur in the OM as trimers of a 36 to 38 kDal monomer, with a quite open structure to the channel formed. The naming of porins is often referred to the *Escherichia coli* Omp (outer membrane protein) nomenclature, because they were first studied systematically in this species. Thus, OmpC and OmpF, together with PhoE are found in *E. coli*, and similar porins are widely distributed among other Gram-negative bacteria. In addition to size limitation, the trimeric pores formed have some preference for the type of small molecules that will be admitted, with OmpC and OmpF selecting molecules with positive or no charge, and PhoE preferring anions. Other porins are synthesized under special circumstances, for example, the LamB porin of *E. coli*, which has specificity for oligosaccharides, is formed under conditions of carbon limitation. Nikaido [9] has presented a brief overview of porin regulation. Although a overview of the regulation of the classical porins is beyond the scope of the present discussion, it is worth noting that OmpF, which has a larger pore diameter than OmpC, is down-regulated in *E. coli* in response to antibiotic exposure, mediated by the global regulator MarA ([10]; see also Miller and Rather, Chapter 3, this volume).

SELF-INDUCED UPTAKE OF AMINOGLYCOSIDE
ANTIBIOTICS THROUGH THE OUTER MEMBRANE
OF GRAM-NEGATIVE BACTERIA

Polycationic antibiotics can traverse the outer membrane by means other than diffusion through pores. As originally outlined by Hancock [2,11,12], aminoglycoside antibiotics such as tobramycin and gentamicin enter Gram-negative bacteria by a self-promoted pathway, involving disruption of LPS-Mg^{++} cross bridges, which as indicated above pose the major barrier to antibiotic entry. Disruption depends on the cationic structure of aminoglycosides, can be effected by other polycations such as polymyxin and protamine, and is mechanistically dependent on displacement of Mg^{++} from cross bridges. The resulting rearrangement of LPS and exposure of the OM bilayer provides sufficient localized destabilization of the bilayer to allow entry of aminoglycoside to the periplasm. Bacterial species such as *Burkholderia cepacia*, which have LPS with a low phosphate and high arabinosamine content, are resistant to polycationic antibiotics, apparently because the *B. cepacia* LPS does not bind polycations effectively [13]. The general phenomenon of self-induced uptake is discussed by Hancock [5] in the context of modification of outer membrane permeability for enhancement of antibiotic efficacy.

RESISTANCE ASSOCIATED WITH SPECIES-SPECIFIC VARIATIONS
AND MUTATIONAL ALTERATION IN ENVELOPE STRUCTURE

Specificity of porins for substrates is sometimes extended to antibiotics; for example, the β-lactam imipenem is specifically taken up by *Pseudomonas aeruginosa* through the substrate-specific porin OprD [14]. OprD formation is highly regulated by nitrogen and carbon sources [15], and its down-regulation results in imipenem resistance [16], as do mutations in OprD [17].

Do altered porins occur in drug-resistant clinical isolates? In some species, the answer would seem to be yes: Mallea et al. [18] described two nosocomial clones of *Enterobacter aerogenes* lacking OmpC and OmpF, and a third having thermolabile porins. This collection of clinically derived strains has been augmented by Bornet et al. [19]; the emergence of imipenem (and multidrug) resistance associated with decreased porin synthesis could be reversed by cessation of imipenem therapy. It appears that at least with *Enterobacter*, the ease with which porin deficiency-associated multidrug resistance arises, in combination with the presence of extended-spectrum β-lactamase, is already creating major difficulties in treatment of this third leading cause of nosocomial respiratory tract infections. A similar situation may be presenting itself in therapy of *Klebsiella pneumoniae* [20,21]. Clinical isolates of this pathogen now are appearing that have lost the OmpK35 porin (a homolog to *E. coli* OmpF), and like *Enterobacter*, this phenotype is seen in combination with the presence of extended-spectrum β-lactamases.

Recently, evidence for a porin-efflux pump collaboration in *Neisseria gonorrhoeae* has been discovered. Olesky et al. [22] previously had identified specific mutations in porin IB that give rise to intermediate-level penicillin and tetracycline resistance (*penB* strains), but only in the presence of an *mtr* mutation, which increased expression of the *mtrCDE*-encoded efflux pump operon. Porins from these *penB* mutant strains were overexpressed, isolated, and their physical properties studied *in vitro* in planar lipid bilayers or liposomes [23]. Conductance and ion selectivities of the porin channels were measured, as well as the flux of a range of sugars and β-lactam antibiotics. The mutant porins did not differ from their wild-type counterparts, consistent with the *in vivo* finding that strains with altered porins did not have altered MICs, unless the MtrC-MtrD-MtrE pump was overexpressed. Olesky et al. [23] suggest two possibilities: (*i*) only a slight decrease in antibiotic permeation through porins is required to confer resistance when expression of the efflux pump is deregulated; and (*ii*) that the porin IB variants may interact directly with the MTR efflux pump and work co-operatively to decrease periplasmic concentrations of antibiotic. Shafer and Folster [24], in their useful commentary on the results of Olesky et al., suggest an additional hypothesis, that the MtrR protein regulates other genes involved in determining levels of penicillin resistance in *N. gonorrhoeae*, and that this regulation may either be dependent on, or independent of, the allelic status of porin IB.

A common pattern may be coming to dominate the emergence of permeability-associated single- or multidrug resistance, at least in Gram-negative pathogens: restricted formation of porins combined with the induction or overexpression of efflux pump(s) with narrow or broad specificity.

ILLICIT UPTAKE: THE USE OF METABOLIC UPTAKE SYSTEMS FOR ANTIBIOTIC ENTRY

If an antibiotic bears a sufficiently close structural relationship to a molecule for which bacteria have specific uptake systems, then the antibiotic may be carried into the cell by that system. Two recent examples of this phenomenon are outlined below.

The work of Raynaud et al. [25] on uptake of the anti-tuberculosis drug pyrazinamide (PZA) by mycobacteria suggests that the drug is transported into the cell by a system normally used for transporting nicotinamide. That this transport system has substantial specificity is supported by the finding that pyrazinoic acid (POA), the intracellularly active form of the drug, has low activity when administered extracellularly; that is, the amide form, but not the acid form, can utilize the nicotinamide transport system.

Measurement of PZA susceptibility of *Mycobacterium tuberculosis* strains is complicated by the low *in vitro* activity of the drug at neutral pH, a problem that is obviated by carrying out susceptibility measurements at acidic pH [26]. The reason for this has been analyzed recently by Zhang et al. [27] as part of a more extensive study on the biochemical basis of PZA susceptibility in *M. tuberculosis* (largely due to defective efflux). The susceptibility measurement effect involves the equilibration of POA—synthesized intracellularly from accumulated PZA—in its protonated form across the cytoplasmic membrane (POA, in common with most weak organic acids, will diffuse across a lipid bilayer). When the extracellular pH is neutral (i.e., similar to the intracellular pH), protonated POA will continue to diffuse out of the cell in an effort to achieve equilibrium with the much larger extracellular volume, and this diffusion will continue as POA is produced from PZA, preventing sufficient accumulation of POA to have an inhibitory effect. Equilibration of protonated POA occurs at a much lower extracellular concentration when the pH outside the cell is acidic, allowing accumulation inside the cell sufficient for inhibition to occur.

The above discussion illustrates some of the general complexities of antimicrobial permeability; in this particular case, while a specific transport system may allow entry of the (pro)drug, conversion to a different chemical form requires an entirely new study of the transmembrane behavior of the active drug.

It seems intuitively likely that bacterial transport systems for nutrients or metabolites will not be easily coopted for entry by naturally occurring antibiotics, since the availability of antibiotics emanating from producer species would reasonably have created a selective pressure for susceptible species to modify (via evolutionary selection) their transport systems to exclude antibiotics. Indeed, the relatively few examples of illicit transport of antimicrobials appear rather to involve synthetic drugs (such as PZA) with antimicrobial activity. However, if the nutrient is sufficiently essential, evolution of the transport system may be constrained, and its exploitation for uptake by a structurally related, naturally occurring antibiotic will remain intact. Such would seem to be the case for the transport of iron-binding molecules (siderophores) across the outer membrane of Gram-negative bacteria [28]. Iron, as Fe^{+3}, is insoluble, and is transported into bacteria as soluble iron complexes, bound to bacterially synthesized, low-molecular-weight siderophores. Outer membrane proteins with requisite specificity bind siderophore-iron complexes, transport them across the outer membrane by energy-dependent processes, and deposit them in the periplasm, where the complexes bind to soluble binding proteins, which in turn deliver them to ABC (ATP-binding cassette) transporters in the cytoplasmic membrane. Gram-positive bacteria have essentially similar systems, but without the outer membrane binding protein component.

Albomycin is a broad-spectrum antibiotic of the Fe^{+3}-binding sideromycin class, produced by a species of *Streptomyces*. It has a low MIC for both Gram-positive and Gram-negative bacterial species, and this high specific activity depends on illicit transport of albomycin by the FhuA-FhuD-FhuB-FhuC system, the physiological activity of which is to transport ferrichrome-Fe^{+3} complexes into the cell. FhuA is the outer membrane binding protein, FhuD is the periplasmic binding protein, and FhuB and FhuC form a cytoplasmic membrane protein complex that, when energized by ATP, catalyzes the entry of ferrichrome-Fe^{+3} or albomycin-Fe^{+3} complexes into the cytoplasmic compartment. Active transport across the outer membrane via FhuA is thought to depend on input of energy from the cytoplasmic membrane via the TonB-ExbB-ExbD complex [29]. When albomycin-Fe^{+3} enters the cytoplasm, a thioribosyl pyrimidine moiety with cell-inhibitory activity must be cleaved by a peptidase from the parent antibiotic molecule. As stressed by Braun [28], this arrangement provides an approach to the design of antibiotics that can be actively transported into bacteria. Structures of Fe^{+3}-hydroxamate and Fe^{+3}-albomycin co-crystals with FhuA provide detailed molecular parameters for the design of such compounds. Interestingly, a rifamycin derivative (CGP 4832), which is not structurally related to either hydroxamates or albomycin (and does not bind iron), also is actively transported across the outer membrane by the FhuA protein energized by the TonB-ExbB-ExbD system, but not involving the other Fhu proteins. This suggests that, after reaching the periplasm, the antibiotic diffuses across the cytoplasmic membrane down its concentration gradient.

Other antibiotic classes have been conjugated to siderophores in order to utilize the siderophore transport systems for delivering active compounds to the bacterial cytoplasm. These are summarized by Braun [28], and include sulfonamides and β-lactams. Results were somewhat mixed; this could be due to difficulties in obtaining intracellular release of the active moiety, or delivery to a cellular compartment that is not optimal for antibiotic action. Although the use of Fe^{+3}-siderophore delivery systems for illicit antibiotic uptake is still in an early phase, the general approach is very promising, for at least two reasons: (i) in infections, circulating Fe^{+3} is commonly extremely low, and pathogen iron uptake systems will be derepressed and attempting to function at high efficiency; (ii) antibiotic resistance by loss of an uptake system would be detrimental to the success of the pathogen. Thus, this would appear to be a promising avenue for the development of semi-synthetic compounds based on siderophore structures. Clearly, knowledge of the intracellular systems that release moieties with antibiotic activity will have to be gained; those systems that are essential to survival of the targeted pathogen would seem to be the best to explore.

ENERGY-DEPENDENT UPTAKE OF ANTIBIOTICS AND RESISTANCE ARISING FROM COMPROMISE OF ENERGY METABOLISM

In accord with the thermodynamics of diffusion processes, bacteria will accumulate an antibiotic in their cytoplasmic compartment at higher concentrations than in the extracellular milieu only if an energy-requiring accumulation process is involved. The exception to this occurs if an intracellular binding site for the antibiotic exists, with a sufficiently strong binding constant such that a significant fraction of intracellular antibiotic is not freely diffusible. In this situation, antibiotic would continue to diffuse into the cell down its concentration gradient.

In general, two modes of providing energy to an antibiotic uptake system can be envisioned: the first is the proton electrochemical gradient, and the second is ATP (as described above for siderophores and albomycin). Either mode requires that a mechanism exists in the CM for coupling utilization of energy to transmembrane movement of the antibiotic. This might, as discussed in the previous section, be coupled to movement across the OM as well as the CM.

The dependence of aminoglycoside antibiotic uptake on energy production by bacterial cells was established four decades ago [30–32]. Uptake of aminoglycosides has been studied more intensively, by more investigators, than any other class of antibiotic. This focus of attention probably has been due to the energy-dependent nature of the uptake process across the CM, and the insights that it might provide to antibiotic uptake in general. This energy dependence has been associated with the membrane potential [33–36], and may reflect a response by positively charged antibiotics to that potential, which carries a negative charge on the interior face of the CM. Results of Miller and associates [37] are consistent with diffusion across the CM through a voltage-gated channel, which closes following uptake due to decreased membrane potential associated with effects of aminoglycosides themselves on the CM. This would suggest that in the absence of a membrane potential, no uptake would occur. However, E. coli [38], and perhaps other bacteria, can adapt to loss of membrane potential and take up aminoglycosides by ATP-dependent processes.

Loss of aminoglycoside uptake associated with energetic deficiencies in clinically important pathogens such as *S. aureus* has been established for some years [39]. Proctor et al. [40–43] have revisited this problem with small-colony variant (SCV) clinical isolates of *S. aureus*. They have stressed the importance of the ability of these variants to persist in chronic infections, in the face of antibiotic treatment [43]. Whereas reduced uptake of cationic antimicrobials accounts for resistance to aminoglycosides, resistance to other antibiotics (e.g., β-lactams) appears to result from changes in growth rate or other more general physiological alterations.

It seems likely that a variety of uptake systems, energized by either a membrane potential or by ATP, will be found to exist in different bacterial species; for example, as discussed by Pym and Cole in this volume (Chapter 13), the anti-tuberculosis drug pyrazinamide recently has been found to enter mycobacteria via an ATP-dependent transport system. Understanding the energetics of antibiotic efflux will require a comparable understanding of the energetics of influx, since approaches to blocking or reducing efflux must not also compromise uptake. In some situations, such as the uptake of rifampin by *E. coli* and *S. aureus* [44], uptake of the antibiotic appears not to be energy dependent at all, as judged by the lack of any effect of inhibitors such as CCCP and dinitrophenol on the process. In this particular example, the saturability of rifampin accumulation appeared to reflect its binding to the intracellular target molecule (RNA polymerase) rather than specific interaction with an envelope component. This may be the situation for many antimicrobial compounds, but to establish this will require study of each individual antibiotic or antibiotic class.

MODIFICATION OF UPTAKE BY ADMINISTRATION OF A SECOND AGENT

Combined antibiotic therapy has of course a venerable history in infectious disease therapy. However, increased understanding of bacterial physiology and structure can provide possibilities for therapeutic use of agents that are not themselves antimicrobial, but have an enhancing effect on the action of a known antibiotic. For example, Rajyaguru and Muszynski [45] showed that susceptibility of *B. cepacia* isolates to several standard antibiotics could be significantly enhanced *in vitro* (four-fold or greater reduction in MIC) by the cationic compounds chlorpromazine and prochlorperazine. The mechanism by which this apparent change in permeability occurred was not clarified, but did not involve changes in outer membrane composition, nor was outer membrane permeability to the fluorescent probe 1*N*-phenylnaphthylamine enhanced. However, electron microscopy revealed a widening of the periplasmic space, suggesting that these cationic agents may alter interactions between outer membrane and cytoplasmic membrane.

There are difficulties in this type of analysis because of the probable involvement of drug efflux systems in most intrinsic antibiotic resistance (see Lomovskaya et al., Chapter 4, this volume). In Gram-negative bacteria, the structure of these efflux systems commonly is characterized by membrane-spanning cytoplasmic membrane and outer membrane proteins, joined by a periplasmic connecting protein.

Thus, modifications of the spatial relationship between cytoplasmic and outer membranes could disrupt the structural integrity of drug efflux systems. The opportunistic pathogen *P. aeruginosa* possesses a highly impermeable outer membrane, which contributes to its intrinsic multidrug resistance. This broad resistance is augmented by several multidrug efflux systems ([46]; Lomovskaya et al., Chapter 4, this volume). The interplay between the *P. aeruginosa* outer membrane and these efflux systems recently was described by Poole's group [47]. In this study, enhancement of outer membrane permeability (using agents whose action was known) could be separated from reduction in efflux, by use of mutants in which the MexA-MexB-OprM efflux pump had been genetically inactivated. The two effects (permeability enhancement and efflux reduction) were found to be synergistic in enhancing antibiotic susceptibility. Interestingly, overproduction of the efflux system by means of multi-copy plasmids containing the *mexA-mexB-oprM* genes overcame the effect of the permeability agents, and increased drug resistance above that of the wild type. Thus, a response of an efflux system-containing bacterial pathogen population to a combined antibiotic/permeabilizing agent exposure might be to select out mutants in which duplication of the efflux genes has occurred.

The area of outer membrane permeabilization has been reviewed by Vaara [48], with a significant focus on cationic agents, such as antimicrobial peptides. More recently, Nikaido [49] has discussed the contributions of outer membrane and efflux pumps to drug resistance in Gram-negative bacteria, in the context of the possibilities for improving drug access. It is clear from the examples cited above, that while the permeabilization approach has promise, it will be necessary to understand a great deal more about the response of bacterial cells to enhanced access of antimicrobials to the Gram-negative periplasmic space and to the cytoplasm of all bacteria. A more promising route to counteracting drug efflux may be to find agents that inhibit the efflux pumps directly, as discussed by Lomovskaya et al. (Chapter 4, this volume).

Miller and Rather (Chapter 3, this volume) discuss the effects of salicylic acid on reduction of Gram-negative antibiotic susceptibility, in the context of the known inducing effect of this agent on the *mar* system and on down-regulation of the OmpF porin. However, in *S. aureus*, salicylate effects are more complex, involving the apparent induction of an antibiotic resistance permeability barrier [50]. Price et al. [51] have provided a general review on salicylate effects on bacteria.

EFFECTS OF BIOFILM FORMATION ON ANTIBIOTIC PERMEATION

Much attention has been focused in recent years on bacterial biofilms, based on the recognition that they occur commonly both at infection sites and during environmental growth of bacterial populations [52,53]. The genetics and physiology of the development of biofilms is being thoroughly explored [54–56]. The role of biofilms in modifying antibiotic susceptibility of bacterial cells within these structures has received its share of attention. Initially, it was supposed that the biofilm acted only as a physical barrier, with the abundant extracellular biopolymers shielding resident bacterial cells from the action of antibiotics by retarding diffusion of the inhibitory agents. However, recent data suggest that other explanations are more likely [57,58].

First, the apparent lack of diffusion of some antibiotics into biofilms can be attributed in some cases to reaction inactivation, that is, the inactivation of the antibiotic by the generally higher concentration of cells at the periphery of the biofilm. An example would be failure of β-lactams to penetrate the biofilm when β-lactamase is present in the biofilm population; mutational loss of the β-lactamase results in ready penetration [57].

Second, it has been suggested repeatedly that biofilms contain a physiologically heterogeneous cell population [52,55], and direct evidence is available to support this contention [55,59]. This heterogeneity includes *persistors*, i.e., a fraction of cells that—in the context of antibiotic killing—are resistant even to high drug concentrations [50,60]. Biofilm cells that survive bactericidal drug treatments are not drug-resistant mutants; instead, they remain wild type in their genetic makeup. Biofilms appear to promote the formation of higher fractions of these persistors than do planktonic (non-biofilm) cultures. Extension of *in vitro* studies to *in vivo* infections involving biofilm formation by bacterial pathogens can be made as follows: antibiotic treatment will eliminate most genetically susceptible members of the population, both inside and outside a biofilm, and cells of the immune system will eliminate the planktonic remainder. However, it is known that immune cells are unable to penetrate the extracellular polysaccharide matrix that characterizes biofilms [61]; therefore, if a residuum of persistors is present inside the biofilm, the potential for regrowth of the pathogen population following termination of therapy remains significant.

OVERCOMING RESISTANCE ASSOCIATED WITH PERMEABILITY BARRIERS

The use of permeability-enhancing agents to improve drug access was discussed above (see Modification of Uptake by Administration of a Second Agent). A second approach might be to circumvent the permeability barrier entirely by presenting antibiotic to the cell interior via a membrane fusion mechanism. For example, in recent studies, drug-resistant mucoid *P. aeruginosa* could be eradicated in an animal model by the use of a suspension of liposomes containing encapsulated bactericidal antibiotic [62]. Mechanistic studies suggested that delivery of the antibiotic to the bacterial cytoplasm occurred by liposome-bacteria fusion. This is particularly noteworthy because of the high degree of intrinsic resistance exhibited by *P. aeruginosa* consequent to the low-permeability characteristics of its outer membrane.

REFERENCES

1. Chopra I, Ball P. Transport of antibiotics into bacteria. *Adv Microb Physiol* 1982; 23: 183–240.
2. Hancock REW, Bell A. Antibiotic uptake in Gram-negative bacteria. *Eur J Clin Microbiol* 1988; 7: 713–720.
3. Nikaido H. Outer membrane barrier as a mechanism of antimicrobial resistance. *Antimicrob Agents Chemother* 1989; 33: 1831–1836.
4. Van JCN, Gutmann L. Antibiotic resistance due to reduced permeability in Gram-negative bacteria. *Presse Med* 1994: 23: 522–531.
5. Hancock REW. The bacterial outer membrane as a drug barrier. *Trends Microbiol* 1997; 5: 37–42.

6. Yethon JA, Heinrichs DE, Monteiro MA, Perry MB, Whitfield C. Involvement of *waaY*, *waaO*, and *waaP* in the modification of *Escherichia coli* lipopolysaccharide and their role in the formation of a stable outer membrane. *J Biol Chem* 1998; 273: 26310–26316.

7. Nikaido H. Prevention of drug access to bacterial targets: permeability barriers and active efflux. *Science* 1994; 264: 382–388.

8. Nikaido H. Porins and specific diffusion channels in bacterial outer membranes. *J Biol Chem* 1994; 269: 3905–3908.

9. Nikaido H. Microdermatology: cell surface in the interaction of microbes with the external world. *J Bacteriol* 1999; 181: 4–8.

10. Alekshun MN, Levy SB. The *mar* regulon: multiple resistance to antibiotics and other toxic chemicals. *Trends Microbiol* 1999; 7: 410–413.

11. Hancock REW, Raffle VJ, Nicas TI. Involvement of the outer membrane in gentamicin and streptomycin uptake and killing in *Pseudomonas aeruginosa*. *Antimicrob Agents Chemother* 1981; 19: 777–785.

12. Hancock REW. Alterations in outer membrane permeability. *Annu Rev Microbiol* 1984; 38: 237–264.

13. Moore RA, Hancock REW. Involvement of outer membrane of *Pseudomonas cepacia* in aminoglycoside and polymyxin resistance. *Antimicrob Agents Chemother* 1986; 30: 923–926.

14. Trias J, Nikaido H. Protein D2 channel of the *Pseudomonas aeruginosa* outer membrane has a binding site for basic amino acids and peptides. *J Biol Chem* 1990; 265: 15680–15684.

15. Ochs MM, Lu CD, Hancock REW, Abdelal AT. Amino acid-mediated induction of the basic amino acid-specific outer membrane porin OprD from *Pseudomonas aeruginosa*. *J Bacteriol* 1999; 181: 5426–5432.

16. Masuda N, Sakagawa E, Ohya S. Outer membrane proteins responsible for multiple drug resistance in *Pseudomonas aeruginosa*. *Antimicrob Agents Chemother* 1995; 39: 645–649.

17. Kohler T, Michea-Hamzehpour M, Epp SF, Pechere JC. Carbapenem activities against *Pseudomonas aeruginosa*: respective contributions of OprD and efflux systems. *Antimicrob Agents Chemother* 1999; 43: 424–427.

18. Mallea M, Chevalier J, Bornet C, Eyraud A, Davin-Regli A, Bollet C, Pages J-M. Porin alteration and active efflux: two *in vivo* drug resistance strategies used by *Enterobacter aerogenes*. *Microbiology* 1998; 144: 3003–3009.

19. Bornet C, Davin-Regli A, Bosi C, Pages J-M, Bollet C. Imipenem resistance of *Enterobacter aerogenes* mediated by outer membrane permeability. *J Clin Microbiol* 2000; 38: 1048–1052.

20. Martinez-Martinez L, Hernandez-Alles S, Alberti S, Tomas JM, Benedi VJ, Jacoby GA. *In vivo* selection of porin deficient mutants of *Klebsiella pneumoniae* with increased resistance to cefoxitin and third generation cephalosporins. *Antimicrob Agents Chemother* 1996; 40: 342–348.

21. Hernandez-Alles S, Alberti S, Alvarez D, Domenech-Sanchez A, Martinez-Martinez L, Gil J, Tomas JM, Benedi VJ. Porin expression in clinical isolates of *Klebsiella pneumoniae*. *Microbiology* 1999; 145: 673–679.

22. Olesky M, Hobbs, M, Nicholas RA. Identification and analysis of amino acid mutations in porin IB that mediate intermediate-level resistance to penicillin and tetracycline in *Neisseria gonorrhoeae*. *Antimicrob Agents Chemother* 2002; 46: 2811–2820.

23. Olesky M, Zhao S, Rosenberg RL, Nicholas RA. Porin-mediated antibiotic resistance in *Neisseria gonorrhoeae*: ion, solute and antibiotic permeation through PIB proteins with *penB* mutations. *J Bacteriol* 2006; 188: 2300–2308.

24. Shafer WM, Folster JP. Towards an understanding of chromosomally mediated penicillin resistance in *Neisseria gonorrhoeae*: evidence for a porin-efflux pump collaboration. *J Bacteriol* 2006; 188: 2297–2299.

25. Raynaud C, Laneele M-A, Senaratne RH, Draper P, Laneele G, Daffe M. Mechanisms of pyrazinamide resistance in mycobacteria: importance of lack of uptake in addition to lack of pyrazinamidase activity. *Microbiology* 1999; 145: 1359–1367.

26. Salfinger M, Heifetz L. Determination of pyrazinamide MICs for *Mycobacterium tuberculosis* at different pHs by the radiometric method. *Antimicrob Agents Chemother* 1988; 32: 1002–1004.

27. Zhang Y, Scorpio A, Nikaido H, Sun Z. Role of acid pH and deficient efflux of pyrazinoic acid in unique susceptibility of *Mycobacterium tuberculosis* to pyrazinamide. *J Bacteriol* 1999; 181: 2044–2049.

28. Braun V. Active transport of siderophore-mimicking antibacterials across the outer membrane. *Drug Resist Updates* 1999; 2: 363–369.

29. Braun V. Energy-coupled transport and signal transduction through the Gram-negative outer membrane via TonB-ExbB-ExbD-dependent receptor proteins. *FEMS Microbiol Rev* 1995; 16: 295–307.

30. Hancock REW. Aminoglycoside uptake and mode of action—with special reference to streptomycin and gentamicin. I. Antagonists and mutants. *J Antimicrob Chemother* 1981; 8: 249–276.

31. Hancock REW. Aminoglycoside uptake and mode action—with special reference to streptomycin and gentamicin. II. Effects of aminoglycosides on cells. *J Antimicrob Chemother* 1981; 8: 429–445.

32. Taber HW, Mueller JP, Miller PF, Arrow AS. Bacterial uptake of aminoglycoside antibiotics. *Microbiol Rev* 1987; 51: 439–457.

33. Bryan LE, Van den Elzen HM. Effects of membrane-energy mutations and cations on streptomycin and gentamicin accumulation by bacteria: a model for entry of streptomycin in susceptible and resistant bacteria. *Antimicrob Agents Chemother* 1979; 12: 163–177.

34. Bryan LE, Kwan S. Roles of ribosomal binding, membrane potential, and electron transport in bacterial uptake of streptomycin and gentamicin. *Antimicrob Agents Chemother* 1983; 23: 835–845.

35. Mates SM, Eisenberg ES, Mandel LJ, Patel L, Kaback HR, Miller MH. Membrane potential and gentamicin uptake in *Staphylococcus aureus*. *Proc Natl Acad Sci USA* 1982; 79: 6693–6697.

36. Eisenberg ES, Mandel LJ, Kaback HR, Miller MH. Quantitative association between electrical potential across the cytoplasmic membrane and early gentamicin uptake and killing in *Staphylococcus aureus*. *J Bacteriol* 1984; 157: 863–867.

37. Leviton IM, Fraimow HS, Carrasco N, Dougherty TJ, Miller MH. Tobramycin uptake in *Escherichia coli* membrane vesicles. *Antimicrob Agents Chemother* 1995; 39: 467–475.

38. Fraimow HS, Greenman JB, Leviton IM, Dougherty TJ, Miller MH. Tobramycin uptake in *Escherichia coli* is driven by either electrical potential or ATP. *J Bacteriol* 1991; 173: 2800–2808.

39. Miller MH, Edberg SC, Mandel LJ, Behar CF, Steigbigel NH. Gentamicin uptake in wild-type and aminoglycoside-resistant small-colony mutants of *Staphylococcus aureus*. *Antimicrob Agents Chemother* 1980; 18: 722–729.

40. Proctor RA, Peters G. Small colony variants in staphylococcal infections: diagnostic and therapeutic implications. *Clin Infec Dis* 1998; 27: 419–423.

41. Proctor RA, Kahl B, von Eiff C, Vaudaux PE, Lew DP, Peters G. Staphylococcal small colony variants have novel mechanisms for antibiotic resistance. *Clin Infec Dis* 1998; 27 (Suppl 1): S68–S74.

42. McNamara PJ, Proctor RA. *Staphylococcus aureus* small colony variants, electron transport and persistent infections. *Internat J Antimicrob Agents* 2000; 14: 117–122.

43. Proctor RA, von Eiff C, Kahl BC, Becker K, McNamara P, Herrman M, Peters G. Small colony variants: a pathogenic form of bacteria that facilitates persistent and recurrent infections. *Nature Rev Microbiol* 2006; 4: 295–305.

44. Williams KJ, Piddock LJV. Accumulation of rifampicin by *Escherichia coli* and *Staphylococcus aureus*. *J Antimicrob Chemother* 1998; 42: 597–603.
45. Rajyaguru JM, Muszynski MJ. Enhancement of *Burkholderia cepacia* antimicrobial susceptibility by cationic compounds. *J Antimicrob Chemother* 1997; 40: 345–351.
46. Li X-Z, Nikaido H, Poole K. Role of *mexA-mexB-pprM* in antibiotic efflux in *Pseudomonas aeruginosa*. *Antimicrob Agents Chemother* 1995; 39: 1948–1953.
47. Li X-Z, Zhang L, Poole K. Interplay between the MexA-MexB-OprM multidrug efflux system and the outer membrane barrier in the antibiotic resistance of *Pseudomonas aeruginosa*. *J Antimicrob Chemother* 2000; 45: 433–443.
48. Vaara M. Agents that increase the permeability of the outer membrane. *Microbiol Rev* 1992; 56: 395–411.
49. Nikaido H. The role of outer membrane and efflux pumps in the resistance of Gram-negative bacteria. Can we improve drug access? *Drug Resist Updates* 1998; 1: 93–98.
50. Price CTD, Kaatz GW, Gustafson JE. The multidrug efflux pump NorA is not required for salicylate-induced reduction in drug accumulation by *Staphylococcus aureus*. *Intern J Antimicrob Ag* 2002; 20: 206–213.
51. Price CTD, Lee IR, Gustafson JE. The effects of salicylate on bacteria. *Intern J Biochem Cell Biol* 2000; 32: 1029–1043.
52. Costerton JW, Lewandowski Z, Caldwell DE, Korber DR, Lappin-Scott HM. Microbial biofilms. *Annu Rev Microbiol* 1995; 49: 711–745.
53. Costerton JW, Stewart PS, Greenberg EP. Bacterial biofilms: a common cause of persistent infections. *Science* 1999; 284: 1318–1322.
54. O'Toole GA, Pratt LA, Watnick PI, Newman DK, Weaver VB, Kolter R. Genetic approaches to study of biofilms. *Meth Enzymol* 1999; 310: 91–109.
55. Pratt LA, Kolter R. Genetic analyses of bacterial biofilm formation. *Curr Opin Microbiol* 1999; 2: 598–603.
56. Lewis K. Programmed death in bacteria. *Microbiol Molec Biol Rev* 2000; 64: 503–514.
57. Anderl JN, Franklin MJ, Stewart PS. Role of antibiotic penetration limitation in *Klebsiella pneumoniae* biofilm resistance to ampicillin and ciprofloxacin. *Antimicrob Agents Chemother* 2000, 44: 1818–1824.
58. Brooun A, Liu S, Lewis K. A dose-response study of antibiotic resistance in *Pseudomonas aeruginosa* biofilms. *Antimicrob Agents Chemother* 2000; 44: 640–644.
59. Huang C-T, Xu KD, McFeters GA, Stewart PS. Spatial patterns of alkaline phosphatase expression within bacterial colonies and biofilms in response to phosphate starvation. *Appl Environ Microbiol* 1998; 64: 1526–1531.
60. Lewis K. Persister cells, dormancy and infectious disease. *Nature Rev Microbiol* 2007; 5: 48–56.
61. Hoyle BD, Jass J, Costerton JW. The biofilm glycocalyx as a resistance factor. *J Antimicrob Chemother* 1990; 26: 1–5.
62. Sachetelli S, Khalil H, Chen T, Beaulac C, Senechal S, Lagace J. Demonstration of a fusion mechanism between a fluid bactericidal liposomal formulation and bacterial cells. *Biochim Biophys Acta-Biomembranes* 2000; 1463: 254–266.

9 Genetic Methods for Detecting Bacterial Resistance Genes

Ad C. Fluit

CONTENTS

The increase in antibiotic resistance and the increasing possibilities and availability of molecular techniques make it only natural to use these techniques to study and detect the genes involved. Current molecular techniques are derived from a limited number of basic principles: probes, polymerase chain reaction (PCR), DNA sequencing, and microarrays. Probe technology is based on the interaction between two DNA strands, which are either identical or show mismatches due to mutations. The polymerase chain reaction (PCR) has nearly become a household name. With widely available computer programs, it is rather easy to design a PCR to detect a single gene, but more complex assays have been developed, which include the concomitant use of probes in a process called real-time PCR. DNA sequencing is long known, but developments in fluorescent dyes and the new process of pyrosequencing have revolutionized this technique. Microarrays are a form of large-scale hybridization that is becoming increasingly popular because of its ability to examine many genes simultaneously. Although the principles behind molecular techniques are simple, the design and implementation of a more advanced assay may be problematic. Potential problems involve contamination, lack of a gold standard to validate the assay, poor analytical sensitivity, and a large number of genes or mutations that need to be covered in order to correctly predict resistance. In addition, the importance for therapy of the presence of a mutation or resistance gene has to be considered. In this review, the principles of the techniques and the problems will be discussed. The (im)possibilities of molecular assays will be illustrated by examples intended for routine diagnostics, one of the most challenging environments for molecular techniques. The examples will focus especially, but not exclusively, on the detection of antibiotic resistance in *Mycobacterium tuberculosis*, methicillin-resistant *Staphylococcus aureus* (MRSA), and antibiotic-resistant *Helicobacter pylori*.

INTRODUCTION

The discovery of the double helix structure of DNA by Watson and Crick opened new avenues for the diagnosis of disease. We should no longer be dependent on subjective phenotypic measurements, but can head straight for the mechanism underlying disease. The diagnosis of infectious disease including the determination of antibiotic resistance is no exception. The invention of the PCR promised a revolution in the diagnosis of infectious diseases and other illnesses. However, 20 years after the invention of the PCR the promised revolution has not been borne out in the diagnosis of resistance determinations. It is more appropriate to speak of an evolution. There are a number of reasons for this. The large variety in genetic mechanisms leading to antibiotic resistance, the low cost of culture in combination with the fact that antibiotics are available to treat most infections effectively with empiric therapy without knowing the species or the antibiotic resistance profile of the infecting bacterium, and the problem of contamination in PCR assays (the ability to detect a few copies of a gene is also its weakness). In this chapter, the factors affecting the successful implementation of molecular assays, the basic types of molecular assays used, and the current state of the art for different bacterial species will be discussed.

PHENOTYPING VERSUS GENOTYPING

The correct detection of antibiotic resistance of clinical isolates is of great importance for optimal antibiotic therapy and successful treatment of patients. This only becomes more important with increasing prevalence of antibiotic-resistant isolates. Testing is not only required for therapy, but also to monitor the spread of antibiotic-resistant organisms or resistance genes through the hospital, community, or other reservoirs, such as animal husbandry.

Before the advent of molecular techniques, and even today, the mainstays of detection of antibiotic resistance were, and remain, phenotypic tests. Relatively simple *in vitro* susceptibility tests have been developed that usually allow a correct prediction of the effectiveness of antibiotic therapy *in vivo*, that is, the successful treatment of a patient. In a few cases, these tests, which monitor the inhibition of growth of clinical isolates by particular antibiotics, are insufficient and additional tests are required. An example is resistance to third generation cephalosporins caused by extended-spectrum β-lactamases (ESBLs). Routine *in vitro* susceptibility tests may show susceptibility to some of these antibiotics, but therapy has a high likelihood of failure and additional *in vitro* tests have been developed to detect ESBLs.

The invention of PCR methodology made possible the detection of each gene whose sequence is known. In fact, for all or nearly all resistance genes PCR tests have been described. For a sequenced gene, design of a PCR-based test using widely available programs that usually propose the correct primers to initiate the polymerization reaction is relatively straightforward. Beyond this simple approach, more elaborate methods have been developed that allow quicker and quality controlled answers. However, the availability of these newer molecular methods does not mean by definition that they are an improvement over the existing phenotypic *in vitro* methods. A major drawback of molecular methods is that the genetic mechanism responsible for the resistance should be known. If this mechanism is not known, no appropriate molecular assay can be developed. Even more worrisome is that when only molecular assays are used, resistance caused by newly emerging mechanisms is not detected. An example of a recently discovered unexpected resistance determinant is the plasmid-encoded *qnrA1* gene, which mediates resistance to fluoroquinolones in Enterobacteriaceae [1,2].

Even if the genetic mechanism of resistance is known, the number of possibilities may be limiting, for example, the existence of a large number of different enzymes that lead to a common resistant phenotype. Examples are the number of different β-lactamases that cause resistance to penicillins and other β-lactam antibiotics, or the number of aminoglycoside-modifying enzymes in Gram-negative bacteria that cause resistance to commonly used antibiotics, such as gentamicin and tobramycin. The required assays become highly complex not only due to the number of genes that must be detected, but also due to the number of positive controls required to validate the result of each individual test. Testing may become even more complex when the presence of an antibiotic induces the expression of already present efflux pumps or a reduction in the number of porins through which antibiotics are transported into the cell. Specialized assays to detect the up- or down-regulation of genes encoding these proteins are then required.

Even when the presence of a gene that encodes resistance is detected, this finding does not necessarily mean that it confers a resistance phenotype. The level of expression of the gene may be too low to have an impact on therapy. Expression of resistance genes may also vary between different species or strains of the same species. The importance for antibiotic prescription is not always clear. For example, *Escherichia coli*, one of the most common causes of bacterial infections, encodes a chromosomally located β-lactamase, but this β-lactamase is never expressed. β-lactam resistance in *E. coli* is caused by the acquisition of other β-lactamases. Resistance genes present in class 1 integrons constitute another example. Class 1 integrons are genetic structures characterized by an integrase that is capable of integrating gene cassettes adjacent to the *int* gene (Figure 9.1). The number of gene cassettes may vary from zero to at least seven. More than 100 different gene cassettes have been described, and the vast majority encode for resistance against antibiotics or disinfectants. Nearly all gene cassettes lack a promoter. Transcription occurs from two promoters (P1 and P2) between the integrase gene and the first cassette. Promoter P2 is seldom functional, whereas promoter P1 varies in strength. In addition, expression of genes further away (downstream) from the promoter is lower. This may result in poor expression and thereby a susceptible phenotype despite the presence of an appropriate resistance-encoding gene [3]. A third example is the *qnrA1* gene encoding resistance against fluoroquinolones. Expression of the gene results in a 16- to 100-fold increase in the minimal inhibitory concentration (MIC) for fluoroquinolones depending on the species and the strain. Because the MIC values for fluoroquinolones for many bacterial species are extremely low (≤0.015 μg/mL) a 16- to 100-fold increase in MIC value does not necessarily result in a value above the breakpoint for resistance as defined by the Clinical Laboratory Standards Institute (CLSI, previously known as the National Committee on Clinical Laboratory Standards [NCCLS]) [4]. A final example is MRSA. Methicillin resistance or in fact resistance to all β-lactams in *S. aureus* is dependent on the expression of the *mecA* gene, which encodes a penicillin binding protein, called PBP2A, that is insensitive to inhibition by β-lactam antibiotics. In most strains the expression of *mecA* is constitutive, but in some strains it is controlled. These latter strains show *in vitro* susceptibility to flucloxacillin, the most commonly used β-lactam antibiotic to treat *S. aureus* infections. However, constitutive expression may be achieved by simple deletion of part of the controlling gene. Therefore, the presence of the chromosomal β-lactamase gene in *E. coli* has never led to problems, but the presence of the *mecA* gene in *S. aureus* is considered the gold standard for β-lactam resistance in this

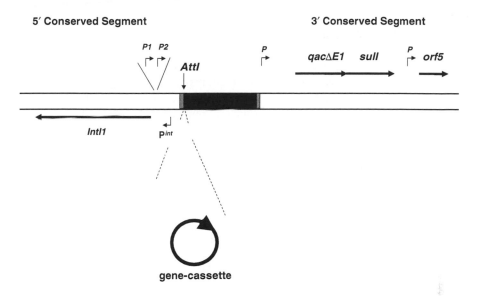

5′ Conserved Segment

3′ Conserved Segment

gene-cassette

FIGURE 9.1 A schematic presentation of a class 1 integron. The genes and their transcription direction are indicated by thick arrows. The *qacΔE1* and *sulI* gene encode resistance to quarternary ammonium compounds and sulfamethoxazole, respectively. The *int* gene encodes the integrase. The function of *orf5* is unknown. The promoters are indicated with small arrows. The gene cassettes that can be excised and circularized or integrated into the *attI*-site are indicated wit black fill. Multiple gene cassettes can be present. The *attI*-site (grey) is split by the integration of a gene cassette. The gene cassettes with a few exceptions are transcribed from promoters P1 and P2, although P2 is often not functional.

species. Consequently, the presence of the *mecA* gene is considered sufficient to consider such an isolate resistant and no β-lactam antibiotics are prescribed.

Thus, knowledge of the genetic mechanism and its expression is paramount for the successful and correct identification of antibiotic resistance by molecular methods. However, despite these limitations molecular assays have been developed, and are routinely used for the detection of antibiotic resistance in diagnostic laboratories for patient care.

MOLECULAR METHODS IN THE DIAGNOSTIC LABORATORY

Introduction

Nearly all molecular methods used to detect resistance encoding genes are based on hybridization, amplification, and DNA sequencing. A thorough understanding of the conditions influencing the results of these tests helps interpret the results, and translate them into optimal antibiotic therapy for the individual patient.

Hybridization

By definition, different genes have different DNA sequences; in principle, it should be possible to select DNA sequences unique for particular (resistance) genes, and utilize hybridization to detect the presence of these genes. These can be either short single-stranded fragments (oligonucleotides) obtained by chemical synthesis, or longer fragments, usually obtained by PCR. For detection, these single-stranded DNA molecules are coupled to a label. Popular labels are fluorescent dyes, enzymes, such as alkaline phosphatase or peroxidase that produce either light or a color stain upon addition of a suitable substrate, or digoxigenin that is subsequently detected by a specific antibody fragment coupled to alkaline phosphatase. These labels generally have replaced the radioactive labels formerly used. The labeled single-stranded DNA molecule is called a probe. Before hybridization can take place, the target gene should be made single-stranded. This is normally achieved by high temperature (>90°C) or hydroxide treatment.

Frequently, the probe is not labeled, but the DNA sequence of interest in the microorganism is labeled in a PCR amplification. The probe is spotted unlabeled on a membrane often in the form of a line; this format is called a reverse line blot. Macroarrays have usually a limited number of DNA fragments on a solid phase, often a nylon membrane. In arrays, a color reaction generally is used to detect hybridization. Arrays may, in fact, be considered a variation of reverse line blot technology.

A perfect match between probe and target gene sequence is not necessary to obtain hybridization. This has two important consequences. First, sequences with a less-than-perfect match also may hybridize, potentially yielding a false-positive result. Second, resistance genes with similar sequences that yield an identical or similar resistance phenotype may be detected with the same probe. In the latter case, the disadvantage is turned into an advantage, since detection can be simplified by the use of a single probe for multiple genes. The specificity of the hybridization can be influenced either by the temperature at which hybridization and subsequent removal of excess probe is carried out, or by the composition of the buffers used. Hybridization becomes more specific when more stringent conditions are used. This means that probe size and hybridization conditions should be tailored to the genetic mechanism responsible for the resistance to be determined.

Heteroduplex formation is a tool that is frequently used to detect point mutations. In this method a hybrid DNA molecule is formed by the hybridization of a wild-type strand with a mutant strand. The mismatch at the position of the mutation influences the temperature or buffer conditions at which the heteroduplex melts to single-stranded molecules. Slab gels commonly have been used in a technique called denaturing gradient gel electrophoresis (DGGE) to detect heteroduplex formation. The amplification products migrate through a gradient of denaturants, and products with a mutation are discerned based on an altered melting temperature. Increasingly, however, capillary or HLPC-based methods are described.

POLYMERASE CHAIN REACTION

PCR is nowadays an integral part of the molecular toolbox. Basically, a primer (an oligonucleotide) is annealed (hybridized) to the target DNA that is made single-stranded at 94°C (denaturation step). The oligonucleotide is used as an initiation point for a heat-stable DNA polymerase (hence the name primer). DNA polymerase extends the primer at a temperature that is usually approximately 72°C in a process called elongation. After these three steps (cycle 1) the procedure is repeated and the newly synthesized DNA strand also becomes a target. Each cycle results in a doubling of the number of strands in optimal conditions. Depending on the amount of input DNA, that is, the number of copies of the target sequence, the efficiency of amplification, and the detection method used, a detectable amount of DNA is produced in less than 3 h after 20 to 40 cycles.

In cases where an extremely low number of copies of the target DNA is present, the PCR procedure can be repeated using the amplification product of the first PCR as a target. When a different set of primers is used, this process is called a nested PCR. A major drawback of this method is its extreme sensitivity and thereby susceptibility to contamination and thus false-positive results.

Frequently, the detection of multiple genes is performed in a single PCR (multiplex PCR) by combining a number of primer sets. However, this is not completely straightforward. Usually, short sequences are more easily amplified than larger fragments; the copy number of the target also plays an important role. A resistance gene present on a plasmid, which has 20 copies per bacterial cell, may amplify more easily than a shorter sequence from a resistance gene present in a single copy on the chromosome of the same bacterial cell. Finally, the base pair composition of the sequence may play a role. Some sequences, for example, GC-rich tracts, are more difficult for DNA polymerase to synthesize than others and this may affect the efficiency of the reaction. The risk of false-negative results increases with an increasing number of primer sets. It is therefore extremely important to optimize these PCRs when new batches of primers are used, because of slight variations between batches. Even a change in thermocycler may affect the outcome.

Although the PCR may yield a detectable fragment (and hence a positive result), the fragment identity should be confirmed. The length of the product may give some indication, but confirmation by a probe or sequencing is usually required, because non-specific products are regularly formed. A commonly used procedure is restriction enzyme analysis. Based on the known target sequence, the position of cuts made by different restriction enzymes can be predicted and thereby the size of the fragments to be expected. Sometimes, the presence or absence of a mutation may lead to the formation or disappearance of a restriction site and thereby different fragment lengths. This technique is known as restriction fragment length polymorphism (RFLP) analysis. A possible drawback is that not all relevant mutations may be covered by restriction enzymes because only a limited number of sequences are recognized by the available restriction enzymes.

A number of variations on the theme of PCR have been developed. The most common is real-time PCR (rtPCR is to be distinguished from reverse-transcriptase PCR, RT-PCR, in which RNA is reverse transcribed into DNA before PCR amplification). In a rtPCR the amplification process can be monitored "live," when either a fluorescent dye or a fluorescently labeled probe is added to the PCR reaction. The dye intercalates into newly formed DNA and fluoresces. The probe interacts with the newly formed product during the annealing step and is then able to fluoresce. Four types of probes can be used: 5′-nuclease (TaqMan) probes, molecular beacons, fluorescence resonance energy transfer (FRET) probes (Figure 9.2), or minor groove binding (MGB) probes. TaqMan probes are short oligonucleotides that

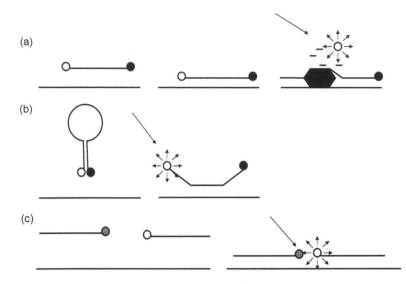

FIGURE 9.2 Schematic presentation of three types of probe used in real-time PCR. Panel a: Taqman probe. The fluorescent dye (white) and the quencher (black) are close together on the probe and no light can be emitted. A DNA polymerase with 5'-exonuclease removes any DNA in front of it during strand synthesis and thereby releases the fluorescent dye. The dye can now be excited (long arrow) and emit light. Panel b: molecular beacon. The fluorescent dye (white) and the quencher (black) are close together when the probe is in the panhandle structure and no light can be emitted. When the probe hybridizes, the fluorescent dye and the quencher are separated and light emission can take place after excitation (long arrow). Panel c: FRET probe. An acceptor molecule (grey) for the excitation (long arrow) of the fluorescent dye (white) is required, but fluorescence can only occur when acceptor and dye are in close proximity, that is, during hybridization.

have a fluorescent dye at one end and a quenching molecule at the other end. As long as fluorescent dye and quencher are in close proximity, no light is emitted by the probe. If the probe anneals to single-stranded DNA of the newly formed product, the *Taq* polymerase will break down the probe during elongation and release the fluorescent dye, which then can emit light that is no longer quenched. Molecular beacons are oligonucleotides that also have a fluorescent dye at one end and a quencher at the other end, but these are attached to short inverted repeats, which allow a "panhandle" to form, which brings dye and quencher close together. When the probe hybridizes, the quencher and dye are separated sufficiently to allow light emission. FRET probes are composed of two oligonucleotides. The first has a light-absorbing molecule at its tail. The second oligonucleotide hybridizes directly behind the first and has the fluorescent dye at its head. Only when the fluorescent dye and the absorbing molecule are next to each other is light emitted. MGB probes can be considered a hybrid between beacons and Taqman probes. The probe is an approximately 15-nucleotide sequence to which fluorescent dye and quencher are attached. In addition, an organic molecule is attached that interacts with the so-called minor groove of the DNA helix. This stabilizes the binding between probe and target, allowing use of shorter probe sequences.

After hybridization, the molecule is still not able to fluoresce; only after release of the fluorescent dye by the exonuclease activity of DNA polymerase during elongation is light emission possible.

In rtPCR, fluorescence is measured during each cycle, and when a certain threshold is reached the PCR is considered positive. A probe can directly confirm the identity of the product formed; when fluorescent dyes are used, a melting curve is usually produced to confirm the identity of the product. By slowly increasing the temperature and measuring fluorescence, the temperature at which the amplification product melts can be determined and compared with the expected melting temperature. For an extensive review of rtPCR see Espy et al. [5].

For all PCRs, appropriate controls should be included. These controls include negative controls to check for contamination and positive controls for both sample preparation and amplification. Amplification controls are particularly important because many compounds can inhibit a PCR amplification. Inhibitory compounds (such as heme, bilirubin, and bile salts) are frequently present in clinical samples. False-positive results may be obtained due to the production of non-specific amplification products when PCR conditions are suboptimal. It is obvious that both false-positive and false-negative results may have serious consequences for the treatment of patients.

When the obstacles of designing an assay that detects all possible mechanisms are overcome another important potential problem should be considered: contamination of a sample by other samples or contamination by amplification products of previous assays. The strength of PCR, its ability to amplify minute amounts of DNA, is also its weakness. Two types of contamination should be discerned: a general contaminant and a sample contaminant. The first type of contaminant will generally affect every sample in an assay. It occurs when reagents, disposables, or the environment are contaminated, but also sample carry-over in pipetting robots. The second type of contaminant only affects a limited number of samples in an assay, for example, due to aerosols from other samples.

The best approach is to prevent contamination because it can be extremely difficult to get rid of once introduced. To reduce the risk of contamination, an inventory of potential sources should be made because some may be quite unexpected. Particular attention should be paid to disposables and reagents; since it is known that many reagents, including high-quality water and *Taq* DNA polymerase preparations, may be contaminated with (bacterial) DNA. Another important source may be present in lytic enzyme preparations that are frequently used to isolate DNA. A completely unexpected source may arise from activities in the laboratory next door. In our laboratory, we had a major problem with a PCR protocol specific for TEM-family β-lactamases, until we found out that colleagues in the laboratory next door were purifying large amounts of recombinant proteins from lysates that contained a high copy number vector (100–200 copies/cell) with an ampicillin resistance marker. The resistance marker was TEM-1 and it is the most common resistance marker used in cloning vectors. Opening tubes with high numbers of copies should also be performed carefully to avoid aerosol formation. A final point is waste disposal. This includes not only reaction tubes, but also other items such as electrophoresis buffer.

Several protocols have been described for decontamination. Surface contaminations are best eliminated with a 0.4% solution of hypochlorite. A commonly used strategy to specifically destroy amplification products is the use of uracil-DNA-glycosylase/dUTP (also known as the ung system). In this system, a nucleotide containing uracil instead of thymine is used. Before an amplification is initiated, the uracil-DNA-glycosylase specifically removes the uracil, which yields a non-functional template.

Another important strategy to prevent contamination is the introduction of "no," "low," and "high" copy DNA work areas. In the "no" copy area, DNA free solutions and buffers are prepared as well as reaction mixtures without target DNA. In the "low" copy area, target DNA is prepared and added to the reaction mixture. In the "high" copy area, amplification and analysis take place. The "no" copy or clean areas should preferably have overpressure to keep contamination out, whereas "high" copy areas should have underpressure to keep DNA in. The workflow should be from "no" copy via "low" copy to "high" copy. Nobody should be allowed to go back from a higher copy level to a lower copy level. The same rule applies to equipment including racks. Although this may seem cumbersome, it will pay off. When a major contamination occurs involving a commercial assay, new primers and/or targets are not readily available. But also for in-house assays, it can be problematic to develop a new assay. For a more extensive review of prevention and destroy techniques see Borst et al. [6].

MICROARRAYS

Microarrays are basically a form of hybridization, but on a large scale. Microarrays use either DNA fragments specific for the genes of interest or oligonucleotides, which are spotted on a solid phase, for example, glass. The (bacterial) DNA to be analyzed for specific genes is then labeled and hybridized with the sequences present on the solid phase. Usually fluorescent labels are used. Frequently, the array is not only hybridized with the labeled DNA of interest, but also with reference DNA that serves as a control for the quality of the spots. The reference DNA uses a label that fluoresces at a different wavelength and thus can be measured independently from the other label. The ratio between the two signals then defines the presence or absence of hybridization and thereby the presence or absence of a gene in the DNA of interest. Although this process seems straightforward, complications in interpretation may arise. As for normal hybridization the stringency of the hybridization conditions influence which variation in the sequence of the gene is considered either positive or negative. Mutations that do not influence resistance may yield a false-negative result, whereas the opposite may also occur. The copy number of the genes also influences the hybridization signal and may influence results.

A variation on the theme is the use of oligonucleotides on a microarray to detect the presence of point mutations that result in a resistant phenotype. In this approach, wild-type and mutant-specific oligonucleotides are usually present.

DNA SEQUENCING

DNA sequencing is still the gold standard for the detection of point mutations. Until recently, DNA sequencing has been based almost exclusively on the method developed by Sanger. In this method, a DNA polymerase elongates a primer, but modified nucleotides specifically block the elongation at either adenines, cytosines, guanines, or thymines. The exact position of termination of elongation for each base is random, but higher concentrations of the modified nucleotides yield shorter sequences. In the early days of sequencing, either the primer or one of the nucleotides was radioactive and a separate incubation was required for each of the four bases. After separation of the elongation products on a gel, an X-ray film was applied and the DNA sequence could be deduced from the resulting image. The use of fluorescently labeled primers made the cumbersome radioactive method less attractive, and an instrument capable of separating and detecting the fluorescent sequencing products became necessary. Besides doing away with radioactivity, the method had the additional advantages of being faster and more amenable to automation. The development of dye terminators, where adenine, cytosine, guanine, and thymine each has its own unique dye emitting light at a different wave length, made the use of single incubations possible, greatly improving efficiency.

A new development is pyrosequencing (Biotage AB, Sweden). The nucleotides are added sequentially in this process. When the first of the four nucleotides is complementary to the template strand, DNA polymerase will incorporate it in the new strand. As a result, one molecule of pyrophosphate is released for each nucleotide molecule incorporated. The pyrophosphate is converted to ATP by sulfurylase. Luciferase uses the ATP as a substrate to generate light, which is detected by a sensitive camera system. The emission or absence of light after the addition of each nucleotide is converted into sequence by a computer program. Using this technique, 40 to 50 nucleotides per fragment can be read.

FINAL REMARKS ON TECHNOLOGY PRINCIPLES

Since the discovery of the double helix structure of DNA, a variety of molecular methods have been developed to identify genetic mechanisms underlying phenotypes. Although the application of these methods seems straightforward (just select the right probe or primers) reality can be more complicated. Variation in patients, type of isolates, frequency of mutations, genes, available technical expertise, costs, and so forth, should be taken into consideration when designing or selecting a particular assay. Because of these variations, all assays should be validated for the situation in which they are used. A complication in the validation of new methods could be that the molecular method is better than the current gold standard. So, compared to the current gold standard, the new method may appear less sensitive or specific. This will require additional determinations to validate the new molecular assay before its implementation in routine diagnostics.

MYCOBACTERIUM TUBERCULOSIS

INTRODUCTION

Tuberculosis is one of the most common infections in many parts of the world, and the treatment of tuberculosis is difficult with only a limited number of antimicrobial agents that show activity against *M. tuberculosis*. This treatment is now under even more pressure with increasing rates of (multi)resistant isolates; therefore, detection of resistance is an important issue. The fact that mycobacteria are slow growing makes them the ideal target for molecular assays to detect resistance. Most often resistance is caused by the development and selection of mutations. A multitude of different assays have been developed and will be discussed. Mutations in *rpoB* are associated with resistance to rifampin, but specific mutations can also be linked to resistance to more novel rifampin analogs [7–9]. Mutations in the *pncA* gene are associated with resistance to pyrazinamide, but alternative mechanisms must exist because phenotypically resistant isolates without mutations, which produce an active pyrazinamidase, have been described [10]. Another complication in interpreting the results of different assays is the occurrence of mixed cultures [11].

PROBE-BASED TECHNOLOGY FOR THE DETECTION OF
RESISTANCE IN *MYCOBACTERIUM TUBERCULOSIS*

A large group of assays is based on the detection of mismatches between the (susceptible) wild-type sequence and the (resistant) mutant sequence. DGGE was used to scan 775 bp of *rpoB* containing relevant mutations. Five primer sets were used, but the primer pair covering the rifampin resistance-determining region (RRDR) detected mutations in 955 of the resistant isolates and in 2% of the susceptible isolates. Two additional areas with mutations were covered by two other sets. The last two primer sets identified two other relevant mutations elsewhere in the gene [12]. The same method has also been applied to the detection of mutations in the *pncA* gene responsible for pyrazinamide resistance. Five sets of primers were used to scan for mutations in a 600 bp region. The assays identified a mutated gene in 82 of 83 resistant isolates and in only 1 of 98 susceptible isolates [13].

Others have used a variant of DGGE, the single-strand conformation polymorphism (SSCP) analysis. The *rpoB* gene was used to detect both *M. tuberculosis* as well as mutations leading to rifampin resistance. The limited number of samples showed that PCR was able to detect *M. tuberculosis* more readily than Ziehl-Neelsen-based microscopy in fine needle aspirates. Mutations in *rpoB* were also detected, but not validated further. One disadvantage of the procedure was that only 285 bp of the gene were interrogated for mutations [14]. The same limitation holds for another paper describing SSCP for mutation detection in *rpoB*, which also only analyses approximately 306 bp. The assays correctly identified the rifampin-resistant and susceptible isolates among a collection of more than 100 isolates. In addition, *Mycobacterium avium* and *Mycobactrium terrae* could be identified [15]. However, the study by Mani et al. [16] showed that PCR-DGGE or PCR-SSCP were inferior to DNA sequencing for detection of *rpoB* mutations.

A variant of DGGE and SSCP is heteroduplex formation. One particular assay described the analysis of mutations in five different genes involved in resistance to different antibiotics in *M. tuberculosis*. The genes are *rpoB*, *katG*, *pncA*, *rpsL*, and *embB*. After amplification of a suitable DNA fragment, the denatured products were allowed to form duplexes with reference DNA molecules, and these were detected by HPLC [17]. Although this holds promise, only a small section of the *rpoB* gene was covered, and in addition, specialized equipment is needed. A comparable assay detected mutations in *gyrA*, but the study was confined to only 20 isolates [18]. In a variation on this approach, an assay has been developed that detects both *pncA* mutations and discriminates between *M. tuberculosis* and *Mycobacterium bovis*. This also required the use of a *pncA* probe specific for each species. The assay performed as expected on a limited number of isolates [19].

Cheng et al. [20] described a marriage between heteroduplex analysis and SSCP. Both methods are used to increase the reliability of detection of mutations. This was demonstrated with mutations in the *gyrA* gene. The profiles obtained for 138 isolates—including 32 with mutations in *gyrA*-correlated with *in vitro* susceptibility to different fluoroquinolones—and were confirmed by DNA sequencing.

In a special form of this technique, heteroduplexes are formed between the (mutated) target DNA formed by a 305 bp RRDR *rpoB* amplification product and a special synthetically generated probe called the Universal Heteroduplex Generator (Genelab, Louisiana State University, Baton Rouge, LA) with deletions and substitutions for optimal detection of mutations in the RRDR. Actual detection is on a gel. The method correctly identified 90 of 97 rifampin-resistant isolates and all 21 susceptible isolates [21]. The method holds promise for the identification of mutations associated with resistance to more improved variants of rifampin. A follow-up with the same PCR and probe but modified with fluorogenic primers and adapted to capillary gel electrophoresis and fluorescence detection was developed. This allowed the detection time to be reduced by several hours, but discrimination was difficult, and the method required specialized equipment not routinely available in clinical microbiology laboratories. In a follow-up study, the system was evaluated on 1892 sputum samples from 394 patients. Compared to sputum smears the assay had a sensitivity and specificity of 99.8% and 82.9%, respectively, but when compared to IS*6110* PCR this was 70.1% and 92.9%, respectively [22].

A different form of analysis for the formation of heteroduplex formation is by chemical cleavage when a mismatch is present. Analysis of the reaction products is by gel electrophoresis. This method is less attractive, because of the hazardous nature of the chemicals and the use of radioactivity for detection [23]. Another variant of heteroduplex detection of mutations uses RNA/RNA duplexes and is quite complex to perform, but needs only a PCR apparatus and gel electrophoresis equipment. Briefly, the RRDR is amplified with PCR, using primers with either the T7 or Sp6 bacteriophage promoter sequences attached. After amplification, T7 and Sp6 RNA polymerases are added and allowed to synthesize RNA initiating at the promoter sequences. The RNA is then allowed to form duplexes with reference RNA. At the position of mismatches the RNA is cleaved by RNase. The cleavage products can then be detected on a gel [24].

Reverse-line blot assays are among the most popular type of probe system used to detect antimicrobial resistance determining mutations among *M. tuberculosis* isolates. The one most commonly used is the INNO-LiPA (Innogenetics, Belgium). In this assay the gene region of interest is first amplified by nested PCR. The inner primers are biotinylated, yielding a labeled amplification product that is denatured and hybridized to oligonucleotide probes applied as a line on a membrane. After hybridization, a color reaction is performed. The hybridization is automated. The assay has a sample inhibition control.

A meta-analysis evaluated 15 studies testing the INNO-LiPA Rif-TB. Eleven of these studies used clinical isolates, one used clinical specimens, and three studies used both. Twelve of the studies analyzing isolates demonstrated a sensitivity greater than 95% and a specificity of 100%. The four studies that analyzed clinical specimens had a sensitivity between 80% and 100% and a specificity of 100%. The authors concluded that the LiPA is highly sensitive and specific, but appears to have a lower sensitivity on clinical material. The authors concluded that more evidence is needed before LiPA can be used to detect MDR-TB among populations at risk in clinical practice [25]. Examples of studies and their main characteristics are provided in Table 9.1. These data indicate that the INNO-LiPA is comparable to classical methods to detect *M. tuberculosis*, but is not as effective as some of the best molecular assays available. At least one study [31] tried to extend the interpretation of the INNO-LiPA to multi-resistance in the *M. tuberculosis* isolates tested. However, this is not straightforward because it is dependent on the prevalence of rifampin mono-resistance in the population tested. An additional disadvantage is that shifts in the resistance pattern are not recognized.

In-house reverse-line blots for the detection of rifampin resistance based on amplification of the RRDR have also been developed. An example is the study by Morcillo et al. [32]. The RRDR was amplified with a biotinylated primer. The amplification products were subsequently hybridized with a set of 11 oligonucleotides on a membrane. More than 250 isolates were tested and 90 of the 97 resistant isolates were correctly identified, as were all susceptible isolates. The costs were substantially less than for commercial assays, but quality control is included in the commercial

TABLE 9.1
Characteristics of INNO-LiPA Rif.TB for the Detection Mutations Associated with Rifampin Resistance in *Mycobacterium tuberculosis*

Samples	No. of Samples	Sensitivity (%)	Specificity (%)	Gold Standard	Footnote	Reference
Respiratory specimen	60	100	100	DNA sequencing	1	26
Smear-positive	281	100	96.9	phenotype	2	27
Isolates	70	95.3	100	phenotype	—	28
Isolates	52	100	92.0	phenotype	3	29
Isolates	41	56.1	—	phenotype	4	30
Isolates	411	98.5	100	phenotype	—	31
Isolates	128	97.3	100	phenotype	5	11

1. One isolate was consistently resistant, but showed no signal in the INNO-LiPa and also showed no mutations after DNA sequencing.
2. 287 yielded amplification product and 281 of these were culture-positive; 12 amplification-negative samples were true negatives, 19 had a low bacterial load, and 32 showed inhibition.
3. The 2 false-positives showed mutations, but were phenotypically susceptible.
4. No sensitive isolates were tested.
5. The 3 isolates that were false-negative had mutations outside the RRDR of *rpoB*.

assays and should also be applied to the in-house test. This requires validation when new batches of primers or oligonucleotides are used or new batches of membranes prepared. In an extension of the previous study, not only mutations in *rpoB* associated with rifampin resistance were detected, but also mutations in the promoter regions of *inhA* and *aphC*, *rpsL*, *rrs*, and *embB*. The first two genes determine resistance to isoniazid, the next two determine resistance to streptomycin, and the last one to ethambutol. Rifampin resistance was identified correctly in 132 of 155 isolates; ethambutol resistance was identified correctly in 28 of 55 isolates, isoniazid resistance in 16.9% and 13.2% of isolates for the promoter region of the *inhA* and *aphC* genes, respectively. Mutations at *rrs*513 and *rps*L188 were detected in 15.1% and 17.0% of the streptomycin-resistant isolates [33]. The authors show the applicability of the method to the detection of a larger collection of mutations covering a greater diversity of resistance genes important in resistance in *M. tuberculosis*. However, the assay will only be valuable when the assignment of the resistance or susceptibility is at least, and preferably better than, 98%.

Another example of a commercial reverse line blot is the Genotype MTBDR (Hain Lifescience, Germany) for the detection of rifampin and isoniazid resistance. The test correctly identified rifampin resistance in 102 of 103 multidrug-resistant isolates, whereas mutations in *katG* were identified correctly in 91 isolates albeit with 2 isolates that were considered resistant because of absence of wild-type hybridization. However, 12 of the 103 isoniazid-resistant isolates apparently had mutations in other genes involved in resistance to isoniazid. All 40 susceptible isolates showed wild-type hybridization patterns [34].

The method of reverse-line blot has also been applied to DNA extracted from Ziehl-Neelsen slides with relatively good sensitivities, but a result was in large part dependent on sufficient bacteria on the slide. With one exception the identification of resistant and susceptible isolates was in accordance with susceptibility testing [35].

Oligonucleotide macroarrays are also a valuable format for the detection of rifampin and isoniazid resistance. One study used oligonucleotides that were directed against the RRDR of *rpoB*, *katG*, and *mabA-inhA*. The target DNA was amplified using biotinylated primers. Isolates with 38 different RRDR genotypes, four *katG* genotypes, and two *mabA-inhA* genotypes were correctly identified, except for one RRDR genotype, which had a nine base pair insertion. All wild-type sequences were also correctly identified [36].

PCR-ELISA starts with amplification using a digoxigenin-labeled primer and the subsequent capture of denatured amplification product utilizing five different biotinylated probes. The biotin-labeled probes are captured in streptavidin-coated wells and any product is detected with a horseradish peroxidase labeled anti-digoxigenin antibody. This method was utilized in a small study with 50 rifampin-resistant isolates and 30 AFB-positive sputum specimens, demonstrating the feasibility of the approach [37].

PCR-Based Methods for the Detection of Antibiotic Resistance in *Mycobacterium tuberculosis*

A number of groups have developed PCRs in which one of the primers is either specific for the mutation or for the wild-type sequence. For elongation to occur, the

3′-nucleotide, where DNA polymerase initiates elongation, has to match perfectly. PCRs based on this principle have been called multiplex allele-specific (MAS) PCR, amplification refractory mutations system (ARMS) PCR, or allele-specific depletory PCR. When a wild-type primer is used and the sequence has a mutation, no elongation occurs and no amplification product is formed. Although this is an ingenious method for the detection of mutations, the large number of amplifications becomes unwieldy when all mutations leading to resistance in *M. tuberculosis* need to be covered. Nevertheless, these types of tests have indeed been developed to cover mutations in all important resistance genes of *M. tuberculosis* [38–45].

Low-stringency single-specific-primer PCR has its own limitations. In this method [46], a specific PCR is first performed. The amplified product is used for a second PCR with only one of the two original primers, under low stringency conditions. The resulting products are analyzed on a polyacrylamide gel. This produces a fingerprint, which is dependent on the mutations in the target. The low stringency conditions may be difficult to reproduce exactly and random variations may lead to variations in the fingerprint. The method was only tested on 12 specimens and although the authors claim that it is highly sensitive and rapid, the method appears rather cumbersome with a high risk of contamination and difficulty in reproducing the fingerprints.

At least two studies have demonstrated that PCR-RFLP methodology can be used to detect resistance mutations in *M. tuberculosis* [40,47]. It was successfully used for detection of the *katG* 315 mutation, the *embB* 306 and 497 mutations, and the *iniA* 501 mutation. A shortcoming of the technique is that not all mutations can be detected. Either no suitable restriction enzymes are available or a large number of restriction digests have to be performed and analyzed on agarose gels.

Efficacy of pyrazinamide treatment is dependent on the *M. tuberculosis* enzyme pyrazinamidase that converts the prodrug to an active form. Measurement of enzyme activity is therefore an effective way to predict susceptibility. However, a large number of bacterial cells are required to obtain sufficient material for such measurements. To overcome this obstacle, a PCR was developed in which the forward primer has a T7 RNA polymerase promoter attached. The resulting amplification product has an active T7 RNA polymerase promoter. This promoter can be used to transcribe the gene after the addition of T7 RNA polymerase and other components. In a coupled system, the mRNA is translated to form pyrazinamidase, and the activity of the enzyme is subsequently determined. All 45 isolates including 30 resistant isolates gave the expected result, namely reduced activity for resistant isolates and normal activity for susceptible isolates [48].

Real-time PCR is a commonly used technique to detect mutations associated with isoniazid, rifampin, and ethambutol resistance in *M. tuberculosis*. The critical parameters of these studies are summarized in Table 9.2. An important conclusion from those studies is that rtPCR assays can readily reach sensitivities and specificities of 100%, when compared with DNA sequencing of the mutations that are targeted. However, the number of samples tested in most studies has been lower than desired. When compared to phenotypes the sensitivity may drop dramatically because not all relevant mutations are targeted by rtPCR assays. This can result in different sensitivities for the same assay depending on the specific mutations present

TABLE 9.2

Characteristics of Real-Time PCR Assays for the Detection Mutations Associated with Antibiotic Resistance in *Mycobacterium tuberculosis*

Gene	Samples	Probe	No. of Samples*	Sensitivity (%)	Specificity	Gold Standard	Analytical Sensitivity (CFU)	Footnote	Reference
katG, ahpC, kasA, inhA	Lysates	Beacons	149	85	100	Phenotype	—	a	49
RRDR rpoB	Lysates	Beacon	149	98	100	Phenotype	—	—	49
Ison	Respiratory	FRET	37	100	100	DNA sequencing	10^3	—	50
Rif	Respiratory	FRET	37	100	100	DNA sequencing	10^3	—	50
inhA	Partially purified lysates	FRET	35	100	100	DNA sequencing	—	—	51
inhA, katG	Lysates	FRET	73	100	100	DNA sequencing	—	—	52
RRDR rpoB	Lysates	FRET	73	100	100	DNA sequencing	—	—	52
katG	ZN-positive sputa	MGB probes	40	70/82	94/100	DNA sequencing	$1 \cdot 10^3$	b	53
katG	Lysates	FRET	56	100	100	DNA sequencing	—	—	54
RRDR rpoB	Lysates	FRET	56	100	100	DNA sequencing	—	—	54
katG	Lysates	FRET	33	100	100	DNA sequencing	—	—	55
RRDR rpoB	Lysates	FRET	33	100	100	DNA sequencing	—	—	55
RRDR rpoB	Lysates	Beacons	148	96.9	100	DNA sequencing	—	c	56
RRDR rpoB	Lysates	FRET	119	92.7	100	Phenotype	—	d	57
RRDR rpoB	Partially purified lysate	Beacons	243	100	100	DNA sequencing	—	e	58
katG	Lysates/sputa	MGB probes	45/27	100/100	100/100	DNA sequencing	—	—	59
RRDR rpoB	Lysates/sputa	MGB probes	45/27	100/100	100/100	DNA sequencing	—	—	59
embB	Lysates/sputa	MGB probes	45/27	100/100	100/100	DNA sequencing	—	—	59
katG, inhA	Lysates	Beacons	196	100	100	DNA sequencing	10–100 fg	f	60
rpoB	Lysates	Beacons	196	100	100	DNA sequencing	10–100 fg	f	60
katG, inhA	Sputa	FRET	205/108	53.8	100	Phenotype	—	g	61
RRDR rpoB	Sputa	FRET	205/108	100	100	Phenotype	—	g	61

continued

TABLE 9.2 (continued)

Characteristics of Real-Time PCR Assays for the Detection Mutations Associated with Antibiotic Resistance in *Mycobacterium tuberculosis*

Gene	Samples	Probe	No. of Samples*	Sensitivity (%)	Specificity	Gold Standard	Analytical Sensitivity (CFU)	Footnote	Reference
katG, inhA, ahpC	Lysates	FRET	150	76	100	Phenotype	—	—	62
katG, inhA	Clinical samples	5′-exonuclease	224	76.6	100	DNA sequencing	1.5×10^3	h	63
rpoB	Clinical samples	5′-exonuclease	224	88.8	100	DNA sequencing	1.5×10^3	h	63

*Where two numbers are provided: no. of samples/no. of patients.

[a]Collections from Madrid and New York show difference in detection rate.

[b]With the wild-type probe 70% sensitivity and 82% specificity; with mutant probe 94% sensitivity and 100% specificity; analytical sensitivity 1 genome with purified DNA and 10^3 CFU/mL on sputa.

[c]Discrepancy for two isolates where genotype and phenotype did not match.

[d]Three resistant isolates were missed on basis of DNA sequencing.

[e]88.8% sensitivity and 99.4% on basis of phenotype.

[f]82.7% sensitivity and 100% for isoniazid based on phenotype and 97.5% sensitivity and 100% specificity for rifampin based on phenotype; the analytical sensitivity is dependent on the probe used.

[g]205 samples from 108 patients; auramine-positive sputa; with the *inhA*-specific amplification 200 of 205 samples positive and 198 of 205 samples with the *katG*-specific amplification, this corresponds to 100% and 98.1% of all patients.

[h]224 samples of 181 patients; 76.6% sensitivity for *inhA* and 76.6% sensitivity for *katG*. Not all possible mutations were addressed in the assay.

in the population tested. This is demonstrated in the study by Piatek et al. [49] who showed different sensitivities when isolates from Madrid, Spain, or New York were tested. So, knowledge of the mutations present in the population to be tested is required to design an adequate test.

PCR-DNA sequencing has been demonstrated by several groups and can be considered a straightforward PCR followed by DNA sequencing [64,65]. As with most PCRs the main problem is efficient extraction of *M. tuberculosis* DNA from the sample.

Pyrosequencing, although relatively new, has already been applied to the detection of mutations in *M. tuberculosis*. It could be considered a type of SNP analysis. The gene region of interest is amplified by PCR and the amplification product sequenced using a reagent kit and pyrosequencing instrumentation. Mutations were detected in the *katG* and *rpoB* genes. In addition, discrimination between closely related species of the *M. tuberculosis* complex was possible [66]. The method promises to be relatively cheap and can be applied to high-throughput situations. Although pyrosequencing requires specialized equipment, it can be easily used for the detection of other mutations in any other gene as well without optimization of the detection step. The feasibility of this approach was also demonstrated by another group, who sequenced mutations in the *rpoB*, *katG*, and *embB* genes. The analytical sensitivity was approximately 45 fg. DNA and sequences were known within 2 h after amplification [67].

MICROARRAYS FOR THE DETECTION OF RESISTANCE IN *MYCOBACTERIUM TUBERCULOSIS*

Both high- and low-density arrays have been used to detect resistance in *M. tuberculosis*. A demonstration of the feasibility of low-density microarrays to detect mutations in *rpoB* with fluorescently labeled DNA amplified from colonies used 53 rifampin-resistant and 15 susceptible isolates. These were successfully tested using 50 different probes specific for the wild-type sequence of the RRDR. Results could be obtained within 1.5 h after amplification [68]. A small study with only 33 isolates showed the feasibility of the method to detect mutations in *pncA* associated with pyrazinamide resistance. The array utilized 79 overlapping oligonucleotide probes covering the gene [69]; another study showed similar results with 57 isolates [70]. The array is built from a set of oligonucleotide probes specific for the wild-type sequence and a set specific for the mutations. A similar approach has been used for the detection of mutations in *rpoB*. The results were generally consistent with sequencing data, but some discrepant results were obtained from mixed cultures [71]. This can be expected, as the ratio of different sequence variants may differ from culture to culture. In addition, the efficacy of binding to the probes may vary somewhat, leading to insufficient signal at low concentrations of the specific variant.

A complex microarray has been devised by Mikhailovich et al. [72,73]. The oligonucleotides are not spotted directly onto a glass slide, but a polyacrylamide gel pad is generated on the slide. The slide is placed on a Peltier element (heating and cooling element) allowing temperature to be controlled. A nested PCR for *rpoB* RRDR with a fluorescent labeled primer in the second PCR and a discriminating set of oligonucleotide probes was developed. For each mutation, a matching probe was present. The labeled amplification product is transferred to the microarray and allowed to hybridize for 14 to 18 h. The use of the gel pads also allowed the use of PCR (or any other reaction) on the slide. The chip correctly identified all isolates with mutant *rpoB* genes using hybridization in one study and 95% of mutations in another [72,73]. In addition, a method based on ligation of two oligonucleotides was demonstrated. This assay utilized the fact that DNA ligase can only join two molecules when they are next to each other and that the ligation site is perfectly double-stranded. Therefore, only homoduplexes can be ligated and are detected [72].

STAPHYLOCOCCUS AUREUS

THE PROBLEM OF MRSA

In *S. aureus*, resistance to methicillin (and thereby by definition all β-lactam antibiotics) is dependent on the presence of the *mecA* gene, which encodes PBP2a, an alternative penicillin-binding protein that despite its name does not bind β-lactam antibiotics and is therefore insensitive to their action. However, not only *S. aureus* may harbor the *mecA* gene. It is also found among coagulase-negative staphylococci (CoNS), including *S. epidermidis*, which is normally a human commensal. Therefore, the detection of MRSA in non-sterile sites cannot be achieved by the simple detection of *S. aureus* and the *mecA* gene. To address the problem of *mecA*-positive CoNS, PCRs have been developed that show a link between the *mecA* gene and the *S. aureus* genome. These PCRs utilize sequences in the larger genetic elements in which *mecA* is present, the Staphylococcal Cassette Chromosome *mec* or SCC*mec*, and *orfX* sequences (into which SCC*mec* is integrated in the staphylococcal chromosome) unique for *S. aureus*.

A problem with this approach is that a number of SCC*mec* variants are known and new ones are regularly discovered, and the sequence of these SCC*mec* types and variants show poor homology near the border with the *orfX*. A PCR using the other border between SCC*mec* and the staphylococcal chromosome is not possible, because both the SCC*mec* sequence at this end and the adjacent staphylococcal chromosome sequence are highly variable. Thus, it is very difficult to design primers that recognize all known and preferably also undiscovered SCC*mec* elements. This may result in false-negatives and the potential spread of a novel strain of MRSA. Another potential problem is the presence of SCC*mec*-like elements that either have lost the *mecA* gene or never possessed the *mecA* gene, but are recognized by the primers. These false-positive reactions may lead to less-than-optimal antibiotic therapy, because vancomycin is often the drug of choice when flucloxacillin cannot be used. Vancomycin is less effective than flucloxacillin and also has more side effects [74,75]. In addition, the patient may be kept in isolation to prevent further spread of the MRSA in the hospital. This is a costly and for the patient unpleasant procedure.

Another issue is whether enrichment cultures are needed before the detection of MRSA in clinical specimens. This will depend in large part on the method chosen. Some methods are only suited for pure cultures and the issue is clear. However, some assays are capable of detecting MRSA present at levels of less than 10 CFU per specimen, and at times the molecular assay appears to be even more sensitive than culture. However, not all specimens will have loads this low and neither are all assays this sensitive. Culture will mean the loss of valuable time, but even the presence of a single CFU means the presence of resistant organisms that may become selected during antibiotic therapy. One of the safest approaches, if feasible, would be to directly perform a molecular assay, and culture additional or leftover material and check the next day for growth from specimens that were negative in the molecular assay.

A large number of straightforward PCR assays for the detection of MRSA have been developed and a number of them have been published [76]. Also, multiplex PCRs for the simultaneous detection of MRSA and for example, mupirocin resistance or other resistance genes have been described, but these methods usually use agarose gel detection of the amplification product and are considered conventional PCRs [76–78]. A more advanced form is the color development–based PCR for the detection of *mecA*. In this PCR, one of the primers is biotinylated and the amplification product is subsequently captured by streptavidin bound in a well. The captured product is then hybridized with a horseradish peroxidase–labeled probe. When substrate is added the enzyme yields color [79]. Here we will discuss the more recent assays based on probe, rtPCR, and microarray technologies.

MRSA DETECTION BY PROBE-BASED METHODS

A number of probe-based methods have been developed. Often these assays rely on a PCR step to obtain sufficient material and thereby analytical sensitivity, but not all assays that have probe-based technology as their main mechanism of recognizing particular sequences utilize PCR. The Velogene™ Rapid MRSA Identification Assay (ID Biomedical Corp., Canada) uses an isothermal cycling process. In this procedure, a biotin-DNA-RNA-DNA-fluorescein probe recognizes the *mecA* gene, and the hybrid target DNA-RNA probe is specifically recognized by RNase H. The probe is cleaved, allowing new probe to bind. The uncleaved probe is captured by streptavidin and detected with an enzyme-labeled antibody directed against fluorescein. When no color develops the sample contains *mecA*. The method yields good results, but is limited to isolates [80].

Another isothermal RNA amplification-based test with a color reaction detection system had an analytical sensitivity equivalent to only 4×10^6 to 2×10^7 CFU/mL of enrichment broth [81]. Unfortunately, this sensitivity is poor in comparison with many of the other methods, especially rtPCR-based methods.

More conventional probe systems are also used. At least one such system is sold commercially and has been evaluated in the international literature. The EVIGENE MRSA Detection Kit (Statens Serum Institut, Denmark) can be used to detect MRSA in positive blood culture bottles. The system uses probes specific for *nuc* (encodes a DNA nuclease specific for *S. aureus*), *mecA*, and 16S rRNA genes. Hybridization, capture and color reaction are all performed in microwells. A result can be obtained within 7 h after the detection of a positive blood culture. A total of 200 blood bottles positive for Gram-positive cocci were evaluated. Eighteen bottles obtained from 12 patients hybridized with the *nuc* and *mecA* probes and 17 of these were also positive in a *mecA* and *femB* (involved in cell-wall synthesis in *S. aureus*) multiplex PCR. Although the remaining sample was negative in the multiplex PCR, it was MRSA-positive in other assays. A few other discrepancies were also noted. Two samples were identified as methicillin-susceptible *S. aureus* (MSSA) based on a barely positive EVIGENE result, but this could not be confirmed by *femB*-specific PCR. One sample contained *S. epidermidis* and *S. capitis*, whereas the other only contained *S. capitis*. Twenty-five of the 200 bottles were negative with the staphylococcal-specific 16S RNA gene probe, which was confirmed in 14 cases, but 11 bottles yielded staphylococci upon culture. Although the authors found the kit user friendly, they did not express a clear opinion about the value of the kit [82]. From the data available, it can be concluded the MRSA were accurately detected, but the identification of staphylococci may be less accurate. Another risk may be the presence of mixed cultures, which can lead to false-positive results. A second study with the same kit using 242 CoNS yielded 237 valid tests in which all *mecA*-positive isolates were correctly identified. In a blood bottle procedure, 67 of 72 *S. aureus* isolates were correctly identified, and the presence or absence of *mecA* was correctly assigned in all cases. Eight of the CoNS containing samples yielded non-valid results, but the others were all correctly identified. The invalid results were the consequence of a positive control under the specified cut-off [83].

Another kit that utilizes hybridization is the GenoType Version 1, MRSA (Hain Lifescience GmbH, Germany). In this system, the *mecA* and a sequence specific for *S. aureus* is amplified and detected by reverse blotting. Testing showed that 138 MRSA were correctly identified, but five MRSA yielded ambiguous results and needed additional testing. The *mecA* probe also recognized, as expected, the *mecA* gene in CoNS [84]. In another test, the kit detected 12 out of 13 MRSA correctly. The analytical sensitivity using blood bottles was 10^4 CFU/mL [85]. This method is therefore not suitable for clinical samples.

REAL-TIME PCR DETECTION OF MRSA

The rtPCR technology in its simplest form amplifies the *mecA* gene and a *Staphylococcus*-specific gene, in this case *femA*. Amplification is detected by the presence of CYBR Green, a dye that fluoresces when intercalated into DNA, and the identity of the amplification products is confirmed by their melting curves. A potential difficulty is the closeness of the melting curves of different staphylococci. Generally, it works quite well [86]. Another simple way to detect MRSA is the use of a selective broth, which allows growth of MRSA but not MSSA. After overnight culture, the *nuc* gene was detected. The positive predictive value was poor (31.8%), but the negative predictive value was high (99.6%). The assay's role in routine diagnostics could be to reduce the workload by eliminating further processing of negative samples [87]. However, this assay has the danger that *mecA*-positive isolates expressing low levels of methicillin resistance, for example, due to poor induction of *mecA* in isolates containing an intact *mecA* regulatory region, are missed. The choice of a good selective medium may reduce that risk. This is not the only assay that does not make use of the full potential of rtPCR; additional assays have been described that have a rather poor analytical sensitivity. A method that also included automated DNA isolation required an input of 0.5 McFarland [88]. This is equivalent to approximately 1.5×10^8 CFU/mL, so the assay is not suited for direct assessment of clinical samples. However, assays that specifically detect MRSA in clinical samples with high analytical sensitivity have been described. Hagen et al.

developed an assay using rtPCR and FRET technology. The PCR detected MRSA in mixed samples and was evaluated with approximately 250 isolates including CoNS. The assay had 98% sensitivity and 100% specificity. The method also showed positivity after overnight culture enrichment for 20 of 27 swabs culture-positive for MRSA among a total of 60 swabs. The analytical sensitivity was less than 10 CFU/swab [89].

Real-time PCR has also been used for the detection of Panton-Valentine leucocidin (PVL)-positive MRSA. PVL is encoded in some *S. aureus* strains, including some MRSA strains and especially community-associated MRSA strains. The presence of PVL is associated with invasive disease, in particular with necrotizing pneumonia, an often fatal disease usually found in teenagers and young adults [90]. With the increasing appearance of community-acquired MRSA, the number of potentially severe infections due to PVL-encoding strains could be increasing and early recognition becomes more important. The assay simultaneously detects the *nuc*, *mecA*, and the two PVL-encoding genes. An analysis of 1552 phenotypically resistant clinical isolates was performed and 103 isolates were used for validation with conventional PCR. This showed complete agreement. The utility of the multiplex assay was further established by concordance of the results of 98 PVL-positive isolates that were identified earlier in an rtPCR assay that only detected PVL. The specificity of the assays was stated as good [91].

The IDI-MRSA Test (Infectio Diagnostic Inc., Canada) is in principle able to identify MRSA in mixed cultures, which include *mecA*-positive CoNS. The test has an *orfX*-specific and a number of SCC*mec*-specific primers. The assay has an internal control and utilizes material directly from a swab. The assay can be performed in 1.5 h under optimal conditions. A first evaluation of isolates clearly proved the principle: 1636 of 1657 MRSA were identified correctly. Twenty-six of 569 MSSA were identified as MRSA. None of 62 non-staphylococcal species and 212 methicillin-resistant CoNS and 74 methicillin-susceptible CoNS was positive [92]. The discrepant results for the MRSA might be explained by sequence variants or novel types of SCC*mec* not covered by the primers in the test. The 26 incorrectly identified MSSA may at least partly be explained by the presence of SCC elements that lack *mecA*, but are recognized by one of the primers. The sensitivity was 91.75%, the specificity 93.5%, the positive predictive value 82.6%, and the negative predictive value 97.1% based on 288 patients [93]. It should be noted these values are dependent on the distribution of SCC*mec* types and variants. Consequently, these values may differ at different geographic locations, for example, countries with high- or low-endemic prevalence of MRSA or high- or low-endemic levels of community-associated MRSA. Based on the first publication, in-house versions have been developed and at least one had been published and showed similar results when compared to the other assays [93].

A simplified version of the IDI MRSA Test was developed by Cuny and Witte. They used an *orfX* primer and a 16-mer primer that shows a maximum of two mismatches with different SCC*mec* types as defined by Hiramatsu [94]. One hundred MRSA representing different SCC*mec* types were detected. None of the 100 MSSA and the 130 CoNS, including 90 *mecA*-positive isolates, was positive in the PCR with the exception of a *mecA*-negative *Staphylococcus delphini* isolate which gave a weak band on agarose gel [95].

A different approach to assess the presence of MRSA in samples with a mixed staphylococcal flora was chosen by Francois et al. [96]. They developed a quantitative PCR for the *mecA* gene, the *femA* gene of *S. aureus* and the *femA* gene of *S. epidermidis*, the CoNS most commonly present on humans. This allows the linkage of the *mecA* gene to either species except when the results indicate that the number of gene copies is approximately similar for all three genes. How often this is the case in practice remains to be seen. The assay uses 5′-exonuclease probe technology together with TaqMan instrumentation, which allows the simultaneous testing of 30 samples in less than 6 h. The assay reliably detected 5 CFU of MRSA even in the presence of a 1000-fold excess of MRSE and was linear over a 6-log range for up to 1 million copies. For optimal performance, the assay uses an immuno-capture procedure with a biotinylated anti-protein A antibody. *S. aureus* cells (only *S. aureus* carries protein A) that bind the antibody are recovered by paramagnetic beads with streptavidin, which recognizes the biotin. Evaluation of 48 clinical samples showed that the immuno-qPCR detected all 23 culture-positive samples, but 16 of the 25 samples considered negative by microbiological methods were also negative in the PCR. The others were positive in the PCR. The authors concluded that these are possibly false-positives in patients treated for MRSA carriership, suggesting that the PCR detects nonviable bacteria [96]. However, these patients relapsed within 2 weeks. So, one may argue that viable bacteria were present and certainly identification of these patients is clinically relevant both in terms of therapy as well as infection prevention.

MICROARRAYS FOR THE DETECTION OF MRSA

The first attempts have been made to produce arrays for the simultaneous detection of several staphylococcal genes including resistance genes. In these arrays, multiple probes are spotted on a solid surface, in this case a membrane, which is hybridized with the DNA from the sample that has been labeled in a PCR reaction with biotin to allow a color reaction. Although microarrays hold great promise, the analytical sensitivity, which is on the order of 10^7 CFU, is still a major problem [97].

DETECTION OF OTHER RESISTANCE DETERMINANTS IN *STAPHYLOCOCCUS AUREUS*

Besides flucloxacillin, other antibiotics play an important role in the treatment of *S. aureus* infections. These antibiotics include fluoroquinolones, aminoglycosides, macrolides, and glycopeptides. It is therefore not surprising that molecular assays have been devised to detect the genes or mutations responsible for some of these antibiotic resistances. Although the majority use conventional PCR, fluoroquinolone and erythromycin resistance have been detected with more advanced methods.

Fluoroquinolone resistance in *S. aureus* is primarily mediated by mutations in codon 80 and 84 of *grlA*, encoding the A subunit of topoisomerase IV, but mutations in *grlB* encoding the B subunit of topoisomerase IV, and *gyrA* and *gyrB* encoding DNA gyrase can be involved as well. The mutations can be detected with rtPCR. One study described the application of rtPCR to the detection of mutations in *gyrA*. The analytical sensitivity of the test, utilizing molecular beacons as probes, was as few as 10 genome copies. The correlation between MIC and the rtPCR results was

98.8% [98]. Another assay demonstrated the applicability of denaturing HLPC to detect mutations in all four genes involved in fluoroquinolone resistance [99]. Unfortunately, both tests did not involve clinical specimens or a large number of clinical isolates.

A microarray has been employed to detect resistance to erythromycin. This microarray consisted of seven oligonucleotides to detect the seven most important genes involved. These genes (*ermA, ermB, ermC, ereA, ereB,* and *msrA/B*) were correctly identified [100]. It should be noted, however, that only 18 clinical isolates were tested. This makes a good assessment impossible.

HELICOBACTER PYLORI

INTRODUCTION

H. pylori is a human-specific pathogen associated with a variety of diseases including gastric and duodenal ulcers [101]. The antibiotics recommended for first-line treatment are clarithromycin and amoxicillin or metronidazole [102]. However, resistance against clarithromycin and metronidazole is increasing and probably leading to treatment failures [103]. Resistance to clarithromycin is the result of point mutations in the peptidyl-transferase region of the 23S ribosomal RNA (rRNA). These point mutations affect the binding of macrolides to the 23S rRNA. Three point mutations have been associated with resistance: A2142G, A2143G, and A2142C [104–107]. In addition, a T2183C mutation has been described, but its influence on the susceptibility to clarithromycin is unknown [108]. Several molecular techniques have been applied to the detection of these mutations. These include analysis of PCR products by restriction enzyme analysis, reverse-line blot assays, oligonucleotide ligation assays, fluorogenic probes, DNA enzyme immunoassay, and denaturing HPLC [106,109–113]. With increasing numbers of patients who fail clarithromycin treatment, other antibiotics such as tetracycline and fluoroquinolones are used, but resistance against these antibiotics also has been documented. High-level tetracycline resistance is caused by mutations at positions 926–928 in the 16S rRNA gene [114]. Resistance to fluoroquinolones is caused by mutations in the gene encoding the GyrA subunit of gyrase [115–117].

PROBE-BASED TECHNOLOGIES

A number of assays have been developed that use the formation of heteroduplexes for the detection of mutations. The formation of such duplexes can be determined by a variety of techniques, including DGGE and solid phase enzyme color reaction schemes.

A DGGE method was demonstrated for the detection of clarithromycin-resistant isolates in gastric biopsies. A total of 23 clarithromycin-susceptible and 19 resistant isolates was used to optimize the procedure, which identified heteroduplex and modified homoduplex molecules for the resistant isolates. The final evaluation was performed on 140 gastric biopsies. The 23S rRNA gene PCR proved to be less sensitive than histology-microscopy, but more sensitive than culture in the detection of *H. pylori* (49.3%, 53.6%, and 39.3%, respectively). The assay detected 25 biopsies with clarithromycin resistance out of a total of 69 PCR-positive biopsies. The mutations and wild-type sequences were confirmed by DNA sequencing. The T2183C mutation alone was also detected, but it yielded a distinct band and could thus easily be recognized. In addition, mixed infections, that is, the simultaneous presence of resistant and susceptible bacterial cells in the same sample, were detected [118]. Unfortunately, the PCR was not sensitive enough to obtain products from all histology-microscopy and/or culture-positive biopsies and the procedure requires overnight electrophoresis.

The preferential homoduplex formation assay (PFHA) is also a heteroduplex assay. In PFHA, a PCR is performed with one primer labeled with biotin and the other with dinitrophenol. After 23S rRNA amplification, the product is denatured and allowed to hybridize with unlabeled wild-type amplification product. The mixture is transferred to a microtiter plate coated with streptavidin, which binds biotin-labeled molecules. The unlabeled wild-type product preferentially recognizes wild-type biotin-labeled product from the sample. Because dinitrophenol is not present in this product, an alkaline phosphatase labeled antibody against it is not recognized and color cannot develop. The two strands from the amplification product from a mutated resistant isolate also preferentially recognize each other. Because this product contains dinitrophenol it is recognized by the antibody and color can develop. The analytical sensitivity was five organisms using purified DNA. Sensitivity of PCR on 254 gastric fluid samples was higher than for culture (95.7% vs. 89.0%). A total of 412 patients had positive samples in the PFHA. Seventy-five samples had isolates with mutations associated with clarithromycin resistance and half of these showed evidence for mixed infections. The mutations were confirmed by sequencing [119]. However, no details were given on the sequence of wild-type strains or the relationship with phenotypic resistance, leaving open the possibility of false-negative results with this assay. Gastric fluid samples proved to be more sensitive than biopsy samples with this assay. Furthermore, the PFHA method was more sensitive in the detection of mixed infections than was PCR-RFLP [119]. This method is significantly improved when compared to an older version by the same group. It has higher sensitivity, improved detection of mixed infections, and a higher throughput; however, at least theoretically, not only mixed infections can yield results indicating the simultaneous presence of susceptible and resistant isolates, but also mutations in only one of the two 23S rRNA genes yield a similar result. This phenomenon has indeed been demonstrated by Elviss et al. [120].

An assay using denaturing HPLC yielded comparable results. An analysis of 81 clinical isolates, including 51 clarithromycin-resistant isolates and 101 *H. pylori*-positive gastric biopsies, was performed. The analytical sensitivity with purified DNA was five organisms. No amplification of DNA from closely related species or other gastric flora was observed. All susceptible isolates yielded homoduplexes and resistant isolates yielded heteroduplexes. Sequencing confirmed the presence of the known mutations. In addition, the novel C2195T mutation was found. This mutation occurred in combination with one of the known mutations. Twenty-five of the gastric biopsies showed heteroduplexes including five culture-negative biopsies. DNA sequencing showed the presence of the expected mutations [108]. As shown by the results, new mutations in the amplified sequences can also be present. The significance for resistance is not clear. Fortunately, the mutation was detected, but when it occurs at the edges of a probe or outside the sequence probed, it may be missed.

Another frequently used hybridization technique for the detection of mutations in *H. pylori* is fluorescence *in situ* hybridization (FISH). FISH is performed with mutation-specific probes on either fresh biopsies or formalin-fixed paraffin-embedded tissue. Published studies including a study of the commercially available SeaFAST *H. pylori* Combi-kit showed specificities between 94% and 100%, whereas sensitivity varied between 97% and 100%. Results are usually obtained in 3 hrs [121–124]. The assays also are able to detect mixed infections. In one of the studies, re-examination of a sample showed that less than 1% of the bacterial cells were resistant, and in another study 11 discrepant results were explained as mixed infections [122,123]. The tests could also be more sensitive in the detection of infection than other methods. The study with fresh biopsies from 83 infected children showed that hybridization detected 77 positive samples, versus 75 with culture and 71 with epsilometer testing. However, six isolates were FISH-negative, but positive with at least three other methods including histology and different urease breath tests [122]. In addition, inactive coccoid forms possibly capable of reversion to vegetative forms were detected in gastric tissue [121].

A PCR-Line Probe assay (LiPA) for the detection of clarithromycin-resistant *H. pylori* also has been developed. The assay was able to detect seven 23S rRNA mutations and for typing purposes four and three genotypes in the *vacA* s-region and m-region, respectively. In total, 299 isolates were tested including 130 resistant to clarithromycin by MIC-testing; 127 had mutations, whereas 167 of the 169 susceptible isolates were wild type [125].

PCR-BASED DETECTION OF MUTATIONS ASSOCIATED WITH ANTIBIOTIC RESISTANCE IN *HELICOBACTER PYLORI*

The PCR-based methods for the detection of antibiotic resistance in *H. pylori* fall into two categories: PCR-RFLP and rtPCR.

The drawback of PCR-RFLP was illustrated in one of the studies, which could only detect the A2143G mutation. Sensitivity using fecal samples was as good as or even slightly better than ELISA for *H. pylori* or culture of gastric biopsies. Samples that were both ELISA- and culture-negative were also PCR-negative [124]. Unfortunately, a nested PCR was required, increasing the workload and the risk of contamination. Another study included more mutations, but had only two samples with clarithromycin-resistant bacteria [126]. The mutations responsible for high-level tetracycline resistance in *H. pylori* could also easily be detected by PCR-RFLP [127]. However, PCR-RFLP failed to detect most mixed infections [128].

A study of 145 isolates combined several PCR-based techniques. The first method was called 3′-mismatched PCR (3M-PCR). In 3M-PCR, the last nucleotide of a primer does not match with either the wild type or mutant, preventing elongation by DNA polymerase in the PCR. The four reactions needed to detect the three mutations associated with clarithromycin resistance were combined in a single reaction. In addition, rtPCR (Table 9.3) and PCR-RFLP were performed. 3M-PCR proved the best to detect resistant isolates. The analytical sensitivity varied for the different mutations and the wild type. In the rtPCR, the analytical sensitivity was 10^2 to 10^4 CFU; for 3M-PCR, 10^2 CFU; and for PCR-RFLP, 5×10^3 to 1×10^4 CFU [120].

rtPCR is a tool that also has been used to detect resistance in *H. pylori* because it can rather easily detect a limited number of mutations. This makes rtPCR suited to detect resistance against clarithromycin, tetracycline, and fluoroquinolones in *H. pylori*, and a number of assays have been developed and tested. The data from these studies are summarized in Table 9.3. It can be concluded in general that rtPCR is both a specific and sensitive method to detect resistant *H. pylori*. However, the occurrence of mixed infections, particularly among patients who have failed initial therapy, is a point of concern. In particular, the presence of false-negatives is a problem because antibiotic treatment may lead to the quick selection of resistant bacteria; however, this is also a problem for the other techniques.

MICROARRAYS FOR THE DETECTION OF RESISTANCE IN *HELICOBACTER PYLORI*

Xing et al. developed a rather complex microelectronic chip array for the detection of *H. pylori* and resistance to clarithromycin and tetracycline. Detailed description of the assay is beyond the scope of this chapter. Briefly, the target is amplified using biotinylated primers and the amplification product is immobilized and hybridized with a stabilizer and a fluorescent detection probe, followed by detection of fluorescence. The assay was specific for *H. pylori*. The analytical sensitivity was 1×10^8 CFU in stool [136]. Although this may be a promising approach, a number of remarks should be made. In the study resistance mutations were detected, but not evaluated on a sufficiently large set of clinical isolates or samples to draw a sound conclusion about its capability. Furthermore, it will require specialized equipment to perform the assays and the availability of a number of different assays and ease of operation may well decide its success. The advantage would be the combination of the detection of several mutations, but other array techniques hold similar promise.

FINAL REMARKS FOR THE DETECTION OF RESISTANCE IN *HELICOBACTER PYLORI*

It should be noted that the interpretation of results obtained with different assays is influenced by the presence of mixed infections. Some studies used patients who failed therapy and especially these patient groups seem to have a higher number of mixed infections. Low numbers of resistant bacteria may be missed in some assays. This may influence the sensitivity and specificity of the test depending on the gold standard used, especially when the gold standard is more effective than the current gold standard. Nevertheless, it can be concluded that molecular assays are a valuable asset for the detection of resistance in this important pathogen.

NEISSERIA GONORRHOEA

Neisseria gonorrhoea is a relatively difficult organism to culture, and antibiotic resistance, especially against fluoroquinolones, is becoming an increasing problem. It is therefore not surprising that a number of molecular tests have been developed to detect fluoroquinolone resistance in this organism. Resistance to fluoroquinolones is mediated by mutations in the *gyrA* and/or *parC* genes encoding subunits of gyrase

TABLE 9.3

Characteristics of Real-Time PCR Assays for the Detection of Mutations Associated with Antibiotic Resistance in *Helicobacter pylori*

Gene	Samples	Probe	No. of Samples	Sensitivity (%)	Specificity (%)	Gold Standard	Analytical Sensitivity (CFU)	Footnote	Reference
23S rRNA	Biopsy	Melting curve	45	100	98	Phenotype	2	a	129
23S rRNA	Stool	Melting curve	45	98	98	Phenotype	2	a	129
23S rRNA	Gastric tissue	Two probes	151	100	100	PCR-RFLP	—*	b	130
23S rRNA	Biopsy	FRET	200	98.4	94.1	Phenotype	—	c	131
23S rRNA	Biopsy	FRET	65	100	98.2	Phenotype	30	d	132
16S rRNA	Isolate	FRET	118	100	100	DNA sequence	—	—	133
	Biopsy	FRET	20	100	100	DNA sequence	—	—	133
16S rRNA	Biopsy	Melting curve	154	100	100	DNA sequence	10	—	134
gyrA	Isolates	FRET	35	100	100	DNA sequence	—	—	135

*No analytical sensitivity given.

a Some resistant isolates missed due to mixed infections. Six resistant isolates only identified by PCR.

b Only mutations at position 2143 and 2144. Mixed infections were found.

c Samples from patients with failed first eradication attempt. Multiple genotypes in 42 cases. In 5 cases phenotypically resistant cultures showed no mutations, 3 of which were confirmed by PCR-RFLP, whereas 4 phenotypically susceptible cultures showed mutations both by Real Time PCR and PCR-RFLP.

d Only one sample gave discrepant results. This was a mixed infection with a ratio of susceptible to resistant bacteria of 1 to 11.5. The assay could be used as a quantitative PCR with a linearity of 6-log starting at 300 CFU.

and topisomerase. In the A subunit of gyrase, the amino acid substitutions Ser91Tyr and Asp95Asn are involved, and in the ParC subunit of topoisomerase IV Asp86Asn, Ser87Ile, and Glu91Gly are associated with fluoroquinolone resistance [137,138]. A variety of molecular techniques have been used to detect the associated mutations.

An rtPCR approach has been described that only focuses on *gyrA* mutations. The test employs FRET technology, had an analytical sensitivity of five genome copies per reaction, and the reaction can be performed in 1 h. The assay was tested on 55 isolates and 36 clinical urethral specimens without bacterial culture. The results were in complete accordance with DNA sequencing data [139]. A second rtPCR has a somewhat different protocol that includes the detection of mutations in *parC*. Its sensitivity was equal to that of the former method, but it uses 5′-exonuclease probes as its detection principle. The same amplification products were also used for direct sequencing with one of the amplification primers as sequencing primer. The results of both methods were in full agreement [140]. A third method relied on DNA sequencing of short sequences. Mutations in *gyrA* were reliably identified [141]. The two latter assays were not demonstrated directly on clinical specimens, making an adequate assessment of their true potential impossible. The *gyrA* mutations alone were also targeted by a mismatch amplification assay (MAMA). Only when the primer matches perfectly is an amplification product formed. A drawback is the time needed for the test because the analysis of the product is performed on agarose gel. Nevertheless, validation of the assay showed excellent results [142].

Denaturing HPLC has been used to detect mutations in both *gyrA* and *parC*. The results obtained with 81 isolates completely agreed with DNA sequencing results [143]. A commercial assay, the Bed-side ICAN NG-QR detection kit (Takara bio, Japan) basically also utilizes hybridization to detect mutations. The method is comparable to the Velogene kit for resistance determination in *M. tuberculosis*, except that a DNA-RNA probe is used instead of a DNA-RNA-DNA probe. The kit yielded a number of false-positive results when compared to MIC data [144]. It can therefore be concluded that the kit performed only reasonably and although the kit is called "Bed-side" unfortunately no data were provided to show this potential ability.

Microarrays to detect mutations associated with fluoroquinolone resistance have been demonstrated. Two microarray approaches to detect mutations in both *gyrA* and *parC* have been described. One of the microarray methods used two different fluorescent labels: one to label the sample DNA and one to label a control. The procedure is a bit unusual in that an approximately 250 bp amplification product is used instead of the usual approximately 50 bp product. The shorter product allows easier detection of a mutation. Nevertheless, the results were completely concordant with DNA sequencing results [145]. Unfortunately, only isolates were tested. In a different microarray approach 87 clinical specimens were tested. The data obtained were completely compatible with sequence analysis, and an analytical sensitivity of five genome copies could be reached using purified DNA [146]. Both studies show the feasibility of designing suitable microarrays for the detection of fluoroquinolone resistance in *N. gonorrhoea*, but tests on clinical specimens are required to use the maximum potential of these tests.

It can be concluded that a number of different approaches were shown to be feasible for the detection of fluoroquinolone resistance in *N. gonorrhoea*. Some of the studies show that analytical sensitivities of five gene copies can be reached, possibly providing sufficient sensitivity for direct detection of resistance in clinical specimens. This will allow the quick and accurate assessment of fluoroquinolone resistance in clinical specimens suspected of harboring this organism.

ENTEROBACTERIACEAE

Enterobacteriaceae are one of the most important groups of pathogens causing a wide range of diseases. They form a large potential reservoir for resistance genes and their close relationship offers the potential for exchange of these resistance genes between different family members. This is exactly what has happened, and continues to occur. Many resistance genes are shared by different species of Enterobacteriaceae, but also a multitude of different resistance genes or variations on existing themes have developed. Examples of the first group are the aminoglycoside-modifying enzymes and the TEM ESBLs are an example of the second group. In addition to gene acquisition, resistance also can be caused by mutations, for example, fluoroquinolone resistance. Enterobacteriaceae are fast-growing organisms, and thus a quick resistance determination may be important in treating severe illness. Effective empiric treatment is still available, and several molecular methods have been developed that can offer the required speed for detection.

In Enterobacteriaceae, fluoroquinolone resistance is an increasing problem. Resistance is mainly caused by mutations in the *gyrA* gene that encodes the A subunit of DNA gyrase, but mutations in *gyrB*, *parC* and *parE*, encoding the B subunit of DNA gyrase and the two subunits of DNA topoisomerase IV, respectively, can also contribute. Mutations in the *gyrA* gene are clustered in the quinolone resistance determining region (QRDR). Several tests have been described that detect fluoroquinolone resistance in *Salmonella* and *Yersinia pestis*. *Y. pestis* has received increased interest because of its potential in biological warfare or bioterrorism.

An rtPCR using FRET technology with three probes (specific for Asp87Asn, Asp87Gly, Ser83Phe) was described for *S. enterica* serovar Typhimurium DT104 *gyrA*. A total of 92 isolates were evaluated and 86 showed expected mutations. Six isolates had a lower melting temperature indicating a mutation, and DNA sequencing showed that five isolates had a mutation different from those expected, but a sixth had no mutation in *gyrA* and its resistance was caused by an undetermined mechanism [147]. The reason for the lower melting temperature of these last isolates remained unexplained. A rtPCR for *Y. pestis* using FRET technology could detect 5 CFU in crude lysates [148]. In another variation, 5'-exonuclease detection technology was used instead of FRET. This method reached an analytical sensitivity of 1 CFU with partially purified lysates [149]. Therefore, both methods are rather comparable. However, the performance on clinical samples was not reported.

Denaturing HPLC is another popular method to detect mutations associated with fluoroquinolone resistance in *S. enterica* and *Y. pestis*. Evaluation of the method for the detection of mutations in *gyrA*, *gyrB*, *parC*, and *parE* using standard HPLC equipment showed that the method correctly predicted the presence or absence of mutations for 50 *Salmonella* isolates when compared to conventional DNA sequencing [150]. A second group used a similar approach, but only investigated *gyrA* mutations.

The method clearly identified the mutations and in addition detected more rare mutations. It was shown that an rtPCR method with mutation-specific probes required additional effort in case no match with one of the probes was found. The authors therefore concluded that denaturing HPLC is easier to perform when rare mutations are present in the population [151]. The same method has also been used to detect mutations in *gyrA* of *Y. pestis*. The method was shown to be satisfactory when tested on nearly 100 isolates and compared to conventional DNA sequencing [152].

Although tetracycline is an older antibiotic, there is still an interest in it. Furthermore, some of the genes conferring resistance to tetracycline are also responsible for resistance against newer tetracyclines including tigecycline [153]. The mechanisms of resistance are efflux, ribosomal protection, and modification of the antibiotic. A large number of different resistance genes encode these mechanisms. A number of molecular techniques have been developed to detect tetracycline resistance, but these are usually limited to a single gene thereby limiting their utility for diagnostic purposes. An example of single resistance determinant assays is an rtPCR assay for *tetR* of Tn*10* [154].

β-Lactam antibiotics form an important class of antimicrobials to treat infections with Enterobacteriaceae. However, resistance to the older members of this class is widespread and is increasing against the newer members. The presence of β-lactamases is the most important mechanism of resistance. Sometimes the activity of these β-lactamases can be blocked by an inhibitor like clavulanic acid or tazobactam. However, β-lactamases that became inhibitor resistant have been described. It is therefore not surprising that molecular tests, such as SCCP, have been described to detect these β-lactamases [155]. These methods are principally of value as epidemiological tools.

Trimethoprim is also a frequently used antibiotic to treat infections with Enterobacteriaceae and it is usually prescribed in combination with sulfamethoxazole. Resistance against trimethoprim is common, however, and can be mediated by at least a score of different genes. This makes these genes an important target for epidemiological studies and molecular tests are useful for this purpose. This led, for example, to the development of a PCR-RFLP to detect up to 16 different trimethoprim resistance encoding genes [156]. Probe-based assays have also been developed for epidemiological studies of multiple resistance determinants [157]. For the same purposes DNA microarrays are being developed. An example is a microarray that detects 25 virulence and 23 antibiotic resistance encoding genes in *Salmonella* and enterovirulent *E. coli*. The array used probes that were amplified by PCR [158]. The same method to generate probes for the microarray was also used in another study, in which a variety of resistance encoding genes were analyzed [159]. A final example of a microarray for epidemiological and surveillance purposes targeted *Vibrio* spp. It tested for a number of markers including resistance genes and was composed of long oligonucleotides [160].

In principle, the use of microarrays holds the possibility to check many resistance-encoding genes simultaneously, but much development has to be performed both in terms of number of genes and mutations covered, as well as costs, before they can compete with conventional phenotypic assays. rtPCR and techniques such as denaturing HPLC may have applications in some specialized niches in which resistance levels are high and speed is of importance, for example, in critically ill intensive-care patients.

CAMPYLOBACTER SPP.

Campylobacter species are an important source of food-borne infections in humans. Fluoroquinolones and macrolides are commonly used antibiotics to treat these infections, but resistance is an increasing problem. Several molecular tests have been developed to detect resistance. rtPCR is a popular choice for the detection of fluoroquinolone resistance in *Campylobacter*. Fluoroquinolone resistance is associated with mutations in the *gyrA* gene. One rtPCR assay using FRET technology easily detected mutations at codon 86 in 36 *Campylobacter coli* isolates [161]. The same group demonstrated similar results for *Campylobacter jejuni* [162]. Another rtPCR also focused on codon 86 of *C. jejuni* and used TaqMan probes. The test had an excellent analytical sensitivity by detecting femtogram levels of target genomes. The test correctly predicted mutations, although confusion over one isolate remained [163]. Mutations in *gyrA* have also correctly been detected by PCR-SSCP in 162 *C. jejuni* isolates [164]. In *C. coli* the use of PCR-RFLP to detect the mutation at codon 86 was demonstrated [165].

An LiPA has been developed to detect both fluoroquinolone and macrolide resistance in *C. jenjuni* and *C. coli*. Macrolide resistance is caused by mutations in the 23S rRNA. The results from the LiPA agreed with the phenotypic resistance determinations for the 42 isolates tested [166]. Another group also tested the LiPA, but only on 25 isolates. Nevertheless, the results were satisfactory [167]. The MAMA-PCR was also successfully applied to the detection of erythromycin resistance in both *C. jejuni* and *C. coli* [168].

PNEUMOCOCCI AND STREPTOCOCCI

With increasing levels of penicillin and erythromycin resistance among pneumococci, fluoroquinolones have become more important for the treatment of patients, but resistance to this class of antibiotics has now been documented. To study the epidemiology of fluoroquinolone resistance, PCR assays have been developed to detect mutations in *parC*, *parE*, and *gyrA* associated with this resistance. One group developed a PCR-RFLP test and a second group developed a PCR-oligonucleotide assay [169,170]. The latter assay detected only a few of the mutations that may contribute to fluoroquinolone resistance.

A PCR-reverse line blot was developed to detect multiple resistance genes in *Streptococcus agalactiae*, an important cause of neonatal and maternal sepsis. The assay involved the macrolide resistance genes *ermA/TR*, *ermB*, and *mefA/E*, the tetracycline resistance determinants *tet(M)* and *tet(O)*, and the aminoglycoside resistance genes *aph-A3* and *aad-6*. Testing of 512 isolates showed that the assay is well qualified for surveillance of antibiotic resistance among group B streptococci [171]. A rtPCR using FRET technology was developed for the surveillance of *erm*- and *mef*-encoded erythromycin resistance among β-hemolytic streptococci (group B, C and G streptococci) with the expected results [172]. However, it should be noted that to be useful for diagnostic purposes all resistance mechanisms need to be covered. The same argument is valid for most surveillance and epidemiological purposes.

OTHER BACTERIAL SPECIES

For the surveillance and monitoring of antimicrobial therapy, the tetracycline resistance determinant *tetQ* was monitored along with *Actinobacillus actinomycetemcomitans*, *Porphyromonas gingivalis*, and *Prevotella intermedia* in dental plaque. For this purpose, a quantitative PCR with TaqMan probes was developed. The PCR was linear over a range of 10 to 10^7 copies for the *tetQ* gene and 10^2 to 10^7 for bacterial cells. The authors noted, however, that sampling is critical since it can heavily influence the outcome of the assay [173]. This example shows that molecular assays may also be useful in infections that are currently less well explored.

CONCLUDING REMARKS

Molecular techniques offer the possibility for more timely antibiotic resistance profiles of both slow-growing microorganisms and microorganisms that are difficult to culture. For severe infections, molecular techniques may provide more rapid determination of resistance profiles. This becomes more important with increasing antibiotic resistance, which compromises adequate options for empiric therapy. Molecular techniques have been described for the detection of antibiotic resistance for a large number of resistance determinants and a wide range of bacterial species. The development of new molecular techniques always led to quick adaptation for the detection of antibiotic resistance. Despite the possibilities offered by molecular techniques, their use is frequently limited to a research setting and implementation in routine diagnostics can be problematic. There are a number of reasons for this: (1) The cost of molecular tests is (considerably) higher than that of phenotypic tests; (2) The number of commercially available tests is limited; (3) The design and validation of a new assay requires considerable technical and microbiologic expertise, especially when the molecular test appears to be more sensitive than the existing gold standard; and (4) Often organisms are multi-resistant and multiple genes or point mutations are involved. Therefore, commercial assays are limited to a number of niche markets such as *M. tuberculosis*, MRSA, and *H. pylori*.

The drawbacks mentioned often limit the application of molecular techniques to epidemiological studies for one or a few genes. These are usually detected by a conventional PCR. Conventional PCR has come within the grasp of more and more laboratories. The technique is straightforward and simple, especially with the development of software programs that help design appropriate primers for any gene for which a sequence is known. This simplicity also represents a danger since the PCR protocol itself may lead to unexpected amplification products and contamination by other samples or previous amplification products. This requires rigorous laboratory procedures and quality control. Unfortunately, the required expertise is not present everywhere and even when present the molecular techniques may provide unexpected challenges. This is underscored by a study by

Noordhoek et al. [174], who demonstrated that a considerable number of laboratories had difficulty in correctly identifying samples containing *M. tuberculosis*. More complicated techniques increase the risk for false-negative and false-positive results. However, new techniques may help reduce the risks by improved concepts and instrumentation.

New techniques also offer possibilities for more effectively coping with large numbers of resistance determinants or point mutations. Advances in two areas—sequencing and microarrays—are potentially important in this respect. Pyrosequencing is likely to become an important tool for the detection of point mutations. Its ability to inexpensively sequence small (40–50 bp) stretches of DNA make it ideal to detect point mutations or single nucleotide polymorphisms (SNPs). Its impact on routine diagnostics is difficult to predict, but the technique has significant potential. Pyrosequencing will certainly become an important tool for research, not only for the detection of SNPs, but in conjunction with the new technology introduced by 454 Life Sciences (U.S.A.) for sequencing of large DNA fragments, such as resistance plasmids. Briefly, DNA is fragmented to random pieces of appropriate size and ligated with primers for amplification and sequencing. The fragments are made single-stranded and bound to microbeads in such a way that only one fragment per bead is obtained. The beads are emulsified in buffer oil. Each bead is encapsulated in its own buffer capsule. The buffer contains all ingredients for amplification, the products of which become bound to the beads. The beads are then prepared for pyrosequencing, which is performed in a massively parallel fashion. Up to 200,000 sequences are generated. Special computer software determines the final sequence based on overlapping fragments. The other technique that has the potential to become more prominent is microarray technology. Microarrays offer the ability to interrogate thousands of genes simultaneously, but relatively large numbers of bacteria are required to obtain sufficient amounts of labeled DNA. Another issue is quality control, which becomes more complicated with increasing numbers of genes.

Currently, rtPCR, reverse line blot, and heteroduplex analyses are the most important techniques for the detection of antibiotic resistance in routine settings. Most studies presented here to detect antibiotic resistance among multidrug-resistant isolates do not cover all possible mechanisms. When epidemiological studies show that certain mechanisms are either absent or have a very low prevalence, it may be a cost-effective approach to ignore rare mechanisms, although the risk exists that some of these mechanisms may become more important. The consequences of non-detection of resistant isolates are not certain, but it may be expected that these will replace other strains and become responsible for new outbreaks of (multi)resistant strains. It will therefore be necessary to design surveillance studies to capture resistance mechanisms not included on a routine basis.

Molecular assays have a place in routine diagnosis of antibiotic resistance. In the future, with the development of new technologies and improvement of current technologies, the importance of molecular assays for routine detection of antibiotic resistance will only increase, although the implementation of these techniques may be slower than desirable.

REFERENCES

1. Martinez-Martinez, L., Pascual, A., and Jacoby, G.A. Quinolone resistance from a transferable plasmid. *Lancet.* 351, 797, 1998.
2. Tran, J.H. and Jacoby, G.A. Mechanism of plasmid-mediated quinolone resistance. *Proc. Natl. Acad. Sci. USA* 99, 5638, 2002.
3. Fluit, A.C. and Schmitz, F.J. Resistance integrons and super-integrons. *Clin. Microbiol. Infect.* 10, 272, 2004.
4. Clinical and Laboratory Standards Institute. Performance Standards for Antimicrobial Susceptibility Testing; Fifteenth Informational Supplement. Document M100-S15. Clinical and Laboratory Standards Institute, Wayne PA, 2005.
5. Espy, M.J. et al. Real-Time PCR in clinical microbiology: applications for routine laboratory testing. *Clin. Microbiol. Rev.* 19, 165, 2006.
6. Borst, A., Box, A.T., and Fluit A.C. False-positive results and contamination in nucleic acid amplification assays: suggestions for a prevent and destroy strategy. *Eur. J. Clin. Microbiol. Infect. Dis.* 23, 289, 2004.
7. Moghazeh, S.L. et al. Comparative antimycobacterial activities of rifampin, rifapentine, and KRM-1648 against a collection of rifampin-resistant *Mycobacterium tuberculosis* isolates with known *rpoB* mutations. *Antimicrob. Agents Chemother.* 40, 2655, 1996.
8. Williams, D.L. et al. Contribution of *rpoB* mutations to development of rifamycin cross-resistance in *Mycobacterium tuberculosis*. *Antimicrob. Agents Chemother.* 42, 1853, 1998.
9. Yang, B. et al. Relationship between antimycobacterial activities of rifampicin, rifabutin and KRM-1648 and *rpoB* mutations of *Mycobacterium tuberculosis*. *J. Antimicrob. Chemother.* 42, 621, 1998. Erratum in: *J. Antimicrob. Chemother.* 43, 613, 1999.
10. Davies, A.P. et al. Comparison of phenotypic and genotypic methods for pyrazinamide susceptibility testing with *Mycobacterium tuberculosis*. *J. Clin. Microbiol.* 38, 3686, 2000.
11. de Oliveira, M.M. et al. Rapid detection of resistance against rifampicin in isolates of *Mycobacterium tuberculosis* from Brazilian patients using a reverse-phase hybridization assay. *J. Microbiol. Methods* 53, 335, 2003.
12. McCammon, M.T. Detection of *rpoB* mutations associated with rifampin resistance in *Mycobacterium tuberculosis* using denaturing gradient gel electrophoresis. *Antimicrob. Agents Chemother.* 49, 2200, 2005.
13. McCammon, M.T. et al. Detection by denaturing gradient gel electrophoresis of *pncA* mutations associated with pyrazinamide resistance in *Mycobacterium tuberculosis* isolates from the United States-Mexico border region. *Antimicrob. Agents Chemother.* 49, 2210, 2005.
14. Gong, G. et al. Nested PCR for diagnosis of tuberculous lymphadenitis and PCR-SSCP for identification of rifampicin resistance in fine-needle aspirates. *Diagn. Cytopathol.* 26, 228, 2002.
15. Kim, B.J. et al. Simultaneous identification of rifampin-resistant *Mycobacterium tuberculosis* and nontuberculous mycobacteria by polymerase chain reaction-single strand conformation polymorphism and sequence analysis of the RNA polymerase gene (*rpoB*). *J. Microbiol. Methods* 58, 111, 2004.
16. Mani, C. et al. Comparison of DNA sequencing, PCR-SSCP and PhaB assays with indirect sensitivity testing for detection of rifampicin resistance in *Mycobacterium tuberculosis*. *Int. J. Tuberc. Lung Dis.* 7, 652, 2003.
17. Shi, R. et al. Temperature-mediated heteroduplex analysis for the detection of drug-resistant gene mutations in clinical isolates of *Mycobacterium tuberculosis* by denaturing HPLC, SURVEYOR nuclease. *Microbes Infect.* 8, 128, 2006.

18. Cooksey, R.C. et al. Temperature-mediated heteroduplex analysis performed by using denaturing high-performance liquid chromatography to identify sequence polymorphisms in *Mycobacterium tuberculosis* complex organisms. *J. Clin. Microbiol.* 40, 1610, 2002.

19. Mohamed, A.M. et al. Temperature-mediated heteroduplex analysis for detection of *pncA* mutations associated with pyrazinamide resistance and differentiation between *Mycobacterium tuberculosis* and *Mycobacterium bovis* by denaturing high-performance liquid chromatography. *J. Clin. Microbiol.* 42, 1016, 2004.

20. Cheng, A.F.B. et al. Multiplex PCR amplimer conformation analysis for rapid detection of *gyrA* mutations in fluoroquinolone-resistant *Mycobacterium tuberculosis* clinical isolates. *Antimicrob. Agents Chemother.* 48, 596, 2004.

21. Saribas, Z. et al. Rapid detection of rifampin resistance in *Mycobacterium tuberculosis* isolates by heteroduplex analysis and determination of rifamycin cross-resistance in rifampin-resistant isolates. *J. Clin. Microbiol.* 41, 816, 2003.

22. Mayta, H. et al. Evaluation of a PCR-based universal heteroduplex generator assay as a tool for rapid detection of multidrug-resistant *Mycobacterium tuberculosis* in Peru. *J. Clin. Microbiol.* 41, 5774, 2003.

23. Bahrmand, A.R. et al. Chemical cleavage of mismatches in heteroduplexes of the *rpoB* gene for detection of mutations associated with resistance of *Mycobacterium tuberculosis* to rifampin. *Scand. J. Infect. Dis.* 32, 395, 2000.

24. Morsczeck, C., Langendorfer, D., and Schierholz, J.M. A quantitative real-time PCR assay for the detection of *tetR* of Tn*10* in *Escherichia coli* using SYBR Green and the Opticon. *J. Biochem. Biophys. Methods* 59, 217, 2004.

25. Morgan, M. et al. A commercial line probe assay for the rapid detection of rifampicin resistance in *Mycobacterium tuberculosis*: a systematic review and meta-analysis. *BMC Infect. Dis.* 5, 62, 2005.

26. Johansen, I.S. et al. Direct detection of multidrug-resistant *Mycobacterium tuberculosis* in clinical specimens in low- and high-incidence countries by line probe assay. *J. Clin. Microbiol.* 41, 4454, 2003.

27. Viveiros, M. et al. Direct application of the INNO-LiPA Rif.TB line-probe assay for rapid identification of *Mycobacterium tuberculosis* complex strains and detection of rifampin resistance in 360 smear-positive respiratory specimens from an area of high incidence of multidrug-resistant tuberculosis. *J. Clin. Microbiol.* 43, 4880, 2005.

28. Somoskovi, A. et al. Use of molecular methods to identify the *Mycobacterium tuberculosis* complex (MTBC) and other mycobacterial species and to detect rifampin resistance in MTBC isolates following growth detection with the BACTEC MGIT 960 system. *J. Clin. Microbiol.* 41, 2822, 2003.

29. Jureen, P., Werngren, J., and Hoffner, S.E. Evaluation of the line probe assay (LiPA) for rapid detection of rifampicin resistance in *Mycobacterium tuberculosis*. *Tuberculosis (Edinb).* 84, 311, 2004.

30. Cavusoglu, C. et al. Characterization of *rpoB* mutations in rifampin-resistant clinical isolates of *Mycobacterium tuberculosis* from Turkey by DNA sequencing and line probe assay. *J. Clin. Microbiol.* 40, 4435, 2002.

31. Traore, H. et al. Detection of rifampicin resistance in *Mycobacterium tuberculosis* isolates from diverse countries by a commercial line probe assay as an initial indicator of multidrug resistance. *Int. J. Tuberc. Lung Dis.* 4, 481, 2000.

32. Morcillo, N. et al. A low cost, home-made, reverse-line blot hybridisation assay for rapid detection of rifampicin resistance in *Mycobacterium tuberculosis*. *Int. J. Tuberc. Lung Dis.* 6, 959, 2002.

33. Mokrousov, I. et al. Multicenter evaluation of reverse line blot assay for detection of drug resistance in *Mycobacterium tuberculosis* clinical isolates. *J. Microbiol. Methods* 57, 323, 2004.

34. Hillemann, D. et al. Use of the genotype MTBDR assay for rapid detection of rifampin and isoniazid resistance in *Mycobacterium tuberculosis* complex isolates. *J. Clin. Microbiol.* 43, 3699, 2005.
35. Van Der Zanden, A.G.M. et al. Use of DNA extracts from Ziehl-Neelsen-stained slides for molecular detection of rifampin resistance and spoligotyping of *Mycobacterium tuberculosis. J. Clin. Microbiol.* 41, 1101, 2003.
36. Brown, T.J. et al. The use of macroarrays for the identification of MDR *Mycobacterium tuberculosis. J. Microbiol. Methods* 65, 294, 2005; [Epub ahead of print] 2005.
37. Garcia, L. et al. Mutations in the *rpoB* gene of rifampin-resistant *Mycobacterium tuberculosis* isolates in Spain and their rapid detection by PCR-enzyme-linked immunosorbent assay. *J. Clin. Microbiol.* 39, 1813, 2001.
38. Mokrousov, I. et al. Detection of isoniazid-resistant *Mycobacterium tuberculosis* strains by a multiplex allele-specific PCR assay targeting *katG* codon 315 variation. *J. Clin. Microbiol.* 40, 2509, 2002.
39. Fan, X.Y. et al. Rapid detection of *rpoB* gene mutations in rifampin-resistant *Mycobacterium tuberculosis* isolates in Shanghai by using the amplification refractory mutation system. *J. Clin. Microbiol.* 41, 993, 2003.
40. Leung, E.T. et al. Detection of the *katG* Ser315Thr substitution in respiratory specimens from patients with isoniazid-resistant *Mycobacterium tuberculosis* using PCR-RFLP. *J. Med. Microbiol.* 52, 999, 2003.
41. Mokrousov, I. et al. Allele-specific *rpoB* PCR assays for detection of rifampin-resistant *Mycobacterium tuberculosis* in sputum smears. *Antimicrob. Agents Chemother.* 47, 2231, 2003.
42. Mokrousov, I. et al. PCR-based methodology for detecting multidrug-resistant strains of *Mycobacterium tuberculosis* Beijing family circulating in Russia. *Eur. J. Clin. Microbiol. Infect. Dis.* 22, 342, 2003.
43. Iwamoto, T. and Sonobe, T. Peptide nucleic acid-mediated competitive PCR clamping for detection of rifampin-resistant *Mycobacterium tuberculosis. Antimicrob. Agents Chemother.* 48, 4023, 2004.
44. Dubiley, S. et al. New PCR-based assay for detection of common mutations associated with rifampin and isoniazid resistance in *Mycobacterium tuberculosis. Clin. Chem.* 51, 447, 2005.
45. Yang, Z. et al. Simultaneous detection of isoniazid, rifampin, and ethambutol resistance of *Mycobacterium tuberculosis* by a single multiplex allele-specific polymerase chain reaction (PCR) assay. *Diagn. Microbiol. Infect. Dis.* 53, 201, 2005.
46. Carvalho, W.S. et al. Low-stringency single-specific-primer PCR as a tool for detection of mutations in the *rpoB* gene of rifampin-resistant *Mycobacterium tuberculosis. J. Clin. Microbiol.* 41, 3384, 2003.
47. Ahmad, S., Mokaddas, E., and Jaber, A.A. Rapid detection of ethambutol-resistant *Mycobacterium tuberculosis* strains by PCR-RFLP targeting *embB* codons 306 and 497 and *iniA* codon 501 mutations. *Mol. Cell. Probes* 18, 299, 2004.
48. Suzuki, Y. et al. Rapid detection of pyrazinamide-resistant *Mycobacterium tuberculosis* by a PCR-based in vitro system. *J. Clin. Microbiol.* 40, 501, 2002.
49. Piatek, A.S. et al. Genotypic analysis of *Mycobacterium tuberculosis* in two distinct populations using molecular beacons: implications for rapid susceptibility testing. *Antimicrob. Agents Chemother.* 44, 103, 2000.
50. Marin, M. et al. Rapid direct detection of multiple rifampin and isoniazid resistance mutations in *Mycobacterium tuberculosis* in respiratory samples by real-time PCR. *Antimicrob. Agents Chemother.* 48, 4293, 2004.
51. Rindi, L. et al. A real-time PCR assay for detection of isoniazid resistance in *Mycobacterium tuberculosis* clinical isolates. *J. Microbiol. Methods* 55, 797, 2003.

52. Torres, M.J. et al. Improved real-time PCR for rapid detection of rifampin and isoniazid resistance in *Mycobacterium tuberculosis* clinical isolates. *Diagn. Microbiol. Infect. Dis.* 45, 207, 2003.

53. van Doorn, H.R. et al. Detection of a point mutation associated with high-level isoniazid resistance in *Mycobacterium tuberculosis* by using real-time PCR technology with 3'-minor groove binder-DNA probes. *J. Clin. Microbiol.* 41, 4630, 2003.

54. Torres, M.J. et al. Use of real-time PCR and fluorimetry for rapid detection of rifampin and isoniazid resistance-associated mutations in *Mycobacterium tuberculosis. J. Clin. Microbiol.* 38, 3194, 2000.

55. Garcia de Viedma, D. et al. New real-time PCR able to detect in a single tube multiple rifampin resistance mutations and high-level isoniazid resistance mutations in *Mycobacterium tuberculosis. J. Clin. Microbiol.* 40, 988, 2002.

56. Edwards, K.J. et al. Detection of *rpoB* mutations in *Mycobacterium tuberculosis* by biprobe analysis. *J. Clin. Microbiol.* 39, 3350, 2001.

57. Kocagoz, T., Saribas, Z., and Alp, A. Rapid determination of rifampin resistance in clinical isolates of *Mycobacterium tuberculosis* by Real-Time PCR. *J. Clin. Microbiol.* 43, 6015, 2005.

58. Varma-Basil, M. et al. Rapid detection of rifampin resistance in *Mycobacterium tuberculosis* isolates from India and Mexico by a molecular beacon assay. *J. Clin. Microbiol.* 42, 5512, 2004.

59. Wada, T. et al. Dual-probe assay for rapid detection of drug-resistant *Mycobacterium tuberculosis* by Real-Time PCR. *J. Clin. Microbiol.* 42, 5277, 2004.

60. Lin, S.Y. et al. Rapid detection of isoniazid and rifampin resistance mutations in *Mycobacterium tuberculosis* complex from cultures or smear-positive sputa by use of molecular beacons. *J. Clin. Microbiol.* 42, 4204, 2004.

61. Ruiz, M. et al. Direct detection of rifampin- and isoniazid-resistant *Mycobacterium tuberculosis* in auramine-rhodamine-positive sputum specimens by Real-Time PCR. *J. Clin. Microbiol.* 42, 1585, 2004.

62. Saribas, Z. et al. Use of fluorescence resonance energy transfer for rapid detection of isoniazid resistance in *Mycobacterium tuberculosis* clinical isolates. *Int. J. Tuberc. Lung Dis.* 9, 181, 2005.

63. Espasa, M. et al. Direct detection in clinical samples of multiple gene mutations causing resistance of *Mycobacterium tuberculosis* to isoniazid and rifampicin using fluorogenic probes. *J. Antimicrob. Chemother.* 55, 860, 2005.

64. Patnaik, M., Liegmann, K., and Peter, J.B. Rapid detection of smear-negative *Mycobacterium tuberculosis* by PCR and sequencing for rifampin resistance with DNA extracted directly from slides. *J. Clin. Microbiol.* 39, 51, 2001.

65. Yam, W.C. et al. Direct detection of rifampin-resistant *Mycobacterium tuberculosis* in respiratory specimens by PCR-DNA sequencing. *J. Clin. Microbiol.* 42, 4438, 2004.

66. Arnold, C. et al. Single-nucleotide polymorphism-based differentiation and drug resistance detection in *Mycobacterium tuberculosis* from isolates or directly from sputum. *Clin. Microbiol. Infect.* 11, 122, 2005.

67. Zhao, J.R. et al. Development of a pyrosequencing approach for rapid screening of rifampin, isoniazid and ethambutol-resistant *Mycobacterium tuberculosis. Int. J. Tuberc. Lung Dis.* 9, 328, 2005.

68. Yue, J. et al. Detection of rifampin-resistant *Mycobacterium tuberculosis* strains by using a specialized oligonucleotide microarray. *Diagn. Microbiol. Infect. Dis.* 48, 47, 2004.

69. Denkin, S. et al. Microarray-based *pncA* genotyping of pyrazinamide-resistant strains of *Mycobacterium tuberculosis. J. Med. Microbiol.* 54, 1127, 2005.

70. Wade, M.M. et al. Accurate mapping of mutations of pyrazinamide-resistant *Mycobacterium tuberculosis* strains with a scanning-frame oligonucleotide microarray. *Diagn. Microbiol. Infect. Dis.* 49, 89, 2004.

71. Sougakoff, W. et al. Use of a high-density DNA probe array for detecting mutations involved in rifampicin resistance in *Mycobacterium tuberculosis*. *Clin. Microbiol. Infect.* 10, 289, 2004.

72. Mikhailovich, V. et al. Identification of rifampin-resistant *Mycobacterium tuberculosis* strains by hybridization, PCR, and ligase detection reaction on oligonucleotide microchips. *J. Clin. Microbiol.* 39, 2531, 2001.

73. Mikhailovich, V.M. et al. Detection of rifampicin-resistant *Mycobacterium tuberculosis* strains by hybridization and polymerase chain reaction on a specialized TB-microchip. *Bull. Exp. Biol. Med.* 131, 94, 2001.

74. Small, P.M. and Chambers, H.F. Vancomycin for *Staphylococcus aureus* endocarditis in intravenous drug users. *Antimicrob. Agents Chemother.* 34, 1227, 1990.

75. Conzalez, C. et al. Bacteremic pneumonia due to *Staphylococcus aureus*: A comparison of disease caused by methicillin-resistant and methicillin-susceptible organisms. *Clin. Infect. Dis.* 29, 1171, 1999.

76. Fluit, A.C., Visser, M.R., and Schmitz, F.-J. Molecular detection of antimicrobial resistance. *Clin. Microbiol. Rev.* 14, 836, 2001.

77. Perez-Roth, E. Multiplex PCR for simultaneous identification of *Staphylococcus aureus* and detection of methicillin and mupirocin resistance. *J. Clin. Microbiol.* 39, 4037, 2001.

78. Strommenger, B. et al. Multiplex PCR assay for simultaneous detection of nine clinically relevant antibiotic resistance genes in *Staphylococcus aureus*. *J. Clin. Microbiol.* 41, 4089, 2003.

79. Krishnan, P.U., Miles, K., and Shetty, N. Detection of methicillin and mupirocin resistance in *Staphylococcus aureus* isolates using conventional and molecular methods: a descriptive study from a burns unit with high prevalence of MRSA. *J. Clin. Pathol.* 55, 745, 2002.

80. Arbique, J. et al. Comparison of the VELOGENE Rapid MRSA Identification Assay, Denka MRSA-Screen Assay, and BBL Crystal MRSA ID System for rapid identification of methicillin-resistant *Staphylococcus aureus*. *Diagn. Microbiol. Infect. Dis.* 40, 5, 2001.

81. Levi, K. et al. Evaluation of an isothermal signal amplification method for rapid detection of methicillin-resistant *Staphylococcus aureus* from patient-screening swabs. *J. Clin. Microbiol.* 41, 3187, 2003.

82. Levi, K. and Towner, K.J. Detection of methicillin-resistant *Staphylococcus aureus* (MRSA) in blood with the EVIGENE MRSA detection kit. *J. Clin. Microbiol.* 41, 3890, 2003.

83. Poulsen, A.B., Skov, R., and Pallesen, L.V. Detection of methicillin resistance in coagulase-negative staphylococci and in staphylococci directly from simulated blood cultures using the EVIGENE MRSA Detection Kit. *J. Antimicrob. Chemother.* 51, 419, 2003.

84. Otte, K.M., Jenner, S., and Wulffen, H.V. Identification of methicillin-resistant *Staphylococcus aureus* (MRSA): Comparison of a new molecular genetic test kit (GenoType MRSA) with standard diagnostic methods. *Clin. Lab.* 51, 389, 2005

85. Eigner, U. et al. Evaluation of a rapid direct assay for identification of bacteria and the *mecA* and *van* genes from positive-testing blood cultures. *J. Clin. Microbiol.* 43, 5256, 2005.

86. Rohrer, S. et al. Improved methods for detection of methicillin-resistant *Staphylococcus aureus*. *Eur. J. Clin. Microbiol. Infect. Dis.* 20, 267, 2001.

87. Fang, H. and Hedin, G. Rapid screening and identification of methicillin-resistant *Staphylococcus aureus* from clinical samples by selective-broth and real-time PCR assay. *J. Clin. Microbiol.* 41, 2894, 2003.

88. Grisold, A.J. et al. Detection of methicillin-resistant *Staphylococcus aureus* and simultaneous confirmation by automated nucleic acid extraction and real-time PCR. *J. Clin. Microbiol.* 40, 2392, 2002.

89. Hagen, R.M. et al. Development of a real-time PCR assay for rapid identification of methicillin-resistant *Staphylococcus aureus* from clinical samples. *Int. J. Med. Microbiol.* 295, 77, 2005.

90. Gillet, Y. et al. Association between *Staphylococcus aureus* strains carrying gene for Panton-Valentine leukocidin and highly lethal necrotising pneumonia in young immunocompetent patients. *Lancet* 359, 753, 2002.

91. McDonald, R.R. et al. Development of a triplex real time PCR assay for detection of Panton-Valentine leukocidin toxin genes in clinical isolates of methicillin-resistant *Staphylococcus aureus. J. Clin. Microbiol.* 43, 6147, 2005.

92. Huletsky, A. et al. New real-time PCR assay for rapid detection of methicillin-resistant *Staphylococcus aureus* directly from specimens containing a mixture of staphylococci. *J. Clin. Microbiol.* 42, 1875, 2004.

93. Warren, D.K. et al. Detection of methicillin-resistant *Staphylococcus aureus* directly from nasal swab specimens by a real-time PCR assay. *J. Clin. Microbiol.* 42, 5578, 2004.

94. Chongtrakool, P. et al. Staphylococcal Cassette Chromosome *mec* (SCC*mec*) typing of methicillin-resistant *Staphylococcus aureus* strains isolated in 11 Asian countries: a proposal for a new nomenclature for SCC *mec* elements. *Antimicrob. Agents Chemother.* 50, 1001, 2006.

95. Cuny, C. and Witte, W. PCR for the identification of methicillin-resistant *Staphylococcus aureus* (MRSA) strains using a single primer pair specific for SCC*mec* elements and the neighbouring chromosome-borne *orfX. Clin. Microbiol. Infect.* 11, 834, 2005.

96. Francois, P. et al. Rapid detection of methicillin-resistant *Staphylococcus aureus* directly from sterile or nonsterile clinical samples by a new molecular assay. *J. Clin. Microbiol.* 41, 254, 2003.

97. Monecke, S. and Ehricht, R. Rapid genotyping of methicillin-resistant *Staphylococcus aureus* (MRSA) isolates using miniaturised oligonucleotide arrays. *Clin. Microbiol. Infect.* 11, 825, 2005.

98. Lapierre, P. et al. Real-time PCR assay for detection of fluoroquinolone resistance associated with *grlA* mutations in *Staphylococcus aureus. J. Clin. Microbiol.* 41, 3246, 2003.

99. Hannachi-M'Zali, F. et al. Examination of single and multiple mutations involved in resistance to quinolones in *Staphylococcus aureus* by a combination of PCR and denaturing high-performance liquid chromatography (DHPLC). *J. Antimicrob. Chemother.* 50, 649, 2002.

100. Volokhov, D. et al. Microarray analysis of erythromycin resistance determinants. *J. Appl. Microbiol.* 95, 787, 2003.

101. Mégraud, F., van Loon, F.P.L., and Thijssen, S.F.T. Curved spiral bacilli, in: *Infectious Diseases*, D. Armstrong and J. Cohen, Eds., Mosby, Philadelphia, 1999, 8.19.4.

102. Malfertheiner, P. et al. Current concepts in the management of *Helicobacter pylori* infection. The Maastricht 2-2000 Consensus Report. *Aliment. Pharmacol. Ther.* 16, 167, 2002.

103. Glupzynski, Y. et al. European multicentre survey of in vitro antimicrobial resistance in *Helicobacter pylori. Eur. J. Clin. Microbiol. Infect. Dis.* 20, 820, 2001.

104. Versalovic, J. et al. Mutations in 23S rRNA are associated with clarithromycin resistance in *Helicobacter pylori. Antimicrob. Agents Chemother.* 40, 477, 1996.

105. Taylor, D.E. et al. Cloning and sequence analysis of two copies of a 23S rRNA gene from *Helicobacter pylori* and association of clarithromycin resistance with 23S rRNA mutations. *Antimicrob. Agents Chemother.* 41, 2621, 1997.

106. van Doorn, L.J. et al. Rapid detection, by PCR and reverse hybridization, of mutations in the *Helicobacter pylori* 23S rRNA gene, associated with macrolide resistance. *Antimicrob. Agents Chemother.* 43, 1779, 1999.

107. Alarcón, T. et al. PCR using 3′-mismatched primers to detect A2142C mutation in 23S rRNA conferring resistance to clarithromycin in *Helicobacter pylori* clinical isolates. *J. Clin. Microbiol.* 38, 923, 2000.

108. Posteraro, P. Rapid detection of clarithromycin resistance in *Helicobacter pylori* using a PCR-based denaturing HPLC assay. *J. Antimicrob. Chemother.* 57, 71, 2006.

109. Occhialini, A. et al. Macrolide resistance in *Helicobacter pylori*: rapid detection of point mutations and assays of macrolide binding to ribosomes. *Antimicrob. Agents Chemother.* 41, 2724, 1997.

110. Stone, G.G. et al. A PCR-oligonucleotide ligation assay to determine the prevalence of 23S rRNA gene mutations in clarithromycin-resistant *Helicobacter pylori*. *Antimicrob. Agents Chemother.* 41, 712, 1997.

111. Marais, A. et al. Direct detection of *Helicobacter pylori* resistance to macrolides by a polymerase chain reaction/DNA enzyme immunoassay in gastric biopsy specimens. *Gut* 44, 447, 1999.

112. Oleastro, M. et al. Real-time PCR assay for rapid and accurate detection of point mutations conferring resistance to clarithromycin in *Helicobacter pylori*. *J. Clin. Microbiol.* 41, 397, 2003.

113. Schabereiter-Gurtner, C. et al. Novel real-time PCR assay for detection of *Helicobacter pylori* infection and simultaneous clarithromycin susceptibility testing of stool and biopsy specimens. *J. Clin. Microbiol.* 42, 4512, 2004.

114. Gerrits, M.M. et al. 16S rRNA mutation-mediated tetracycline resistance in *Helicobacter pylori*. *Antimicrob. Agents Chemother.* 46, 2996, 2002.

115. Moore, R.A. et al. Nucleotide sequence of the *gyrA* gene and characterization of ciprofloxacin-resistant mutants of *Helicobacter pylori*. *Antimicrob. Agents Chemother.* 39, 107, 1995.

116. Megraud, F. Epidemiology and mechanism of antibiotic resistance in *Helicobacter pylori*. *Gastroenterology* 115, 1278, 1998.

117. Wang, G. et al. Spontaneous mutations that confer antibiotic resistance in *Helicobacter pylori*. *Antimicrob. Agents Chemother.* 45, 727, 2001.

118. Scarpellini, P. et al. Direct detection of *Helicobacter pylori* mutations associated with macrolide resistance in gastric biopsy material taken from human immunodeficiency virus-infected subjects. *J. Clin. Microbiol.* 40, 2234, 2002.

119. Maeda, S. et al. Detection of clarithromycin-resistant *Helicobacter pylori* strains by a preferential homoduplex formation assay. *J. Clin. Microbiol.* 38, 210, 2000.

120. Elviss, N.C., Lawson, A.J., and Owen R.J. Application of 3′-mismatched reverse primer PCR compared with real-time PCR and PCR-RFLP for the rapid detection of 23S rDNA mutations associated with clarithromycin resistance in *Helicobacter pylori*. *Int. J. Antimicrob. Agents* 23, 349, 2004.

121. Trebesius, K. et al. Rapid and specific detection of *Helicobacter pylori* macrolide resistance in gastric tissue by fluorescent *in situ* hybridisation. *Gut* 46, 608, 2000.

122. Feydt-Schmidt, A. et al. Fluorescence *in situ* hybridization vs. epsilometer test for detection of clarithromycin-susceptible and clarithromycin-resistant *Helicobacter pylori* strains in gastric biopsies from children. *Aliment. Pharmacol. Ther.* 16, 2073, 2002.

123. Juttner, S. et al. Reliable detection of macrolide-resistant *Helicobacter pylori* via fluorescence *in situ* hybridization in formalin-fixed tissue. *Mod. Pathol.* 17, 684, 2004.

124. Morris, J.M. et al. Evaluation of seaFAST, a rapid fluorescent *in situ* hybridization test, for detection of *Helicobacter pylori* and resistance to clarithromycin in paraffin-embedded biopsy sections. *J. Clin. Microbiol.* 43, 3494, 2005.

125. van Doorn, L.J. et al. Accurate prediction of macrolide resistance in *Helicobacter pylori* by a PCR line probe assay for detection of mutations in the 23S rRNA gene: multicenter validation study. *Antimicrob. Agents Chemother.* 45, 1500, 2001.

126. Fontana, C. et al. Detection of clarithromycin-resistant *Helicobacter pylori* in stool samples. *J. Clin. Microbiol.* 41, 3636, 2003.

127. Ribeiro, M.L. et al. Detection of high-level tetracycline resistance in clinical isolates of *Helicobacter pylori* using PCR-RFLP. *FEMS Immunol. Med. Microbiol.* 40, 57, 2004.

128. Maeda, S. Detection of clarithromycin-resistant *Helicobacter pylori* strains by a preferential homoduplex formation assay. *J. Clin. Microbiol.* 38, 210, 2000.

129. Schabereiter-Gurtner, C. et al. Novel real-time PCR assay for detection of *Helicobacter pylori* infection and simultaneous clarithromycin susceptibility testing of stool and biopsy specimens. *J. Clin. Microbiol.* 42, 4512, 2004.

130. Matsumura, M. et al. Rapid detection of mutations in the 23S rRNA gene of *Helicobacter pylori* that confers resistance to clarithromycin treatment to the bacterium. *J. Clin. Microbiol.* 39, 691, 2001.

131. Oleastro, M. et al. Real-time PCR assay for rapid and accurate detection of point mutations conferring resistance to clarithromycin in *Helicobacter pylori*. *J. Clin. Microbiol.* 41, 397, 2003.

132. Lascols, C. et al. Fast and accurate quantitative detection *of Helicobacter pylori* and identification of clarithromycin resistance mutations in *H. pylori* isolates from gastric biopsy specimens by real-time PCR. *J. Clin. Microbiol.* 41, 4573, 2003.

133. Lawson, A.J., Elviss, N.C., and Owen, R.J. Real-time PCR detection and frequency of 16S rDNA mutations associated with resistance and reduced susceptibility to tetracycline in *Helicobacter pylori* from England and Wales. *J. Antimicrob. Chemother.* 56, 282, 2005.

134. Glocker, E. et al. Real-time PCR screening for 16S rRNA mutations associated with resistance to tetracycline in *Helicobacter pylori*. *Antimicrob. Agents Chemother.* 49, 3166, 2005.

135. Glocker, E. and Kist, M. Rapid detection of point mutations in the *gyrA* gene of *Helicobacter pylori* conferring resistance to ciprofloxacin by a fluorescence resonance energy transfer-based real-time PCR approach. *J. Clin. Microbiol.* 42, 2241, 2004.

136. Xing, J.Z. et al. Development of a microelectronic chip array for high-throughput genotyping of *Helicobacter* species and screening for antimicrobial resistance. *J. Biomol. Screen.* 10, 235, 2005.

137. Deguchi, T. et al. DNA gyrase mutations in quinolone-resistant clinical isolates of *Neisseria gonorrhoeae*. *Antimicrob. Agents Chemother.* 39, 561, 1995.

138. Deguchi, T. et al. Quinolone-resistant *Neisseria gonorrhoeae*: correlation of alterations in the GyrA subunit of DNA gyrase and the ParC subunit of topoisomerase IV with antimicrobial susceptibility profiles. *Antimicrob. Agents Chemother.* 40, 1020, 1996.

139. Rimbara, E. et al. Development of a highly sensitive method for detection of clarithromycin-resistant *Helicobacter pylori* from human feces. *Curr. Microbiol.* 51, 1, 2005.

140. Giles, J. et al. Use of applied biosystems 7900HT sequence detection system and Taqman assay for detection of quinolone-resistant *Neisseria gonorrhoeae*. *J. Clin. Microbiol.* 42, 3281, 2004. Erratum in: *J. Clin. Microbiol.* 42, 4916, 2004.

141. Gharizadeh, B. et al. Detection of *gyrA* mutations associated with ciprofloxacin resistance in *Neisseria gonorrhoeae* by rapid and reliable pre-programmed short DNA sequencing. *Int. J. Antimicrob. Agents* 26, 486, 2005.

142. Sultan, Z. et al. Comparison of mismatch amplification mutation assay with DNA sequencing for characterization of fluoroquinolone resistance in *Neisseria gonorrhoeae*. *J. Clin. Microbiol.* 42, 591, 2004.

143. Shigemura, K. et al. Rapid detection of *gyrA* and *parC* mutations in fluoroquinolone-resistant *Neisseria gonorrhoeae* by denaturing high-performance liquid chromatography. *J. Microbiol. Methods* 59, 415, 2004.

144. Horii, T. et al. Rapid detection of fluoroquinolone resistance by isothermal chimeric primer-initiated amplification of nucleic acids from clinical isolates of *Neisseria gonorrhoeae*. *J. Microbiol. Methods* 65, 557, 2006; [Epub ahead of print] 2005.

145. Booth, S.A. et al. Design of oligonucleotide arrays to detect point mutations: molecular typing of antibiotic resistant strains of *Neisseria gonorrhoeae* and hantavirus infected deer mice. *Mol. Cell. Probes* 17, 77, 2003.

146. Zhou, W. et al. Detection of *gyrA* and *parC* mutations associated with ciprofloxacin resistance in *Neisseria gonorrhoeae* by use of oligonucleotide biochip technology. *J. Clin. Microbiol.* 42, 5819, 2004.

147. Walker, R.A. et al. Use of a LightCycler *gyrA* mutation assay for rapid identification of mutations conferring decreased susceptibility to ciprofloxacin in multiresistant *Salmonella enterica* serotype Typhimurium DT104 isolates. *J. Clin. Microbiol.* 39, 1443, 2001.

148. Lindler, L.E., Fan, W., and Jahan, N. Detection of ciprofloxacin-resistant *Yersinia pestis* by fluorogenic PCR using the LightCycler. *J. Clin. Microbiol.* 39, 3649, 2001.

149. Lindler, L.E. and Fan, W. Development of a 5' nuclease assay to detect ciprofloxacin resistant isolates of the biowarfare agent *Yersinia pestis*. *Mol. Cell. Probes* 17, 41, 2003.

150. Randall, L.P., Coldham, N.G., and Woodward, M.J. Detection of mutations in *Salmonella enterica gyrA, gyrB, parC* and *parE* genes by denaturing high performance liquid chromatography (DHPLC) using standard HPLC instrumentation. *J. Antimicrob. Chemother.* 56, 619, 2005.

151. Eaves, D.J. et al. Detection of *gyrA* mutations in quinolone-resistant *Salmonella enterica* by denaturing high-performance liquid chromatography. *J. Clin. Microbiol.* 40, 4121, 2002.

152. Hurtle, W. et al. Detection and identification of ciprofloxacin-resistant *Yersinia pestis* by denaturing high-performance liquid chromatography. *J. Clin. Microbiol.* 41, 3273, 2003.

153. Fluit, A.C. et al. The presence of tetracycline resistance determinants and the susceptibility for tigecycline and minocycline. *Antimicrob. Agents Chemother.* 49, 1636, 1638, 2005.

154. Mokrousov, I. et al. Molecular characterization of multiple-drug-resistant *Mycobacterium tuberculosis* isolates from northwestern Russia and analysis of rifampin resistance using RNA/RNA mismatch analysis as compared to the line probe assay and sequencing of the *rpoB* gene. *Res. Microbiol.* 153, 213, 2002.

155. Winstanley, T.G. et al. Phenotypic detection of β-lactamase-mediated resistance to oxyimino-cephalosporins in Enterobacteriaceae: evaluation of the Mastascan Elite Expert System. *J. Antimicrob. Chemother.* 56, 292, 2005.

156. Navia, M.M. et al. Detection of dihydrofolate reductase genes by PCR and RFLP. *Diagn. Microbiol. Infect. Dis.* 46, 295, 2003.

157. Gauthier, M. and Blais, B.W. Cloth-based hybridization array system for the detection of multiple antibiotic resistance genes in *Salmonella enterica* subsp. *enterica* serotype Typhimurium DT104. *Lett. Appl. Microbiol.* 38, 265, 2004.

158. Chen, S. et al. A DNA microarray for identification of virulence and antimicrobial resistance genes in *Salmonella* serovars and *Escherichia coli*. *Mol. Cell. Probes* 19, 195, 2005.

159. Call, D.R. et al. Identifying antimicrobial resistance genes with DNA microarrays. *Antimicrob. Agents Chemother.* 47, 3290, 2003.

160. Vora, G.J. et al. Microarray-based detection of genetic heterogeneity, antimicrobial resistance, and the viable but nonculturable state in human pathogenic *Vibrio* spp. *Proc. Natl. Acad. Sci. USA* 102, 19109, 2005.

161. Carattoli, A., Dionisi, A., and Luzzi, I. Use of a LightCycler *gyrA* mutation assay for identification of ciprofloxacin-resistant *Campylobacter coli*. *FEMS Microbiol. Lett.* 214, 87, 2002.

162. Dionisi, A.M., Luzzi, I., and Carattoli, A. Identification of ciprofloxacin-resistant *Campylobacter jejuni* and analysis of the *gyrA* gene by the LightCycler mutation assay. *Mol. Cell. Probes* 18, 255, 2004.
163. Wilson, D.L. et al. Identification of ciprofloxacin-resistant *Campylobacter jejuni* by use of a fluorogenic PCR assay. *J. Clin. Microbiol.* 38, 3971, 2000.
164. Hakanen, A. et al. *gyrA* polymorphism in *Campylobacter jejuni*: detection of *gyrA* mutations in 162 *C. jejuni* isolates by single-strand conformation polymorphism and DNA sequencing. *Antimicrob. Agents Chemother.* 46, 2644, 2002.
165. Alonso, R. et al. PCR-restriction fragment length polymorphism assay for detection of *gyrA* mutations associated with fluoroquinolone resistance in *Campylobacter coli*. *Antimicrob. Agents Chemother.* 48, 4886, 2004.
166. Niwa, H. et al. Simultaneous detection of mutations associated with resistance to macrolides and quinolones in *Campylobacter jejuni* and *C. coli* using a PCR-line probe assay. *Int. J. Antimicrob. Agents* 22, 374, 2003.
167. Niwa, H. et al. Rapid detection of mutations associated with resistance to erythromycin in *Campylobacter jejuni/coli* by PCR and line probe assay. *Int. J. Antimicrob. Agents* 18, 359, 2001. Erratum in: *Int. J. Antimicrob. Agents* 22, 461, 2001.
168. Alonso, R. et al. MAMA-PCR assay for the detection of point mutations associated with high-level erythromycin resistance in *Campylobacter jejuni* and *Campylobacter coli* strains. *J. Microbiol. Methods* 63, 99, 2005.
169. Bui, M.H. et al. PCR-oligonucleotide ligation assay for detection of point mutations associated with quinolone resistance in *Streptococcus pneumoniae*. *Antimicrob. Agents Chemother.* 47, 1456, 2003.
170. Alonso, R., Galimand, M., and Courvalin P. An extended PCR-RFLP assay for detection of *parC*, *parE* and *gyrA* mutations in fluoroquinolone-resistant *Streptococcus pneumoniae*. *J. Antimicrob. Chemother.* 53, 682, 2004.
171. Zeng, X. et al. Simultaneous detection of nine antibiotic resistance-related genes in *Streptococcus agalactiae* using multiplex PCR and reverse line blot hybridization assay. *Antimicrob. Agents Chemother.* 50, 204, 2006.
172. Twagira, M.F. et al. Development of a real-time PCR assay on the Roche Light-Cycler for the detection of *erm* and *mef* erythromycin resistance genes in beta-haemolytic streptococci. *J. Antimicrob. Chemother.* 56, 793, 2005.
173. Maeda, H. et al. Quantitative real-time PCR using TaqMan and SYBR Green for *Actinobacillus actinomycetemcomitans*, *Porphyromonas gingivalis*, *Prevotella intermedia*, *tetQ* gene and total bacteria. *FEMS Immunol. Med. Microbiol.* 39, 81, 2003.
174. Noordhoek, G.T. et al. Multicentre quality control study for detection of *Mycobacterium tuberculosis* in clinical samples by nucleic amplification methods. *Clin. Microbiol. Infect.* 10, 295, 2004.

10 Evolution and Epidemiology of Antibiotic-Resistant Pneumococci

Christopher Gerard Dowson
and Krzysztof Trzcinski

CONTENTS

Streptococcus pneumoniae is still an important human bacterial pathogen. The past two decades have witnessed the global spread of resistance to major groups of anti-pneumococcal drugs and there are no countries free of multidrug-resistant strains. In this naturally transformable organism resistance to antibiotics can arise by both inter- and intra-species recombination events, enabling resistance acquired by horizontal gene transfer or from point mutation to spread throughout the population. However, there is also strong evidence for clonal expansion by the international spread of multidrug-resistant strains. Among such successful clones, the Spain[23F]-1, Spain[6B]-2, and Spain[9V]-3 in particular have reached the status of pandemic clones. All three have emerged as being penicillin-non-susceptible (PNSP) and often resistant to tetracyclines, macrolides, chloramphenicol, or co-trimoxazole. The presence of the most successful, the Spain[23F]-1 clone, and its other capsular variants, has been documented in 42 countries all over the world. The susceptibility-testing results survey done for this study showed that prevalence of PNSP was above 40% in 25 out of 96 countries and below 5% in only 8 of them. This chapter offers insight into the mechanisms of the antibiotic resistance acquisition in pneumococci, their evolution, and the epidemiology of multidrug-resistant strains.

INTRODUCTION

Streptococcus pneumoniae (the Pneumococcus) is the causative agent of pneumonia, otitis media, meningitis, and bacteraemia, and a major cause of morbidity and mortality worldwide particularly among the young, elderly, and immunocompromised [1,2]. The past two decades have witnessed the acquisition and global spread of chromosomal and transposon-encoded resistance to the major groups of effective antibiotics [3–6]. There is therefore increasing pressure to develop novel therapeutic agents. However, in order to understand fully the spread of resistance, we need to look jointly at the mechanisms of resistance and the evolutionary processes involved in their acquisition and dissemination. For this, we also need a clear picture of the population structure of carried and invasive isolates of this naturally transformable organism.

The past 60 years of selection by diverse antimicrobials has revealed an extensive range of resistance mechanisms, many of which are dealt with elsewhere in this volume. Therefore, in looking ahead to the selection of novel stable targets for chemotherapy or vaccination, we need to take account of the processes involved in the development of resistance during the past decades. The Pneumococcus and other naturally transformable organisms such as *Neisseria* spp. that evolve by intra- and inter-species recombination are perhaps among the most difficult to deal with. Many of their loci are effectively moving targets [7–9], not only transferring freely from one strain to another, but also able to evolve by acquiring highly divergent blocks of nucleotides from related species that will generate novel proteins with altered catalytic activities or different antigenic profiles [10–15]. The following gives some insight into the role of horizontal gene transfer in the evolution and epidemiology of antibiotic-resistant pneumococci.

EVOLUTION OF ANTIBIOTIC RESISTANCE

β-LACTAM RESISTANCE AND CLINICAL CONSEQUENCES

Since its detection in 1967, penicillin resistance in *S. pneumoniae* has become increasingly prevalent worldwide [16]. A *S. pneumoniae* isolate is considered to lack susceptibility when the minimal inhibitory concentration (MIC) of penicillin is greater than 0.06 mg/L [17] and is treated as a PNSP. Isolates for which penicillin MICs ranged from 0.12 to 1 mg/L fit the category of intermediate susceptibility and high-level resistance to penicillin, when the MIC is greater than 1 mg/L [17]. With few exceptions, infections caused by strains intermediately susceptible to penicillin can be successfully treated with other anti-pneumococcal β-lactams, such as amoxicillin or the broad-spectrum third generation cephalosporins, cefotaxime, and ceftriaxone [18,19]. However, highly penicillin-resistant isolates are invariably cross resistant to a range of β-lactam antibiotics including cefotaxime and ceftriaxone [20], posing reduced therapeutic options. In countries such as Spain, Hungary, and South Africa lack of susceptibility to penicillin among *S. pneumoniae* is not only found among a high proportion of all pneumococci isolated [21,22], but isolates possess levels of resistance to penicillin of 1 to 4 mg/L and occasionally up to 8 to 16 mg/L [23]. Recent clinical studies have shown that β-lactams are generally still useful for the treatment of pneumococcal infections that do not involve cerebrospinal fluid [24,25] as pneumococcal bacteremia caused by PNSP is not associated with increased morbidity or mortality [26]; and there is no poorer outcome for pneumococcal pneumonia caused by intermediately resistant strains, when patients were treated with amoxicillin [27]. It is less clear whether this is also the case for otitis media [28]. Nevertheless, β-lactam resistance is associated with increased costs of health care [29] and alternative therapy may well be necessary for organisms expressing high-level resistance [30]. The existence and spread of highly penicillin-resistant strains [31–34] is potentially a major concern as pneumococci of this phenotype are not only more difficult or inadequately treated with β-lactams, but also frequently non-susceptible to several other anti-pneumococcal drugs [35,36].

Role of Penicillin-Binding Proteins in β-Lactam Resistance

Lack of susceptibility to penicillin in clinical isolates of *S. pneumoniae* is due to the presence of high-molecular-weight penicillin-binding proteins (PBPs) that have a greatly reduced affinity for the β-lactam antibiotics [23,37]. No other mechanisms of acquired resistance to β-lactams have been described in clinical isolates of pneumococci to date. Although there are numerous alterations to the genes encoding low affinity PBPs [38,39], several changes within the transpeptidase domain of different PBPs have been identified as important in resistance [40,41]. It would appear for PBP2X that resistance is due to amino acid substitutions within a buried cavity near the catalytic site, which contains a structural water molecule [42]. The examination of β-lactam-resistant laboratory mutants has shown that, in addition to PBPs, mutations in *ciaR/H, cpoA,* and *murM/N* genes could potentially influence resistance. Although currently these alternative determinants have not been found to be directly responsible for increased resistance among clinical isolates [43–45], the functional inactivation of *murM/N* genes has been shown to obliterate high-level penicillin resistance in clinical isolates [46]. MurM/N are involved in the sequential addition of Ala or Ser, and Ala, respectively, to the Lys residue, which is found on the pentapeptide branch of the carbohydrate backbone of peptidoglycan; however, the enzymology of these reactions has only just been characterized (Lloyd, dePascale, Bugg, Roper, and Dowson, unpublished data). The prevalence of side branches varies from strain to strain, and although reported to be associated with strains exhibiting high levels of penicillin resistance, high levels of branching are also found in the susceptible isolate R6 [47]. Interestingly R6 was originally selected for laboratory use because of its ability to be transformed to penicillin resistance at high frequency. The enzymological basis for these differences in crosslinking between strains and its possible role in high-level penicillin resistance are the subject of current study (Lloyd et al., unpublished data).

The primary target of a β-lactam antibiotic is the essential PBP [48] with the highest affinity for that particular antibiotic, and for many clinically important β-lactams this is PBP2X [49]. However, the use of primary target in this context does not pre-suppose that this is the only killing target, but rather that which influences MIC due to the differential affinities of PBPs for different β-lactam antibiotics. For clinical isolates of *S. pneumoniae* challenged by different β-lactams, either as the result of treatment of a pneumococcal infection or during asymptomatic carriage, when a different organism is the desired target, there may be selection for the acquisition of different permutations of low affinity PBPs. High-level resistance to oxacillin requires low affinity forms of PBPs 2X and 2B [50], cephalosporin-resistant PBPs 2X and 1A [20,51], and penicillin-resistant PBPs 2X, 1A, and 2B [37]; in laboratory-derived mutants and recently in a clinical isolate PBP2A has also been implicated in resistance [52]. The inevitability of highly penicillin-resistant clinical isolates being cross-resistant to other groups of β-lactams now becomes obvious.

Role of Oral Streptococci in the Formation of Mosaic *pbp* Genes

Low affinity forms of PBP1A, PBP2B, and PBP2X have arisen initially by the horizontal transfer and recombination of homologous chromosomally encoded *pbp* genes from closely related species of streptococci. *S. mitis* and *S. oralis* have been identified as two of the species responsible for contributing genetic material for the formation of a low affinity PBP2B in many penicillin-resistant isolates of *S. pneumoniae* [40,53]. However, analysis of *pbp* genes from a diverse collection of resistant isolates has revealed that several additional, as yet unidentified species also have been involved in the evolution of these mosaic genes [40]. Recent analysis of the population structure of pneumococci and the closely related oral streptococci has revealed that isolates identified as *S. mitis* represent a highly divergent group of organisms. In addition, there is a previously unidentified group of organisms that lie between *S. mitis* and pneumococci [54]. These are being investigated as alternative DNA donors involved in the evolution of PBPs and a range of pneumococcal virulence determinants. Experimentally it has been shown that oral streptococci with MICs for penicillin as high as 64 mg/L can transform pneumococci to this level of resistance, although this requires the acquisition of altered forms of PBPs 2A and 1B from *S. oralis* together with 2X, 1A, and 2B [55,56].

Work examining the degree of sexual isolation between pneumococci and the related oral streptococci [57] has revealed, as found previously for *Bacillus* [58], a log linear relationship between nucleotide divergence and sexual isolation.

Apart from the acquisition of novel PBPs by recombination, there is now also evidence that mosaic *pbp* genes have further evolved, by spontaneous mutation, altering levels of cross resistance to penicillin and cephalosporins, presumably in response to clinical exposure to these different classes of β-lactams [20].

DEVELOPMENT OF MULTIPLE ANTIBIOTIC RESISTANCE

Resistance of pneumococci to tetracyclines [59–61], chloramphenicol [62], and macrolides [63,64] is due to acquisition of the highly mobile conjugative transposon Tn1545, or related transposons, which may carry one or more of these and other resistance determinants [65,66]. These transposons possess an integration/excision system, encoded by the genes *int/xis*, and terminally associated (host-derived) integration sequences [67]. Transfer of the transposon from the donor to the recipient chromosome involves excision of the element from the host chromosome, formation of a covalently closed circular intermediate, entry into the recipient cell, and subsequent integration into its chromosome [67]. The site of integration may be determined by DNA topology rather than sequence specificity.

The stability of transposon-encoded resistance within pneumococci has not been determined. However, it is clear that members of some multiply-resistant pneumococcal clones do differ in their resistance profiles, that many more allelic variants of the tetracycline resistance gene (*tetM*) are found within pneumococci than previously described [59], and that different *tetM* alleles can be found among members of the same clonal group [68]. In general, the *tetM* positive isolates are resistant to all clinically available tetracyclines [69] except tigecycline [70]; however, isolates with MICs of tetracycline 2 to 4 mg/L (susceptible or intermediate-susceptible [17]), which gave positive hybridization signals with *tetM* probes, have been also described [71]. Apart from *tetM*, presence of two other tetracycline resistance determinants have been documented in *S. pneumoniae*, namely *tetO*, coding for ribosomal protection similar to TetM [72], and the *tetK* coded efflux mechanism [73].

Two mechanisms of resistance to macrolides have been described in pneumococci thus far. Active efflux due to the acquisition of the *mefA* gene was identified in isolates expressing a low-level resistance to erythromycin (MICs ranged from 1 to 32 mg/L). Such isolates were once treated as macrolide resistant, but susceptible to lincosamides and streptogramins (M phenotype) [74,75]. The second mechanism described is based on ribosomal protection due to acquisition of the *ermB* gene, where *ermB* positive isolates are resistant to macrolides, lincosamides, and streptogramins B (MLS$_B$ phenotype) [74,75] and exhibit high-level resistance to erythromycin (MICs above 32 mg/L) [71,76]. MLS$_B$ and M phenotypes were also described in *ermB* negative and *mefE* negative strains indicating the presence of novel genes or allelic variants of already identified genes [77]. Finally, the macrolide-streptogramin-resistant but lincosamide-susceptible *S. pneumoniae* (so-called MS phenotype) also has been described [77].

Pneumococcal resistance to trimethoprim and the sulfonamides, which inhibit bacterial purine synthesis, has also been identified [78–80]; however, this is clearly chromosomally encoded, and involves alterations to housekeeping *sulA* (dihydropteroate synthase) and *dfr* (dihydrofolate reductase) genes within the pneumococcal genome. Although point mutations and codon duplications are frequently associated with resistance there is also some evidence that inter-species recombination has played a role in the evolution of resistance [78]. A similar situation is found in the evolution of pneumococcal resistance to rifampicin, where there is evidence of resistance arising due to recombination rather than the more frequently occurring point mutations within the gene encoding the β subunit of RNA polymerase (*rpoB*) [81].

The use of fluoroquinolones in the treatment of pneumococcal infections has resulted in decreased susceptibility [82]. This appears to be due to target alterations in DNA gyrase (GyrA) and topoisomerase IV (ParC) [83–85] or to the action of an efflux pump encoded by *pmrA* [86,87]. Although alterations in GyrA and ParC appear to have evolved by point mutations in *S. pneumoniae*, it is clear that high-level quinolone-resistant viridans streptococci also have evolved [88] and may, if resistance becomes prevalent, act as a source of resistance genes for pneumococci. Recent investigations do show evidence that interspecies recombination has also played a role in the evolution of fluoroquinolone resistance in clinical isolates of *S. pneumoniae* [89]. To date, there are no reports of vancomycin resistance in clinical isolates of *S. pneumoniae*; however, it has been shown that loss of function of the VncS histidine kinase of a two-component sensor-regulator system in laboratory strains of *S. pneumoniae* produced tolerance to vancomycin and other classes of antibiotic, indicating that this may be a precursor to the evolution of vancomycin resistance in the community [90].

EPIDEMIOLOGY OF *S. PNEUMONIAE*

POPULATION STRUCTURE OF *S. PNEUMONIAE*

Asymptomatic carriage of pneumococci in the throat or nasopharynx is widespread, with carriage rates being especially high in children [91–93]. There is also clear evidence of spread among families [94], and colonization by multiple pneumococcal capsular types has also been reported [91]. Some serotypes are particularly associated with disease in children [95] or adults [96]) and others with carriage [97] or HIV infection [98]. However, it is only just becoming apparent that among isolates associated with invasive disease there are important virulent pneumococcal clones that are responsible for many cases of disease around the world [8] and that these clones are also frequently carried asymptomatically [7].

There is clear evidence from population genetic analysis that the pneumococcal chromosome is at linkage equilibrium, that is, freely recombining, and that recombination by transformation and possibly transduction may introduce blocks of nucleotides from other *S. pneumoniae* strains or other species ranging in size from tens of base pairs [99] to tens of kilobase pairs [100,101]. This can result in alterations to single loci or whole operons. One therefore has to be careful in epidemiological analyses not to rely upon single markers in strain identification, especially if those markers are immunologically reactive and liable to change under the selective pressure of the human immune system.

Capsular serotyping has been the cornerstone of pneumococcal epidemiology for many years. However, this is a fairly blunt instrument when trying to understand the movement and evolution of specific pneumococcal clones, especially now that serotype exchange among clones is well documented [100,102,103], and the current best estimate suggests that serotype exchange may occur among 4% to 6% of isolates [7]. Therefore, tracking the spread of prevalent susceptible or resistant clones requires the use of techniques, such as pulse field gel electrophoresis (PFGE) [104], restriction fragment end labelling (RFEL) with PBP genotyping [105] or the more recently developed multi-locus sequence typing (MLST) [8], or multi-locus restriction typing (MLRT) [7]. Clearly, transportability, access to reference strains, and composite databases are important for positive strain identification. A database also showing clonal variants is especially important for organisms, such as pneumococci, in which clones initially sufficiently stable to track do start to break down due to the ongoing process of recombination.

Apart from tracking the clonal spread of organisms, it is also possible to examine the horizontal spread of resistance genes. This has been undertaken successfully for the dissemination of *pbp* genes by restriction fragment length polymorphism (RFLP) analysis of amplified *pbp* gene fragments [102,106], and using similar techniques for *tetM* [68,107].

INTERCONTINENTAL SPREAD OF RESISTANT CLONES

Numerous multidrug-resistant pneumococcal clones have been identified [8], with at least five of these shown to be major pandemic clones (http://spneumoniae.mlst.net; http://www.sph.emory.edu/PMEN/pmen_ww_spread_clones.html). The oldest and most prevalent is pandemic or Spain[23F]-1. This clone has been reported in 42 countries (Figure 10.1) across all continents. Isolates of this clone are usually resistant to a wide range of anti-pneumococcal drugs, including tetracyclines, co-trimoxazole, chloramphenicol, and often macrolides. MICs for penicillin are generally 1 to 2 mg/L [108], but may reach 8 mg/L [109,110]. Originally of serotype 23F this clone has acquired at least eight distinct capsular type variants: 3 [111], 6B [109], 9V [108,111], 7 [112], 11 (http://spneumoniae.mlst.net), 14 [109,111,113–115], 19A (http://spneumoniae.mlst.net), and 19F [102], with 19F being the most prevalent variant reported [71,108,111,113–118]. The early spread of the Spain[23F]-1 clone from Spain to the United Kingdom (Figure 10.1) was highly correlated to holidaymakers returning from Spain. Each year approximately 52 million people visit Spain as a holiday destination plus a large number of migrant workers. Together these most likely represent the predominant means by which clones originating in Spain have spread worldwide. It is not clear whether major international sporting events, such as the 1992 Barcelona Olympics, further contributed to this. However, there is clear evidence of other pathogens spreading during crowded international gatherings, such as the hajj in Mecca, which resulted in an outbreak of meningococcal disease, and measles at the International Special Olympic Games in the United States [119]. Given that the population of Barcelona is approximately 1.5 million and that the tickets sold for the 1992 Barcelona Olympic Games totalled three million (http://www.olympic. org/uk/organisation/facts/programme/ticketing_uk.asp), it would not be surprising that the close proximity of different nationalities at this event may well have played some role in the global transmission of the Spain[23F]-1 clone illustrated in Figure 10.1.

Second in temporal sequence of isolation is the multidrug-resistant Spain[6B]-2 clone [116]. This clone spread across Western Europe at the end of the 1980s and the beginning of the 1990s [110,116,120,121], and is now present in North and South America [112,122–124], Asia [118,125], and Australia (http://spneumoniae.mlst.net).

The third multidrug-resistant pandemic strain is the major penicillin-resistant Spain[9V]-3 clone. This strain was originally intermediately susceptible to penicillin and additionally resistant to co-trimoxazole; however, by the mid-1990s, members of this clone had acquired resistance to macrolides and chloramphenicol. It is now also clear that serotype 14 variants of this clone are widely distributed in France [110], Denmark, Spain, Uruguay [100], Poland [100,126], Portugal [127], the Netherlands

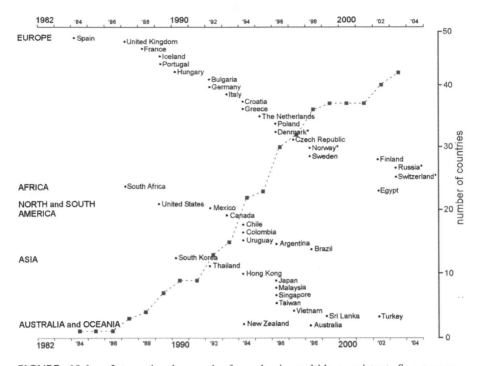

FIGURE 10.1 International spread of pandemic multidrug-resistant *Streptococcus pneumoniae* clone Spain[23F]-1. *Dots* indicate the year of the strain appearance in a particular country (timeline at the top and the bottom of the graph). *Asterisks* indicate countries where single locus variants of the dominant Spain[23F]-1 type ST81 (but not ST81 itself) were identified by MLST. Squares show cumulative number of countries (right Y-axis) in which presence of the clone was reported up to the particular year. Countries in alphabetical order: Argentina [153], Australia (http://spneumoniae.mlst.net), Brazil [112], Bulgaria [76], Canada [154], Chile [155], Colombia [128], Croatia [117], Czech Republic (http://spneumoniae.mlst.net), Denmark (http://spneumoniae.mlst.net), Egypt (http://spneumoniae.mlst.net), Finland (http://spneumoniae.mlst.net), France [156], Germany [121], Greece [113], Hong Kong [115], Hungary [121], Iceland [157], Italy [71], Japan [158], Malaysia [158], Mexico [159], New Zealand [160], Norway (http://spneumoniae.mlst.net), Poland [126], Portugal [131], Singapore [158], Spain [116], Sri Lanka [161], Russia (http://spneumoniae.mlst.net), South Africa [116], South Korea [117], Sweden [162], Switzerland (http://spneumoniae.mlst.net), Taiwan [158], Thailand [132], Turkey [163], the Netherlands [164], Uruguay [165], United Kingdom [102], United States [4], and Vietnam [166].

[113], Mexico [109], and Colombia [128]. 23F variants of this clone have been described in Germany [121], and 11A in Israel [129]. Presence of the Spain [9V]-3 has been reported on all continents except Australia (http://www.sph.emory.edu/PMEN/pmen_ww_spread_clones.html).

Intercontinental spread of at least two other clones indicate their pandemic potential, namely England[14]-9 [130] reported thus far in five European countries, the United States, Argentina, and Australia, and Taiwan[19F]-14 [118] reported in three

Asian countries, South Africa, the United Kingdom, Greece, the United States, and Australia (http://www.sph.emory.edu/PMEN/pmen_ww_spread_clones.html).

Despite the fact that there is some degree of similarity observed in resistance profiles of particular pandemic clones, different genes or even mechanisms of resistance might be responsible for similar phenotypes. For example, among the Spain[23F]-1 clone isolates collected in the United States in 1996 to 1997, both *ermB* and *mefE* genes coding for macrolide resistance were observed [114]. Isolates of MLS_B and M phenotypes were observed among Taiwanese PNSP of the same clone isolated in 1996 to 1997 [118]. Moreover, early Bulgarian [76], Italian [71], and Portuguese [113,127,131] isolates of this pandemic clone were susceptible to macrolides. This might indicate that antibiotic resistance profiles vary in particular clones, rather than exhibiting an immutable pandemic pattern. Fluidity in resistance profile would enable strains to respond to local or national variations in prescribing policy.

There are also several currently more geographically restricted national clones of multidrug-resistant pneumococci (http://www.sph.emory.edu/PMEN/pmen_ww_spread_clones.html), most of them expressing intermediate susceptibility to penicillin [108,126–128,132]. One of the best described is the 19A Hungarian clone [121,133], which has been responsible for one of the highest frequencies of resistance to penicillin observed worldwide [134]. Perhaps surprisingly, the spread of this clone appears to have been restricted to the Czech Republic [133]. This was possibly due to the socio-economic situation in Europe prior to the end of the 1980s, when traveling and mass migration were restricted in former Eastern Bloc countries.

Multidrug resistance in pneumococci is not only observed in PNSP. Penicillin-susceptible serotype 3 strains that are resistant to macrolides, lincosamides (MLS_B phenotype) and tetracyclines have been isolated in South Africa [135], and penicillin-susceptible serogroup 6 strains resistant to macrolides and lincosamides, tetracyclines, co-trimoxazole, and chloramphenicol have been isolated in Greece [136]. Penicillin-susceptible, multiply-resistant serotype 5, 6, 11, and 23 strains have also been isolated in Colombia [137], Portugal [127], and Hong Kong [138].

PREVALENCE OF PNSP WORLDWIDE

Lack of susceptibility to penicillin is the most often analyzed mechanism of resistance in pneumococci, and for this reason, it is an accepted marker of overall non-susceptibility despite the fact that it is not always a dominating mechanism of resistance. Analysis of reports for which data for more than three different antimicrobial group drugs were available revealed that lack of susceptibility to penicillin was only the third most common mechanism of resistance, after lack of susceptibility to tetracycline or co-trimoxazole.

A compilation of published data for the prevalence of PNSP in 96 countries is presented in Figure 10.2. A map for the year 1999 published in the first edition of this book [139] presents data collected mostly in the 1990s. The map for the year 2007 presents data published since then, with the exception of Bangladesh, Papua New Guinea, Pakistan, Serbia, and Zambia, for which no new reports were available. Surveys did not necessarily cover the same period. When more than one source of data was available for the particular country, the most recent or largest dataset is cited. Results for invasive pneumococcal diseases had priority over non-invasive and

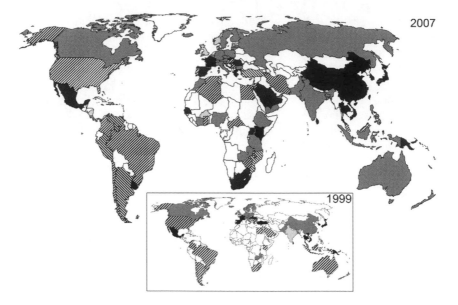

FIGURE 10.2 The worldwide prevalence of penicillin-nonsusceptible pneumococci (PNSP, penicillin MIC >0.1 mg/L). Countries shown in light gray represent those with <5% PNSP (category A), in dark gray those with 5% to 20% PNSP (category B), in black-white stripes with 20% to 40% PNSP (category C), and black with over 40% PNSP (category D). No data were available for unshaded areas. The small framed map shows the prevalence according to data published up to the year 1999 [139]; the main map shows results of the surveillances published between 1999 and 2007. In 2007 within category A were: Burkina Faso [167], Denmark [168], Estonia [169], Jamaica [171], the Netherlands [172], Nicaragua [173], Norway [174], and the United Kingdom [175,176]. Within category B were: Australia [177,178], Austria [179], Bangladesh [180], Belgium [181,182], Bulgaria [183], Canada [184], Costa Rica [185], the Czech Republic [186,187], Ethiopia [188], Fiji [189], Finland (www.ktl.fi/extras/fire), Gambia [190], Germany [182], Iceland [191], Indonesia [192,193], India [161,170,194], Ireland [195], Italy [182,196], Latvia [187], Lithuania [187,197], Morocco [198,199], Mozambique [200], New Zealand [201], Nigeria [202], the Philippines [161,194], Pakistan [193], Poland [203,204], Russia [205], Rwanda [206], Slovenia [207,208], Sweden [209], Switzerland [141,182], Tanzania [210], United Arab Emirates [211], Yemen [212], Zambia [213]. Within category C were: Algeria [199], Argentina [214,215], Brazil [216], Central Africa Republic [217], Chile [215,218], Colombia [215], Côte d'Ivoire [198,219], Croatia [187,220], Dominican Republic [221], Egypt [199,222], Ghana [223], Hungary [224], Iran [225], Israel [226,227], Jordan [199], Malawi [228], Malaysia [194], Panama [229], Peru [230], Portugal [231], Trinidad and Tobago [232], Tunisia [199,233], Turkey [234,235], the United States of America [142,144], Venezuela [215,229], Yugoslavia/Serbia [236], Zimbabwe [32]. Within category D were: China [194,237,238], France [33,182], Greece [239], Hong Kong [194], Japan [240,241], Kenya [242], Kuwait [243,244], Lebanon [245], Mexico [215,246,247], Papua New Guinea [140], Qatar [248], Romania [187,249], Saudi Arabia [161,250], Senegal [198], Singapore [161,194,251], Slovak Republic [187], South Africa [252], South Korea [161,194], Spain [182,253], Sri Lanka [161,194], Taiwan [161,194,254], Thailand [161,194,255], Uganda [256,257], Uruguay [215], Vietnam [161,194,258].

the carriage data were presented when no other were available. Only in a few countries were ongoing nationwide surveys on PNSP prevalence conducted. For many places results from limited geographic areas (mainly highly urbanized), short periods of time or only pre-selected groups of patients were available. Since it is documented that the prevalence of PNSP depends upon type of infection [140,141], patient group (e.g., HIV positive vs. HIV negative) [32], and group age [142], results of this analysis should be treated with caution, as each may not represent a broad picture of PNSP distribution.

For purposes of this study, the prevalence of PNSP was described by four arbitrarily chosen categories based on the percentage of PNSP among clinical or carried *S. pneumoniae* isolates [139]: category A—countries for which percentage of PNSP was below 5%; category B—prevalence of PNSP ranged from 5% to 20%; category C—PNSP ranged from 20% to 40%; and category D—prevalence of PNSP was above 40%.

The following countries changed category between 1999 and 2007 to a higher prevalence of PNSP: Malaysia, China (two categories up), Greece, India, Kenya, Saudi Arabia, Singapore, Slovak Republic, South Africa, and Uruguay; or to a lower PNSP prevalence: Norway, Australia, Bulgaria, Croatia, Hungary, Indonesia, Rwanda, Turkey, and the United Kingdom. The significance of these changes is uncertain, since for the majority of countries the category remained the same over a decade. Changes in the category may reflect different populations sampled, but can also show the effect of positive or negative selection. Increase in resistance, usually attributed to selective pressure applied by antimicrobial therapies, is well documented for pneumococci. There is also evidence that intervention can lead to an increase in the susceptibility of *S. pneumoniae* population under selective pressure. In the United States, PNSP were associated with serotypes targeted by the pneumococcal conjugated vaccine licensed in 2000. Here the incidence of antibiotic-resistant invasive disease declined substantially after vaccine introduction into the routine childhood immunization program [143,144]. The impact of vaccination was also significant for some other forms of resistance [144].

SUMMARY AND FUTURE PROSPECTS

There has been a substantial increase in antibiotic resistance observed in pneumococci within the last decade. This has been directly connected with the spread of particular pandemic clones of multidrug-resistant strains, and the development of local epidemic strains. There are no countries that are free of multidrug-resistant PNSP; however, there are pronounced differences observed in the frequencies of PNSP even between neighboring countries. As to whether this is due to differences in antibiotic usage policies, to vaccination strategies, or to other factors is unclear. Interestingly, the development of multivalent conjugate pneumococcal vaccines may have a significant impact upon the prevalence of antibiotic resistance. Included within the polyvalent vaccines are those childhood-associated serotypes 6B, 9V, 14, and 23F that represent the burden of pandemic multi-resistant clones. Eradication of these serotypes from the vaccinated population will hopefully reduce the frequency of their occurrence among other non-vaccinated members of the population [145];

however, this is little comfort to countries unable to afford or implement mass vaccination programs. Moreover, it is unclear in the mid- to long term whether restricted valency vaccines will select for new pandemic clones from serotypes beyond the scope of the proposed vaccines or lead to the evolution of novel capsular types. Furthermore, modeling the impact of conjugate vaccines upon the prevalence of penicillin resistance strongly suggests that vaccination alone may not be successful in controlling selection for resistance in *S. pneumoniae* [146]. It is clear, however, that continued selection for resistance will continue and that relative levels of antibiotic consumption are important [147–149]. To lengthen the lifespan of existing and future agents, it has been proposed that national drug prescribing policies combination therapy or cycling may prove useful [150]. For pneumococci, it is clear that increased doses of amoxicillin are effective against disease and the spread of PNSP and strains with intermediate levels of resistance [151]. However, modeling the population impact of upgrading the dose of amoxicillin suggests that the use of high doses may well facilitate the spread of highly penicillin-resistant strains [152].

Time will tell whether the commencement of mass conjugate vaccination and upgrading doses of antibiotic ushers in a decline in global pneumococcal infection or just another phase in the evolution of this highly adaptable organism.

REFERENCES

1. Feldman C. Pneumonia associated with HIV infection. *Curr Opin Infect Dis* 2005;18:165–70.
2. Feldman C. Pneumonia in the elderly. *Med Clin North Am* 2001;85:1441–59.
3. Appelbaum PC. Emerging resistance to antimicrobial agents in Gram-positive bacteria. Pneumococci. *Drugs* 1996;51 Suppl 1:1–5.
4. Munoz R, Coffey TJ, Daniels M, et al. Intercontinental spread of a multiresistant clone of serotype 23F *Streptococcus pneumoniae*. *J Infect Dis* 1991;164:302–6.
5. McDougal LK, Facklam R, Reeves M, et al. Analysis of multiply antimicrobial-resistant isolates of *Streptococcus pneumoniae* from the United States. *Antimicrob Agents Chemother* 1992;36:2176–84.
6. Baquero F. Gram-positive resistance: challenge for the development of new antibiotics. *J Antimicrob Chemother* 1997;39 Suppl A:1–6.
7. Muller-Graf CD, Whatmore AM, King SJ, et al. Population biology of *Streptococcus pneumoniae* isolated from oropharyngeal carriage and invasive disease. *Microbiology* 1999;145 (Pt 11):3283–93.
8. Enright MC and Spratt BG. A multilocus sequence typing scheme for *Streptococcus pneumoniae*: identification of clones associated with serious invasive disease. *Microbiology* 1998;144 (Pt 11):3049–60.
9. Hanage WP, Kaijalainen T, Herva E, Saukkoriipi A, Syrjanen R, and Spratt BG. Using multilocus sequence data to define the pneumococcus. *J Bacteriol* 2005;187:6223–30.
10. Smith JM, Dowson CG, and Spratt BG. Localized sex in bacteria. Nature 1991;349:29–31.
11. Dowson CG, Barcus V, King S, Pickerill P, Whatmore A, and Yeo M. Horizontal gene transfer and the evolution of resistance and virulence determinants in *Streptococcus*. *Soc Appl Bacteriol Symp Ser* 1997;26:42S–51S.
12. McDaniel LS, McDaniel DO, Hollingshead SK, and Briles DE. Comparison of the PspA sequence from *Streptococcus pneumoniae* EF5668 to the previously identified PspA sequence from strain Rx1 and ability of PspA from EF5668 to elicit protection against pneumococci of different capsular types. *Infect Immun* 1998;66:4748–54.

13. Whatmore AM, Barcus VA, and Dowson CG. Genetic diversity of the streptococcal competence (com) gene locus. *J Bacteriol* 1999;181:3144–54.
14. Iannelli F, Oggioni MR, and Pozzi G. Allelic variation in the highly polymorphic locus pspC of *Streptococcus pneumoniae*. *Gene* 2002;284:63–71.
15. King SJ, Whatmore AM, and Dowson CG. NanA, a neuraminidase from *Streptococcus pneumoniae*, shows high levels of sequence diversity, at least in part through recombination with *Streptococcus oralis*. *J Bacteriol* 2005;187:5376–86.
16. Dowson CG, Coffey TJ, and Spratt BG. Origin and molecular epidemiology of penicillin-binding-protein-mediated resistance to β-lactam antibiotics. *Trends Microbiol* 1994; 2:361–6.
17. NCCLS. Methods for dilution antimicrobial susceptibility tests for bacteria that grow aerobically; Approved standard—Fifth edition. Vol. NCCLS document M7-A5. 2000 (NCCLS, ed.).
18. Gehanno P, Nguyen L, Barry B, et al. Eradication by ceftriaxone of *Streptococcus pneumoniae* isolates with increased resistance to penicillin in cases of acute otitis media. *Antimicrob Agents Chemother* 1999;43:16–20.
19. Silverstein M, Bachur R, and Harper MB. Clinical implications of penicillin and ceftriaxone resistance among children with pneumococcal bacteremia. *Pediatr Infect Dis J* 1999;18:35–41.
20. Coffey TJ, Daniels M, McDougal LK, Dowson CG, Tenover FC, and Spratt BG. Genetic analysis of clinical isolates of *Streptococcus pneumoniae* with high-level resistance to expanded-spectrum cephalosporins. *Antimicrob Agents Chemother* 1995;39:1306–13.
21. Fenoll A, Martin Bourgon C, Munoz R, Vicioso D, and Casal J. Serotype distribution and antimicrobial resistance of *Streptococcus pneumoniae* isolates causing systemic infections in Spain, 1979–1989. *Rev Infect Dis* 1991;13:56–60.
22. Marton A. Pneumococcal antimicrobial resistance: the problem in Hungary. *Clin Infect Dis* 1992;15:106–11.
23. Markiewicz Z and Tomasz A. Variation in penicillin-binding protein patterns of penicillin-resistant clinical isolates of pneumococci. *J Clin Microbiol* 1989;27:405–10.
24. Yu VL, Chiou CC, Feldman C, et al. An international prospective study of pneumococcal bacteremia: correlation with *in vitro* resistance, antibiotics administered, and clinical outcome. *Clin Infect Dis* 2003;37:230–7.
25. Feldman C. Clinical relevance of antimicrobial resistance in the management of pneumococcal community-acquired pneumonia. *J Lab Clin Med* 2004;143:269–83.
26. Ho PL, Que TL, Ng TK, Chiu SS, Yung RW, and Tsang KW. Clinical outcomes of bacteremic pneumococcal infections in an area with high resistance. *Eur J Clin Microbiol Infect Dis* 2006;25:323–7.
27. Falco V, Almirante B, Jordano Q, et al. Influence of penicillin resistance on outcome in adult patients with invasive pneumococcal pneumonia: is penicillin useful against intermediately resistant strains? *J Antimicrob Chemother* 2004;54:481–8.
28. Barry B, Gehanno P, Blumen M, and Boucot I. Clinical outcome of acute otitis media caused by pneumococci with decreased susceptibility to penicillin. *Scand J Infect Dis* 1994;26:446–52.
29. Klepser ME, Klepser DG, Ernst EJ, et al. Health care resource utilization associated with treatment of penicillin-susceptible and -nonsusceptible isolates of *Streptococcus pneumoniae*. *Pharmacotherapy* 2003;23:349–59.
30. Reechaipichitkul W, Assawasanti K, and Chaimanee P. Risk factors and clinical outcomes of penicillin resistant *S. pneumoniae* community-acquired pneumonia in Khon Kaen, Thailand. *Southeast Asian J Trop Med Public Health* 2006;37:320–6.
31. Doern GV, Brueggemann AB, Blocker M, et al. Clonal relationships among high-level penicillin-resistant *Streptococcus pneumoniae* in the United States. *Clin Infect Dis* 1998;27:757–61.

32. Gwanzura L, Pasi C, Nathoo KJ, et al. Rapid emergence of resistance to penicillin and trimethoprim-sulphamethoxazole in invasive *Streptococcus pneumoniae* in Zimbabwe. *Int J Antimicrob Agents* 2003;21:557–61.

33. Demachy MC, Vernet-Garnier V, Cottin J, et al. Antimicrobial resistance data on 16,756 *Streptococcus pneumoniae* isolates in 1999: A Pan-Regional Multicenter Surveillance Study in France. *Microb Drug Resist* 2005;11:323–9.

34. Cafini F, del Campo R, Alou L, et al. Alterations of the penicillin-binding proteins and murM alleles of clinical *Streptococcus pneumoniae* isolates with high-level resistance to amoxicillin in Spain. *J Antimicrob Chemother* 2006;57:224–9.

35. Lister PD. Multiply-resistant pneumococcus: therapeutic problems in the management of serious infections. *Eur J Clin Microbiol Infect Dis* 1995;14 Suppl 1: S18–S25.

36. McGowan JE, Jr. and Metchock BG. Penicillin-resistant pneumococci—an emerging threat to successful therapy. *J Hosp Infect* 1995;30 Suppl:472–82.

37. Barcus VA, Ghanekar K, Yeo M, Coffey TJ, and Dowson CG. Genetics of high level penicillin resistance in clinical isolates of *Streptococcus pneumoniae*. *FEMS Microbiol Lett* 1995;126:299–303.

38. Dowson CG, Hutchison A, Brannigan JA, et al. Horizontal transfer of penicillin-binding protein genes in penicillin-resistant clinical isolates of *Streptococcus pneumoniae*. *Proc Natl Acad Sci USA* 1989;86:8842–6.

39. Martin C, Sibold C, and Hakenbeck R. Relatedness of penicillin-binding protein 1a genes from different clones of penicillin-resistant *Streptococcus pneumoniae* isolated in South Africa and Spain. *EMBO J* 1992;11:3831–6.

40. Dowson CG, Coffey TJ, Kell C, and Whiley RA. Evolution of penicillin resistance in *Streptococcus pneumoniae*; the role of *Streptococcus mitis* in the formation of a low affinity PBP2B in *S. pneumoniae*. *Mol Microbiol* 1993;9:635–43.

41. Smith AM and Klugman KP. Alterations in penicillin-binding protein 2B from penicillin-resistant wild-type strains of *Streptococcus pneumoniae*. *Antimicrob Agents Chemother* 1995;39:859–67.

42. Mouz N, Gordon E, Di Guilmi AM, et al. Identification of a structural determinant for resistance to β-lactam antibiotics in Gram-positive bacteria. *Proc Natl Acad Sci USA* 1998;95:13403–6.

43. Guenzi E, Gasc AM, Sicard MA, and Hakenbeck R. A two-component signal-transducing system is involved in competence and penicillin susceptibility in laboratory mutants of *Streptococcus pneumoniae*. *Mol Microbiol* 1994;12:505–15.

44. Zahner D, Grebe T, Guenzi E, et al. Resistance determinants for β-lactam antibiotics in laboratory mutants of *Streptococcus pneumoniae* that are involved in genetic competence. *Microb Drug Resist* 1996;2:187–91.

45. Grebe T, Paik J, and Hakenbeck R. A novel resistance mechanism against β-lactams in *Streptococcus pneumoniae* involves CpoA, a putative glycosyltransferase. *J Bacteriol* 1997;179:3342–9.

46. Filipe SR and Tomasz A. Inhibition of the expression of penicillin resistance in *Streptococcus pneumoniae* by inactivation of cell wall muropeptide branching genes. *Proc Natl Acad Sci USA* 2000;97:4891–6.

47. Fiser A, Filipe SR, and Tomasz A. Cell wall branches, penicillin resistance and the secrets of the MurM protein. *Trends Microbiol* 2003;11:547–53.

48. Kell CM, Sharma UK, Dowson CG, Town C, Balganesh TS, and Spratt BG. Deletion analysis of the essentiality of penicillin-binding proteins 1A, 2B and 2X of *Streptococcus pneumoniae*. *FEMS Microbiol Lett* 1993;106:171–5.

49. Grebe T and Hakenbeck R. Penicillin-binding proteins 2b and 2x of *Streptococcus pneumoniae* are primary resistance determinants for different classes of β-lactam antibiotics. *Antimicrob Agents Chemother* 1996;40:829–34.

50. Dowson CG, Johnson AP, Cercenado E, and George RC. Genetics of oxacillin resistance in clinical isolates of *Streptococcus pneumoniae* that are oxacillin resistant and penicillin susceptible. *Antimicrob Agents Chemother* 1994;38:49–53.

51. Munoz R, Dowson CG, Daniels M, et al. Genetics of resistance to third-generation cephalosporins in clinical isolates of *Streptococcus pneumoniae*. *Mol Microbiol* 1992;6:2461–5.

52. Smith AM, Feldman C, Massidda O, McCarthy K, Ndiweni D, and Klugman KP. Altered PBP 2A and its role in the development of penicillin, cefotaxime, and ceftriaxone resistance in a clinical isolate of *Streptococcus pneumoniae*. *Antimicrob Agents Chemother* 2005;49:2002–7.

53. Sibold C, Henrichsen J, Konig A, Martin C, Chalkley L, and Hakenbeck R. Mosaic pbpX genes of major clones of penicillin-resistant *Streptococcus pneumoniae* have evolved from pbpX genes of a penicillin-sensitive *Streptococcus oralis*. *Mol Microbiol* 1994;12:1013–23.

54. Whatmore AM, Efstratiou A, Pickerill AP, et al. Genetic relationships between clinical isolates of *Streptococcus pneumoniae*, *Streptococcus oralis*, and *Streptococcus mitis*: characterization of "Atypical" pneumococci and organisms allied to *S. mitis* harboring *S. pneumoniae* virulence factor-encoding genes. *Infect Immun* 2000;68:1374–82.

55. Konig A, Reinert RR, and Hakenbeck R. *Streptococcus mitis* with unusually high level resistance to β-lactam antibiotics. *Microb Drug Resist* 1998;4:45–9.

56. Hakenbeck R, Konig A, Kern I, et al. Acquisition of five high-Mr penicillin-binding protein variants during transfer of high-level β-lactam resistance from *Streptococcus mitis* to *Streptococcus pneumoniae*. *J Bacteriol* 1998;180:1831–40.

57. Majewski J, Zawadzki P, Pickerill P, Cohan FM, and Dowson CG. Barriers to genetic exchange between bacterial species: *Streptococcus pneumoniae* transformation. *J Bacteriol* 2000;182:1016–23.

58. Zawadzki P, Roberts MS, and Cohan FM. The log-linear relationship between sexual isolation and sequence divergence in *Bacillus* transformation is robust. *Genetics* 1995;140:917–32.

59. Oggioni MR, Dowson CG, Smith JM, Provvedi R, and Pozzi G. The tetracycline resistance gene *tet*(M) exhibits mosaic structure. *Plasmid* 1996;35:156–63.

60. Widdowson CA, Klugman KP, and Hanslo D. Identification of the tetracycline resistance gene, *tet*(O), in *Streptococcus pneumoniae*. *Antimicrob Agents Chemother* 1996;40:2891–3.

61. Widdowson CA and Klugman KP. The molecular mechanisms of tetracycline resistance in the pneumococcus. *Microb Drug Resist* 1998;4:79–84.

62. Friedland IR and Klugman KP. Failure of chloramphenicol therapy in penicillin-resistant pneumococcal meningitis. *Lancet* 1992;339:405–8.

63. Trieu-Cuot P, Poyart-Salmeron C, Carlier C, and Courvalin P. Nucleotide sequence of the erythromycin resistance gene of the conjugative transposon Tn*1545*. *Nucleic Acids Res* 1990;18:3660.

64. Widdowson CA and Klugman KP. Emergence of the M phenotype of erythromycin-resistant pneumococci in South Africa. *Emerg Infect Dis* 1998;4:277–81.

65. Linares J and Tubau F. Pneumococcal meningitis and third generation cephalosporins. *Enferm Infecc Microbiol Clin* 1996;14:1–6.

66. Clewell DB, Flannagan SE, and Jaworski DD. Unconstrained bacterial promiscuity: the Tn*916*-Tn*1545* family of conjugative transposons. *Trends Microbiol* 1995;3:229–36.

67. Scott JR and Churchward GG. Conjugative transposition. *Annu Rev Microbiol* 1995;49:367–97.

68. Doherty N, Trzcinski K, Pickerill P, Zawadzki P, and Dowson CG. Genetic diversity of the *tet*(M) gene in tetracycline-resistant clonal lineages of *Streptococcus pneumoniae*. *Antimicrob Agents Chemother* 2000;44:2979–84.

69. Poyart-Salmeron C, Trieu-Cuot P, Carlier C, and Courvalin P. Nucleotide sequences specific for Tn*1545*-like conjugative transposons in pneumococci and staphylococci resistant to tetracycline. *Antimicrob Agents Chemother* 1991;35:1657–60.

70. Izdebski R, Sadowy E, Fiett J, Grzesiowski P, Gniadkowski M, and Hryniewicz W. Clonal diversity and resistance mechanisms in *Streptococcus pneumoniae* nonsusceptible to tetracyclines in Poland. *Antimicrob Agents Chemother* 2007.

71. Marchese A, Ramirez M, Schito GC, and Tomasz A. Molecular epidemiology of penicillin-resistant *Streptococcus pneumoniae* isolates recovered in Italy from 1993 to 1996. *J Clin Microbiol* 1998;36:2944–9.

72. Luna VA and Roberts MC. The presence of the *tetO* gene in a variety of tetracycline-resistant *Streptococcus pneumoniae* serotypes from Washington State. *J Antimicrob Chemother* 1998;42:613–9.

73. Mendonca-Souza CR, Carvalho Mda G, Barros RR, et al. Occurrence and characteristics of erythromycin-resistant *Streptococcus pneumoniae* strains isolated in three major Brazilian states. *Microb Drug Resist* 2004;10:313–20.

74. Sutcliffe J, Tait-Kamradt A, and Wondrack L. *Streptococcus pneumoniae* and *Streptococcus pyogenes* resistant to macrolides but sensitive to clindamycin: a common resistance pattern mediated by an efflux system. *Antimicrob Agents Chemother* 1996;40:1817–24.

75. Barry AL, Fuchs PC, and Brown SD. Antipneumococcal activities of a ketolide (HMR 3647), a streptogramin (quinupristin-dalfopristin), a macrolide (erythromycin), and a lincosamide (clindamycin). *Antimicrob Agents Chemother* 1998;42:945–6.

76. Setchanova L and Tomasz A. Molecular characterization of penicillin-resistant *Streptococcus pneumoniae* isolates from Bulgaria. *J Clin Microbiol* 1999;37:638–48.

77. Johnston NJ, De Azavedo JC, Kellner JD, and Low DE. Prevalence and characterization of the mechanisms of macrolide, lincosamide, and streptogramin resistance in isolates of *Streptococcus pneumoniae*. *Antimicrob Agents Chemother* 1998;42:2425–6.

78. Maskell JP, Sefton AM, and Hall LM. Mechanism of sulfonamide resistance in clinical isolates of *Streptococcus pneumoniae*. *Antimicrob Agents Chemother* 1997;41:2121–6.

79. Pikis A, Donkersloot JA, Rodriguez WJ, and Keith JM. A conservative amino acid mutation in the chromosome-encoded dihydrofolate reductase confers trimethoprim resistance in *Streptococcus pneumoniae*. *J Infect Dis* 1998;178:700–6.

80. Vinnicombe HG and Derrick JP. Dihydropteroate synthase from *Streptococcus pneumoniae*: characterization of substrate binding order and sulfonamide inhibition. *Biochem Biophys Res Commun* 1999;258:752–7.

81. Enright M, Zawadski P, Pickerill P, and Dowson CG. Molecular evolution of rifampicin resistance in *Streptococcus pneumoniae*. *Microb Drug Resist* 1998;4:65–70.

82. Doern GV, Pfaller MA, Erwin ME, Brueggemann AB, and Jones RN. The prevalence of fluoroquinolone resistance among clinically significant respiratory tract isolates of *Streptococcus pneumoniae* in the United States and Canada—1997 results from the SENTRY Antimicrobial Surveillance Program. *Diagn Microbiol Infect Dis* 1998;32:313–6.

83. Choi H, Lee HJ, and Lee Y. A mutation in QRDR in the ParC subunit of topoisomerase IV was responsible for fluoroquinolone resistance in clinical isolates of *Streptococcus pneumoniae*. *Yonsei Med J* 1998;39:541–5.

84. Taba H and Kusano N. Sparfloxacin resistance in clinical isolates of *Streptococcus pneumoniae*: involvement of multiple mutations in *gyrA* and *parC* genes. *Antimicrob Agents Chemother* 1998;42:2193–6.

85. Barry AL, Brown SD, and Fuchs PC. Fluoroquinolone resistance among recent clinical isolates of *Streptococcus pneumoniae*. *J Antimicrob Chemother* 1999;43:428–9.

86. Brenwald NP, Gill MJ, and Wise R. Prevalence of a putative efflux mechanism among fluoroquinolone-resistant clinical isolates of *Streptococcus pneumoniae*. *Antimicrob Agents Chemother* 1998;42:2032–5.

87. Gill MJ, Brenwald NP, and Wise R. Identification of an efflux pump gene, *pmrA*, associated with fluoroquinolone resistance in *Streptococcus pneumoniae*. *Antimicrob Agents Chemother* 1999;43:187–9.

88. Gonzalez I, Georgiou M, Alcaide F, Balas D, Linares J, and de la Campa AG. Fluoroquinolone resistance mutations in the *parC*, *parE*, and *gyrA* genes of clinical isolates of viridans group streptococci. *Antimicrob Agents Chemother* 1998;42:2792–8.

89. de la Campa AG, Balsalobre L, Ardanuy C, Fenoll A, Perez-Trallero E, and Linares J. Fluoroquinolone resistance in penicillin-resistant *Streptococcus pneumoniae* clones, Spain. *Emerg Infect Dis* 2004;10:1751–9.

90. Novak R, Henriques B, Charpentier E, Normark S, and Tuomanen E. Emergence of vancomycin tolerance in *Streptococcus pneumoniae*. *Nature* 1999;399:590–3.

91. Austrian R. Some aspects of the pneumococcal carrier state. *J Antimicrob Chemother* 1986;18 Suppl A:35–45.

92. Appelbaum PC, Gladkova C, Hryniewicz W, et al. Carriage of antibiotic-resistant *Streptococcus pneumoniae* by children in eastern and central Europe—a multicenter study with use of standardized methods. *Clin Infect Dis* 1996;23:712–7.

93. Sung RY, Cheng AF, Chan RC, Tam JS, and Oppenheimer SJ. Epidemiology and etiology of pneumonia in children in Hong Kong. *Clin Infect Dis* 1993;17:894–6.

94. Hendley JO, Sande MA, Stewart PM, and Gwaltney JM, Jr. Spread of *Streptococcus pneumoniae* in families. I. Carriage rates and distribution of types. *J Infect Dis* 1975;132:55–61.

95. Austrian R. Epidemiology of pneumococcal capsular types causing pediatric infections. *Pediatr Infect Dis J* 1989;8:S21–2.

96. Scott JA, Hall AJ, Dagan R, et al. Serogroup-specific epidemiology of *Streptococcus pneumoniae*: associations with age, sex, and geography in 7,000 episodes of invasive disease. *Clin Infect Dis* 1996;22:973–81.

97. Hoeprich PD. Bacterial pneumonias. In: Hoeprich PD, Jordan MC, and Ronald AR, Eds. *Infectious diseases*. New York: Lippincott, 1994:421–433.

98. Crewe-Brown HH, Karstaedt AS, Saunders GL, et al. *Streptococcus pneumoniae* blood culture isolates from patients with and without human immunodeficiency virus infection: alterations in penicillin susceptibilities and in serogroups or serotypes. *Clin Infect Dis* 1997;25:1165–72.

99. Whatmore AM and Dowson CG. The autolysin-encoding gene (*lytA*) of *Streptococcus pneumoniae* displays restricted allelic variation despite localized recombination events with genes of pneumococcal bacteriophage encoding cell wall lytic enzymes. *Infect Immun* 1999;67:4551–6.

100. Coffey TJ, Daniels M, Enright MC, and Spratt BG. Serotype 14 variants of the Spanish penicillin-resistant serotype 9V clone of *Streptococcus pneumoniae* arose by large recombinational replacements of the *cpsA-pbp1a* region. *Microbiology* 1999;145 (Pt 8):2023–31.

101. Trzcinski K, Thompson CM, and Lipsitch M. Single-step capsular transformation and acquisition of penicillin resistance in *Streptococcus pneumoniae*. *J Bacteriol* 2004;186:3447–52.

102. Coffey TJ, Dowson CG, Daniels M, et al. Horizontal transfer of multiple penicillin-binding protein genes, and capsular biosynthetic genes, in natural populations of *Streptococcus pneumoniae*. *Mol Microbiol* 1991;5:2255–60.

103. Coffey TJ, Enright MC, Daniels M, et al. Recombinational exchanges at the capsular polysaccharide biosynthetic locus lead to frequent serotype changes among natural isolates of *Streptococcus pneumoniae*. *Mol Microbiol* 1998;27:73–83.

104. Tomasz A, Corso A, Severina EP, et al. Molecular epidemiologic characterization of penicillin-resistant *Streptococcus pneumoniae* invasive pediatric isolates recovered in six Latin-American countries: an overview. PAHO/Rockefeller University Workshop. Pan American Health Organization. *Microb Drug Resist* 1998;4:195–207.

105. Hermans PW, Sluijter M, Hoogenboezem T, Heersma H, van Belkum A, and de Groot R. Comparative study of five different DNA fingerprint techniques for molecular typing of *Streptococcus pneumoniae* strains. *J Clin Microbiol* 1995;33:1606–12.
106. Coffey TJ, Dowson CG, Daniels M, and Spratt BG. Horizontal spread of an altered penicillin-binding protein 2B gene between *Streptococcus pneumoniae* and *Streptococcus oralis*. *FEMS Microbiol Lett* 1993;110:335–9.
107. Dzierzanowska-Fangrat K, Semczuk K, Gorska P, et al. Evidence for tetracycline resistance determinant *tet*(M) allele replacement in a *Streptococcus pneumoniae* population of limited geographical origin. *Int J Antimicrob Agents* 2006;27:159–64.
108. Coffey TJ, Berron S, Daniels M, et al. Multiply antibiotic-resistant *Streptococcus pneumoniae* recovered from Spanish hospitals (1988–1994): novel major clones of serotypes 14, 19F and 15F. *Microbiology* 1996;142 (Pt 10):2747–57.
109. Echaniz-Aviles G and Velazquez-Meza ME, Carnalla-Barajas MN, et al. Predominance of the multiresistant 23F international clone of *Streptococcus pneumoniae* among isolates from Mexico. *Microb Drug Resist* 1998;4:241–6.
110. Doit C, Loukil C, Fitoussi F, Geslin P, and Bingen E. Emergence in France of multiple clones of clinical *Streptococcus pneumoniae* isolates with high-level resistance to amoxicillin. *Antimicrob Agents Chemother* 1999;43:1480–3.
111. Nesin M, Ramirez M, and Tomasz A. Capsular transformation of a multidrug-resistant *Streptococcus pneumoniae in vivo*. *J Infect Dis* 1998;177:707–13.
112. Wolf B, Rey LC, Brisse S, et al. Molecular epidemiology of penicillin-resistant *Streptococcus pneumoniae* colonizing children with community-acquired pneumonia and children attending day-care centres in Fortaleza, Brazil. *J Antimicrob Chemother* 2000;46:757–65.
113. Hermans PW, Sluijter M, Dejsirilert S, et al. Molecular epidemiology of drug-resistant pneumococci: toward an international approach. *Microb Drug Resist* 1997;3:243–51.
114. Corso A, Severina EP, Petruk VF, Mauriz YR, and Tomasz A. Molecular characterization of penicillin-resistant *Streptococcus pneumoniae* isolates causing respiratory disease in the United States. *Microb Drug Resist* 1998;4:325–37.
115. Ip M, Lyon DJ, Yung RW, Chan C, and Cheng AF. Evidence of clonal dissemination of multidrug-resistant *Streptococcus pneumoniae* in Hong Kong. *J Clin Microbiol* 1999;37:2834–9.
116. Sibold C, Wang J, Henrichsen J, and Hakenbeck R. Genetic relationships of penicillin-susceptible and -resistant *Streptococcus pneumoniae* strains isolated on different continents. *Infect Immun* 1992;60:4119–26.
117. Tarasi A, Chong Y, Lee K, and Tomasz A. Spread of the serotype 23F multidrug-resistant *Streptococcus pneumoniae* clone to South Korea. *Microb Drug Resist* 1997;3:105–9.
118. Shi ZY, Enright MC, Wilkinson P, Griffiths D, and Spratt BG. Identification of three major clones of multiply antibiotic-resistant *Streptococcus pneumoniae* in Taiwanese hospitals by multilocus sequence typing. *J Clin Microbiol* 1998;36:3514–9.
119. Thackway SV, Delpech VC, Jorm LR, McAnulty JM, and Visotina M. Monitoring acute diseases during the Sydney 2000 Olympic and Paralympic Games. *Med J Aust* 2000;173:318–21.
120. Soares S, Kristinsson KG, Musser JM, and Tomasz A. Evidence for the introduction of a multiresistant clone of serotype 6B *Streptococcus pneumoniae* from Spain to Iceland in the late 1980s. *J Infect Dis* 1993;168:158–63.
121. Reichmann P, Varon E, Gunther E, et al. Penicillin-resistant *Streptococcus pneumoniae* in Germany: genetic relationship to clones from other European countries. *J Med Microbiol* 1995;43:377–85.
122. Versalovic J, Kapur V, Mason EO, Jr., et al. Penicillin-resistant *Streptococcus pneumoniae* strains recovered in Houston: identification and molecular characterization of multiple clones. *J Infect Dis* 1993;167:850–6.

123. Gherardi G, Whitney CG, Facklam RR, and Beall B. Major related sets of antibiotic-resistant Pneumococci in the United States as determined by pulsed-field gel electrophoresis and *pbp1a-pbp2b-pbp2x-dhf* restriction profiles. *J Infect Dis* 2000;181:216–29.
124. Vela MC, Fonseca N, Di Fabio JL, and Castaneda E. Presence of international multi-resistant clones of *Streptococcus pneumoniae* in Colombia. *Microb Drug Resist* 2001;7:153–64.
125. Greenberg D, Dagan R, Muallem M, and Porat N. Antibiotic-resistant invasive pediatric *Streptococcus pneumoniae* clones in Israel. *J Clin Microbiol* 2003;41:5541–5.
126. Overweg K, Hermans PW, Trzcinski K, Sluijter M, de Groot R, and Hryniewicz W, Multidrug-resistant *Streptococcus pneumoniae* in Poland: identification of emerging clones. *J Clin Microbiol* 1999;37:1739–45.
127. De Lencastre H, Kristinsson KG, Brito-Avo A, et al. Carriage of respiratory tract pathogens and molecular epidemiology of *Streptococcus pneumoniae* colonization in healthy children attending day care centers in Lisbon, Portugal. *Microb Drug Resist* 1999;5:19–29.
128. Castaneda E, Penuela I, Vela MC, and Tomasz A. Penicillin-resistant *Streptococcus pneumoniae* in Colombia: presence of international epidemic clones. Colombian pneumococcal study group. *Microb Drug Resist* 1998;4:233–9.
129. Porat N, Arguedas A, Spratt BG, et al. Emergence of penicillin-nonsusceptible *Streptococcus pneumoniae* clones expressing serotypes not present in the antipneumococcal conjugate vaccine. *J Infect Dis* 2004;190:2154–61.
130. Hall LM, Whiley RA, Duke B, George RC, and Efstratiou A. Genetic relatedness within and between serotypes of *Streptococcus pneumoniae* from the United Kingdom: analysis of multilocus enzyme electrophoresis, pulsed-field gel electrophoresis, and antimicrobial resistance patterns. *J Clin Microbiol* 1996;34:853–9.
131. Pato MV, Carvalho CB, and Tomasz A. Antibiotic susceptibility of *Streptococcus pneumoniae* isolates in Portugal. A multicenter study between 1989 and 1993. *Microb Drug Resist* 1995;1:59–69.
132. Dejsirilert S, Overweg K, Sluijter M, et al. Nasopharyngeal carriage of penicillin-resistant *Streptococcus pneumoniae* among children with acute respiratory tract infections in Thailand: a molecular epidemiological survey. *J Clin Microbiol* 1999;37:1832–8.
133. Figueiredo AM, Austrian R, Urbaskova P, Teixeira LA, and Tomasz A. Novel penicillin-resistant clones of *Streptococcus pneumoniae* in the Czech Republic and in Slovakia. *Microb Drug Resist* 1995;1:71–8.
134. Marton A. Epidemiology of resistant pneumococci in Hungary. *Microb Drug Resist* 1995;1:127–30.
135. Lawrenson JB, Klugman KP, Eidelman JI, Wasas A, Miller SD, and Lipman J. Fatal infection caused by a multiply resistant type 3 pneumococcus. *J Clin Microbiol* 1988;26:1590–1.
136. Syrogiannopoulos GA, Grivea IN, Beratis NG, et al. Resistance patterns of *Streptococcus pneumoniae* from carriers attending day-care centers in southwestern Greece. *Clin Infect Dis* 1997;25:188–94.
137. Tamayo M, Sa-Leao R, Santos Sanches I, Castaneda E, and de Lencastre H. Dissemination of a chloramphenicol- and tetracycline-resistant but penicillin-susceptible invasive clone of serotype 5 *Streptococcus pneumoniae* in Colombia. *J Clin Microbiol* 1999;37:2337–42.
138. Luey KY and Kam KM. Vaccine coverage of *Streptococcus pneumoniae* in Hong Kong with attention to the multiple-antibiotic-resistant strains. *Vaccine* 1996;14:1573–80.
139. Dowson CG and Trzcinski K. Evolution and epidemiology of antibiotic-resistant pneumococci. In: Lewis K, Salyers AA, Taber HW and Wax RG, eds. *Bacterial Resistance to Antimicrobials.* New York, Basel: Marcel Dekker, Inc., 2002:265–293.

140. Lehmann D, Gratten M, and Montgomery J. Susceptibility of pneumococcal carriage isolates to penicillin provides a conservative estimate of susceptibility of invasive pneumococci. *Pediatr Infect Dis J* 1997;16:297–305.
141. Kronenberg A, Zucs P, Droz S, and Muhlemann K. Distribution and invasiveness of *Streptococcus pneumoniae* serotypes in Switzerland, a country with low antibiotic selection pressure, from 2001 to 2004. *J Clin Microbiol* 2006;44:2032–8.
142. Brown SD and Farrell DJ. Antibacterial susceptibility among *Streptococcus pneumoniae* isolated from paediatric and adult patients as part of the PROTEKT US study in 2001–2002. *J Antimicrob Chemother* 2004;54 Suppl 1:i23–9.
143. Talbot TR, Poehling KA, Hartert TV, et al. Reduction in high rates of antibiotic-nonsusceptible invasive pneumococcal disease in Tennessee after introduction of the pneumococcal conjugate vaccine. *Clin Infect Dis* 2004;39:641–8.
144. Kyaw MH, Lynfield R, Schaffner W, et al. Effect of introduction of the pneumococcal conjugate vaccine on drug-resistant *Streptococcus pneumoniae*. *N Engl J Med* 2006; 354:1455–63.
145. Kayhty H, Auranen K, Nohynek H, Dagan R, and Makela H. Nasopharyngeal colonization: a target for pneumococcal vaccination. *Expert Rev Vaccines* 2006;5:651–67.
146. Temime L, Guillemot D, and Boelle PY. Short- and long-term effects of pneumococcal conjugate vaccination of children on penicillin resistance. *Antimicrob Agents Chemother* 2004;48:2206–13.
147. Baquero F, Martinez-Beltran J, and Loza E. A review of antibiotic resistance patterns of *Streptococcus pneumoniae* in Europe. *J Antimicrob Chemother* 1991;28 Suppl C:31–8.
148. Jones ME, Karlowsky JA, Blosser-Middleton R, Critchley I, Thornsberry C, and Sahm DF. Relationship between antibiotic resistance in *Streptococcus pneumoniae* and that in *Haemophilus influenzae*: evidence for common selective pressure. *Antimicrob Agents Chemother* 2002;46:3106–7.
149. Jones RN, Biedenbach DJ, and Beach ML. Influence of patient age on the susceptibility patterns of *Streptococcus pneumoniae* isolates in North America (2000–2001): report from the SENTRY Antimicrobial Surveillance Program. *Diagn Microbiol Infect Dis* 2003;46:77–80.
150. Okeke IN, Klugman KP, Bhutta ZA, et al. Antimicrobial resistance in developing countries. Part II: strategies for containment. *Lancet Infect Dis* 2005;5:568–80.
151. Dagan R and Lipsitch M. Changing the ecology of pneumococci with antibiotics and vaccines. In: Toumanen EI, Mitchell TJ, Morrison DA and Spratt BG, eds. *The pneumococcus*. 1 ed. Washington, D.C.: ASM Press, 2004:283–331.
152. Wang YC and Lipsitch M. Upgrading antibiotic use within a class: tradeoff between resistance and treatment success. *Proc Natl Acad Sci USA* 2006;103:9655–60.
153. Rossi A, Corso A, Pace J, Regueira M, and Tomasz A. Penicillin-resistant *Streptococcus pneumoniae* in Argentina: frequent occurrence of an internationally spread serotype 14 clone. *Microb Drug Resist* 1998;4:225–31.
154. Louie M, Louie L, Papia G, Talbot J, Lovgren M, and Simor AE. Molecular analysis of the genetic variation among penicillin-susceptible and penicillin-resistant *Streptococcus pneumoniae* serotypes in Canada. *J Infect Dis* 1999;179:892–900.
155. Gherardi G, Inostrozo JS, O'Ryan M, et al. Genotypic survey of recent β-lactam-resistant pneumococcal nasopharyngeal isolates from asymptomatic children in Chile. *J Clin Microbiol* 1999;37:3725–30.
156. Doit C, Denamur E, Picard B, Geslin P, Elion J, and Bingen E. Mechanisms of the spread of penicillin resistance in *Streptococcus pneumoniae* strains causing meningitis in children in France. *J Infect Dis* 1996;174:520–8.
157. Sa-Leao R, Vilhelmsson SE, de Lencastre H, Kristinsson KG, and Tomasz A. Diversity of penicillin-nonsusceptible *Streptococcus pneumoniae* circulating in Iceland after the introduction of penicillin-resistant clone Spain(6B)-2. *J Infect Dis* 2002;186:966–75.

158. Song JH, Lee NY, Ichiyama S, et al. Spread of drug-resistant *Streptococcus pneumoniae* in Asian countries: Asian Network for Surveillance of Resistant Pathogens (ANSORP) Study. *Clin Infect Dis* 1999;28:1206–11.

159. Calderon-Jaimes E, Echaniz-Aviles G, Conde-Gonzalez C, et al. [The resistance and serotyping of 83 strains of *Streptococcus pneumoniae* isolated from asymptomatic carriers and ill children]. *Bol Med Hosp Infant Mex* 1993;50:854–60.

160. Bean DC and Klena JD. Characterization of major clones of antibiotic-resistant *Streptococcus pneumoniae* in New Zealand by multilocus sequence typing. *J Antimicrob Chemother* 2005;55:375–8.

161. Lee NY, Song JH, Kim S, et al. Carriage of antibiotic-resistant pneumococci among Asian children: a multinational surveillance by the Asian Network for Surveillance of Resistant Pathogens (ANSORP). *Clin Infect Dis* 2001;32:1463–9.

162. Henriqus Normark B, Christensson B, Sandgren A, et al. Clonal analysis of *Streptococcus pneumoniae* nonsusceptible to penicillin at day-care centers with index cases, in a region with low incidence of resistance: emergence of an invasive type 35B clone among carriers. *Microb Drug Resist* 2003;9:337–44.

163. Sener B, McGee L, Pinar A, and Eser O. Genomic backgrounds of drug-resistant *Streptococcus pneumoniae* in Ankara, Turkey: Identification of emerging new clones. *Microb Drug Resist* 2006;12:109–14.

164. Hermans PW, Sluijter M, Elzenaar K, et al. Penicillin-resistant *Streptococcus pneumoniae* in the Netherlands: results of a 1-year molecular epidemiologic survey. *J Infect Dis* 1997;175:1413–22.

165. Camou T, Hortal M, and Tomasz A. The apparent importation of penicillin-resistant capsular type 14 Spanish/French clone of *Streptococcus pneumoniae* into Uruguay in the early 1990s. *Microb Drug Resist* 1998;4:219–24.

166. Parry CM, Duong NM, Zhou J, et al. Emergence in Vietnam of *Streptococcus pneumoniae* resistant to multiple antimicrobial agents as a result of dissemination of the multiresistant Spain(23F)-1 clone. *Antimicrob Agents Chemother* 2002;46:3512–7.

167. Yaro S, Lourd M, Traore Y, et al. Epidemiological and molecular characteristics of a highly lethal pneumococcal meningitis epidemic in Burkina Faso. *Clin Infect Dis* 2006;43:693–700.

168. Konradsen HB and Kaltoft MS. Invasive pneumococcal infections in Denmark from 1995 to 1999: epidemiology, serotypes, and resistance. *Clin Diagn Lab Immunol* 2002;9:358–65.

169. Altraja A, Naaber P, Tamm E, Meriste S, Kullamaa A, and Leesik H. Antimicrobial susceptibility of common pathogens from community-acquired lower respiratory tract infections in Estonia. *J Chemother* 2006;18:603–9.

170. Jain A, Kumar P, and Awasthi S. High nasopharyngeal carriage of drug resistant *Streptococcus pneumoniae* and *Haemophilus influenzae* in North Indian school-children. *Trop Med Int Health* 2005;10:234–9.

171. Allen UD, Thomas S, Carapetis J, et al. Serotypes of respiratory tract isolates of *Streptococcus pneumoniae* from Jamaican children. *Int J Infect Dis* 2003;7:29–35.

172. Bogaert D, Sluijter M, Toom NL, et al. Dynamics of pneumococcal colonization in healthy Dutch children. *Microbiology* 2006;152:377–85.

173. Matute AJ, Brouwer WP, Hak E, Delgado E, Alonso E, and Hoepelman IM. Aetiology and resistance patterns of community-acquired pneumonia in Leon, Nicaragua. *Int J Antimicrob Agents* 2006;28:423–7.

174. Pedersen MK, Hoiby EA, Froholm LO, Hasseltvedt V, Lermark G, and Caugant DA. Systemic pneumococcal disease in Norway 1995–2001: capsular serotypes and anti-microbial resistance. *Epidemiol Infect* 2004;132:167–75.

175. Johnson AP, Potz N, Waight P, et al. Susceptibility of pneumococci causing meningitis in England and Wales to first-line antimicrobial agents, 2001–2004. *J Antimicrob Chemother* 2005;56:1181–2.

176. Denham BC and Clarke SC. Serotype incidence and antibiotic susceptibility of *Streptococcus pneumoniae* causing invasive disease in Scotland, 1999–2002. *J Med Microbiol* 2005;54:327–31.

177. Watson M, Roche P, Bayley K, et al. Laboratory surveillance of invasive pneumococcal disease in Australia, 2003 predicting the future impact of the universal childhood conjugate vaccine program. *Commun Dis Intell* 2004;28:455–64.

178. Watson M, Bayley K, Bell JM, et al. Laboratory surveillance of invasive pneumococcal disease in Australia in 2001 to 2002—implications for vaccine serotype coverage. *Commun Dis Intell* 2003;27:478–87.

179. Buxbaum A, Forsthuber S, Graninger W, and Georgopoulos A. Serotype distribution and antimicrobial resistance of *Streptococcus pneumoniae* in Austria. *J Antimicrob Chemother* 2004;54:247–50.

180. Saha SK, Rikitomi N, Ruhulamin M, et al. Antimicrobial resistance and serotype distribution of *Streptococcus pneumoniae* strains causing childhood infections in Bangladesh, 1993 to 1997. *J Clin Microbiol* 1999;37:798–800.

181. Flamaing J, Verhaegen J, and Peetermans WE. *Streptococcus pneumoniae* bacteraemia in Belgium: differential characteristics in children and the elderly population and implications for vaccine use. *J Antimicrob Chemother* 2002;50:43–50.

182. Reinert RR, Reinert S, van der Linden M, Cil MY, Al-Lahham A, and Appelbaum P. Antimicrobial susceptibility of *Streptococcus pneumoniae* in eight European countries from 2001 to 2003. *Antimicrob Agents Chemother* 2005;49:2903–13.

183. Petrov M, Hadjieva N, Kantardjiev T, Velinov T, and Bachvarova A. Surveillance of antimicrobial resistance in Bulgaria—a synopsis from BulSTAR 2003. *Euro Surveill* 2005;10.

184. Powis J, McGeer A, Green K, et al. *In vitro* antimicrobial susceptibilities of *Streptococcus pneumoniae* clinical isolates obtained in Canada in 2002. *Antimicrob Agents Chemother* 2004;48:3305–11.

185. Ulloa-Gutierrez R, Avila-Aguero ML, Herrera ML, Herrera JF and Arguedas A. Invasive pneumococcal disease in Costa Rican children: a seven year survey. *Pediatr Infect Dis J* 2003;22:1069–74.

186. Urbaskova P, Motlova J, and Zemlickova H. Antibiotic resistance in invasive pneumococci and their serotypes in the Czech Republic. *Cas Lek Cesk* 2004; 143:178–83.

187. Nagai K, Appelbaum PC, Davies TA, et al. Susceptibility to telithromycin in 1,011 *Streptococcus pyogenes* isolates from 10 central and Eastern European countries. *Antimicrob Agents Chemother* 2002;46:546–9.

188. Gebreselassie S. Patterns of isolation of common Gram-positive bacterial pathogens and their susceptibilities to antimicrobial agents in Jimma Hospital. *Ethiop Med J* 2002;40:115–27.

189. Russell FM, Carapetis JR, Ketaiwai S, et al. Pneumococcal nasopharyngeal carriage and patterns of penicillin resistance in young children in Fiji. *Ann Trop Paediatr* 2006;26:187–97.

190. Hill PC, Akisanya A, Sankareh K, et al. Nasopharyngeal carriage of *Streptococcus pneumoniae* in Gambian villagers. *Clin Infect Dis* 2006;43:673–9.

191. Arason VA, Sigurdsson JA, Erlendsdottir H, Gudmundsson S, and Kristinsson KG. The role of antimicrobial use in the epidemiology of resistant pneumococci: A 10-year follow up. *Microb Drug Resist* 2006;12:169–76.

192. Soewignjo S, Gessner BD, Sutanto A, et al. *Streptococcus pneumoniae* nasopharyngeal carriage prevalence, serotype distribution, and resistance patterns among children on Lombok Island, Indonesia. *Clin Infect Dis* 2001;32:1039–43.

193. Rehman N and Ahmad SI. Antibiotic susceptibility of *Streptococcus pneumoniae* in Karachi. *J Pak Med Assoc* 2000;50:58–60.

194. Song JH, Jung SI, Ko KS, et al. High prevalence of antimicrobial resistance among clinical *Streptococcus pneumoniae* isolates in Asia (an ANSORP study). *Antimicrob Agents Chemother* 2004;48:2101–7.

195. Murphy OM, Murchan S, Whyte D, et al. Impact of the European Antimicrobial Resistance Surveillance System on the development of a national programme to monitor resistance in *Staphylococcus aureus* and *Streptococcus pneumoniae* in Ireland, 1999–2003. *Eur J Clin Microbiol Infect Dis* 2005;24:480–3.

196. Monaco M, Camilli R, D'Ambrosio F, Del Grosso M, and Pantosti A. Evolution of erythromycin resistance in *Streptococcus pneumoniae* in Italy. *J Antimicrob Chemother* 2005;55:256–9.

197. Hjaltested FK, Bornatoniene J, Erlendsdottir H, et al. Resistance in respiratory tract pathogens and antimicrobial use in Icelandic and Lithuanian children. *Scand J Infect Dis* 2003;35:21–6.

198. Benbachir M, Benredjeb S, Boye CS, et al. Two-year surveillance of antibiotic resistance in *Streptococcus pneumoniae* in four African cities. *Antimicrob Agents Chemother* 2001;45:627–9.

199. Borg M, Scicluna E, De Kraker M, et al. Antibiotic resistance in the southeastern Mediterranean—preliminary results from the ARMed project. Euro Surveill 2006;11.

200. Roca A, Sigauque B, Quinto L, et al. Invasive pneumococcal disease in children <5 years of age in rural Mozambique. *Trop Med Int Health* 2006;11:1422–31.

201. Bean DC, Ikram RB, and Klena JD. Molecular characterization of penicillin non-susceptible *Streptococcus pneumoniae* in Christchurch, New Zealand. *J Antimicrob Chemother* 2004;54:122–9.

202. Johnson WB, Adedoyin OT, Abdulkarim AA, and Olanrewaju WI. Bacterial pathogens and outcome determinants of childhood pyogenic meningitis in Ilorin, Nigeria. *Afr J Med Med Sci* 2001;30:295–303.

203. Sadowy E, Izdebski R, Skoczynska A, Grzesiowski P, Gniadkowski M, and Hryniewicz W. Phenotypic and molecular analysis of penicillin-nonsusceptible *Streptococcus pneumoniae* isolates in Poland. *Antimicrob Agents Chemother* 2007;51:40–7.

204. Semczuk K, Dzierzanowska-Fangrat K, Lopaciuk U, Gabinska E, Jozwiak P, and Dzierzanowska D. Antimicrobial resistance of *Streptococcus pneumoniae* and *Haemophilus influenzae* isolated from children with community-acquired respiratory tract infections in Central Poland. *Int J Antimicrob Agents* 2004;23:39–43.

205. Stratchounski LS, Kozlov RS, Appelbaum PC, Kretchikova OI, and Kosowska-Shick K. Antimicrobial resistance of nasopharyngeal pneumococci from children from day-care centres and orphanages in Russia: results of a unique prospective multicentre study. *Clin Microbiol Infect* 2006;12:853–66.

206. Bogaerts J, Lepage P, Taelman H, et al. Antimicrobial susceptibility and serotype distribution of *Streptococcus pneumoniae* from Rwanda, 1984–1990. *J Infect* 1993; 27:157–68.

207. Cizman M, Pokorn M, and Paragi M. Antimicrobial resistance of invasive *Streptococcus pneumoniae* in Slovenia from 1997 to 2000. *J Antimicrob Chemother* 2002;49:582–4.

208. Cizman M, Srovin T, Pokorn M, Cad Pecar S, and Battelino S. Analysis of the causes and consequences of decreased antibiotic consumption over the last 5 years in Slovenia. *J Antimicrob Chemother* 2005;55:758–63.

209. Nilsson P and Laurell MH. A 10-year follow-up study of penicillin-non-susceptible *S. pneumoniae* during an intervention programme in Malmo, Sweden. *Scand J Infect Dis* 2006;38:838–44.

210. Batt SL, Charalambous BM, Solomon AW, et al. Impact of azithromycin administration for trachoma control on the carriage of antibiotic-resistant *Streptococcus pneumoniae*. *Antimicrob Agents Chemother* 2003;47:2765–9.

211. Mahmoud R, Mahmoud M, Badrinath P, Sheek-Hussein M, Alwash R, and Nicol AG. Pattern of meningitis in Al-Ain medical district, United Arab Emirates—a decadal experience (1990–99). *J Infect* 2002;44:22–5.

212. Al Khorasani A and Banajeh S. Bacterial profile and clinical outcome of childhood meningitis in rural Yemen: a 2-year hospital-based study. *J Infect* 2006;53:228–34.

213. Woolfson A, Huebner R, Wasas A, Chola S, Godfrey-Faussett P, and Klugman K. Nasopharyngeal carriage of community-acquired, antibiotic-resistant *Streptococcus pneumoniae* in a Zambian paediatric population. *Bull World Health Organ* 1997; 75:453–62.

214. Grenon S, von Specht M, Corso A, Pace J, and Regueira M. Distribution of serotypes and antibiotic susceptibility patterns of *Streptococcus pneumoniae* strains isolated from children in Misiones, Argentina. *Enferm Infec Microbiol Clin* 2005;23:10–4.

215. Castanheira M, Gales AC, Mendes RE, Jones RN, and Sader HS. Antimicrobial susceptibility of *Streptococcus pneumoniae* in Latin America: results from five years of the SENTRY Antimicrobial Surveillance Program. *Clin Microbiol Infect* 2004; 10:645–51.

216. Brandileone MC, Casagrande ST, Guerra ML, Zanella RC, Andrade AL, and Di Fabio JL. Increase in numbers of β-lactam-resistant invasive *Streptococcus pneumoniae* in Brazil and the impact of conjugate vaccine coverage. *J Med Microbiol* 2006;55: 567–74.

217. Rowe AK, Schwartz B, Wasas A, and Klugman KP. Evaluation of the Etest as a means of determining the antibiotic susceptibilities of isolates of *Streptococcus pneumoniae* and *Haemophilus influenzae* from children in the Central African Republic. *J Antimicrob Chemother* 2000;45:132–3.

218. Diaz A, Alvarez M, Callejas C, Rosso R, Schnettler K, and Saldias F. Clinical picture and prognostic factors for severe community-acquired pneumonia in adults admitted to the intensive care unit. *Arch Bronconeumol* 2005;41:20–6.

219. Tanon-Anoh MJ, Kacou-Ndouba A, Yoda M, Ette-Akre E, Sanogo D, and Kouassi B. Particularities of bacterial ecology of acute otitis media in an African subtropical country (Cote d'Ivoire). *Int J Pediatr Otorhinolaryngol* 2006;70:817–22.

220. Pankuch GA, Bozdogan B, Nagai K, et al. Incidence, epidemiology, and characteristics of quinolone-nonsusceptible *Streptococcus pneumoniae* in Croatia. *Antimicrob Agents Chemother* 2002;46:2671–5.

221. Schrag SJ, Pena C, Fernandez J, et al. Effect of short-course, high-dose amoxicillin therapy on resistant pneumococcal carriage: a randomized trial. *JAMA* 2001; 286:49–56.

222. El Kholy A, Baseem H, Hall GS, Procop GW, and Longworth DL. Antimicrobial resistance in Cairo, Egypt 1999–2000: a survey of five hospitals. *J Antimicrob Chemother* 2003;51:625–30.

223. Adjei O and Agbemadzo T. Susceptibility of *Streptococcus pneumoniae* strains isolated from cerebrospinal fluid in Ghana. *J Antimicrob Chemother* 1996;38:746–7.

224. Dobay O, Rozgonyi F, Hajdu E, Nagy E, Knausz M, and Amyes SG. Antibiotic susceptibility and serotypes of *Streptococcus pneumoniae* isolates from Hungary. *J Antimicrob Chemother* 2003;51:887–93.

225. Kohanteb J and Sadeghi E. Penicillin-resistant *Streptococcus pneumoniae* in Iran. *Med Princ Pract* 2007;16:29–33.

226. Dagan R, Givon-Lavi N, Zamir O, and Fraser D. Effect of a nonavalent conjugate vaccine on carriage of antibiotic-resistant *Streptococcus pneumoniae* in day-care centers. *Pediatr Infect Dis J* 2003;22:532–40.

227. Regev-Yochay G, Raz M, Dagan R, et al. Nasopharyngeal carriage of *Streptococcus pneumoniae* by adults and children in community and family settings. *Clin Infect Dis* 2004;38:632–9.

228. Feikin DR, Davis M, Nwanyanwu OC, et al. Antibiotic resistance and serotype distribution of *Streptococcus pneumoniae* colonizing rural Malawian children. *Pediatr Infect Dis J* 2003;22:564–7.
229. Jacobs MR and Appelbaum PC. Susceptibility of 1100 *Streptococcus pneumoniae* strains isolated in 1997 from seven Latin American and Caribbean countries. Laser Study Group. *Int J Antimicrob Agents* 2000;16:17–24.
230. Ochoa TJ, Rupa R, Guerra H, et al. Penicillin resistance and serotypes/serogroups of *Streptococcus pneumoniae* in nasopharyngeal carrier children younger than 2 years in Lima, Peru. *Diagn Microbiol Infect Dis* 2005;52:59–64.
231. Dias R, Louro D, and Canica M. Antimicrobial susceptibility of invasive *Streptococcus pneumoniae* isolates in Portugal over an 11-year period. *Antimicrob Agents Chemother* 2006;50:2098–105.
232. Orrett FA and Changoor E. Bacteremia in children at a regional hospital in Trinidad. *Int J Infect Dis* 2006.
233. Maalej SM, Kassis M, Rhimi FM, Damak J, and Hammami A. Bacteriology of community acquired meningitis in Sfax, Tunisia (1993–2001). *Med Mal Infect* 2006;36:105–10.
234. Yalcin I, Gurler N, Alhan E, et al. Serotype distribution and antibiotic susceptibility of invasive *Streptococcus pneumoniae* disease isolates from children in Turkey, 2001–2004. *Eur J Pediatr* 2006;165:654–7.
235. Erdem H and Pahsa A. Antibiotic resistance in pathogenic *Streptococcus pneumoniae* isolates in Turkey. *J Chemother* 2005;17:25–30.
236. Petreska-Sibinovska D, Mraovic M, and Lazarcvic G. The development of resistance of *Streptococcus pneumoniae* strains isolated from children. *Clin Microbiol Infect* 1999;5:82.
237. Tiemei Z, Xiangqun F, and Youning L. Resistance phenotypes and genotypes of erythromycin-resistant *Streptococcus pneumoniae* isolates in Beijing and Shenyang, China. *Antimicrob Agents Chemother* 2004;48:4040–1.
238. Ho PL, Lam KF, Chow FK, et al. Serotype distribution and antimicrobial resistance patterns of nasopharyngeal and invasive *Streptococcus pneumoniae* isolates in Hong Kong children. *Vaccine* 2004;22:3334–9.
239. Paraskakis I, Kafetzis DA, Chrisakis A, et al. Serotypes and antimicrobial susceptibilities of 1033 pneumococci isolated from children in Greece during 2001–2004. *Clin Microbiol Infect* 2006;12:490–3.
240. Working Group of Tokai Anti-biogram Study G, Mitsuyama J, Yamaoka K, et al. Sensitivity surveillance of *Streptococcus pneumoniae* isolates for several antibiotics in Gifu prefecture (2004). *Jpn J Antibiot* 2006;59:137–51.
241. Ubukata K, Chiba N, Hasegawa K, Kobayashi R, Iwata S, and Sunakawa K. Antibiotic susceptibility in relation to penicillin-binding protein genes and serotype distribution of *Streptococcus pneumoniae* strains responsible for meningitis in Japan, 1999 to 2002. *Antimicrob Agents Chemother* 2004;48:1488–94.
242. Kariuki S, Muyodi J, Mirza B, Mwatu W, and Daniels JJ. Antimicrobial susceptibility in community-acquired bacterial pneumonia in adults. *East Afr Med J* 2003;80:213–7.
243. Al Sweih N, Jamal W, and Rotimi VO. Spectrum and antibiotic resistance of uropathogens isolated from hospital and community patients with urinary tract infections in two large hospitals in Kuwait. *Med Princ Pract* 2005;14:401–7.
244. Mokaddas EM, Wilson S, and Sanyal SC. Prevalence of penicillin-resistant *Streptococcus pneumoniae* in Kuwait. *J Chemother* 2001;13:154–60.
245. Karam Sarkis D, Hajj A, and Adaime A. Evolution of the antibiotic resistance of *Streptococcus pneumoniae* from 1997 to 2004 at Hotel-Dieu de France, a university hospital in Lebanon. *Pathol Biol (Paris)* 2006;54:591–5.

246. Quinones-Falconi F, Calva JJ, Lopez-Vidal Y, Galicia-Velazco M, Jimenez-Martinez ME, and Larios-Mondragon L. Antimicrobial susceptibility patterns of *Streptococcus pneumoniae* in Mexico. *Diagn Microbiol Infect Dis* 2004;49:53–8.

247. Calva-Mercado JJ, Castillo G, and Lopez-Vidal Y. Fluoroquinolone activity in clinical isolates of *Streptococcus pneumoniae* with different susceptibility to penicillin: an epidemiological study in five cities of Mexico. *Gac Med Mex* 2005;141:253–8.

248. Elsaid MF, Flamerzi AA, Bessisso MS, and Elshafie SS. Acute bacterial meningitis in Qatar. *Saudi Med J* 2006;27:198–204.

249. Porat N, Leibovitz E, Dagan R, et al. Molecular typing of *Streptococcus pneumoniae* in northeastern Romania: unique clones of *S. pneumoniae* isolated from children hospitalized for infections and from healthy and human immunodeficiency virus-infected children in the community. *J Infect Dis* 2000;181:966–74.

250. Al-Mazrou A, Twum-Danso K, Al Zamil F, and Kambal A. *Streptococcus pneumoniae* serotypes/serogroups causing invasive disease in Riyadh, Saudi Arabia: extent of coverage by pneumococcal vaccines. *Ann Saudi Med* 2005;25:94–9.

251. Chiang WC, Teoh OH, Chong CY, Goh A, Tang JP, and Chay OM. Epidemiology, clinical characteristics and antimicrobial resistance patterns of community-acquired pneumonia in 1702 hospitalized children in Singapore. *Respirology* 2007;12:254–61.

252. Liebowitz LD, Slabbert M, and Huisamen A. National surveillance programme on susceptibility patterns of respiratory pathogens in South Africa: moxifloxacin compared with eight other antimicrobial agents. *J Clin Pathol* 2003;56:344–7.

253. Perez-Trallero E, Garcia-de-la-Fuente C, Garcia-Rey C, et al. Geographical and ecological analysis of resistance, coresistance, and coupled resistance to antimicrobials in respiratory pathogenic bacteria in Spain. *Antimicrob Agents Chemother* 2005; 49:1965–72.

254. Lauderdale TL, Wagener MM, Lin HM, et al. Serotype and antimicrobial resistance patterns of *Streptococcus pneumoniae* isolated from Taiwanese children: comparison of nasopharyngeal and clinical isolates. *Diagn Microbiol Infect Dis* 2006;56:421–6.

255. Watanabe H, Asoh N, Hoshino K, et al. Antimicrobial susceptibility and serotype distribution of *Streptococcus pneumoniae* and molecular characterization of multidrug-resistant serotype 19F, 6B, and 23F Pneumococci in northern Thailand. *J Clin Microbiol* 2003;41:4178–83.

256. Joloba ML, Bajaksouzian S, Palavecino E, Whalen C, and Jacobs MR. High prevalence of carriage of antibiotic-resistant *Streptococcus pneumoniae* in children in Kampala, Uganda. *Int J Antimicrob Agents* 2001;17:395–400.

257. Yoshimine H, Oishi K, Mubiru F, et al. Community-acquired pneumonia in Ugandan adults: short-term parenteral ampicillin therapy for bacterial pneumonia. *Am J Trop Med Hyg* 2001;64:172–7.

258. Schultsz C, Vien LM, Campbell JI, et al. Changes in the nasal carriage of drug-resistant *Streptococcus pneumoniae* in urban and rural Vietnamese schoolchildren. *Trans R Soc Trop Med Hyg* 2007;101:484–92.

11 Antimicrobial Resistance in the *Enterococcus*

George M. Eliopoulos

CONTENTS

INTRODUCTION

Enterococcal infections can be among the most challenging problems encountered in the practice of clinical infectious diseases. The *Enterococcus* is the third most common pathogen causing left-sided, native valve infective endocarditis, after streptococci and *Staphylococcus aureus* [1]. Because enterococci are typically resistant to killing by penicillins or glycopeptides alone [2], successful treatment has usually required combinations of cell wall-active agents with aminoglycosides that achieve synergistic bactericidal activity *in vitro* [3]. Such regimens, when they can be employed, may be associated with significant toxicities. However, in some cases, resistance to either the cell wall antibiotic or to the aminoglycoside, or to both components, can preclude the possibility of even attaining bactericidal synergism with such combination regimens [4].

Ominously, in recent years, enterococci resistant to multiple antimicrobial agents have become increasingly prevalent in the hospital environment. In one worldwide surveillance study from 2000 to 2004, *Enterococcus* spp. accounted for 12.3% of more than 9000 bacterial isolates collected from intensive care unit (ICU) patients in 29 countries, second in frequency only to *S. aureus* [5]. More than half of these enterococcal isolates were resistant to tetracycline, levofloxacin, and quinupristin-dalfopristin; 28% were resistant to ampicillin; and approximately 20% were non-susceptible to vancomycin. From ICUs in the United States, even higher rates of vancomycin resistance have been reported. Vancomycin-resistant strains accounted for 28.5% of enterococcal isolates identified in 2003 as nosocomial pathogens, which represented a 12% increase over the previous five-year average [6]. Such vancomycin-resistant enterococci (VRE), which are predominantly *Enterococcus faecium*, are commonly resistant to multiple older antibiotics [7], necessitating the increased use of newer antimicrobials for therapy. Because infections caused by VRE often occur in patients who have significant underlying medical conditions, who are immuno-compromised, or who have undergone surgical procedures, the additional clinical burden of these infections is high [8–11]. Costs associated with these infections and with their treatment contribute to escalating health care expenditures.

Vancomycin resistance genes originating in enterococci have now been found in several clinical isolates of *S. aureus* [12]. This validates concerns expressed more than a decade ago that VRE may serve as a reservoir of genes that could confer upon staphylococci resistance to glycopeptides [13], the principal antibiotics for treatment of infections caused by methicillin-resistant strains.

RESISTANCE TO ANTIBIOTICS ACTIVE ON THE CELL WALL OR CELL MEMBRANE

β-Lactam Antibiotics

Target-Based Resistance: Low-Affinity Penicillin-Binding Proteins

Resistance in Enterococcus faecalis
The inhibitory activity of penicillin against various enterococcal species is influenced by the amounts and relative binding affinities for penicillin of penicillin-binding proteins (PBPs) found on the cell membrane of these organisms [14–18]. *Enterococcus faecalis* are generally susceptible to ampicillin, with minimal inhibitory concentrations for 90% of isolates (MIC_{90}s) at 1 to 2 µg/mL [19]. The presence of a high-molecular-weight (74 kDa) PBP, designated PBP5, contributes to the relatively high MICs of penicillins in this species and results in high-level resistance to ceftriaxone. Deletion of *pbp5* results in a >1000-fold reduction in MICs of ceftriaxone, and a more modest four-fold reduction in MICs of ampicillin from 2 to 0.5 µg/mL [20].

However, the presence of PBP5 per se may not be sufficient for resistance to β-lactam antibiotics in *E. faecalis*. Despite the normal expression of PBP5, deletion of the *croRS* locus, which encodes a two-component regulatory system responsive to the presence of ceftriaxone in growth media, resulted in a 4000-fold increase in susceptibility to ceftriaxone and a four-fold increase in susceptibility to ampicillin, changes identical in magnitude to those observed with deletion of *pbp5* [21]. The *croRS* deletion also resulted in large decreases in MICs of cefuroxime and cefepime, and more modest reductions in MICs of cephalothin, imipenem, and other β-lactam antibiotics. The mechanisms involved are unknown, as there were no apparent changes in PBP patterns, cell wall precursors, or muropeptides produced [21]. A later study showed that a *croR* mutant was more susceptible than the parent *E. faecalis* not only to ampicillin and cefotaxime, but also to the non-β-lactam cell wall-active antimicrobial, D-cycloserine [22].

Two isolates of *E. faecalis* with ampicillin MICs of 32 to 64 µg/mL were described, which had commensurate increases in resistance to penicillin (MIC, 64 µg/mL) and imipenem (MIC, 16 µg/mL) [23]. These isolates were β-lactamase-negative, but demonstrated increased amounts of PBP5 and relative reductions in binding of penicillin to PBPs 1 and 6. Ono et al. [24] studied a series of *E. faecalis* urine isolates with ampicillin MICs of 1, 8, or 16 µg/mL (imipenem MICs of 0.5 to 1, 4, and 32 µg/mL, respectively). As measured by competition assays, the affinity of ampicillin and imipenem for PBP4 diminished in parallel with increases in MICs. Sequencing of the *pbp4* gene revealed that in comparison with an ampicillin-susceptible clinical isolate, single amino acid substitutions (Tyr605His) were found in PBP4 for isolates with ampicillin MICs of 8 µg/mL, and double substitutions (Tyr605His, Pro520Ser) in those with MICs of 16 µg/mL [24].

Recently, attention has been drawn to another phenotype of *E. faecalis*. These isolates are susceptible to ampicillin (MICs, ≤2 µg/mL), but disproportionately resistant to penicillin (MICs, ≥16 µg/mL) and imipenem (MICs, 4 to 16 µg/mL) [25]. Mechanisms accounting for this dissociated resistance are under investigation. However, the existence of such isolates is important to recognize, so that imipenem or penicillin susceptibilities can be determined directly (i.e., not just inferred from ampicillin susceptibility test results) if these antibiotics are used to treat *E. faecalis* infections.

Resistance in *Enterococcus faecium*

As a species, clinical isolates of *E. faecium* are considerably more resistant to penicillins than are *E. faecalis*. Grayson et al. [26] examined a collection of *E. faecium* recovered from clinical specimens in Boston from 1989 to 1990 and reported an ampicillin MIC_{90} of 128 µg/mL. Such high levels of resistance to penicillins are a relatively recent development in this species. For isolates collected over the preceeding 20 years (1969 to 1988) in the same laboratory, the MIC_{90} of ampicillin was 32 µg/mL. Twenty-four isolates collected from an antibiotic-naïve population on the Solomon Islands in 1968 were substantially more susceptible to ampicillin, with an MIC_{90} of 2 µg/mL [26].

Resistance to penicillins in *E. faecium* has been associated with production of a low-affinity penicillin-binding protein, PBP5. Loss of this non-essential PBP can render strains highly susceptible to penicillins, reducing ampicillin MICs to as low as 0.03 to 0.06 μg/mL [27,28]. Examination of clinical isolates with varying degrees of penicillin resistance found a correlation between modestly elevated MICs up to 64 μg/mL and increased expression of PBP5. At higher MICs, point mutations in *pbp5* were observed, and these were associated with further decreases in affinity of the PBP for penicillins [29–32]. The role of specific mutations in *pbp5* was investigated by introducing mutated copies of the gene situated on plasmids into a strain of *E. faecium* that had spontaneously lost its native copy of the gene [33]. This study showed that individual mutations found in clinical isolates modestly raised ampicillin MICs. A combination of three amino acid substitutions, together with a Ser466′ insertion, near the active site raised the ampicillin MIC to 185 μg/mL. Reduced affinities of the mutant PBPs for penicillin correlated with increased ampicillin MICs against the strains [33].

Rice et al. [34] have shown that the *pbp5* gene located on the *E. faecium* chromosome is a transferable element. By conjugation experiments, several ampicillin-resistant strains from humans or turkeys were shown to transfer ampicillin resistance into a *pbp5*-deleted recipient at frequencies from 5×10^{-10} to 3×10^{-7}. The donor *pbp5* was preferentially inserted into the region from which the recipient copy had previously been excised. They postulated that such transferable resistance genes may have contributed to the rapid emergence of ampicillin-resistant *E. faecium* in the United States [34]. The authors had shown in an earlier paper that resistance to ampicillin and to vancomycin could be co-transferred from *E. faecium* in which the VanB mobile element Tn*5382/1549* was genetically linked to *pbp5* [35,36].

Another target-based mechanism of β-lactam resistance has now been described in *E. faecium*, which is distinct from that attributable to low-affinity PBPs. Mainardi et al. [37] selected ampicillin-resistant mutants (MIC, >2000 μg/mL) from a highly susceptible parent strain lacking pbp5 (MIC, 0.06 μg/mL). Examination of the peptidoglycan produced demonstrated that the native D-Ala → D-Asp-L-Lys (or D-Asn-L-Lys) crosslinks, formed by DD-transpeptidases that are inhibited by penicillin, had been completely replaced in the mutant by L-Lys → D-Asp-L-Lys (or D-Asn-L-Lys) crosslinks formed by a penicillin-insensitive LD-transpeptidation reaction. Study of mutants derived by step-wise selection demonstrated that the native peptidoglycan was gradually replaced by the novel crosslinks [38]. The LD-transpeptidase could be detected in the parent as well as in the mutants; however, resistance was accompanied by the appearance of large amounts of the tetrapeptide precursors for the LD-transpeptidation. This suggested that a β-lactam-insensitive DD-carboxypeptidase that was not detectable in the parent determined the extent to which novel crosslinks were formed [38]. This enzyme would cleave the terminal D-Ala from the usual pentapeptide substrate for DD-transpeptidation, tipping the balance toward the penicillin-insensitive cell wall components. PBPs of the highly ampicillin-resistant mutant and their affinities for β-lactams were similar to those of the parent strain. Homologs of the ampicillin-insensitive LD-transpeptidase, designated $\mathrm{Ldt_{fm}}$, were detected in other Gram-positive species, including *E. faecalis* and *Bacillus anthracis* [39]. The crystal structure of a catalytically active fragment of this transpeptidase has been reported [40].

β-Lactamase Production

Investigators have long searched for β-lactamases as an explanation of the relative resistance of enterococci to penicillins. β-Lactamase-producing *E. faecalis* were first identified in the early 1980s in Texas and Pennsylvania [41,42] but are very rare. Isolates have also been characterized from elsewhere in the eastern United States, Argentina, and Lebanon [43,44]. A single isolate of a β-lactamase-producing *E. faecium* was detected in a medical center in Virginia where β-lactamase-producing *E. faecalis* were endemic [45]. More recently, a β-lactamase-producing *E. faecalis* isolate was recovered in Australia, [46] and β-lactamase-positive isolates of both *E. faecalis* and *E. faecium* were reported from a liver transplant center in China [47].

Exactly why β-lactamase-producing enterococci remain so uncommon is not clear because large outbreaks of colonization or infection or both have been reported [48,49]. These isolates cannot be detected reliably by susceptibility testing to penicillin or ampicillin. The level of β-lactamase produced is low, so that MICs determined under standard testing conditions will usually not exceed those of non-β-lactamase-producing isolates. High inoculum testing can demonstrate a large increase in the MIC of these penicillins [42]; however, direct detection of the β-lactamase by nitrocefin hydrolysis is preferred.

Initial studies with β-lactamase-producing enterococci showed that the β-lactamase gene, *blaZ*, was of staphylococcal origin, and typically on transferable plasmids that also encoded high-level gentamicin resistance [50–52]. Enterococci express the β-lactamase gene constitutively. This has been attributed for the most part to absence or alteration of the regulatory genes found in staphylococci. However, transfer of genes from a strain of *E. faecalis* that constitutively produced enzyme despite having intact regulatory genes, resulted in inducible expression of the enzyme in a staphylococcal recipient, suggesting that additional, unknown host factors influence β-lactamase production [52,53].

β-Lactamase genes have been located on the chromosome of several *E. faecalis* isolates [54]. Rice et al. [55] determined that *blaZ* was on an approximately 65-kb chromosomal composite element, Tn*5385*, containing transposons and insertion sequences (of both staphylococcal and enterococcal origins) and encoding resistance to erythromycin (*ermAM*), tetracyclines [*tet*(M)], mercury (*merRAB*), streptomycin (*aadE*), and gentamicin (*aac6'-aph2"*).

The first reported β-lactamase-producing *E. faecalis* from Houston, as well as isolates from hospital outbreaks in Virginia and North Carolina, belong to a clonal cluster that also includes the first vancomycin-resistant *E. faecalis* detected in the United States, and a number of other highly pathogenic clinical isolates [56]. β-Lactamase-producing *E. faecalis* recovered in Connecticut and Argentina belong to an unrelated clonal group, while the lineage of strains from Boston and Lebanon appears unrelated to either [56].

GLYCOPEPTIDES

The glycopeptides provide important alternatives to penicillins for patients intolerant of those agents or who have infections due to enterococci that are resistant to the penicillins. In the mid-1980s, reports of enterococci demonstrating resistance to vancomycin began to emerge, primarily as nosocomial pathogens. A decade later, more than 15% of U.S. nosocomial bloodstream isolates of enterococci were vancomycin resistant [57]. The National Nosocomial Infections Surveillance System reported that in 2003, 28.5% of enterococci recovered from infections in patients hospitalized in U.S. intensive care units were vancomycin resistant [6]. An independent surveillance project of hospital laboratories across the United States undertaken in 2004 reported that 72% of *E. faecium* isolates were by that time resistant to vancomycin [58].

Certain strains of VRE appear to be particularly well suited to persist within healthcare institution environments. In the mid-1990s, within months of its introduction into a Boston hospital from an affiliated medical center where it had been endemic, one clonal type of VRE became dominant over the dozen or so strains circulating at the time [59]. Multi-locus sequence typing studies have shown that vancomycin-resistant *E. faecium* belonging to a globally distributed clonal lineage (designated clonal complex-17) appear to be particularly well adapted to the hospital environment [60]. These organisms are characterized by ampicillin resistance and by the presence of a putative pathogenicity island.

The appearance of VRE in hospitals has major repercussions on health care resources [61,62]. Environmental contamination of the hospital environment can be extensive. Some hospital rooms become so widely contaminated that extraordinary cleaning procedures (e.g., taking up to four hours) are necessary to decontaminate a vacated room successfully, in order to minimize the risk of VRE acquisition by subsequent occupants [63]. Some patients who acquire VRE may harbor these organisms in their gastrointestinal tracts for years [64].

Phenotypic Descriptions of Glycopeptide Resistance Classes

Glycopeptide resistance in enterococci results from modification of peptidoglycan precursors, such that effective binding by the glycopeptide with resulting inhibition of cell wall synthesis is prevented. Following early reports describing VRE, phenotypic classification schemes were developed based on species identity and levels of resistance to vancomycin and teicoplanin [65].

The VANA phenotype described isolates with inducible, high-level (MIC typically greater than 64 µg/mL) resistance to vancomycin and resistance to teicoplanin. These were generally *E. faecium* or *E. faecalis*. The VANB designation was applied to strains of these species that were resistant to vancomycin, but susceptible to teicoplanin [65]. Although vancomycin MICs were generally lower than those of VANA strains, the range was broad (4 to >1000 µg/mL), sometimes overlapping the MIC range of susceptible isolates [66]. Limitations of phenotypic classification schemes soon became evident. For example, mutants derived from VANB enterococci were described that had become resistant to teicoplanin [67]. Such isolates thus resembled VANA strains. *Enterococcus gallinarum* and *Enterococcus casseliflavus,* which inherently display low-level resistance to vancomycin and remain susceptible to teicoplanin, comprised the VANC phenotype. Isolates of these species can, however, acquire additional vancomycin resistance genes, which result in phenotypes consistent with VANA or VANB [68–70].

Genotypic Classification of Glycopeptide Resistance

A concise description of vancomycin resistance mechanisms in enterococci and their genotypic classification is presented in a recent review by Courvalin [71].

Resistance Associated with the vanA Determinant

Leclercq et al. [72,73] described the plasmid-borne resistance determinant Tn*1546*, responsible for vancomycin and teicoplanin resistance in an isolate of *E. faecium*. This transposon is capable of transfer from donor strains (of animal origin) to recipient strains of human origin in the intestines of human volunteers [74]. Resistance has now been shown to result from the cooperative effects of several enzymes mediated by genes carried on this transposon. The *vanA* gene encodes production of a ligase that results in the synthesis of D-alanine-D-lactate. As D-Ala-D-Lac becomes incorporated into peptidoglycan precursors in preference to D-Ala-D-Ala, the resulting pentadepsipeptide has 1000-fold lower binding affinity for vancomycin than does the usual peptidoglycan precursor terminating in D-Ala-D-Ala [75]. D-Lactate for this reaction is derived from pyruvate through the action of a dehydrogenase encoded by the *vanH* gene. The D,D-dipeptidase encoded by *vanX* reduces levels of D-alanine-D-alanine formed by the native enterococcal ligase. Transcription of the *vanHAX* operon is under control of a two-component regulatory system comprised of a histidine kinase sensor (VanS), which modulates phosphorylation of a transcriptional regulator (VanR); these are encoded by *vanS* and *vanR* genes of the transposon [75–81]. Cleavage of terminal D-alanine by a D,D-carboxypeptidase, VanY, further reduces the availability of glycopeptide-inhibitable pentapeptide target that might be formed and contributes to resistance [82]. Finally, the *vanZ* gene of Tn*1546* can confer low-level resistance to teicoplanin, but mechanisms involved remain elusive [83].

The origins of these vancomycin resistance gene clusters are unknown [71]. They may have evolved from the self-protective mechanisms of glycopeptide-producing bacteria [84,85] or originated in other bacterial species. A gene cluster homologous with the *vanA* gene cluster (designated *vanF* and including *vanH, X, Y, Z, R,* and *S* homologs) has been found in *Paenibacillus popilliae*, a biopesticide used in the control of Japanese beetle larvae [86,87]. These genes have been detected in stored samples of these highly vancomycin-resistant organisms (MIC, >1000 µg/mL) that antedate the clinical introduction of vancomycin. A sequence upstream of $vanR_F$ showed 95% identity with portions of the Tn*1546* transposase [87]. Gene clusters with homology to VanA (or VanB) of enterococci have been discovered in other *Paenibacillus* spp. and in *Rhodococcus* spp. and other organisms [88,89].

Resistance Associated with the vanB Determinant

The *vanB* gene cluster also results in production of peptidoglycan precursors terminating in D-alanine-D-lactate, to which vancomycin binds poorly. Three alleles of *vanB* have been described based on nucleotide sequences [90]. Genes homologous with *vanH* and *vanX* are designated $vanH_B$ and $vanX_B$; $vanY_B$ is found in some isolates [71]. Unlike VanR–VanS, the $VanR_B$–$VanS_B$ regulatory system is not inducible by teicoplanin [91]. However, mutations in the regulatory system can result in resistance to teicoplanin [92,93]. There are no genes homologous with *vanZ*, but another gene of unknown function, *vanW*, is present [91].

Resistance determinants of the VanB type can be plasmid mediated, but have also been found on large, mobile chromosomal elements [55,94,95]. A 27-kb *vanB* transposon, Tn*5382*, has been found to be genetically linked to *pbp5* in *E. faecium* [35]. The *vanB* and *pbp5* genes, conferring vancomycin and penicillin resistance, were shown to be co-transferred from several VanB strains into recipients strains by conjugation [36]. Close linkage between Tn*5382* and *pbp5* was demonstrated in 30 of 32 vancomycin-resistant *E. faecium* from Taiwan. These strains were of diverse pulsed-field gel electrophoresis (PFGE) types [96].

Grayson's group has detected VanB operons in several species of vancomycin-resistant Gram-positive intestinal anaerobes, carried on elements similar to enterococcal transposons Tn*5382* or Tn*1549* [97,98]. Conjugal transfer of a *vanB2* Tn*1549*-like element from *Clostridium symbiosum* into both *E. faecium* and *E. faecalis* could be demonstrated in the gastrointestinal tracts of gnotobiotic mice [99]. This work suggests the possible role of anaerobic gut flora as an additional reservoir of vancomycin resistance genes.

Resistance Associated with the vanC Determinant
Low-level resistance to vancomycin, with susceptibility to teicoplanin, results from the presence of VanC-type resistance determinants intrinsic to *E. gallinarum* (*vanC-1*) and *E. casseliflavus* (*vanC-2*)/*Enterococcus flavescens* (*vanC-3*) [100]. The mechanisms involved have recently been reviewed by Reynolds and Courvalin [101]. The VanC gene cluster differs from those of VanA and VanB in several respects. The altered target consists of peptidoglycan precursors terminating in D-alanine-D-serine (in contrast to D-alanine-D-lactate), which also demonstrates reduced binding affinity for vancomycin [102]. The cluster also contains a gene encoding a serine racemase, VanT, which ensures the availability of sufficient amounts of D-serine substrate for the VanC ligase [103]. *E. gallinarum* BM4174 contains not only the ligase encoded by *vanC* and a native D-Ala-D-Ala ligase, but also a second ligase with D-Ala-D-Ala activity, encoded by *ddl2* present on the gene cluster; the role of this third ligase is uncertain [104].

Another difference between the VanC gene cluster and those of VanA and VanB is that the (VanX) D,D-dipeptidase and the (VanY) D,D-carboxypeptidase functions of the latter are assumed by a single protein, VanXY$_C$, encoded by *vanXY$_C$* [105]. Regulatory genes *vanR$_C$* and *vanS$_C$* are present. Resistance may be inducible or constitutive, the latter most likely resulting from amino acid substitutions in VanS$_C$ [106].

E. gallinarum or *E. casseliflavus*, species with intrinsic VanC resistance mechanisms, can also acquire *vanA* or *vanB* resistance genes [107–110]. If this occurs, higher levels of resistance to glycopeptides than is characteristic for VanC species is likely to be observed.

E. gallinarum and *E. casseliflavus/flavescens* are less commonly recognized as human pathogens than are *E. faecalis* or *E. faecium*. Nevertheless, both groups are occasionally isolated from bloodstream infections, biliary tract sepsis, or infections associated with medical interventions [111–115].

vanD and beyond
Additional gene clusters that confer glycopeptide resistance in enterococci have been discovered. A protocol to detect *vanA-vanE* and *vanG* ligase genes by

multiplex PCR has been described [116]. The VanD cluster, like VanA and VanB, results in peptidoglycan precursors terminating in D-alanine-D-lactate and has been found in both *E. faecalis* and *E. faecium* [71]. Alleles of *vanD* have been described in clinical isolates [117]. Descriptions of early isolates of *E. faecium* reported MICs of vancomycin ≥64 µg/mL, and moderate resistance to teicoplanin (MICs, 4 µg/mL) [118,119]. *E. faecalis* strains more recently characterized were less resistant to vancomycin (MICs, 16 µg/mL) and were susceptible to teicoplanin [120]. *vanD* has also been found in a highly vancomycin-resistant isolate of *E. gallinarum* (MIC, 256 µg/mL) [121].

The VanD cluster contains genes *vanH$_D$*, *vanX$_D$*, and *vanY$_D$* like those in VanA-type strains, but there is no gene homologous with *vanZ* [120]. The activity of VanX$_D$ is variable; however, D,D-dipeptidase activity appears to be of less importance here because several strains examined produced non-functional intrinsic *ddl* ligases. The observation that these strains grow in the absence of glycopeptide induction can be attributed to mutations, deletions or insertions in the *vanS$_D$* gene of the VanS$_D$-VanR$_D$ regulatory system, resulting in constitutive expression of resistance [120]. In contrast to the penicillin-insensitive D,D-carboxypeptidases of the VanA and VanB clusters, VanY$_D$ is a penicillin-inhibitable enzyme localized to the cell membrane (a PBP), saturated at penicillin concentrations <1 µg/mL [122].

Fines et al. [123] described a gene cluster that was assigned the designation VanE in *E. faecalis* BM4405, an isolate from Chicago. This organism produced D-Ala-D-Ser peptidoglycan precursors conferring low-level resistance to vancomycin (MIC, 16 µg/mL), but not to teicoplanin (MIC, 0.5 µg/mL). Phylogenetically, based on alignment of ligase sequences, VanE was most closely related to the VanC group. The new operon, consisting of the genes *vanE*, *vanXY$_E$*, *vanT$_E$*, *vanR$_E$*, and *vanS$_E$*, was localized to the bacterial chromosome [124]. A stop codon within the sequence of *vanS$_E$* in *E. faecalis* BM4405 and a deletion in that gene in one of three distinct VanE-type isolates of *E. faecalis* recovered in Australia suggest that the sensor protein of this cluster is not always functional [124,125].

The VanF designation was assigned to the glycopeptide resistance gene cluster of *Paenibacillus popilliae*, as discussed previously (see the section Resistance Associated with the *vanA* Determinant). The VanG cluster was encountered in four moderately vancomycin-resistant (MICs, 12 to 16 µg/mL), teicoplanin-susceptible isolates from one hospital in Brisbane, Australia [126]. These and other isolates were subsequently studied in detail, with the following results [127]. As is the case for VanC and VanE, the product of the VanG ligase is D-alanine-D-serine. The cluster contains genes for VanT$_G$, a serine racemase, and VanW$_G$, of unknown function. VanXY$_G$, with approximately 40% identity with the bifunctional D,D-dipeptidase/ D,D-carboxypeptidases VanXY$_C$ and VanXY$_E$, showed very weak D-Ala-D-Ala dipeptidase activity. Despite the presence of the bifunctional enzyme and of a gene for a D,D-carboxypeptidase, *vanY$_G$* (which contains a frameshift mutation predicting a non-functional protein), the bacteria showed minimal carboxypeptidase activity. As a result of these two weak enzymatic activities, peptidoglycan precursors terminating in D-Ala as well as D-Ser were found in substantial amounts. The *vanR$_G$*- *vanS$_G$* regulatory cluster includes *vanU$_G$*, encoding a predicted transcriptional activator. Finally, Depardieu et al. [127] demonstrated that the VanG resistance cluster could be transferred by conjugation at low frequencies on a large chromosomal element containing also *erm*(B), permitting selection of transconjugants by acquisition of resistance to erythromycin.

A novel mechanism of glycopeptide resistance has recently been described, based on the discovery that peptidoglycan crosslinks can occur via a penicillin-insensitive L,D-transpeptidase as described in the section Resistance in *Enterococcus faecium*, above. This crosslinking reaction utilizes precursors lacking the terminal D-alanine components, so are not targets of glycopeptide action. Cremniter et al. [128] generated mutants of *E. faecium* after serial passage on increasing concentrations of glycopeptides or on glycopeptides after initially selecting for ampicillin resistance [37,39]. Mutants were recovered that were highly resistant to both vancomycin and teicoplanin (MIC of vancomycin, 1000 µg/mL; MIC of teicoplanin, 250 to 1000 µg/mL) [128]. Analysis of peptidoglycan from one of the most resistant mutants demonstrated that UDP-MurNAc-tetrapeptide (a vancomycin-insensitive target) accounted for >99% of the detectable peptidoglycan precursor. The authors found peptidoglycan crosslinks between L-Lys (the third amino acid attached to UDP-MurNAc) of one stem peptide and the D-*iso*-aspartyl or D-*iso*-asparaginyl side chain (attached to the L-Lys^3) of another stem peptide, reflecting the activity of the glycopeptide-insensitive L,D-transpeptidation reaction [128].

RESISTANCE TO LIPOPEPTIDE ANTIBIOTICS

In the United States, daptomycin was approved in 2003 for treatment of complicated skin and skin structure infections caused by a number of pathogens, including vancomycin-susceptible *E. faecalis*. The mechanisms of action of this cyclic lipopeptide antibiotic are not fully understood. However, the initial effects of daptomycin appear to result from its insertion into the bacterial cell membrane in a calcium-dependent process, formation of oligomers that disrupt membrane integrity, and subsequently depolarization of the bacterial cell [129]. From a large collection of clinical isolates, MIC$_{90}$s of daptomycin against *E. faecalis* and *E. faecium* were 2 and 4 µg/mL, respectively [130].

Since the introduction of daptomycin into clinical practice, there have been isolated reports of clinical failure with off-label use of the drug. Some of the enterococci recovered in this setting demonstrated decreased susceptibility to daptomycin. For example, from one patient treated with daptomycin and amikacin for a bloodstream infection caused by a daptomycin-susceptible *E. faecalis* (MIC, 1 µg/mL), a non-susceptible (MIC, 16 µg/mL) isolate was recovered that was indistinguishable from the initial strain by PFGE [131]. From another patient, a daptomycin non-susceptible *E. faecium* (MIC, ≥32 µg/mL) was recovered from blood cultures after treatment with daptomycin; the isolate was highly related by PFGE to an antecedent, susceptible urinary isolate (MIC, 2 µg/mL) [132].

Mechanisms of resistance to daptomycin have not yet been reported in enterococci. Studies in staphylococci suggest that mutations predicted to affect cell membrane composition, with the potential to influence activity of cationic antibiotics, are the first to appear in the step-wise selection of mutants with elevated daptomycin MICs [133].

RESISTANCE TO AGENTS ACTING ON THE BACTERIAL RIBOSOME

AMINOGLYCOSIDES

As single agents, the aminoglycosides lack useful activity against enterococci. Nevertheless, they constitute a critical component of regimens designed to achieve the bactericidal action sought for treatment of enterococcal endocarditis [3]. Enterococci are typically tolerant of the bactericidal effects of cell wall-active agents, such as penicillins and glycopeptides [2]. As a result, administration of penicillin alone in earlier years produced only modest results in the treatment of enterococcal endocarditis [134]. The observation that empirical therapy with penicillin together with streptomycin resulted in surprisingly favorable results was subsequently validated by several clinical studies (later with gentamicin instead of streptomycin). As a result, combination regimens consisting of a cell wall-active agent and an aminoglycoside have now become standard for the treatment of enterococcal endocarditis [3]. Such combinations result in synergistic killing, which can be demonstrated by time-kill studies *in vitro*. Experiments carried out more than 35 years ago showed that the otherwise limited intracellular uptake of aminoglycosides by enterococci can be substantially enhanced in the presence of a cell wall-active agent, allowing the former to reach intracellular concentrations lethal to the bacterium [135].

Resistance to Synergistic Killing

The synergistic bactericidal effect of penicillin-aminoglycoside combinations is dependent upon the ability of the latter to interact with the bacterial ribosome, perturbing its function. Resistance to the bactericidal action of the aminoglycosides can result from target insensitivity (streptomycin) or from the presence of enzymes that modify the drugs (streptomycin or 2-deoxystreptamine aminoglycosides). Such enzymes modify aminoglycosides by adenylylation, acetylation, or phosphorylation. Aminoglycoside-modifying enzymes have been the subject of a recent review [136]. Although streptomycin is uncommonly used for treatment of enterococcal infections today because the drug is not affected by mechanisms that inactivate gentamicin or other 2-deoxystreptamines, it is sometimes useful when there is resistance limited to the more commonly used agents.

High-Level Resistance to Streptomycin
Some enterococci were discovered to be so highly resistant to streptomycin (MICs, >2000 μg/mL) that even exposed to the drug in combination with penicillins, synergistic killing was not achievable. Laboratory mutants exhibiting high-level streptomycin resistance were shown to be resistant to synergistic killing even though uptake of streptomycin was enhanced in the presence of penicillin [135]. Ribosomal resistance to streptomycin explains resistance to synergism among some clinical isolates of *E. faecalis*. Ribosomal polypeptide synthesis by crude 30S extracts obtained from two such isolates was hardly affected by streptomycin at concentrations as high as 100 μg/mL; in contrast, concentrations as low as 1 μg/mL inhibited synthesis by ribosomes from a normally susceptible strain [137]. For six isolates of *E. faecalis* proven or suspected to be ribosomally resistant to the drug, streptomycin MICs were 128,000 μg/mL [137].

The presence of aminoglycoside-modifying enzymes is, however, a more common mechanism of resistance to streptomycin. Expression of streptomycin adenylylating enzymes typically results in streptomycin MICs of 4000 to 16,000 µg/mL [137,138]. The gene *aadE* encodes the enzyme ANT(6)-I that confers resistance to streptomycin [139]. Another gene, *aadA,* that was previously recognized in *S. aureus* and *E. coli* and more recently confirmed in enterococci, encodes the enzyme ANT(3″)-Ia (alternatively called ANT(3″) [9]), which confers resistance to both streptomycin and spectinomycin [138,140]. An adenylylating enzyme with 80% amino acid identity to ANT(6)-I and 13.9% identity to ANT(3″)-Ia was encountered in an isolate of *E. casseliflavus,* and subsequently detected in *E. faecium*; this was identical to an enzyme previously reported in *Lactococcus* [141].

Enzymes That Modify 2-Deoxystreptamine Aminoglycosides
Resistance to high levels of kanamycin (MICs, >2000 µg/mL) with loss of penicillin–kanamycin synergism was already common by the 1970s [142]. Resistance was caused by phosphorylation at the 3′-OH position of the aminoglycoside by the action of APH(3′)-IIIa [143]. Modification of amikacin by this enzyme has a curious result: the MIC is little affected (i.e., there is no high-level resistance), but any killing effect is lost; thus, amikacin may actually antagonize whatever killing may result from the penicillin alone [143,144]. Examination of more than 500 enterococcal isolates from patients at the Sapporo Medical University Hospital in Japan collected from 1997 to 2003 found *aph(3′)-IIIa* in approximately 50% of *E. faecalis* and 60% of *E. faecium* [145]. Tobramycin, which lacks the 3′-OH group, is not phosphorylated by the enzyme. Modification of kanamycin, tobramycin, and amikacin, with resistance to synergistic killing in the presence of a cell wall-active agent, can also arise from the action of a 4′,4″-nucleotidyltransferase, ANT(4′)-Ia [146]. Both *aph(3′)-IIIa* and *ant(4′)-Ia* can be found on transferable elements. Neither enzyme generated affects the potential synergistic activity of gentamicin.

An aminoglycoside acetylating enzyme, AAC(6′)-Ii, is chromosomally determined and present in *E. faecium* uniquely, although produced in low concentration, which does not necessarily result in high-level resistance [142,147]. The enzyme modifies kanamycin and tobramycin; thus, *E. faecium* are considered resistant to synergism by combinations involving these aminoglycosides. Although amikacin contains a 6′-amino group that is in principle susceptible to modification, this group is protected by the molecule's bulky 2-aminohydroxybutyryl moiety, and thus (in the absence of other enzymes) can participate in synergistic killing [142]. The C1 component of gentamicin is not acetylated by this enzyme; as a result, the clinical preparation of gentamicin retains the capacity for synergistic killing of *E. faecium* [147]. Other genes encoding aminoglycoside acetyltransferase activity have been encountered in *E. hirae* [designated *aac(6′)-Iih*] and in *E. durans* [designated *aac(6′)-Iid*]. At the amino acid level, these sequences shared 65% and 68% identity with AAC(6′)-Ii [148]. Like the latter, their presence precluded synergy between penicillin and kanamycin or tobramycin, but did not affect the synergistic activity of gentamicin. These genes were not detected in strains of *E. faecalis* or *E. faecium* examined.

Because gentamicin resists inactivation by the aforementioned modifying enzymes, it remained for many years the aminoglycoside in widest use to achieve synergistic killing of enterococci, irrespective of species and enzyme complement. In 1979, however, Horodniceanu et al. [149] reported high-level gentamicin resistance in *E. faecalis* from France. Within a decade of this description, in some U.S. healthcare institutions more than 25% of isolates displayed high-level resistance [150]. High-level resistance to gentamicin was first observed in *E. faecium* in Boston in 1988 [151]; over the next two years, more than 60% of *E. faecium* isolates demonstrated this trait [26].

High-level gentamicin resistance resulted from a single enzyme with both 2″-phosphorylating and 6′-acetylating activities [152]. This broad-spectrum enzyme can modify all 2-deoxystreptamine aminoglycosides available in the United States [153]. The result is that none of these drugs serves as an effective component of regimens intended to achieve bactericidal synergism when the enzyme is produced. The bifunctional enzyme is encoded by a gene, designated *aac(6′)-Ie-aph(2″)-Ia* [140], which has now been found in several enterococcal species and in isolates recovered from human samples, animal sources, and from food [151,154,155]. The gene is transposon-encoded and has been demonstrated in chromosomal elements as well as on diverse plasmids [54,156–160].

Although the *aac(6′)-Ie-aph(2″)-Ia* gene is by far the most common determinant of high-level gentamicin resistance (defined by MIC > 500 µg/mL), other phosphotransferases have now been implicated in a smaller proportion of isolates [140]. The gene *aph(2″)-Ib* was detected in a strain of *E. faecium* resistant to high levels of gentamicin and to synergistic killing by ampicillin–gentamicin combinations [161]. The gene was present in 5% of 121 high-level gentamicin resistant enterococci from Detroit hospitals, but five of the six isolates were clonal. In nine *E. faecium* studied, *aph(2″)-Ib* was found in close proximity to an acetyltransferase, *aac(6′)-Im* [162]. Crude extracts prepared from *E. coli* into which this gene had been cloned revealed acetyltransferase activity against kanamycin, tobramycin, and amikacin, but not gentamicin [162]. The contribution of this enzyme to clinically important aminoglycoside resistance in enterococci is uncertain.

APH(2″)-Ic was originally detected in *E. gallinarum*, but is also found in *E. faecalis* and *E. faecium* [155,163]. Although the presence of this enzyme negates ampicillin–gentamicin synergistic killing, gentamicin MICs may only reach 256 to 512 µg/mL, so resistance may escape detection by standard screening tests [163,164]. On the other hand, because amikacin retains synergistic potential against some isolates with this enzyme [164], isolates with faint growth in screening concentrations of gentamicin may be incorrectly assumed to possess *aac(6′)-Ie-aph(2″)-Ia* and to be resistant to synergy with all 2-deoxystreptamines, including amikacin [140]. APH(2″)-Id, initially observed in *E. casseliflavus*, produced high-level resistance to kanamycin, tobramycin, and gentamicin, but not to amikacin [165]. The *aph(2″)-Id* gene is also found in *E. faecium* [155]. Although synergistic killing by ampicillin plus amikacin was seen in the *E. casseliflavus* isolate, this was not the case for two *E. faecium* isolates [165]. Another phosphotransferase gene, *aph(2″)-Ie*, conferring high-level gentamicin resistance in *E. faecium* [145] and *E. casseliflavus* [141], has been reported from Japan and China, respectively. Vakulenko et al. [166] have described a multiplex PCR for the detection of six aminoglycoside-modifying enzyme genes from lysed colony suspensions of enterococci.

Other Mechanisms of Resistance to Gentamicin Synergism

Moellering et al. [167] reported resistance to penicillin–gentamicin bactericidal synergism for an isolate of *E. faecalis* that was relatively susceptible to gentamicin (MIC, 8 μg/mL) and which lacked detectable aminoglycoside-modifying enzymes. The patient from whom the isolate was recovered had relapsed after two courses of ampicillin plus gentamicin for treatment of enterococcal endocarditis, but was subsequently cured with a regimen of ampicillin plus tobramycin (MIC, 16 μg/mL). Time-kill studies *in vitro* confirmed bactericidal synergism for penicillin combined with tobramycin, but not gentamicin. Penicillin exposure enhanced the uptake of radiolabeled tobramycin, but not gentamicin. These results, together with the observation of unusually small colonies on agar, led the authors to consider a defect in aminoglycoside uptake specific to gentamicin [167]. Studies by Aslangul et al. [168,169] also point to the likelihood of impaired gentamicin uptake in *E. faecalis* passaged *in vitro* on gentamicin. They were able to select mutants with gentamicin MICs up to 400 μg/mL. For the most resistant mutant in the series, synergism could be shown between amoxicillin and gentamicin only with very high (i.e., not clinically achievable) concentrations of gentamicin, and the combination did not show an enhanced effect in an animal model. The authors excluded mutations affecting 16S rRNA or the ribosomal protein L6, and they excluded the presence of known modifying enzymes [168,169].

MACROLIDES, LINCOSAMIDES, AND STREPTOGRAMINS

Macrolides

Resistance to erythromycin and other macrolide antibiotics is very common among enterococci [170]. A survey of isolates collected at European university hospitals from 1997 to 1998 determined that only 14.8% of 403 *E. faecalis* and 6.6% of 86 *E. faecium* were susceptible to erythromycin [171]. A study by Jones et al. [172] of more than 1900 enterococcal isolates recovered from laboratories across the United States in 1992 found even lower rates of susceptibility, with only 2.9% of isolates susceptible to erythromycin. However, macrolide resistance is not intrinsic to the species. Atkinson et al. [173] examined 220 enterococcal isolates that had been collected in the Washington, D.C. area in 1953 to 1954. Only six of these organisms were resistant to erythromycin: three to this drug alone, and three to both erythromycin and clindamycin. The latter three isolates yielded DNA that hybridized with a probe for *erm*(AM), a gene of the *erm* class b family of ribosomal methylases [174].

The *erm*(B) genes are found on a common enterococcal transposon, Tn*917*, that has been detected in both plasmid and chromosomal locations [175–177]. Similar elements (Tn*917*-like) have been detected as components of large, transferable, chromosomal multidrug resistant elements in enterococci [55,178]. Using primers for *erm*(B), Portillo et al. [179] found evidence for this determinant in 39 of 40 enterococci from Spain with erythromycin MICs > 128 μg/mL; the one highly resistant isolate lacking *erm*(B) amplified with primers for the related gene, *erm*(A). Expression of *erm* genes in enterococci is inducible in some isolates. Alterations in the regulatory leader sequence have been found in other isolates with constitutive expression of the enzyme [180]. In one isolate of *E. faecalis*, a single amino acid change in the leader peptide of a ribosomal methylase gene led to the unusual situation that induction by the 16-membered macrolide tylosin was stronger than that by erythromycin [181].

In various Gram-positive bacteria, macrolide resistance (without resistance to lincosamides or to streptogramin A compounds) results from efflux mediated by *mef* genes [174]. Luna et al. [182] examined 32 erythromycin-resistant enterococcal isolates using DNA probes for *mef* and *erm*(B) genes. Nineteen percent were positive for *erm*(B) and 22% hybridized with *mef*; none was positive for both, and the remaining isolates hybridized with neither probe. As with *S. pneumoniae* carrying the *mefE* gene, levels of resistance to erythromycin were modest (MICs, ≤16 µg/mL) in the *mef*-positive enterococci. The authors also demonstrated conjugal transfer of *mef* genes into and from *E. faecalis* [182].

In 2000, Portillo et al. [179] reported the presence of the gene *msr*(C) in all 23 *E. faecium* isolates examined, and in no strains representing other species. Southern hybridizations localized the genes to the bacterial chromosome, not plasmids. Although detected by primers designed for the staphylococcal gene *msr*(A) associated with erythromycin resistance, the sequences of *msr*(C) and *msr*(A) shared only 62% identity at the DNA level. The *msr*(C) gene was present irrespective of erythromycin susceptibility (MICs, ≤0.125 to >128 µg/mL). Singh et al. [183] subsequently reported the complete sequence of *msr*(C), confirmed its presence in all 233 *E. faecium* isolates but in none of the other enterococcal species examined, and demonstrated that disruption of the gene resulted in modest increases in susceptibility to 14-, 15-, and 16-membered macrolides and to quinupristin, a streptogramin B antibiotic. Further support for a role of this gene in macrolide resistance comes from studies that showed that transfer of enterococcal *msr*(C) into an erythromycin-susceptible (MIC, 0.25 µg/mL) *S. aureus* substantially increased the level of erythromycin resistance in the recipient (MICs, 16 to 64 µg/mL). *E. faecium* isolates, mostly from animal sources or sewage, have now been reported that lack *msr*(C), suggesting that while the gene is widespread in the species, it is not essential [184]. A gene encoding an amino acid sequence with 40% identity to MsrC was found in *E. faecalis* V583; in a microarray system, gene expression was strongly up-regulated in cells grown in the presence of erythromycin [185]. The mechanism by which MsrC contributes to macrolide resistance has not yet been established. Although the staphylococcal MsrA belongs to a family of ABC (ATP-binding cassette) transporters, the protein has no transmembrane domains, so whether drug efflux is actually involved is unclear [186,187].

Lincosamides

As a species, *E. faecalis* is resistant to the lincosamides [188]. Among 403 isolates of this species collected in Europe in the late 1990s, only 4.4% were susceptible to clindamycin [171]. In contrast, a substantial minority of *E. faecium* isolates are susceptible to this drug [171,188]. Ribosomal methylase (*erm*) genes are prevalent among enterococci (see Macrolides section) [189]; methylation of rRNA leads to resistance to macrolides, lincosamides, and streptogramin B drugs. Constitutive expression of the enzyme due to deletions or truncations in regulatory elements could result in lincosamide resistance [180].

Singh et al. [190] provided an alternative explanation for the characteristic resistance of *E. faecalis* to both lincosamides and dalfopristin (a streptogramin A antibiotic). They detected the *lsa* gene in 180 of 180 isolates of *E. faecalis* examined, and not in any of the almost 200 isolates of other species tested. Presence of *lsa* was associated with resistance to clindamycin, even in *erm*(B)-negative *E. faecalis* isolates; disruption of the *lsa* gene resulted in susceptibility to clindamycin and dalfopristin. Two clinical isolates of *E. faecalis* that were susceptible to clindamycin and dalfopristin were shown to have mutations that introduced stop codons into the gene sequence [191]. The predicted amino acid sequence of Lsa was similar to those of ABC transporter proteins [190]; however, the putative Lsa protein lacks evidence of transmembrane domains, and thus it is not clear that resistance is due to efflux or perhaps to other mechanisms [186].

Enzymatic modification of lincosamides as a mechanism of resistance has also been described. Bozdogan et al. [192] discovered a 3-lincosamide *O*-nucleotidyltransferase, which adenylates the 3-hydroxyl groups of lincomycin and clindamycin. The gene, *linB*, encoding production of this enzyme was detected in all 14 *E. faecium* strains that (among 110 isolates tested) inactivated clindamycin by Gots' test.

Streptogramins

Quinupristin–dalfopristin is a combination drug consisting of semi-synthetic derivatives of natural antibiotics of the streptogramin B and streptogramin A classes, respectively. It has been approved to treat infections due to *E. faecium* [193]; however, isolates of *E. faecalis* are almost universally resistant to the drug [194]. As mentioned previously (see the section Lincosamides), the product of the *lsa* gene found in the latter species confers resistance to streptogramin A antibiotics [190]. Dina et al. [191] described two urine isolates of *E. faecalis* that were uncharacteristically susceptible to clindamycin, dalfopristin, and to quinupristin–dalfopristin. The gene sequence of *lsa* contained mutations that generated premature stop codons in both strains.

The streptogramin A component is susceptible to inactivation by acetyltransferases, mediated by genes *vat*(D) (previously *satA*) and *vat*(E) (previously *satG*) found in enterococci from human, animal, food, or sewage sources [195–200]. Other mechanisms of dalfopristin resistance have been described in staphylococci, but not yet in enterococci. These include *vga*(A) and *vga*(B), which mediate resistance to streptogramin A compounds by putative ATP-binding cassette efflux mechanisms [201,202]. VgaA belongs to the same sub-family of ABC systems as MsrA [187]. The rRNA methyltransferase encoded by *cfr* results in methylation of 23S rRNA at position A2503 in the peptidyltransferase center of the ribosome, conferring resistance to chloramphenicol, lincosamides, pleuromutilins, oxazolidinones, and streptogramin A compounds in staphylococci [203].

The expression of *erm*(B) genes with resulting methylation of the ribosomal target confers resistance to the streptogramin B class, in addition to macrolides and lincosamides. The gene *vgb*, encoding a streptogramin B hydrolyzing enzyme, has been found in enterococci [204]. One isolate of *E. faecium* that was resistant to both vancomycin and quinupristin–dalfopristin was studied intensively. The organism contained the genes *vanA*, *vat*(D), *erm*(B), and *vgb* [205,206]. The first three of these resistance traits were co-transferred on a single plasmid.

The two components of quinupristin–dalfopristin interact synergistically [207]. In *E. faecium*, resistance to quinupristin alone abolishes bactericidal activity of the drug; inhibitory activity is maintained unless the organism is resistant to dalfopristin as well as to quinupristin [208]. In pneumococci and *S. aureus*, mutations affecting ribosomal proteins can confer resistance to quinupristin–dalfopristin [209,210].

CHLORAMPHENICOL

Although chloramphenicol is not widely used in the United States, the drug is active *in vitro* against many isolates of multiply drug-resistant VRE [7], and it has been employed with some success to treat enterococcal bloodstream infections and meningitis [211–215]. Susceptibility of enterococci to chloramphenicol varies by geographic region. Of more than 1000 enterococcal isolates collected in the year 2000 in North America, 87% were susceptible to chloramphenicol; the same year, approximately 70% of enterococci from Europe and Latin America were susceptible [216]. A Philadelphia group noted an increase in the proportion of VRE resistant to chloramphenicol in the decade of the 1990s, from 0% to 12% [217]. Prior chloramphenicol use and prior fluoroquinolone use were risk factors for bloodstream infection with a chloramphenicol-resistant VRE.

Chloramphenicol is inactivated by chloramphenicol acetyltransferases, which can be inducible, and encoded by genes on conjugative or non-conjugative resistance plasmids or on the chromosome of enterococci [177,218,219]. Lynch et al. [220] provided evidence for energy-driven efflux of chloramphenicol in *E. faecalis* and *E. faecium*, even for strains with MICs in the susceptible range.

TETRACYCLINES

Resistance to tetracycline is common among enterococci. Examination of more than 3000 isolates collected between 2000 and 2004 revealed that only 38.4% were susceptible to tetracycline [221]. Resistance is most commonly associated with *tet*(M), usually found on the conjugative transposon, Tn*916* [222,223]. By a ribosomal protection mechanism, Tet(M) confers resistance to both tetracycline and to minocycline. Tn*916*-like structures have been found on large chromosomal elements of *E. faecium*, capable of co-transferring several antibiotic resistance traits, including vancomycin and ampicillin resistance [35]. The ribosomal protection mechanism genes *tet*(O) and *tet*(S) have also been found in enterococci, but appear to be rare [224,225]. The *tet*(S) gene has been found on a chromosomally situated conjugative transposon in *E. faecium* [226].

Another mechanism of tetracycline resistance in enterococci is drug efflux [227]. Most commonly, efflux is mediated by the *tet*(L) gene, although *tet*(K) has been found [225,228]. Combinations of efflux genes with one or more ribosomal protection genes [e.g., *tet*(L) + *tet*(M) + *tet*(O)] have also been documented [225]. In contrast to tetracycline and minocycline, tigecycline, which is a member of the glycylcycline class, is not significantly affected by ribosomal protection or by tetracycline efflux mechanisms; it thus inhibits enterococci resistant to the earlier compounds [229]. In the collection of more than 3000 enterococcal isolates mentioned above, approximately 93% were susceptible to tigecycline [221]. Mechanisms accounting for the apparently reduced susceptibility of the remaining 7% of isolates have not yet been reported.

OXAZOLIDINONES

Currently, only one member of this antimicrobial class, linezolid, is approved for clinical use. It is often employed for treatment of infections due to hospital-associated strains of VRE, which are typically resistant to penicillins as well as to glycopeptides. Large surveillance studies document that resistance to linezolid among enterococci is rare [58,230–233]. Nevertheless, resistance has emerged among clinical isolates, and nosocomial transmission has been well described [234–239]. This topic has been reviewed by Meka and Gold [240].

Prystowsky et al. [241] studied linezolid-resistant enterococci selected *in vitro* during serial passage on the antimicrobial. Resistance was selected in five of five isolates of *E. faecalis*, with MICs rising from 1 µg/mL to 128 µg/mL. For *E. faecium*, the MIC rose from 1 µg/mL to 16 µg/mL in one of five strains and from 1 µg/mL to 8 µg/mL in two others. In the gene for 23S rRNA in the *E. faecalis* mutant, a $G \rightarrow U$ transversion at bp 2576 was documented, and in the most resistant *E. faecium* mutant a $G \rightarrow A$ transition at bp 2505 was seen. Both of these mutations are in domain V of the ribosomal peptidyltransferase center. Among clinical isolates of linezolid-resistant *E. faecium*, however, the point mutation at bp 2576 is the one that has been observed [236,242–247].

Enterococci contain multiple copies of 23S rRNA genes, and examination of clinical isolates of *E. faecalis* with varying degrees of resistance to linezolid has demonstrated a rough correlation between the proportion of copies bearing a G2576T mutation and the level of oxazolidinone resistance [243]. Rather than representing independent mutational events, the presence of multiple copies of the mutant gene is thought to result from gene conversion, that is, the exchange of mutated copies for wild-type copies under selective antimicrobial pressure. To test this hypothesis, Lobritz et al. [248] attempted to select linezolid-resistant mutants from a strain of *E. faecalis* and from a related, recombination-defective strain. From the recombination-proficient strain, mutants were easily derived with four of four copies of the 23S rRNA gene demonstrating a point mutation at bp 2576. The resulting linezolid MIC of the mutants was 128 µg/mL. From the recombination-defective strain, growth was achieved only on linezolid concentrations of 8 µg/mL or less. A point mutation, G2505A, was present in only one copy of the 23S rRNA gene, indicating that in this setting of defective recombination, the mutated copy could not serve as a template to increase the proportion of resistant copies by gene conversion.

Woodford et al. [249] used real-time PCR to detect an enterococcal strain that was susceptible to linezolid (MIC, 4 µg/mL), but nevertheless contained a G2576T polymorphism. His group subsequently confirmed that result by pyrosequencing, showing that one to two mutant copies were present in this linezolid-susceptible isolate of *E. faecium* [250]. The significance of this observation is that the presence of a single mutant 23S rRNA gene copy may escape detection, because the isolate remains susceptible; however, the organism retains a template for recombination that could result in a higher proportion of mutant copies and clinical resistance upon exposure to the drug.

RESISTANCE TO FLUOROQUINOLONES

A surveillance study carried out shortly after the introduction of ciprofloxacin revealed that approximately 88% of *E. faecalis* and 50% of *E. faecium* were susceptible to the fluoroquinolone at concentrations ≤1 μg/mL [251]. By the year 2000, only 39% of more than 1000 isolates from North America were susceptible at this concentration [216]. Gatifloxacin was somewhat more active, with 51% of strains in that collection susceptible to the drug. In a collection of more than 3000 isolates recovered between 2000 and 2004, only 50.8% were susceptible to levofloxacin [221]. Clearly, resistance to fluoroquinolone antimicrobials is widespread in the genus *Enterococcus*.

Although additional mechanisms of fluoroquinolone resistance have been described in Gram-negative bacteria, in enterococci resistance is caused by mutations in genes encoding the topoisomerase target proteins, by drug efflux, or by the combination of these mechanisms. Mutations affecting *parC*, encoding a subunit of topoisomerase IV, and *gyrA*, encoding the A subunit of DNA gyrase, are the ones most commonly associated with resistance to this class [252–256]. These mutations are not random, but tend to cluster in certain areas of the genes, referred to as quinolone resistance determining regions. The most resistant isolates typically have one or more mutations affecting genes for both the topoisomerase IV and DNA gyrase enzymes [255,256].

Lynch et al. [220] demonstrated active efflux of norfloxacin from both *E. faecalis* and *E. faecium*. An enterococcal efflux pump encoded by a homolog of the staphylococcal *norA* gene, designated *emeA*, was identified in *E. faecalis* [257]. Deletion of *emeA* resulted in an approximately two-fold increase in susceptibility to norfloxacin and ciprofloxacin, among other substances. Genes from *E. faecalis*, designated *efrAB*, encode an ABC efflux pump, the presence of which results in about a four-fold increase in MICs of norfloxacin and ciprofloxacin when expressed in *E. coli* [258]. It is likely that both target mutations and one or more drug efflux pumps contribute to the ultimate level of fluoroquinolone resistance in any one enterococcal isolate [259,260].

CONCLUSIONS

As the preceding discussion illustrates, enterococci have amply demonstrated a remarkable repertoire of resistance mechanisms offering protection against new and old antimicrobial agents alike. Davis et al. [261], comparing sequences from the *E. faecalis* V583 genome with amino acid sequences of known multidrug resistance transporters, identified 34 candidate genes for multidrug efflux pumps, 23 of which belonged to the ABC family of transporters. Another analysis of the same genome by Paulsen et al. [262] concluded that more than a quarter of the genetic complement represents mobile or exogenous DNA, including three plasmids, and a variety of transposons, phage genes, and integrated plasmids. From such observations, one might reasonably conclude that there may be endless possibilities for enterococci to develop mechanisms ensuring survival under the most hostile conditions, including antimicrobial therapy. At the same time, study of antimicrobial resistance in enterococci continues to reveal previously unknown and most intriguing mechanisms of resistance, which will very likely give rise over time to new drug targets and insightful approaches for treating patients suffering from infections due to these organisms.

REFERENCES

1. McDonald JR, Olaison L, Anderson DJ, et al. Enterococcal endocarditis: 107 cases from the international collaboration on endocarditis merged database. *Am J Med* 2005;118:759–66.
2. Storch GA and Krogstad DJ. Antibiotic-induced lysis of enterococci. *J Clin Invest* 1981;68:639–45.
3. Baddour LM, Wilson WR, Bayer AS, et al. Infective endocarditis: diagnosis, antimicrobial therapy, and management of complications: a statement for healthcare professionals from the Committee on Rheumatic Fever, Endocarditis, and Kawasaki Disease, Council on Cardiovascular Disease in the Young, and the Councils on Clinical Cardiology, Stroke, and Cardiovascular Surgery and Anesthesia, American Heart Association: endorsed by the Infectious Diseases Society of America. *Circulation* 2005;111: e394–434.
4. Eliopoulos GM. Aminoglycoside resistant enterococcal endocarditis. *Infect Dis Clin North Am* 1993;7:117–33.
5. Sader HS, Jones RN, Dowzicky MJ, and Fritsche TR. Antimicrobial activity of tigecycline tested against nosocomial bacterial pathogens from patients hospitalized in the intensive care unit. *Diagn Microbiol Infect Dis* 2005;52:203–8.
6. National Nosocomial Infections Surveillance (NNIS) System Report, data summary from January 1992 through June 2004, issued October 2004. *Am J Infect Control* 2004;32:470–85.
7. Eliopoulos GM, Wennersten CB, Gold HS, et al. Characterization of vancomycin-resistant *Enterococcus faecium* isolates from the United States and their susceptibility *in vitro* to dalfopristin-quinupristin. *Antimicrob Agents Chemother* 1998;42:1088–92.
8. McNeil SA, Malani PN, Chenoweth CE, et al. Vancomycin-resistant enterococcal colonization and infection in liver transplant candidates and recipients: a prospective surveillance study. *Clin Infect Dis* 2006;42:195–203.
9. Avery R, Kalaycio M, Pohlman B, et al. Early vancomycin-resistant enterococcus (VRE) bacteremia after allogeneic bone marrow transplantation is associated with a rapidly deteriorating clinical course. *Bone Marrow Transplant* 2005;35:497–9.
10. Salgado CD and Farr BM. Outcomes associated with vancomycin-resistant enterococci: a meta-analysis. *Infect Control Hosp Epidemiol* 2003;24:690–8.
11. DiazGranados CA and Jernigan JA. Impact of vancomycin resistance on mortality among patients with neutropenia and enterococcal bloodstream infection. *J Infect Dis* 2005;191:588–95.
12. Appelbaum PC. The emergence of vancomycin-intermediate and vancomycin-resistant *Staphylococcus aureus*. *Clin Microbiol Infect* 2006;12 Suppl 1:16–23.
13. Noble WC, Virani Z, and Cree RG. Co-transfer of vancomycin and other resistance genes from *Enterococcus faecalis* NCTC 12201 to *Staphylococcus aureus*. *FEMS Microbiol Lett* 1992;72:195–8.
14. Williamson R, le Bouguenec C, Gutmann L, and Horaud T. One or two low affinity penicillin-binding proteins may be responsible for the range of susceptibility of *Enterococcus faecium* to benzylpenicillin. *J Gen Microbiol* 1985;131:1933–40.
15. al-Obeid S, Gutmann L, and Williamson R. Modification of penicillin-binding proteins of penicillin-resistant mutants of different species of enterococci. *J Antimicrob Chemother* 1990;26:613–8.

16. Grayson ML, Eliopoulos GM, Wennersten CB, et al. Comparison of *Enterococcus raffinosus* with *Enterococcus avium* on the basis of penicillin susceptibility, penicillin-binding protein analysis, and high-level aminoglycoside resistance. *Antimicrob Agents Chemother* 1991;35:1408–12.

17. Fontana R, Cerini R, Longoni P, Grossato A, and Canepari P. Identification of a streptococcal penicillin-binding protein that reacts very slowly with penicillin. *J Bacteriol* 1983;155:1343–50.

18. Piras G, el Kharroubi A, van Beeumen J, Coeme E, Coyette J, and Ghuysen JM. Characterization of an Enterococcus hirae penicillin-binding protein 3 with low penicillin affinity. *J Bacteriol* 1990;172;6856–62.

19. Fliopoulos GM, Wennersten CB, Gold HS, and Moellering RC, Jr. In vitro activities in new oxazolidinone antimicrobial agents against enterococci. *Antimicrob Agents Chemother* 1996;40:1745–7.

20. Arbeloa A, Segal H, Hugonnet JE, et al. Role of class A penicillin-binding proteins in PBP5-mediated β-lactam resistance in *Enterococcus faecalis*. *J Bacteriol* 2004; 186:1221–8.

21. Comenge Y, Quintiliani R, Jr., Li L, et al. The CroRS two-component regulatory system is required for intrinsic β-lactam resistance in *Enterococcus faecalis*. *J Bacteriol* 2003;185:7184–92.

22. Muller C, Le Breton Y, Morin T, Benachour A, Auffray Y, and Rince A. The response regulator CroR modulates expression of the secreted stress-induced SalB protein in *Enterococcus faecalis*. *J Bacteriol* 2006;188:2636–45.

23. Cercenado E, Vicente MF, Diaz MD, Sanchez-Carrillo C, and Sanchez-Rubiales M. Characterization of clinical isolates of β-lactamase-negative, highly ampicillin-resistant *Enterococcus faecalis*. *Antimicrob Agents Chemother* 1996;40:2420–2.

24. Ono S, Muratani T, and Matsumoto T. Mechanisms of resistance to imipenem and ampicillin in *Enterococcus faecalis*. *Antimicrob Agents Chemother* 2005;49: 2954–8.

25. Metzidie E, Manolis EN, Pournaras S, Sofianou D, and Tsakris A. Spread of an unusual penicillin- and imipenem-resistant but ampicillin-susceptible phenotype among *Enterococcus faecalis* clinical isolates. *J Antimicrob Chemother* 2006;57:158–60.

26. Grayson ML, Eliopoulos GM, Wennersten CB, et al. Increasing resistance to β-lactam antibiotics among clinical isolates of *Enterococcus faecium*: a 22-year review at one institution. *Antimicrob Agents Chemother* 1991;35:2180–4.

27. Sifaoui F, Arthur M, Rice L, and Gutmann L. Role of penicillin-binding protein 5 in expression of ampicillin resistance and peptidoglycan structure in *Enterococcus faecium*. *Antimicrob Agents Chemother* 2001;45:2594–7.

28. Rice LB, Carias LL, Hutton-Thomas R, Sifaoui F, Gutmann L, and Rudin SD. Penicillin-binding protein 5 and expression of ampicillin resistance in *Enterococcus faecium*. *Antimicrob Agents Chemother* 2001;45:1480–6.

29. Zorzi W, Zhou XY, Dardenne O, et al. Structure of the low-affinity penicillin-binding protein 5 PBP5fm in wild-type and highly penicillin-resistant strains of *Enterococcus faecium*. *J Bacteriol* 1996;178:4948–57.

30. Fontana R, Aldegheri M, Ligozzi M, Lopez H, Sucari A, and Satta G. Overproduction of a low-affinity penicillin-binding protein and high-level ampicillin resistance in *Enterococcus faecium*. *Antimicrob Agents Chemother* 1994;38:1980–3.

31. Rybkine T, Mainardi JL, Sougakoff W, Collatz E, and Gutmann L. Penicillin-binding protein 5 sequence alterations in clinical isolates of *Enterococcus faecium* with different levels of β-lactam resistance. *J Infect Dis* 1998;178:159–63.

32. Ligozzi M, Pittaluga F, and Fontana R. Modification of penicillin-binding protein 5 associated with high-level ampicillin resistance in *Enterococcus faecium*. *Antimicrob Agents Chemother* 1996;40:354–7.

33. Rice LB, Bellais S, Carias LL, et al. Impact of specific pbp5 mutations on expression of β-lactam resistance in *Enterococcus faecium. Antimicrob Agents Chemother* 2004;48:3028–32.

34. Rice LB, Carias LL, Rudin S, Lakticova V, Wood A, and Hutton-Thomas R. *Enterococcus faecium* low-affinity pbp5 is a transferable determinant. *Antimicrob Agents Chemother* 2005;49:5007–12.

35. Carias LL, Rudin SD, Donskey CJ, and Rice LB. Genetic linkage and cotransfer of a novel, vanB-containing transposon (Tn*5382*) and a low-affinity penicillin-binding protein 5 gene in a clinical vancomycin-resistant *Enterococcus faecium* isolate. *J Bacteriol* 1998;180:4426–34.

36. Hanrahan J, Hoyen C, and Rice LB. Geographic distribution of a large mobile element that transfers ampicillin and vancomycin resistance between *Enterococcus faecium* strains. *Antimicrob Agents Chemother* 2000;44:1349–51.

37. Mainardi JL, Legrand R, Arthur M, Schoot B, van Heijenoort J, and Gutmann L. Novel mechanism of β-lactam resistance due to bypass of DD-transpeptidation in *Enterococcus faecium. J Biol Chem* 2000;275:16490–6.

38. Mainardi JL, Morel V, Fourgeaud M, et al. Balance between two transpeptidation mechanisms determines the expression of β-lactam resistance in *Enterococcus faecium. J Biol Chem* 2002;277:35801–7.

39. Mainardi JL, Fourgeaud M, Hugonnet JE, et al. A novel peptidoglycan cross-linking enzyme for a β-lactam-resistant transpeptidation pathway. *J Biol Chem* 2005; 280:38146–52.

40. Biarrotte-Sorin S, Hugonnet JE, Delfosse V, et al. Crystal structure of a novel β-lactam-insensitive peptidoglycan transpeptidase. *J Mol Biol* 2006;359:533–8.

41. Murray BE and Mederski-Samaroj B. Transferable β-lactamase. A new mechanism for *in vitro* penicillin resistance in *Streptococcus faecalis. J Clin Invest* 1983;72:1168–71.

42. Murray BE, Church DA, Wanger A, et al. Comparison of two β-lactamase-producing strains of *Streptococcus faecalis. Antimicrob Agents Chemother* 1986;30:861–4.

43. Murray BE, Singh KV, Markowitz SM, et al. Evidence for clonal spread of a single strain of β-lactamase-producing *Enterococcus (Streptococcus) faecalis* to six hospitals in five states. *J Infect Dis* 1991;163:780–5.

44. Murray BE, Lopardo HA, Rubeglio EA, Frosolono M, and Singh KV. Intrahospital spread of a single gentamicin-resistant, β-lactamase-producing strain of *Enterococcus faecalis* in Argentina. *Antimicrob Agents Chemother* 1992;36:230–2.

45. Coudron PE, Markowitz SM, and Wong ES. Isolation of a β-lactamase-producing, aminoglycoside-resistant strain of *Enterococcus faecium. Antimicrob Agents Chemother* 1992;36:1125–6.

46. McAlister T, George N, Faoagali J, and Bell J. Isolation of β-lactamase positive vancomycin resistant *Enterococcus faecalis*; first case in Australia. *Commun Dis Intell* 1999;23:237–9.

47. Zhou JD, Guo JJ, Zhang Q, Chen Y, Zhu SH, and Peng HY. Drug resistance of infectious pathogens after liver transplantation. *Hepatobiliary Pancreat Dis Int* 2006;5:190–4.

48. Rhinehart E, Smith NE, Wennersten C, et al. Rapid dissemination of β-lactamase-producing, aminoglycoside-resistant *Enterococcus faecalis* among patients and staff on an infant-toddler surgical ward. *N Engl J Med* 1990;323:1814–8.

49. Markowitz SM, Wells VD, Williams DS, Stuart CG, Coudron PE, and Wong ES. Antimicrobial susceptibility and molecular epidemiology of β-lactamase-producing, aminoglycoside-resistant isolates of *Enterococcus faecalis. Antimicrob Agents Chemother* 1991;35:1075–80.

50. Murray BE, Mederski-Samoraj B, Foster SK, Brunton JL, and Harford P. *In vitro* studies of plasmid-mediated penicillinase from *Streptococcus faecalis* suggest a staphylococcal origin. *J Clin Invest* 1986;77:289–93.

51. Murray BE, An FY, and Clewell DB. Plasmids and pheromone response of the β-lactamase producer *Streptococcus (Enterococcus) faecalis* HH22. *Antimicrob Agents Chemother* 1988;32:547–51.
52. Tomayko JF, Zscheck KK, Singh KV, and Murray BE. Comparison of the β-lactamase gene cluster in clonally distinct strains of *Enterococcus faecalis*. *Antimicrob Agents Chemother* 1996;40:1170–4.
53. Okamoto R, Okubo T, and Inoue M. Detection of genes regulating β-lactamase production in *Enterococcus faecalis* and *Staphylococcus aureus*. *Antimicrob Agents Chemother* 1996;40:2550–4.
54. Rice LB, Eliopoulos GM, Wennersten C, Goldmann D, Jacoby GA, and Moellering RC, Jr. Chromosomally mediated β-lactamase production and gentamicin resistance in *Enterococcus faecalis*. *Antimicrob Agents Chemother* 1991;35:272–6.
55. Rice LB and Carias LL. Transfer of Tn*5385*, a composite, multiresistance chromosomal element from *Enterococcus faecalis*. *J Bacteriol* 1998;180:714–21.
56. Nallapareddy SR, Wenxiang H, Weinstock GM, and Murray BE. Molecular characterization of a widespread, pathogenic, and antibiotic resistance-receptive *Enterococcus faecalis* lineage and dissemination of its putative pathogenicity island. *J Bacteriol* 2005;187:5709–18.
57. Edmond MB, Wallace SE, McClish DK, Pfaller MA, Jones RN, and Wenzel RP. Nosocomial bloodstream infections in United States hospitals: a three-year analysis. *Clin Infect Dis* 1999;29:239–44.
58. Draghi DC, Sheehan DJ, Hogan P, and Sahm DF. *In vitro* activity of linezolid against key Gram-positive organisms isolated in the United States: results of the LEADER 2004 surveillance program. *Antimicrob Agents Chemother* 2005;49:5024–32.
59. Fridkin SK, Yokoe DS, Whitney CG, Onderdonk A, and Hooper DC. Epidemiology of a dominant clonal strain of vancomycin-resistant *Enterococcus faecium* at separate hospitals in Boston, Massachusetts. *J Clin Microbiol* 1998;36:965–70.
60. Willems RJ, Top J, van Santen M, et al. Global spread of vancomycin-resistant *Enterococcus faecium* from distinct nosocomial genetic complex. *Emerg Infect Dis* 2005; 11:821–8.
61. Hayden MK. Insights into the epidemiology and control of infection with vancomycin-resistant enterococci. *Clin Infect Dis* 2000;31:1058–65.
62. Zirakzadeh A and Patel R. Vancomycin-resistant enterococci: colonization, infection, detection, and treatment. *Mayo Clin Proc* 2006;81:529–36.
63. Martinez JA, Ruthazer R, Hansjosten K, Barefoot L, and Snydman DR. Role of environmental contamination as a risk factor for acquisition of vancomycin-resistant enterococci in patients treated in a medical intensive care unit. *Arch Intern Med* 2003; 163:1905–12.
64. Baden LR, Thiemke W, Skolnik A, et al. Prolonged colonization with vancomycin-resistant *Enterococcus faecium* in long-term care patients and the significance of "clearance." *Clin Infect Dis* 2001;33:1654–60.
65. Leclercq R. Enterococci acquire new kinds of resistance. *Clin Infect Dis* 1997;24 Suppl 1:S80–4.
66. Quintiliani R, Jr., Evers S, and Courvalin P. The *vanB* gene confers various levels of self-transferable resistance to vancomycin in enterococci. *J Infect Dis* 1993; 167:1220–3.
67. Hayden MK, Trenholme GM, Schultz JE, and Sahm DF. *In vivo* development of teicoplanin resistance in a VanB *Enterococcus faecium* isolate. *J Infect Dis* 1993; 167:1224–7.
68. Dutka-Malen S, Blaimont B, Wauters G, and Courvalin P. Emergence of high-level resistance to glycopeptides in *Enterococcus gallinarum* and *Enterococcus casseliflavus*. *Antimicrob Agents Chemother* 1994;38:1675–7.

69. Liassine N, Frei R, Jan I, and Auckenthaler R. Characterization of glycopeptide-resistant enterococci from a Swiss hospital. *J Clin Microbiol* 1998;36:1853–8.
70. Poulsen RL, Pallesen LV, Frimodt-Moller N, and Espersen F. Detection of clinical vancomycin-resistant enterococci in Denmark by multiplex PCR and sandwich hybridization. *Apmis* 1999;107:404–12.
71. Courvalin P. Vancomycin resistance in Gram-positive cocci. *Clin Infect Dis* 2006;42 Suppl 1:S25–34.
72. Leclercq R, Derlot E, Duval J, and Courvalin P. Plasmid-mediated resistance to vancomycin and teicoplanin in *Enterococcus faecium*. *N Engl J Med* 1988; 319:157–61.
73. Leclercq R and Courvalin P. Resistance to glycopeptides in enterococci. *Clin Infect Dis* 1997;24:545–54.
74. Lester CH, Frimodt-Moller N, Sorensen TL, Monnet DL, and Hammerum AM. *In vivo* transfer of the *vanA* resistance gene from an *Enterococcus faecium* isolate of animal origin to an *E. faecium* isolate of human origin in the intestines of human volunteers. *Antimicrob Agents Chemother* 2006;50:596–9.
75. Bugg TD, Wright GD, Dutka-Malen S, Arthur M, Courvalin P, and Walsh CT. Molecular basis for vancomycin resistance in *Enterococcus faecium* BM4147: biosynthesis of a depsipeptide peptidoglycan precursor by vancomycin resistance proteins VanH and VanA. *Biochemistry* 1991;30:10408–15.
76. Reynolds PE, Depardieu F, Dutka-Malen S, Arthur M, and Courvalin P. Glycopeptide resistance mediated by enterococcal transposon Tn*1546* requires production of VanX for hydrolysis of D-alanyl-D-alanine. *Mol Microbiol* 1994;13:1065–70.
77. Wu Z, Wright GD, and Walsh CT. Overexpression, purification, and characterization of VanX, a D-, D-dipeptidase which is essential for vancomycin resistance in *Enterococcus faecium* BM4147. *Biochemistry* 1995;34:2455–63.
78. Arthur M, Depardieu F, Gerbaud G, Galimand M, Leclercq R, and Courvalin P. The VanS sensor negatively controls VanR-mediated transcriptional activation of glycopeptide resistance genes of Tn*1546* and related elements in the absence of induction. *J Bacteriol* 1997;179:97–106.
79. Arthur M, Depardieu F, Reynolds P, and Courvalin P. Quantitative analysis of the metabolism of soluble cytoplasmic peptidoglycan precursors of glycopeptide-resistant enterococci. *Mol Microbiol* 1996;21:33–44.
80. Holman TR, Wu Z, Wanner BL, and Walsh CT. Identification of the DNA-binding site for the phosphorylated VanR protein required for vancomycin resistance in *Enterococcus faecium*. *Biochemistry* 1994;33:4625–31.
81. Haldimann A, Fisher SL, Daniels LL, Walsh CT, and Wanner BL. Transcriptional regulation of the *Enterococcus faecium* BM4147 vancomycin resistance gene cluster by the VanS-VanR two-component regulatory system in *Escherichia coli* K-12. *J Bacteriol* 1997;179:5903–13.
82. Arthur M, Depardieu F, Snaith HA, Reynolds PE, and Courvalin P. Contribution of VanY D,D-carboxypeptidase to glycopeptide resistance in *Enterococcus faecalis* by hydrolysis of peptidoglycan precursors. *Antimicrob Agents Chemother* 1994; 38:1899–903.
83. Arthur M, Depardieu F, Molinas C, Reynolds P, and Courvalin P. The *vanZ* gene of Tn*1546* from *Enterococcus faecium* BM4147 confers resistance to teicoplanin. *Gene* 1995;154:87–92.
84. Pootoolal J, Thomas MG, Marshall CG, et al. Assembling the glycopeptide antibiotic scaffold: The biosynthesis of A47934 from *Streptomyces toyocaensis* NRRL15009. *Proc Natl Acad Sci USA* 2002;99:8962–7.
85. Patel R. Enterococcal-type glycopeptide resistance genes in non-enterococcal organisms. *FEMS Microbiol Lett* 2000;185:1–7.

86. Patel R, Piper K, Cockerill FR, 3rd, Steckelberg JM, and Yousten AA. The biopesticide *Paenibacillus popilliae* has a vancomycin resistance gene cluster homologous to the enterococcal VanA vancomycin resistance gene cluster. *Antimicrob Agents Chemother* 2000;44:705–9.

87. Fraimow H, Knob C, Herrero IA, and Patel R. Putative VanRS-like two-component regulatory system associated with the inducible glycopeptide resistance cluster of *Paenibacillus popilliae*. *Antimicrob Agents Chemother* 2005;49:2625–33.

88. Guardabassi L, Christensen H, Hasman H, and Dalsgaard A. Members of the genera *Paenibacillus* and *Rhodococcus* harbor genes homologous to enterococcal glycopeptide resistance genes *vanA* and *vanB*. *Antimicrob Agents Chemother* 2004;48:4915–8.

89. Guardabassi I., Perichon B, van Heijenoort J, Blanot D, and Courvalin P. Glycopeptide resistance vanA operons in *Paenibacillus* strains isolated from soil. *Antimicrob Agents Chemother* 2005;49:4227–33.

90. Dahl KH, Simonsen GS, Olsvik O, and Sundsfjord A. Heterogeneity in the vanB gene cluster of genomically diverse clinical strains of vancomycin-resistant enterococci. *Antimicrob Agents Chemother* 1999;43:1105–10.

91. Evers S, Courvalin P. Regulation of VanB-type vancomycin resistance gene expression by the VanS(B)-VanR (B) two-component regulatory system in *Enterococcus faecalis* V583. *J Bacteriol* 1996;178:1302–9.

92. Baptista M, Depardieu F, Courvalin P, and Arthur M. Specificity of induction of glycopeptide resistance genes in *Enterococcus faecalis*. *Antimicrob Agents Chemother* 1996;40:2291–5.

93. Baptista M, Rodrigues P, Depardieu F, Courvalin P, and Arthur M. Single-cell analysis of glycopeptide resistance gene expression in teicoplanin-resistant mutants of a VanB-type *Enterococcus faecalis*. *Mol Microbiol* 1999;32:17–28.

94. Quintiliani R, Jr., Courvalin P. Conjugal transfer of the vancomycin resistance determinant vanB between enterococci involves the movement of large genetic elements from chromosome to chromosome. *FEMS Microbiol Lett* 1994;119:359–63.

95. Quintiliani R, Jr. and Courvalin P. Characterization of Tn*1547*, a composite transposon flanked by the IS16 and IS256-like elements, that confers vancomycin resistance in *Enterococcus faecalis* BM4281. *Gene* 1996;172:1–8.

96. Lu JJ, Chang TY, Perng CL, and Lee SY. The *vanB2* gene cluster of the majority of vancomycin-resistant *Enterococcus faecium* isolates from Taiwan is associated with the *pbp5* gene and is carried by Tn*5382* containing a novel insertion sequence. *Antimicrob Agents Chemother* 2005;49:3937–9.

97. Stinear TP, Olden DC, Johnson PD, Davies JK, and Grayson ML. Enterococcal *vanB* resistance locus in anaerobic bacteria in human faeces. *Lancet* 2001;357:855–6.

98. Ballard SA, Pertile KK, Lim M, Johnson PD, and Grayson ML. Molecular characterization of *vanB* elements in naturally occurring gut anaerobes. *Antimicrob Agents Chemother* 2005;49:1688–94.

99. Launay A, Ballard SA, Johnson PD, Grayson ML, and Lambert T. Transfer of vancomycin resistance transposon Tn*1549* from *Clostridium symbiosum* to *Enterococcus* spp. in the gut of gnotobiotic mice. *Antimicrob Agents Chemother* 2006;50:1054–62.

100. Patel R, Uhl JR, Kohner P, Hopkins MK, and Cockerill FR, 3rd. Multiplex PCR detection of *vanA*, *vanB*, *vanC-1*, and *vanC-2/3* genes in enterococci. *J Clin Microbiol* 1997;35:703–7.

101. Reynolds PE and Courvalin P. Vancomycin resistance in enterococci due to synthesis of precursors terminating in D-alanyl-D-serine. *Antimicrob Agents Chemother* 2005;49:21–5.

102. Reynolds PE, Snaith HA, Maguire AJ, Dutka-Malen S, and Courvalin P. Analysis of peptidoglycan precursors in vancomycin-resistant *Enterococcus gallinarum* BM4174. *Biochem J* 1994;301 (Pt 1):5–8.

103. Arias CA, Martin-Martinez M, Blundell TL, Arthur M, Courvalin P, and Reynolds PE. Characterization and modelling of VanT: a novel, membrane-bound, serine racemase from vancomycin-resistant *Enterococcus gallinarum* BM4174. *Mol Microbiol* 1999; 31:1653–64.

104. Ambur OH, Reynolds PE, and Arias CA. D-Ala:D-Ala ligase gene flanking the *vanC* cluster: evidence for presence of three ligase genes in vancomycin-resistant *Enterococcus gallinarum* BM4174. *Antimicrob Agents Chemother* 2002;46:95–100.

105. Reynolds PE, Arias CA, and Courvalin P. Gene *vanXYC* encodes D,D-dipeptidase (VanX) and D,D-carboxypeptidase (VanY) activities in vancomycin-resistant *Enterococcus gallinarum* BM4174. *Mol Microbiol* 1999;34:341–9.

106. Panesso D, Abadia-Patino L, Vanegas N, Reynolds PE, Courvalin P, and Arias CA. Transcriptional analysis of the *vanC* cluster from *Enterococcus gallinarum* strains with constitutive and inducible vancomycin resistance. *Antimicrob Agents Chemother* 2005;49:1060–6.

107. Corso A, Faccone D, Gagetti P, et al. First report of VanA *Enterococcus gallinarum* dissemination within an intensive care unit in Argentina. *Int J Antimicrob Agents* 2005;25:51–6.

108. Camargo IL, Barth AL, Pilger K, Seligman BG, Machado AR, and Darini AL. *Enterococcus gallinarum* carrying the *vanA* gene cluster: first report in Brazil. *Braz J Med Biol Res* 2004;37:1669–71.

109. Mammina C, Di Noto AM, Costa A, and Nastasi A. VanB-VanC1 *Enterococcus gallinarum*, Italy. *Emerg Infect Dis* 2005;11:1491–2.

110. Schooneveldt JM, Marriott RK, and Nimmo GR. Detection of a *vanB* determinant in *Enterococcus gallinarum* in Australia. *J Clin Microbiol* 2000;38:3902.

111. Iaria C, Stassi G, Costa GB, Di Leo R, Toscano A, and Cascio A. Enterococcal meningitis caused by *Enterococcus casseliflavus*. First case report. *BMC Infect Dis* 2005;5:3.

112. Takayama Y, Sunakawa K, and Akahoshi T. Meningitis caused by *Enterococcus gallinarum* in patients with ventriculoperitoneal shunts. *J Infect Chemother* 2003; 9:348–50.

113. Choi SH, Lee SO, Kim TH, et al. Clinical features and outcomes of bacteremia caused by *Enterococcus casseliflavus* and *Enterococcus gallinarum*: analysis of 56 cases. *Clin Infect Dis* 2004;38:53–61.

114. Toye B, Shymanski J, Bobrowska M, Woods W, and Ramotar K. Clinical and epidemiological significance of enterococci intrinsically resistant to vancomycin (possessing the *vanC* genotype). *J Clin Microbiol* 1997;35:3166–70.

115. Dargere S, Vergnaud M, Verdon R, et al. *Enterococcus gallinarum* endocarditis occurring on native heart valves. *J Clin Microbiol* 2002;40:2308–10.

116. Depardieu F, Perichon B, and Courvalin P. Detection of the van alphabet and identification of enterococci and staphylococci at the species level by multiplex PCR. *J Clin Microbiol* 2004;42:5857–60.

117. Woodford N. Epidemiology of the genetic elements responsible for acquired glycopeptide resistance in enterococci. *Microb Drug Resist* 2001;7:229–36.

118. Perichon B, Reynolds P, and Courvalin P. VanD-type glycopeptide-resistant *Enterococcus faecium* BM4339. *Antimicrob Agents Chemother* 1997;41:2016–8.

119. Ostrowsky BE, Clark NC, Thauvin-Eliopoulos C, et al. A cluster of VanD vancomycin-resistant *Enterococcus faecium*: molecular characterization and clinical epidemiology. *J Infect Dis* 1999;180:1177–85.

120. Depardieu F, Kolbert M, Pruul H, Bell J, and Courvalin P. VanD-type vancomycin-resistant *Enterococcus faecium* and *Enterococcus faecalis*. *Antimicrob Agents Chemother* 2004;48:3892–904.

121. Boyd DA, Miller MA, and Mulvey MR. *Enterococcus gallinarum* N04–0414 harbors a VanD-type vancomycin resistance operon and does not contain a D-alanine:D-alanine 2 (ddl2) gene. *Antimicrob Agents Chemother* 2006;50:1067–70.
122. Reynolds PE, Ambur OH, Casadewall B, and Courvalin P. The VanY(D) DD-carboxy-peptidase of *Enterococcus faecium* BM4339 is a penicillin-binding protein. *Microbiology* 2001;147:2571–8.
123. Fines M, Perichon B, Reynolds P, Sahm DF, and Courvalin P. VanE, a new type of acquired glycopeptide resistance in *Enterococcus faecalis* BM4405. *Antimicrob Agents Chemother* 1999;43:2161–4.
124. Abadia Patino L, Courvalin P, and Perichon B. *vanE* gene cluster of vancomycin-resistant *Enterococcus faecalis* BM4405. *J Bacteriol* 2002;184:6457–64.
125. Abadia-Patino L, Christiansen K, Bell J, Courvalin P, and Perichon B. *vanE*-type vancomycin-resistant *Enterococcus faecalis* clinical isolates from Australia. *Antimicrob Agents Chemother* 2004;48:4882–5.
126. McKessar SJ, Berry AM, Bell JM, Turnidge JD, and Paton JC. Genetic characterization of *vanG*, a novel vancomycin resistance locus of *Enterococcus faecalis*. *Antimicrob Agents Chemother* 2000;44:3224–8.
127. Depardieu F, Bonora MG, Reynolds PE, and Courvalin P. The *vanG* glycopeptide resistance operon from *Enterococcus faecalis* revisited. *Mol Microbiol* 2003;50:931–48.
128. Cremniter J, Mainardi JL, Josseaume N, et al. Novel mechanism of resistance to glyco-peptide antibiotics in *Enterococcus faecium*. *J Biol Chem* 2006;281:32254–62.
129. Steenbergen JN, Alder J, Thorne GM, and Tally FP. Daptomycin: a lipopeptide anti-biotic for the treatment of serious Gram-positive infections. *J Antimicrob Chemother* 2005;55:283–8.
130. Critchley IA, Blosser-Middleton RS, Jones ME, Thornsberry C, Sahm DF, and Karlowsky JA. Baseline study to determine *in vitro* activities of daptomycin against Gram-positive pathogens isolated in the United States in 2000–2001. *Antimicrob Agents Chemother* 2003;47:1689–93.
131. Munoz-Price LS, Lolans K, and Quinn JP. Emergence of resistance to daptomycin during treatment of vancomycin-resistant *Enterococcus faecalis* infection. *Clin Infect Dis* 2005;41:565–6.
132. Lewis JS, 2nd, Owens A, Cadena J, Sabol K, Patterson JE, and Jorgensen JH. Emergence of daptomycin resistance in *Enterococcus faecium* during daptomycin therapy. *Antimicrob Agents Chemother* 2005;49:1664–5.
133. Friedman L, Alder JD, and Silverman JA. Genetic changes that correlate with reduced susceptibility to daptomycin in *Staphylococcus aureus*. *Antimicrob Agents Chemother* 2006;50:2137–45.
134. Geraci JE and Martin WJ. Antibiotic therapy of bacterial endocarditis. VI. Subacute enterococcal endocarditis; clinical, pathologic and therapeutic consideration of 33 cases. *Circulation* 1954;10:173–94.
135. Moellering RC, Jr. and Weinberg AN. Studies on antibiotic syngerism against entero-cocci. II. Effect of various antibiotics on the uptake of 14 C-labeled streptomycin by enterococci. *J Clin Invest* 1971;50:2580–4.
136. Vakulenko SB and Mobashery S. Versatility of aminoglycosides and prospects for their future. *Clin Microbiol Rev* 2003;16:430–50.
137. Eliopoulos GM, Farber BF, Murray BE, Wennersten C, and Moellering RC, Jr. Ribo-somal resistance of clinical enterococcal to streptomycin isolates. *Antimicrob Agents Chemother* 1984;25:398–9.
138. Clark NC, Olsvik O, Swenson JM, Spiegel CA, and Tenover FC. Detection of a strepto-mycin/spectinomycin adenylyltransferase gene (*aadA*) in *Enterococcus faecalis*. *Antimicrob Agents Chemother* 1999;43:157–60.

282   Bacterial Resistance to Antimicrobials

139. Krogstad DJ, Korfhagen TR, Moellering RC, Jr., Wennersten C, and Swartz MN. Aminoglycoside-inactivating enzymes in clinical isolates of *Streptococcus faecalis*. An explanation for resistance to antibiotic synergism. *J Clin Invest* 1978; 62:480–6.
140. Chow JW. Aminoglycoside resistance in enterococci. *Clin Infect Dis* 2000;31:586–9.
141. Chen YG, Qu TT, Yu YS, Zhou JY, and Li LJ. Insertion sequence ISEcp1-like element connected with a novel *aph(2″)* allele [*aph(2″)-Ie*] conferring high-level gentamicin resistance and a novel streptomycin adenylyltransferase gene in Enterococcus. *J Med Microbiol* 2006;55:1521–5.
142. Moellering RC, Jr., Korzeniowski OM, Sande MA, and Wennersten CB. Species-specific resistance to antimocrobial synergism in *Streptococcus faecium* and *Streptococcus faecalis*. *J Infect Dis* 1979;140:203–8.
143. Calderwood SB, Wennersten C, and Moellering RC, Jr. Resistance to antibiotic synergism in *Streptococcus faecalis*: further studies with amikacin and with a new amikacin derivative, 4′-deoxy, 6′-N-methylamikacin. *Antimicrob Agents Chemother* 1981;19:549–55.
144. Thauvin C, Eliopoulos GM, Wennersten C, and Moellering RC, Jr. Antagonistic effect of penicillin-amikacin combinations against enterococci. *Antimicrob Agents Chemother* 1985;28:78–83.
145. Mahbub Alam M, Kobayashi N, Ishino M, et al. Detection of a novel *aph(2″)* allele [*aph(2″)-Ie*] conferring high-level gentamicin resistance and a spectinomycin resistance gene ant(9)-Ia (aad 9) in clinical isolates of enterococci. *Microb Drug Resist* 2005;11:239–47.
146. Carlier C and Courvalin P. Emergence of 4′,4″-aminoglycoside nucleotidyltransferase in enterococci. *Antimicrob Agents Chemother* 1990;34:1565–9.
147. Costa Y, Galimand M, Leclercq R, Duval J, and Courvalin P. Characterization of the chromosomal *aac(6′)-Ii* gene specific for *Enterococcus faecium*. *Antimicrob Agents Chemother* 1993;37:1896–903.
148. Del Campo R, Galan JC, Tenorio C, et al. New *aac(6′)-I* genes in *Enterococcus hirae* and *Enterococcus durans*: effect on (β)-lactam/aminoglycoside synergy. *J Antimicrob Chemother* 2005;55:1053–5.
149. Horodniceanu T, Bougueleret L, El-Solh N, Bieth G, and Delbos F. High-level, plasmid-borne resistance to gentamicin in *Streptococcus faecalis* subsp. zymogenes. *Antimicrob Agents Chemother* 1979;16:686–9.
150. Eliopoulos GM. Antibiotic resistance in *Enterococcus* species: an update. *Curr Clin Top Infect Dis* 1996;16:21–51.
151. Eliopoulos GM, Wennersten C, Zighelboim-Daum S, Reiszner E, Goldmann D, and Moellering RC, Jr. High-level resistance to gentamicin in clinical isolates of *Streptococcus (Enterococcus) faecium*. *Antimicrob Agents Chemother* 1988;32:1528–32.
152. Ferretti JJ, Gilmore KS, and Courvalin P. Nucleotide sequence analysis of the gene specifying the bifunctional 6′-aminoglycoside acetyltransferase 2″-aminoglycoside phosphotransferase enzyme in *Streptococcus faecalis* and identification and cloning of gene regions specifying the two activities. *J Bacteriol* 1986;167:631–8.
153. Daigle DM, Hughes DW, and Wright GD. Prodigious substrate specificity of AAC(6′)-APH(2″), an aminoglycoside antibiotic resistance determinant in enterococci and staphylococci. *Chem Biol* 1999;6:99–110.
154. Straut M, de Cespedes G, and Horaud T. Plasmid-borne high-level resistance to gentamicin in *Enterococcus hirae*, *Enterococcus avium*, and *Enterococcus raffinosus*. *Antimicrob Agents Chemother* 1996;40:1263–5.
155. Donabedian SM, Thal LA, Hershberger E, et al. Molecular characterization of gentamicin-resistant Enterococci in the United States: evidence of spread from animals to humans through food. *J Clin Microbiol* 2003;41:1109–13.

156. Hodel-Christian SL and Murray BE. Characterization of the gentamicin resistance transposon Tn*5281* from *Enterococcus faecalis* and comparison to staphylococcal transposons Tn*4001* and Tn*4031*. *Antimicrob Agents Chemother* 1991;35:1147–52.

157. Thal LA, Chow JW, Clewell DB, and Zervos MJ. Tn*924*, a chromosome-borne transposon encoding high-level gentamicin resistance in *Enterococcus faecalis*. *Antimicrob Agents Chemother* 1994;38:1152–6.

158. Rice LB, Carias LL, and Marshall SH. Tn*5384*, a composite enterococcal mobile element conferring resistance to erythromycin and gentamicin whose ends are directly repeated copies of IS*256*. *Antimicrob Agents Chemother* 1995;39:1147–53.

159. Patterson JE, Masecar BL, Kauffman CA, Schaberg DR, Hierholzer WJ, Jr., and Zervos MJ. Gentamicin resistance plasmids of enterococci from diverse geographic areas are heterogeneous. *J Infect Dis* 1988;158:212–6.

160. Casetta A, Hoi AB, de Cespedes G, and Horaud T. Diversity of structures carrying the high-level gentamicin resistance gene (*aac6-aph2*) in *Enterococcus faecalis* strains isolated in France. *Antimicrob Agents Chemother* 1998;42:2889–92.

161. Kao SJ, You I, Clewell DB, et al. Detection of the high-level aminoglycoside resistance gene *aph(2″)-Ib* in *Enterococcus faecium*. *Antimicrob Agents Chemother* 2000; 44:2876–9.

162. Chow JW, Kak V, You I, et al. Aminoglycoside resistance genes *aph(2″)-Ib* and *aac(6′)-Im* detected together in strains of both *Escherichia coli* and *Enterococcus faecium*. *Antimicrob Agents Chemother* 2001;45:2691–4.

163. Chow JW, Zervos MJ, Lerner SA, et al. A novel gentamicin resistance gene in *Enterococcus*. *Antimicrob Agents Chemother* 1997;41:511–4.

164. Chow JW, Donabedian SM, Clewell DB, Sahm DF, and Zervos MJ. *In vitro* susceptibility and molecular analysis of gentamicin-resistant enterococci. *Diagn Microbiol Infect Dis* 1998;32:141–6.

165. Tsai SF, Zervos MJ, Clewell DB, Donabedian SM, Sahm DF, and Chow JW. A new high-level gentamicin resistance gene, *aph(2″)-Id*, in *Enterococcus* spp. *Antimicrob Agents Chemother* 1998;42:1229–32.

166. Vakulenko SB, Donabedian SM, Voskresenskiy AM, Zervos MJ, Lerner SA, and Chow JW. Multiplex PCR for detection of aminoglycoside resistance genes in enterococci. *Antimicrob Agents Chemother* 2003;47:1423–6.

167. Moellering RC, Jr., Murray BE, Schoenbaum SC, Adler J, and Wennersten CB. A novel mechanism of resistance to penicillin-gentamicin synergism in *Streptococcus faecalis*. *J Infect Dis* 1980;141:81–6.

168. Aslangul E, Ruimy R, Chau F, Garry L, Andremont A, and Fantin B. Relationship between the level of acquired resistance to gentamicin and synergism with amoxicillin in *Enterococcus faecalis*. *Antimicrob Agents Chemother* 2005;49:4144–8.

169. Aslangul E, Massias L, Meulemans A, et al. Acquired gentamicin resistance by permeability impairment in *Enterococcus faecalis*. *Antimicrob Agents Chemother* 2006; 50:3615–21.

170. Schulin T, Wennersten CB, Moellering RC, Jr., and Eliopoulos GM. *In vitro* activity of RU 64004, a new ketolide antibiotic, against Gram-positive bacteria. *Antimicrob Agents Chemother* 1997;41:1196–202.

171. Schmitz FJ, Verhoef J, and Fluit AC. Prevalence of resistance to MLS antibiotics in 20 European university hospitals participating in the European SENTRY surveillance programme. Sentry Participants Group. *J Antimicrob Chemother* 1999; 43:783–92.

172. Jones RN, Sader HS, Erwin ME, and Anderson SC. Emerging multiply resistant enterococci among clinical isolates. I. Prevalence data from 97 medical center surveillance study in the United States. Enterococcus Study Group. *Diagn Microbiol Infect Dis* 1995;21:85–93.

173. Atkinson BA, Abu-Al-Jaibat A, and LeBlanc DJ. Antibiotic resistance among entero-cocci isolated from clinical specimens between 1953 and 1954. *Antimicrob Agents Chemother* 1997;41:1598–600.
174. Roberts MC, Sutcliffe J, Courvalin P, Jensen LB, Rood J, and Seppala H. Nomenclature for macrolide and macrolide-lincosamide-streptogramin B resistance determinants. *Antimicrob Agents Chemother* 1999;43:2823–30.
175. Shaw JH and Clewell DB. Complete nucleotide sequence of macrolide-lincosamide-streptogramin B-resistance transposon Tn*917* in *Streptococcus faecalis*. *J Bacteriol* 1985;164:782–96.
176. Leclercq R and Courvalin P. Bacterial resistance to macrolide, lincosamide, and strepto-gramin antibiotics by target modification. *Antimicrob Agents Chemother* 1991;35:1267–72.
177. Pepper K, Horaud T, Le Bouguenec C, and de Cespedes G. Location of antibiotic resistance markers in clinical isolates of *Enterococcus faecalis* with similar anti-biotypes. *Antimicrob Agents Chemother* 1987;31:1394–402.
178. Bonafede ME, Carias LL, and Rice LB. Enterococcal transposon Tn*5384*: evolution of a composite transposon through cointegration of enterococcal and staphylococcal plasmids. *Antimicrob Agents Chemother* 1997;41:1854–8.
179. Portillo A, Ruiz-Larrea F, Zarazaga M, Alonso A, Martinez JL, and Torres C. Macro-lide resistance genes in *Enterococcus* spp. *Antimicrob Agents Chemother* 2000; 44:967–71.
180. Rosato A, Vicarini H, and Leclercq R. Inducible or constitutive expression of resistance in clinical isolates of streptococci and enterococci cross-resistant to erythromycin and lincomycin. *J Antimicrob Chemother* 1999;43:559–62.
181. Oh TG, Kwon AR, and Choi EC. Induction of *ermAMR* from a clinical strain of *Enterococcus faecalis* by 16-membered-ring macrolide antibiotics. *J Bacteriol* 1998; 180:5788–91.
182. Luna VA, Coates P, Eady EA, Cove JH, Nguyen TT, and Roberts MC. A variety of Gram-positive bacteria carry mobile *mef* genes. *J Antimicrob Chemother* 1999; 44:19–25.
183. Singh KV, Malathum K, and Murray BE. Disruption of an *Enterococcus faecium* species-specific gene, a homologue of acquired macrolide resistance genes of staphylo-cocci, is associated with an increase in macrolide susceptibility. *Antimicrob Agents Chemother* 2001;45:263–6.
184. Werner G, Hildebrandt B, and Witte W. The newly described *msrC* gene is not equally distributed among all isolates of *Enterococcus faecium*. *Antimicrob Agents Chemother* 2001;45:3672–3.
185. Aakra A, Vebo H, Snipen L, et al. Transcriptional response of *Enterococcus faecalis* V583 to erythromycin. *Antimicrob Agents Chemother* 2005;49:2246–59.
186. Singh KV and Murray BE. Differences in the *Enterococcus faecalis lsa* locus that influence susceptibility to quinupristin-dalfopristin and clindamycin. *Antimicrob Agents Chemother* 2005;49:32–9.
187. Dassa E and Bouige P. The ABC of ABCS: a phylogenetic and functional classification of ABC systems in living organisms. *Res Microbiol* 2001;152:211–29.
188. Malathum K, Coque TM, Singh KV, and Murray BE. *In vitro* activities of two ketolides, HMR 3647 and HMR 3004, against Gram-positive bacteria. *Antimicrob Agents Chemother* 1999;43:930–6.
189. Lim JA, Kwon AR, Kim SK, Chong Y, Lee K, and Choi EC. Prevalence of resistance to macrolide, lincosamide and streptogramin antibiotics in Gram-positive cocci iso-lated in a Korean hospital. *J Antimicrob Chemother* 2002;49:489–95.
190. Singh KV, Weinstock GM, and Murray BE. An *Enterococcus faecalis* ABC homologue (Lsa) is required for the resistance of this species to clindamycin and quinupristin-dalfopristin. *Antimicrob Agents Chemother* 2002;46:1845–50.

191. Dina J, Malbruny B, and Leclercq R. Nonsense mutations in the *lsa*-like gene in *Enterococcus faecalis* isolates susceptible to lincosamides and Streptogramins A. *Antimicrob Agents Chemother* 2003;47:2307–9.

192. Bozdogan B, Berrezouga L, Kuo MS, et al. A new resistance gene, *linB*, conferring resistance to lincosamides by nucleotidylation in *Enterococcus faecium* HM1025. *Antimicrob Agents Chemother* 1999;43:925–9.

193. Eliopoulos GM. Quinupristin-dalfopristin and linezolid: evidence and opinion. *Clin Infect Dis* 2003;36:473–81.

194. Jones RN, Ballow CH, Biedenbach DJ, Deinhart JA, and Schentag JJ. Antimicrobial activity of quinupristin-dalfopristin (RP 59500, Synercid) tested against over 28,000 recent clinical isolates from 200 medical centers in the United States and Canada. *Diagn Microbiol Infect Dis* 1998;31:437–51.

195. Rende-Fournier R, Leclercq R, Galimand M, Duval J, and Courvalin P. Identification of the *satA* gene encoding a streptogramin A acetyltransferase in *Enterococcus faecium* BM4145. *Antimicrob Agents Chemother* 1993;37:2119–25.

196. Hammerum AM, Jensen LB, and Aarestrup FM. Detection of the *satA* gene and transferability of virginiamycin resistance in *Enterococcus faecium* from food-animals. *FEMS Microbiol Lett* 1998;168:145–51.

197. Werner G, Klare I, and Witte W. Association between quinupristin/dalfopristin resistance in glycopeptide-resistant *Enterococcus faecium* and the use of additives in animal feed. *Eur J Clin Microbiol Infect Dis* 1998;17:401–2.

198. Werner G and Witte W. Characterization of a new enterococcal gene, *satG*, encoding a putative acetyltransferase conferring resistance to Streptogramin A compounds. *Antimicrob Agents Chemother* 1999;43:1813–4.

199. Soltani M, Beighton D, Philpott-Howard J, and Woodford N. Mechanisms of resistance to quinupristin-dalfopristin among isolates of *Enterococcus faecium* from animals, raw meat, and hospital patients in Western Europe. *Antimicrob Agents Chemother* 2000;44:433–6.

200. Donabedian SM, Perri MB, Vager D, et al. Quinupristin-dalfopristin resistance in *Enterococcus faecium* isolates from humans, farm animals, and grocery store meat in the United States. *J Clin Microbiol* 2006;44:3361–5.

201. Allignet J and El Solh N. Characterization of a new staphylococcal gene, *vgaB*, encoding a putative ABC transporter conferring resistance to streptogramin A and related compounds. *Gene* 1997;202:133–8.

202. Lina G, Quaglia A, Reverdy ME, Leclercq R, Vandenesch F, and Etienne J. Distribution of genes encoding resistance to macrolides, lincosamides, and streptogramins among staphylococci. *Antimicrob Agents Chemother* 1999;43:1062–6.

203. Long KS, Poehlsgaard J, Kehrenberg C, Schwarz S, and Vester B. The Cfr rRNA methyltransferase confers resistance to Phenicols, Lincosamides, Oxazolidinones, Pleuromutilins, and Streptogramin A antibiotics. *Antimicrob Agents Chemother* 2006;50:2500–5.

204. Jensen LB, Hammerum AM, Aerestrup FM, van den Bogaard AE, and Stobberingh EE. Occurrence of *satA* and *vgb* genes in streptogramin-resistant *Enterococcus faecium* isolates of animal and human origins in the Netherlands. *Antimicrob Agents Chemother* 1998;42:3330–1.

205. Bozdogan B and Leclercq R. Effects of genes encoding resistance to streptogramins A and B on the activity of quinupristin-dalfopristin against *Enterococcus faecium*. *Antimicrob Agents Chemother* 1999;43:2720–5.

206. Bozdogan B, Leclercq R, Lozniewski A, and Weber M. Plasmid-mediated coresistance to streptogramins and vancomycin in *Enterococcus faecium* HM1032. *Antimicrob Agents Chemother* 1999;43:2097–8.

207. Barriere JC, Berthaud N, Beyer D, Dutka-Malen S, Paris JM, and Desnottes JF. Recent developments in streptogramin research. *Curr Pharm Des* 1998;4:155–80.

208. Caron F, Gold HS, Wennersten CB, Farris MG, Moellering RC, Jr., and Eliopoulos GM. Influence of erythromycin resistance, inoculum growth phase, and incubation time on assessment of the bactericidal activity of RP 59500 (quinupristin-dalfopristin) against vancomycin-resistant *Enterococcus faecium*. *Antimicrob Agents Chemother* 1997;41:2749–53.
209. Jones RN, Farrell DJ, and Morrissey I. Quinupristin-dalfopristin resistance in *Streptococcus pneumoniae*: novel L22 ribosomal protein mutation in two clinical isolates from the SENTRY antimicrobial surveillance program. *Antimicrob Agents Chemother* 2003;47:2696–8.
210. Malbruny B, Canu A, Bozdogan B, et al. Resistance to quinupristin-dalfopristin due to mutation of L22 ribosomal protein in *Staphylococcus aureus*. *Antimicrob Agents Chemother* 2002;46:2200–7.
211. Lautenbach E, Schuster MG, Bilker WB, and Brennan PJ. The role of chloramphenicol in the treatment of bloodstream infection due to vancomycin-resistant *Enterococcus*. *Clin Infect Dis* 1998;27:1259–65.
212. Norris AH, Reilly JP, Edelstein PH, Brennan PJ, and Schuster MG. Chloramphenicol for the treatment of vancomycin-resistant enterococcal infections. *Clin Infect Dis* 1995;20:1137–44.
213. Scapellato PG, Ormazabal C, Scapellato JL, and Bottaro EG. Meningitis due to vancomycin-resistant *Enterococcus faecium* successfully treated with combined intravenous and intraventricular chloramphenicol. *J Clin Microbiol* 2005;43:3578–9.
214. Perez Mato S, Robinson S, and Begue RE. Vancomycin-resistant *Enterococcus faecium* meningitis successfully treated with chloramphenicol. *Pediatr Infect Dis J* 1999; 18:483–4.
215. Ricaurte JC, Boucher HW, Turett GS, Moellering RC, Labombardi VJ, and Kislak JW. Chloramphenicol treatment for vancomycin-resistant *Enterococcus faecium* bacteremia. *Clin Microbiol Infect* 2001;7:17–21.
216. Mutnick AH, Biedenbach DJ, and Jones RN. Geographic variations and trends in antimicrobial resistance among *Enterococcus faecalis* and *Enterococcus faecium* in the SENTRY Antimicrobial Surveillance Program (1997–2000). *Diagn Microbiol Infect Dis* 2003;46:63–8.
217. Gould CV, Fishman NO, Nachamkin I, and Lautenbach E. Chloramphenicol resistance in vancomycin-resistant enterococcal bacteremia: impact of prior fluoroquinolone use? *Infect Control Hosp Epidemiol* 2004;25:138–45.
218. Courvalin PM, Shaw WV, and Jacob AE. Plasmid-mediated mechanisms of resistance to aminoglycoside-aminocyclitol antibiotics and to chloramphenicol in group D streptococci. *Antimicrob Agents Chemother* 1978;13:716–25.
219. Trieu-Cuot P, de Cespedes G, Bentorcha F, Delbos F, Gaspar E, and Horaud T. Study of heterogeneity of chloramphenicol acetyltransferase (CAT) genes in streptococci and enterococci by polymerase chain reaction: characterization of a new CAT determinant. *Antimicrob Agents Chemother* 1993;37:2593–8.
220. Lynch C, Courvalin P, and Nikaido H. Active efflux of antimicrobial agents in wild-type strains of enterococci. *Antimicrob Agents Chemother* 1997;41:869–71.
221. Sader HS, Jones RN, Stilwell MG, Dowzicky MJ, and Fritsche TR. Tigecycline activity tested against 26,474 bloodstream infection isolates: a collection from 6 continents. *Diagn Microbiol Infect Dis* 2005;52:181–6.
222. Flannagan SE, Zitzow LA, Su YA, and Clewell DB. Nucleotide sequence of the 18-kb conjugative transposon Tn916 from *Enterococcus faecalis*. *Plasmid* 1994; 32:350–4.
223. Franke AE and Clewell DB. Evidence for a chromosome-borne resistance transposon (Tn916) in *Streptococcus faecalis* that is capable of "conjugal" transfer in the absence of a conjugative plasmid. *J Bacteriol* 1981;145:494–502.

224. Nishimoto Y, Kobayashi N, Alam MM, Ishino M, Uehara N, and Watanabe N. Analysis of the prevalence of tetracycline resistance genes in clinical isolates of *Enterococcus faecalis* and *Enterococcus faecium* in a Japanese hospital. *Microb Drug Resist* 2005; 11:146–53.
225. Charpentier E, Gerbaud G, and Courvalin P. Presence of the *Listeria* tetracycline resistance gene *tet(S)* in *Enterococcus faecalis*. *Antimicrob Agents Chemother* 1994;38:2330–5.
226. Roberts AP, Davis IJ, Seville L, Villedieu A, and Mullany P. Characterization of the ends and target site of a novel tetracycline resistance-encoding conjugative transposon from *Enterococcus faecium* 664.1H1 *J Bacteriol* 2006,188.4356–61.
227. McMurry LM, Park BH, Burdett V, and Levy SB. Energy-dependent efflux mediated by class L (TetL) tetracycline resistance determinant from streptococci. *Antimicrob Agents Chemother* 1987;31:1648–50.
228. Bentorcha F, De Cespedes G, and Horaud T. Tetracycline resistance heterogeneity in *Enterococcus faecium*. *Antimicrob Agents Chemother* 1991;35:808–12.
229. Boucher HW, Wennersten CB, and Eliopoulos GM. In vitro activities of the glycylcycline GAR-936 against Gram-positive bacteria. *Antimicrob Agents Chemother* 2000;44:2225–9.
230. Jones RN, Ross JE, Fritsche TR, and Sader HS. Oxazolidinone susceptibility patterns in 2004: report from the Zyvox Annual Appraisal of Potency and Spectrum (ZAAPS) Program assessing isolates from 16 nations. *J Antimicrob Chemother* 2006;57:279–87.
231. Hoban DJ, Bouchillon SK, Johnson BM, Johnson JL, and Dowzicky MJ. *In vitro* activity of tigecycline against 6792 Gram-negative and Gram-positive clinical isolates from the global Tigecycline Evaluation and Surveillance Trial (TEST Program, 2004). *Diagn Microbiol Infect Dis* 2005;52:215–27.
232. Streit JM, Sader HS, Fritsche TR, and Jones RN. Dalbavancin activity against selected populations of antimicrobial-resistant Gram-positive pathogens. *Diagn Microbiol Infect Dis* 2005;53:307–10.
233. Bouchillon SK, Hoban DJ, Johnson BM, et al. *In vitro* evaluation of tigecycline and comparative agents in 3049 clinical isolates: 2001 to 2002. *Diagn Microbiol Infect Dis* 2005;51:291–5.
234. Gonzales RD, Schreckenberger PC, Graham MB, Kelkar S, DenBesten K, and Quinn JP. Infections due to vancomycin-resistant *Enterococcus faecium* resistant to linezolid. *Lancet* 2001;357:1179.
235. Pai MP, Rodvold KA, Schreckenberger PC, Gonzales RD, Petrolatti JM, and Quinn JP. Risk factors associated with the development of infection with linezolid- and vancomycin-resistant *Enterococcus faecium*. *Clin Infect Dis* 2002;35:1269–72.
236. Ruggero KA, Schroeder LK, Schreckenberger PC, Mankin AS, and Quinn JP. Nosocomial superinfections due to linezolid-resistant *Enterococcus faecalis*: evidence for a gene dosage effect on linezolid MICs. *Diagn Microbiol Infect Dis* 2003;47:511–3.
237. Dobbs TE, Patel M, Waites KB, Moser SA, Stamm AM, and Hoesley CJ. Nosocomial spread of *Enterococcus faecium* resistant to vancomycin and linezolid in a tertiary care medical center. *J Clin Microbiol* 2006;44:3368–70.
238. Herrero IA, Issa NC, and Patel R. Nosocomial spread of linezolid-resistant, vancomycin-resistant *Enterococcus faecium*. *N Engl J Med* 2002;346:867–9.
239. Bonora MG, Ligozzi M, Luzzani A, Solbiati M, Stepan E, and Fontana R. Emergence of linezolid resistance in *Enterococcus faecium* not dependent on linezolid treatment. *Eur J Clin Microbiol Infect Dis* 2006;25:197–8.
240. Meka VG, and Gold HS. Antimicrobial resistance to linezolid. *Clin Infect Dis* 2004; 39:1010–5.
241. Prystowsky J, Siddiqui F, Chosay J, et al. Resistance to linezolid: characterization of mutations in rRNA and comparison of their occurrences in vancomycin-resistant enterococci. *Antimicrob Agents Chemother* 2001;45:2154–6.

242. Johnson AP, Tysall L, Stockdale MV, et al. Emerging linezolid-resistant *Enterococcus faecalis* and *Enterococcus faecium* isolated from two Austrian patients in the same intensive care unit. *Eur J Clin Microbiol Infect Dis* 2002;21:751–4.
243. Marshall SH, Donskey CJ, Hutton-Thomas R, Salata RA, and Rice LB. Gene dosage and linezolid resistance in *Enterococcus faecium* and *Enterococcus faecalis*. *Antimicrob Agents Chemother* 2002;46:3334–6.
244. Jones RN, Della-Latta P, Lee LV, and Biedenbach DJ. Linezolid-resistant *Enterococcus faecium* isolated from a patient without prior exposure to an oxazolidinone: report from the SENTRY Antimicrobial Surveillance Program. *Diagn Microbiol Infect Dis* 2002; 42:137–9.
245. Dibo I, Pillai SK, Gold HS, et al. Linezolid-resistant *Enterococcus faecalis* isolated from a cord blood transplant recipient. *J Clin Microbiol* 2004;42:1843–5.
246. Klare I, Konstabel C, Mueller-Bertling S, et al. Spread of ampicillin/vancomycin-resistant *Enterococcus faecium* of the epidemic-virulent clonal complex-17 carrying the genes *esp* and *hyl* in German hospitals. *Eur J Clin Microbiol Infect Dis* 2005; 24:815–25.
247. Bae HG, Sung H, Kim MN, Lee EJ, and Koo Lee S. First report of a linezolid- and vancomycin-resistant *Enterococcus faecium* strain in Korea. *Scand J Infect Dis* 2006;38:383–6.
248. Lobritz M, Hutton-Thomas R, Marshall S, and Rice LB. Recombination proficiency influences frequency and locus of mutational resistance to linezolid in *Enterococcus faecalis*. *Antimicrob Agents Chemother* 2003;47:3318–20.
249. Woodford N, Tysall L, Auckland C, et al. Detection of oxazolidinone-resistant *Enterococcus faecalis* and *Enterococcus faecium* strains by real-time PCR and PCR-restriction fragment length polymorphism analysis. *J Clin Microbiol* 2002;40:4298–300.
250. Sinclair A, Arnold C, and Woodford N. Rapid detection and estimation by pyrosequencing of 23S rRNA genes with a single nucleotide polymorphism conferring linezolid resistance in Enterococci. *Antimicrob Agents Chemother* 2003;47:3620–2.
251. Barry AL, Jones RN, Thornsberry C, Ayers LW, Gerlach EH, and Sommers HM. Antibacterial activities of ciprofloxacin, norfloxacin, oxolinic acid, cinoxacin, and nalidixic acid. *Antimicrob Agents Chemother* 1984;25:633–7.
252. Tankovic J, Mahjoubi F, Courvalin P, Duval J, and Leclercq R. Development of fluoroquinolone resistance in *Enterococcus faecalis* and role of mutations in the DNA gyrase *gyrA* gene. *Antimicrob Agents Chemother* 1996;40:2558–61.
253. Eliopoulos GM. Activity of newer fluoroquinolones *in vitro* against Gram-positive bacteria. *Drugs* 1999;58 Suppl 2:23–8.
254. Korten V, Huang WM, and Murray BE. Analysis by PCR and direct DNA sequencing of *gyrA* mutations associated with fluoroquinolone resistance in *Enterococcus faecalis*. *Antimicrob Agents Chemother* 1994;38:2091–4.
255. Kanematsu E, Deguchi T, Yasuda M, Kawamura T, Nishino Y, and Kawada Y. Alterations in the GyrA subunit of DNA gyrase and the ParC subunit of DNA topoisomerase IV associated with quinolone resistance in *Enterococcus faecalis*. *Antimicrob Agents Chemother* 1998;42:433–5.
256. el Amin NA, Jalal S, and Wretlind B. Alterations in GyrA and ParC associated with fluoroquinolone resistance in *Enterococcus faecium*. *Antimicrob Agents Chemother* 1999;43:947–9.
257. Jonas BM, Murray BE, and Weinstock GM. Characterization of *emeA*, a NorA homolog and multidrug resistance efflux pump, in *Enterococcus faecalis*. *Antimicrob Agents Chemother* 2001;45:3574–9.
258. Lee EW, Huda MN, Kuroda T, Mizushima T, and Tsuchiya T. EfrAB, an ABC multidrug efflux pump in *Enterococcus faecalis*. *Antimicrob Agents Chemother* 2003; 47:3733–8.

259. Oyamada Y, Ito H, Fujimoto K, et al. Combination of known and unknown mechanisms confers high-level resistance to fluoroquinolones in *Enterococcus faecium. J Med Microbiol* 2006;55:729–36.
260. Oyamada Y, Ito H, Inoue M, and Yamagishi J. Topoisomerase mutations and efflux are associated with fluoroquinolone resistance in *Enterococcus faecalis. J Med Microbiol* 2006;55:1395–401.
261. Davis DR, McAlpine JB, Pazoles CJ, et al. *Enterococcus faecalis* multi-drug resistance transporters: application for antibiotic discovery. *J Mol Microbiol Biotechnol* 2001; 3:179–84.
262. Paulsen IT, Banerjei L, Myers GS, et al. Role of mobile DNA in the evolution of vancomycin-resistant *Enterococcus faecalis. Science* 2003;299:2071–4.

12 Methicillin Resistance in *Staphylococcus aureus*

Keeta S. Gilmore, Michael S. Gilmore,
and Daniel F. Sahm

CONTENTS

Before the advent of antibiotic therapy, invasive staphylococcal infection was often fatal. The bacterium *Staphylococcus aureus* has since demonstrated a remarkable ability to adapt to antibiotic pressure. Methicillin-resistant strains of *S. aureus,* termed MRSA, are those strains that have acquired the ability to grow in the presence of methylpenicillins and derivatives, including methicillin, oxacillin, and nafcillin. This methicillin resistance is mediated by the acquisition and expression of an altered penicillin-binding protein, PBP2a (PBP2′), which exhibits a decreased affinity for β-lactam antibiotics [1,2]. Penicillin-binding proteins (PBPs) are essential enzymes that catalyze transpeptidation crosslinking of peptidoglycan in the bacterial cell wall and are the targets of the antibiotic methicillin in sensitive strains of *S. aureus.* Inhibition of this reaction with methicillin results in the arrest of cell wall biosynthesis, triggering death of the organism through induction of the autolytic response [3]. MRSA possess a 21- to 67-kb DNA sequence that encodes, among other things, PBP2a and genes for regulation of its expression. Methicillin-susceptible strains are inhibited by oxacillin at concentrations of 4 μg/mL or methicillin at 8 μg/mL. In contrast, MRSA grow in the presence of 16 μg/mL to over 2000 μg/mL of methicillin.

INTRODUCTION

Staphylococci cause a variety of infections, ranging from skin and soft-tissue infections to bloodstream infections and endocarditis, and the pathogenesis of these infections is well described [4–6]. The purpose of this chapter is to review the present and future challenge to health care specifically posed by methicillin-resistant strains of S. aureus. In particular, the subjects of this chapter are the origins and nature of methicillin resistance, its epidemiology among nosocomial and community-acquired strains, and the consequences of this resistance in limiting therapeutic options and its impact on health care costs.

Methicillin resistance in S. aureus was initially detected in Europe in the 1960s shortly after the introduction of methicillin. Today, MRSA are present in the hospitals of most countries and are often resistant to several antibiotics. Clinical infections are most common in patients in hospital intensive care units, nursing homes, and other chronic care facilities; however, MRSA are emerging as an important community-acquired pathogen as well. Currently, most MRSA are susceptible to the glyco-peptides, such as vancomycin and teicoplanin; however, as resistance to these agents increases, some staphylococcal infections could be untreatable.

EMERGENCE OF MRSA

Since the introduction of antibiotics into clinical use in the mid-1940s, microorganisms have shown a remarkable ability to protect themselves by developing and acquiring antibiotic resistance. By 1942, penicillin resistance was reported in S. aureus after only months of limited clinical trials [5]. By 1953, as use of penicillin became more widespread, 64% to 80% of S. aureus isolates were resistant to penicillin, with development of resistance to tetracycline, erythromycin, and other classes of antibiotics beginning to emerge [5]. By 1960, despite using aggressive infection control measures, antibiotic-resistant staphylococci had become the most common cause of hospital-acquired infection worldwide [5,7]. Still, penicillin-resistant S. aureus was largely a nosocomial problem until the 1970s, when it became apparent that penicillin resistance was prevalent among community-acquired isolates as well. By this time, the rates of penicillin resistant S. aureus were about the same for both hospital and community-acquired isolates [8].

Methicillin, a β-lactam effective against penicillin-resistant S. aureus strains, became widely available in 1960. Like the development of penicillin resistance, within a year of its introduction, MRSA were reported in the United Kingdom [9,10]. Sporadic reports of clinical isolates of MRSA were soon observed in the United States [11], with the first well-documented outbreak in the United States in 1968 [12]. These MRSA were resistant to the entire class of β-lactams.

MOLECULAR BASIS FOR METHICILLIN RESISTANCE

The early introduction of β-lactam antibiotics quickly selected for the outgrowth of *S. aureus* strains possessing, or having acquired, the ability to express β-lactamases, achieving a resistance rate of 75% as early as 1952 [13]. The outgrowth of β-lactamase-producing *S. aureus* prompted the commercial development of β-lactamase-resistant derivatives of penicillin, such as methicillin, oxacillin, and nafcillin, which possess an acyl side chain that prevents hydrolysis of the β-lactam ring. The narrow-spectrum staphylococcal β-lactamases exhibit little activity against these semisynthetic penicillins [14].

MECHANISMS OF RESISTANCE TO METHICILLIN

Under new selective pressure, *S. aureus* developed multiple mechanisms of resistance to modified penicillins, including methicillin. Although methicillin is resistant to hydrolysis by small quantities of staphylococcal β-lactamase, strains of *S. aureus* have been isolated that are capable of producing increased levels of β-lactamase [15]. These hyper-producers resist methicillin through limited hydrolysis of the antibiotic, resulting in a phenotype that, with respect to methicillin, is intermediate between susceptible and resistant [15].

Methicillin resistance in *S. aureus* is achieved primarily by the acquisition of an altered PBP, PBP2a (also known as PBP'), which confers resistance to all β-lactams and their derivatives. *S. aureus* natively expresses four other PBPs, designated PBP1, 2, 3, and 4, that are all sensitive to β-lactam antibiotics [16]. The β-lactam antibiotics serve as substrate analogs that covalently bind PBPs, inactivating them at concentrations close to the MIC. PBPs are essential proteins that are anchored to the cytoplasmic membrane and, under normal circumstances, catalyze the transpeptidation reaction that crosslinks bacterial cell wall peptidoglycan. Inhibition of this reaction by the binding of a β-lactam is lethal [14]. Low-level resistance to β-lactam antibiotics has been observed to result from a decrease in the binding affinities of PBPs for penicillins, an increase in the production of PBPs, or both [16,17]. However, the most prevalent means for achieving methicillin resistance is the acquisition of an element termed the staphylococcal cassette chromosome *mec* (SCC*mec*) [18] containing the *mecA* gene encoding PBP2a.

SCC*MEC*

SCC*mec* DNA is a large, 21- to 67-kb, unique class of mobile genetic element always located at a fixed site in the *S. aureus* chromosome near the origin of replication (Figure 12.1) [14,19,20]. Unlike conjugative transposons, SCC*mec* does not contain the *tra* gene complex. SCC*mec* contains *mecA*, the structural gene for PBP2a and regulatory elements, *mecI* and *mecRl*, which control *mecA* transcription. Downstream from *mecA* is a variable segment of DNA that ends with an insertion-like element, IS*431* [21], that serves as a target for homologous recombination for other resistance determinants flanked by similar IS elements [4,22]. Therefore, *mecA* and its associated DNA act as a trap for integration of other determinants, including genes for resistance to fluoroquinolones, aminoglycosides, tetracyclines, macrolides,

mec DNA

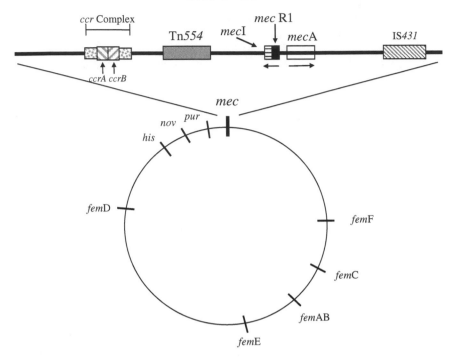

FIGURE 12.1 Organization of the SCC*mec* region of DNA and chromosomal location. SCC*mec* DNA is 21–67 kb containing the PBP2a structural gene, *mec*A and its upstream regulatory elements, *mec*I-*mec*R1 and the *ccr* complex. The regulatory genes are divergently transcribed from *mec*A as indicated by the arrows. Further upstream from *mec*A is Tn*554* and downstream from *mec*A is a variable region ending with IS*431*. (Adapted from Chambers HF, *J Infect Dis* 179, 1999; Hiramatsu K, *Microbiol Immunol* 39, 1995; Katayama Y, Ito T, Hiramatsu K, *Antimicrob Agents Chemother* 44, 2000.)

and trimethoprim-sulfamethoxazole [14]. In addition, the transposon Tn*554* containing *ermA*, the gene encoding for inducible erythromycin resistance, is located upstream from *mec*A in over 90% of MRSA [23].

Two genes have been identified for mobilization of SCC*mec*, *ccr*A and *B* (cassette chromosome recombinase genes A and B), which encode DNA recombinases of the invertase/resolvase family [18,20]. These two genes catalyze the precise integration of SSC*mec* into the chromosome in the correct orientation and its precise excision from the chromosome.

Several types of SCC*mec* elements have now been described [18,24–28] and a typing system has been proposed, which classifies variations of SCC*mec* into five types (I to V) based on the *mec* DNA structure and the *ccr*A and *B* genes [20,24]. Types I, II, and III compose the majority of nosocomial strains of MRSA, whereas types IV and V are found in community-acquired MRSA [20,24]. Interestingly, SCC*mec* type V was found to contain a single new site-specific recombinase gene (*ccrC*), which carries out both integration and excision.

mecA

Greater than 90% of MRSA harbor *mecA* [16]. The *mecA* gene is inducible and encodes the high-molecular-weight, 78-kD PBP2a polypeptide. It occurs in both MRSA and methicillin-resistant coagulase-negative staphylococci, and is highly conserved [29–32]. Analysis of the nucleotide sequence of *mecA* and its operator region revealed that sequences contained within the 5′ end were similar to sequences within the β-lactamase gene, *blaZ*, of *S. aureus*. The remainder of the structural gene exhibits sequence similarity to the PBP2 and PBP3 genes of *Escherichia coli* [16,33].

PBP2a

The native PBPs in *S. aureus*, PBP1, 2, 3, and 4 are essential for cell growth and survival of susceptible strains. These PBPs have a high affinity for most β-lactam antibiotics, which bind to the transpeptidase domain preventing crosslinking and new septum initiation [34–36]. PBP2a binds β-lactams with much lower affinity, allowing the organisms to grow in drug concentrations that would otherwise inactivate native PBP and inhibit growth. Initially, PBP2a was thought to substitute for the essential functions of the native PBPs at lethal concentrations of antibiotics [1,14] since it includes both transpeptidase (TPase) and what appeared to be transglycolase (TGase) domains [4,37]. However, while native PBPs produce highly cross-linked peptidoglycan, PBP2a appears to be limited in activity to linking two monomers, and is incapable of generating highly cross-linked oligomers that are typical products of the normal cell wall synthetic machinery [38]. Further, Pinho et al. [37,39] have shown that the concerted action of both PBP2a and native PBP2 is essential for optimal methicillin resistance even when the TPase domain of PBP2 is fully acylated. They found that when the structural gene for PBP2 was inactivated in a highly methicillin-resistant strain, there was a several-fold reduction in the methicillin MIC (from 800 µg/mL to 12 µg/mL) [37]. Additionally, they were able to show that the TGase domain of native PBP2 was insensitive to the presence of β-lactam antibiotics and functions in the presence of β-lactams for cell wall synthesis [37]. Therefore, high-level resistance to methicillin requires the TPase domain of PBP2a in concert with the penicillin-insensitive TGase domain of native PBP2 [39].

Regulation of *mecA*

Expression of PBP2a is controlled by two sets of regulator genes. The first set, which includes *mecR1* and *mecI*, is located within the *mec* DNA immediately upstream of *mecA* and is divergently transcribed from *mecA* [14]. The second set of regulators that affect *mecA* expression, *blaR1* and *blaI*, are chromosomally encoded and also serve to regulate *blaZ*, the staphylococcal penicillinase gene [4,40,41]. Strains that contain functional *mecR1-mecI* regulatory elements are strongly repressed and produce PBP2a only after induction [4,42].

MecI and BlaI are repressor proteins and both can repress *mecA* and *blaZ* [43,44]. Repression by BlaI is weaker than by MecI, and as a result, some PBP2a is produced in uninduced strains. Induction of BlaI-repressed *mecA* by methicillin is as rapid as induction of BlaI-repressed β-lactamase synthesis [4,42]. In contrast, MecI is a strong repressor of *mec*A and leads to an extremely slow induction of PBP2a. As a result, methicillin resistance is established slowly and may only appear after 48 h on methicillin-containing plates, making these strains appear initially falsely susceptible at 24 h [4,42,45].

The MecR1 and BlaR1 proteins are sensor-transducer molecules that are specific for their corresponding repressors, MecI and BlaI, respectively, and cannot substitute for each other in the presence of β-lactam antibiotics [14]. Like BlaR1, MecR1 is a transmembrane protein consisting of an extracellular sensor domain and an intracellular metalloprotease domain [46,47].

Although *mec*A is present in all MRSA, there is considerable variation in the presence of the other genes [48]. *mecR1-mecI* is present in 60% to 95% of *mec*A-positive *S. aureus* [49–51]. Because *mecI* is such a strong repressor, it has been concluded that phenotypically resistant *mec*A-positive *S. aureus* strains either do not possess *mecI*, or have mutations within *mecI*, which prevent it from functioning [16,49,50,52]; or that they have mutations within the *mec*A promoter region corresponding to a presumptive operator of *mec*A, the binding site of the repressor protein [16,50]. Inactivation of *mecI*, by either deletion or mutation, is an essential step in the production of PBP2a and expression of methicillin resistance [53,54]. Two point mutations are frequently detected in the *mecI* gene: a substitution transition at nucleotide position 202 (C to T) or a transversion at position 260 (T to A), either of which generates an in-frame stop codon in the middle of the *mecI* gene [16,49,50,52]. In these strains, a functional repressor protein is not produced, resulting in maximal expression of methicillin resistance [50]. Point mutations in the operator region of the *mec*A promoter that result in derepression have also been identified [16,33].

A small number of *S. aureus* strains have been isolated that carry intact *mecI* and *mecRl,* together with *mec*A, and these strains have been termed pre-MRSA, as typified by prototype *S. aureus* strain N315 [16,50,52]. Pre-MRSA are phenotypically susceptible to methicillin as routinely assayed [45,49,50]. In these strains, the expression of methicillin resistance is fully repressed by *mecI* and is not induced by the presence of methicillin. However, when grown on selective media, resistant cells arise at a high frequency (10^{-5} to 10^{-6}) resulting from point mutations in the *mecI* gene [49,50], circumventing the *mecI*-mediated repression of *mec*A. In the absence of both the *blaR1-blaI* and *mecR1-mecI* regulatory elements, PBP2a is produced constitutively but this does not always correlate with high-level resistance [4,44], leading to the conclusion that other genes also contribute to resistance.

CHROMOSOMAL ELEMENTS AFFECTING METHICILLIN RESISTANCE LEVELS

The observation that PBP2a levels do not correlate directly with resistance levels [4,44,55] led to the search for other factors that influence expression of methicillin resistance. Transposon-mediated insertional inactivation of chromosomal genes identified several that affected methicillin resistance [55–58]. It is now appreciated that methicillin resistance in *S. aureus* is complex and involves auxiliary genes (*aux* genes) or *fem* genes (factors essential for the expression of methicillin resistance) [59,60]. The *fem* genes are primarily housekeeping genes, located throughout the staphylococcal genome and are essential for maximum resistance [4,58,61,62]. Over 20 *fem* genes have been identified (Table 12.1) [4,14,54,63–65].

The *fem* genes occur in both MRSA and methicillin-susceptible *S. aureus* (MSSA), and many encode or regulate the activity of enzymes catalyzing reactions at different stages in peptidoglycan biosynthesis or turnover. However, none has been shown to affect PBP2a expression [44]. Inactivation of *fem* genes, especially those genes involved in cell wall precursor formation, leads to a reduction in methicillin resistance. The function of many *fem* gene encoded proteins is still unknown.

The *fem* genes with the most influence on resistance are *fmhD*, *femA*, and *femB*, which lead to formation of the pentaglycine bridge that crosslinks staphylococcal peptidoglycan (Figure 12.2) [4,14,47,66,67]. FmhB is responsible for incorporation of the first glycyl residue of the pentaglycine bridge. FemA and FemB are responsible for the addition of residues 2–3 and 4–5, respectively, into the bridge [4,14,68]. The pentaglycine bridge has been shown to be essential for PBP2a-mediated resistance, and shortening its length leads to hypersusceptibility to β-lactams as well as other classes of antibiotics [14,44]. Inactivation of *fmhB* is lethal [67].

Disruption of *femC* reduces the basal level of methicillin resistance in MRSA, but still allows formation of a highly resistant subpopulation [4,14]. Mutation in *femC* results in a metabolic block in glutamine production. This block affects peptidoglycan composition by reducing the amidation of isoglutamate in the peptidoglycan stem pentapeptide, resulting in a reduction in the extent of crosslinking in the peptidoglycan. Addition of glutamine to the culture medium restores both isoglutamate amidation and methicillin resistance [4,14].

femD catalyzes the conversion of glucosamine-6-phosphate to glucosamine-1-phosphate, a reaction key to peptidoglycan precursor formation [69]. Inhibition of FemD leads to a reduction in methicillin resistance [70,71]. It has been observed that in cultures of both *femC* mutants and *femD* mutants, spontaneous methicillin-resistant suppressor mutants can be found that render cells highly resistant to methicillin [71].

TABLE 12.1
Partial List of Chromosomal Factors Affecting Methicillin Resistance

Gene	Function	Effect on Methicillin Resistance
fmhB	Addition of 1st glycine to the peptidoglycan pentaglycine bridge	Inactivation is lethal
femA	Addition of 2nd and 3rd glycine to the peptidoglycan pentaglycine bridge	Mutants are methicillin susceptible
femB	Addition of 4th and 5th glycine to the peptidoglycan pentaglycine bridge	Inactivation reduces methicillin resistance
femC	Glutamine synthase repressor	Inactivation reduces methicillin resistance
femD	Phosphoglucosamine mutase crucial for precursor formation	Inactivation reduces methicillin resistance
femF	Catalyzes incorporation of lysine into peptidoglycan stem	Inactivation reduces methicillin resistance
lytH	Autolytic enzyme	Inactivation increases methicillin resistance

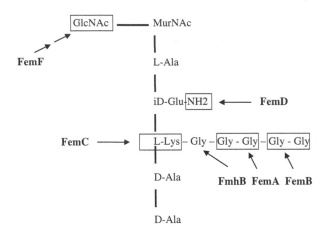

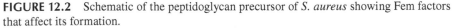

FIGURE 12.2 Schematic of the peptidoglycan precursor of *S. aureus* showing Fem factors that affect its formation.

It is evident from the growing list of auxiliary factors involved in methicillin resistance that disruption of peptidoglycan or membrane biosynthesis has the potential to reduce the optimal function of PBP2a. Methicillin resistance in *S. aureus* involves the concerted actions of *mecRI*, *mecI*, and *mecA* genes together with many metabolic functions of the organism [4].

HETEROGENEOUS METHICILLIN RESISTANCE

An interesting characteristic of methicillin resistance in *S. aureus* is the phenomenon known as heteroresistance, in which subpopulations of cells (10^{-8} to 10^{-4}) in a methicillin-resistant strain, all producing PBP2a, vary markedly in the phenotypic expression of resistance. That is, in clinical isolates of MRSA where the majority of the population is relatively susceptible to β-lactam antibiotics, a small proportion of cells express resistance to high levels of methicillin [61,69]. Although all the cells in an MRSA population have the potential to express resistance to methicillin, the population does not behave in a homogeneous manner [72–74]. The proportion of cells expressing higher resistance levels is strain specific and a reproducible property [75]. The level of resistance in heterogeneous MRSA does not correlate to the quantity of PBP2a present [62,69,76]. In some strains, the highly resistant sub-population will maintain the high level of resistance among descendants of this subpopulation [77]. Among other isolates, however, the highly resistant subclones return to their original resistance upon re-growth from a single colony in drug-free medium [4,77]. Strains consistently producing populations of high-level resistant cells are termed homogeneous expression strains. Even though the subpopulation of highly resistant MRSA within a heterogeneous strain occurs at a low frequency, it can overgrow a culture under conditions of antibiotic pressure [75]. The practical implication is that every MRSA strain, irrespective of whether expression is hetero-geneous or homogeneous, may lead to treatment failure *in vivo* [75].

This phenomenon of heterogeneous resistance makes it necessary for clinical laboratories to use special methods to ensure detection of MRSA. While the genetic cause of heterogeneous resistance is poorly understood, it can be overcome by lower incubation temperatures (30°C to 35°C) and the incorporation of higher salt concentrations (2% to 4% NaCl) in the medium, which are conditions that favor enhanced expression of resistance [72,78].

Origins of Methicillin Resistance

Nucleotide sequencing revealed that the *mecA* gene is composed of separate domains exhibiting sequence similarity to two distinct genes: the 5' region of the *mecA* gene is similar to the penicillinase gene (*blaZ*) of *S. aureus,* and the rest of the gene is related to *E. coli* PBP2 and PBP3 [16,33]. Several theories on the origins of this gene have been proposed that: (*i*) *mecA* emerged by homologous recombination between PBP and a β-lactamase gene in an unknown organism [16,33]; or (*ii*) *mecA* originated in a coagulase-negative staphylococcal species, perhaps a close evolutionary relative of *S. sciuri* [14,48]. When bacterial isolates belonging to over 15 species of staphylococci were examined for reactivity with a DNA probe internal to the *mecA* of an MRSA strain, only one species, *S. sciuri* was invariantly positive for all isolates [79]. However, the *mecA* homolog in *S. sciuri* appears to be silent, as *S. sciuri* isolates express no detectable resistance to either methicillin or penicillin [79,80]. It has been suggested that the *mecA* homolog is native in this bacterium, where it performs some physiological function, such as cell wall biosynthesis [79]. The product of *S. sciuri mecA* possesses a putative transglycosylase (TGase) domain with an *N*-terminal membrane anchor sequence and a putative transpeptidase (TPase) domain, similar to other high-molecular-weight PBPs. The *mecA* of *S. sciuri* exhibits an overall inferred amino acid sequence similarity of 88% and identity of 80% when compared to *mecA* of the MRSA [79]. The *S. sciuri mecA* homolog is by far the most closely related of known genes to *mecA* of MRSA, and it appears certain that both genes share a common evolutionary ancestry, with intermediates most likely occurring elsewhere within the genus [79]. Additional evidence for an evolutionary link was demonstrated when *S. sciuri* mutants, selected for by increasing the concentration of methicillin, were shown to have increased rates of transcription of the *S. sciuri mecA* homolog due to a point mutation in the promoter region [81]. Additionally, when this mutated gene was introduced into methicillin-susceptible strains of *S. aureus*, it conferred resistance [81].

Within *S. aureus*, two theories prevail as to the origin of *mecA* and methicillin resistance. The earliest MRSA isolates may have descended from a single methicillin-resistant clone [82] and then entered other phylogenetic lineages of *S. aureus*. Alternately, the *mec* determinant was acquired from a source outside of the species at different times by different strains [4]. Clonal analysis of MRSA [1,83] and of *mec* determinants stemming from *Staphylococcus* spp. other than *S. aureus* (mainly from *S. haemolyticus* and *S. epidermidis*) support the view that the *mec* determinant was disseminated by horizontal transfer, with coagulase-negative staphylococci possibly serving as the intermediary of the *mec* determinant for *S. aureus* [4,48]. Whereas β-lactamases were rapidly and widely disseminated and are now present in about

300   Bacterial Resistance to Antimicrobials

80% of all staphylococci, the *mec* determinant is still largely restricted to discrete clonal lineages, and seems to favor clonal over horizontal spread [4]. Using DNA microarray analysis, MRSA strains were shown to fall into five distinct chromosomal genotypic groups that are highly divergent relative to one another [84], concluding that MRSA strains have arisen multiple independent times by lateral transfer of the *mec* element into methicillin-susceptible precursors [84].

EPIDEMIOLOGY OF MRSA

The anterior nares are a natural human reservoir for *S. aureus* where it can be isolated from 10% to 40% of healthy adults [5,85]. From the nares, spread to the skin (especially eczematous lesions) and then to surgical wounds, foreign bodies (e.g., indwelling devices), burns, and the upper respiratory tract [72,85,86] is common, with the hands being the major mode of transmission [85,87]. That a common cause of frequently severe infections is carried asymptomatically by a large proportion of the population in an accessible site, such as the anterior nares, challenges paradigms of what constitutes a pathogen. Between 20% and 35% of the population are persistent *S. aureus* carriers, and 30% to 70% are intermittent carriers [88,89]. Identification of carriers is an important key to containment, because strains associated with nasal colonization have been observed to account for 40% to 100% of staphylococcal sepsis, and surgical infection is 2 to 17 times more common among carriers than non-carriers [88].

Nosocomial Infections

S. aureus is now the leading cause of nosocomial pneumonia and surgical site infections [90], and is the second leading cause of nosocomial bloodstream infections behind coagulase-negative staphylococci [90,91]. Infections and outbreaks are common throughout the world in nursing homes [92,93] and among outpatient populations [72,94], in addition to those reported in hospitals [85,86]. Infection with MRSA is especially prominent in intensive care units (ICUs) [85].

MRSA is introduced into an institution primarily by admission of an infected or colonized patient who serves as a reservoir [72,95,96]. Less frequently, MRSA can be introduced by colonized or infected health care workers who disseminate the organism directly to patients [72,97]. The principal mode of transmission of MRSA within the hospital is via transiently colonized hands of health care workers, who acquire the organism after close contact with colonized patients, contaminated equipment, or their own flora [72,85,86,96,98,99]. More rarely, patients can acquire MRSA via airborne transmission, as has been observed in burn units [72,85,98,100–102].

Several risk factors for the acquisition of MRSA have been identified. These include prior hospitalization, admission to an intensive care unit (ICU) or burn unit, invasive procedures, skin lesions, age, and previous antimicrobial treatment [103–107]. Current guidelines for MRSA control in hospitals focus on measures to limit MRSA cross contamination and colonization [103]. These guidelines include measures such as hand washing, the identification of human reservoirs, decontamination of the environment, patient isolation, and notification of known carriers when a patient is transferred to another institution [103]. Despite these procedures, MRSA continues to spread in most institutions, and has become endemic rather than epidemic [53].

Currently, approximately two million hospitalizations annually result in nosocomial infection [91]. Surveillance databases, such as The Surveillance Network (TSN), electronically collect and compile data daily from more than 300 clinical laboratories across the United States, identify potential laboratory testing errors, and detect emergence of resistance profiles and mechanisms that pose a public health threat (e.g., vancomycin-resistant staphylococci) [108–110]. In 1991, using data from the National Nosocomial Infections Surveillance (NNIS) system, it was noted that the percentage of MRSA was greatest from hospitals reporting from the southeastern region of the United States [111]. Using data collected from 1998 through 1999 by TSN U.S.A. (Eurofin Medinet, Herndon, Virginia, U.S.A.), this trend continued with the Southeast reporting 45.5% of *S. aureus* isolates to be MRSA compared to a national average of 35.7%. Data collected from TSN from 1998 through 2005 report all regions except New England above 50%, with the Southeast still reporting the highest rates of MRSA, and within this region Kentucky, Tennessee, Alabama, and Mississippi reporting 63% in both inpatient and outpatient populations. These most recent data also show that multidrug resistance rates remain highest among nosocomial strains as compared to community-acquired strains. It should also be noted that in this study of 14,635 MRSA strains, no resistance to vancomycin was noted and only three strains (0.02%) were resistant to linezolid. So, while there have been reports of resistance to these three agents [110,112,113], it seems for now to be extremely rare.

Today, *S. aureus* is still the most common bacterial species isolated from inpatient specimens and the second most common from outpatient specimens (18.7% and 14.7%, respectively) [110]. The MRSA problem arose initially in large tertiary care hospitals [86,114] with patients in burn [83,98,115,116], postoperative [83,95,98], and ICU [95,96]. Increased risk of MRSA infection was associated with use of multiple broad-spectrum antibiotics [5,83,95,96,117], indwelling devices [95,117], ventilatory support [95], severity of underlying disease [83,5,97,117], and length of hospital stay [83,97,117]. Between 1975 and 1996, the NNIS system reported the percentage of MRSA among nosocomial isolates in the United States had increased from 2.4% to 35% of staphylococcal isolates [111,118]. Rates of MRSA have continued to rise. In 2005, the TSN reported MRSA rates were 55% for ICU patients and 59.2% for non-ICU patients [110]. Nosocomial MRSA tend to possess one of three types of SCC*mecA*: I, II, and III. Types II and III code not only for methicillin resistance, but also resistance to multiple non-β-lactam antibiotics [20,24].

COMMUNITY-ACQUIRED MRSA

Based on experience studying penicillin resistance patterns in *S. aureus*, it is not surprising that the epidemiology of MRSA has shifted from that of an almost exclusively nosocomial problem to now being transmitted within the community with increasing frequency [119–121]. Community-acquired MRSA was first described during the 1980–1981 outbreak of MRSA infections in Detroit [120,122], where approximately two-thirds of the patients affected were injection drug users. Consequently, early studies on community-acquired MRSA in the United States described mainly infections in intravenous drug abusers [123,124] and individuals with recognized predisposing risk factors such as persistent carriage, recent hospital stay (within the last 12 months), serious underlying diseases, previous antibiotic therapy, or residence in a long-term care facility [5,119]. By the mid-1990s, community-acquired MRSA infections were beginning to be described in individuals without these identifiable risk factors [120]. While it remained difficult to distinguish between nosocomial and community-acquired MRSA infections, certain trends were beginning to emerge. It was observed that community-acquired MRSA strains tended to remain susceptible to most classes of antibiotics including clindamycin, macrolides, fluoroquinolones, trimethoprim-sulfamethoxazole, and aminoglycosides, resistant only to the β-lactams, in contrast to the multidrug-resistant pattern of nosocomial strains [120]. This more restricted set of antibiotic resistances was also observed in studies of community-acquired MRSA strains among intravenous drug abusers compared with nosocomially acquired MRSA isolates [120,125,126]. Hospital surveys in the United States and Canada documented a substantial proportion of MRSA infection identified on admission to the hospital [104,119,127] revealing that community MRSA infection was more common than expected, and that the majority of isolates represented distinct strains rather than recent descendants of a single strain [119]. Additionally, while previous reports of community-acquired MRSA infections were generally limited to infections among intravenous drug users and individuals with health care associated risk factors [72,123,124,127], these studies revealed cases of MRSA colonization and infection acquired in the community by individuals lacking those predisposing factors. By the late 1990s, clinicians also observed that the community-acquired MRSA strains had a predilection for skin and soft tissue infection [128].

In 2002, a novel SCC type (type IV, described above) was isolated from a community-acquired strain [27] and has been found to be present in 89% of community-acquired isolates [20,129]. This genetic element carries only the *mecA* resistance gene, consistent with the finding that community-acquired strains tend to be susceptible to non-β-lactam antibiotics. In addition, community-acquired MRSA were associated with exotoxin genes, including the Panton-Valentine leukocidin (PVL) [129]. PVL is a two-component staphylococcal membrane toxin that targets leukocytes, and is found in about 2% of all *S. aureus* clinical isolates, including both MRSA and MSSA [130]. Contact between PVL and human neutrophils, monocytes, macrophages, or erythrocytes, results in pore formation and cell lysis through osmotic rupture [130]. Isolates containing the PVL genes are often associated with recurrent and often severe primary skin infections and severe necrotizing pneumonia [131,132].

COST ATTRIBUTABLE TO MRSA

Over nearly five decades, methicillin resistance has represented a major therapeutic, management, and epidemiological problem throughout the world [83,133]. MRSA colonization and infection has been shown to increase morbidity, length of hospital stay, and hospital cost. Nosocomial bloodstream infection with MRSA was found to prolong hospitalization an average of eight days over similar infections caused by MSSA, resulting in an approximately three-fold increase in direct costs [134]. Studies have shown that treating an MRSA infection can cost 6% to 10% more than treating a methicillin-sensitive infection [91]. This difference does not reflect greater virulence of MRSA; rather, it reflects the increased cost of vancomycin treatment, longer hospital stay, and the cost of patient isolation and infection-control measures. In addition to increasing costs, the mortality rate attributable to MRSA infections has been observed in some studies to be more than 2.5 times higher than that attributable to MSSA infections (21% vs. 8%) [91]. Although it should be noted that some of the death rate difference may be related to the underlying condition of patients who become infected with MRSA, such as older patients and patients previously exposed to antibiotics, as well as the lack of effectiveness of vancomycin to cure MRSA [91].

EXPECTATIONS FOR THE FUTURE

Currently, more than 95% of patients with *S. aureus* infections worldwide do not respond to first-line antibiotics, such as penicillin or ampicillin [91,135]. Moreover, MRSA are increasingly found in the community, including individuals who have never been hospitalized [103,104,136]. Many multidrug-resistant MRSA strains are presently only susceptible to a single class of clinically available bactericidal antibiotic, the glycopeptides (vancomycin and teicoplanin), and the widespread acquisition of the *vanA* or *vanB* determinants from enterococci would be a potential public health disaster. Currently, intravenous vancomycin is the standard antibiotic for empirical therapy. But as more vancomycin-intermediate resistant strains of *S. aureus* are isolated, this line of therapy may be compromised. Linezolid, from the new class of antibiotics, oxazolidinones, is available for intravenous and oral administration, but costs $100 to $1000 more per treatment course, which may limit its use [137].

Several studies suggest that reduction of antibiotic use within the hospital could decrease nosocomial acquisition of multi-resistant bacteria [85,138–140], and scheduled rotation of antibiotic use has also been suggested [85,141]. In addition to prudent use of antibiotics, strict compliance with infection control policies can aid in the reduction of nosocomial spread of multidrug-resistant MRSA. This may, however, be harder to effect than decreasing antibiotic use, since studies have shown that compliance with simple hand washing in ICUs varies from only 20 to 40% [85,142–145].

Characterization of the interactions between PBP2a and β-lactams may elucidate the basis for the extremely low affinity for β-lactam antibiotics and contribute to the rationale to design better PBP2a inhibitors, leading to more effective antibacterial agents for MRSA and other bacteria [146]. PBP2a has already been utilized as a screening target for discovery of new β-lactam antibiotics with enhanced affinity and improved activity against MRSA [146,147].

There is an obvious need for more effective antibiotic therapy for infections with MRSA. Reports describing treatment failure of vancomycin for multidrug-resistant MRSA infections have raised concern for the emergence of strains of MRSA for which there will be no effective, affordable therapy. However, new therapeutic agents alone will not provide a long-term solution, and our attention to prevention must remain constant. Strict adherence to hospital infection-control practices, as well as appropriate use of antibiotics and improved surveillance systems to track the emergence of resistance patterns, are of primary importance as we look to the future usefulness of antibiotic therapy against this extremely adaptive organism.

REFERENCES

1. Hartman BJ, Tomasz A. Low-affinity penicillin-binding protein associated with β-lactam resistance in *Staphylococcus aureus*. *J Bacteriol* 1984; 158:513–516.
2. Reynolds PE, Fuller C. Methicillin-resistant strains of *Staphylococcus aureus*; presence of identical additional penicillin-binding protein in all strains examined. *FEMS Microbiol Lett* 1986; 33:251–254.
3. Wise EM, Park JT. Penicillin: its basic site of action as an inhibitor of a peptide cross-linking reaction in cell wall mucopeptide synthesis. *Proc Natl Acad Sci USA* 1965; 54:75–81.
4. Berger-Bächi B. Resistance not mediated by β-lactamase (methicillin resistance). In: Crossley KB, Archer GL, eds. *The Staphylococci in Human Disease*. New York: Churchill Livingstone, 1997:158–167.
5. Bradley SF. Methicillin-resistant *Staphylococcus aureus* infection. *Clin Geriatr Med* 1992; 8553–868.
6. Bamberger DM, Boyd SE. Management of *Staphylococcus aureus* infections. *Am Fam Phys* 2005; 72:2474–2481.
7. Wise RI, Ossman EA, Littlefield DR. Personal reflections on nosocomial staphylococcal infections and the development of hospital surveillance. *Rev Infect Dis* 1989; 11(6):1005–1019.
8. Chambers HF. The changing epidemiology of *Staphylococcus aureus*? *Emerg Infect Dis* 2001; 7:178–182.
9. Hiramatsu K. The emergence of *Staphylococcus aureus* with reduced susceptibility to vancomycin in Japan. *Am J Med* 1998; 104:7S–10S.
10. Barbar M. Methicillin-resistant staphylococci. *J Clin Pathol* 1961; 14:385–393.
11. Bulger RJ. A methicillin-resistant strain of *Staphylococcus aureus*: clinical and laboratory experience. *Annal Intern Med* 1967; 67:81–89.
12. Barrett FF, McGehee RF, Finland M. Methicillin-resistant *Staphylococcus aureus* at Boston City Hospital. Bacteriologic and epidemiologic observations. *N Engl J Med* 1968; 279:441–448.
13. Finland M. Changing patterns of resistance of certain common pathogenic bacteria to antimicrobial agents. *N Engl J Med* 1955; 252:570–580.
14. Chambers HF. Penicillin-binding protein-mediated resistance in pneumococci and staphylococci. *J Infect Dis* 1999; 179:S353–S359.
15. McDougal LK, Thornsberry C. The role of β-lactamase in staphylococcal resistance to penicillinase-resistant penicillins and cephalosporins. *J Clin Microbiol* 1986; 23:832–839.
16. Hiramatsu K. Molecular evolution of MRSA. *Microbiol Immunol* 1995; 39(8): 531–543.
17. Tomasz A, Drugeon HB, de Lencastre HM, Jabes D, McDougal L, Bille J. New mechanism for methicillin resistance in *Staphylococcus aureus:* clinical isolates that lack the PBP 2a gene and contain normal penicillin-binding proteins with modified penicillin-binding capacity. *Antimicrob Agents Chemother* 1989; 33:1869–1874.
18. Katayama Y, Ito T, Hiramatsu K. A new class of genetic element, staphylococcus cassette chromosome *mec,* encodes methicillin resistance in *Staphylococcus aureus*. *Antimicrob Agents Chemother* 2000; 44:1549–1555.
19. Kuhl SA, Pattee PA, Baldwin JN. Chromosomal map location of the methicillin resistance determinant in *Staphylococcus aureus*. *J Bacteriol* 1978; 135: 460–465.
20. Hiramatsu K, Cui L, Kuroda M, Ito T. The emergence and evolution of methicillin-resistant *Staphylococcus aureus*. *Trends Microbiol* 2001; 9:486–493.
21. Archer GL, Niemeyer DM. Origin and evolution of DNA associated with resistance to methicillin in staphylococci. *Trends Microbiol* 1994; 2:343–347.

22. Stewart PR, Dubin DT, Chikramane SG, Inglis B, Matthews PR, Poston SM. IS257 and small plasmid insertions in the *mec* region of the chromosome of *Staphylococcus aureus*. *Plasmid* 1994; 31:12–20.

23. Kreiswirth B, Kornblum J, Arbeit RD, Eisner W, Maslow JN, McGeer A, Low DE, Novick RP. Evidence for a clonal origin of methicillin resistance in *Staphylococcus aureus*. *Science* 1993; 259:227–230.

24. Ito T, Ma XX, Takeuchi F, Okuma K, Yuzawa H, Hiramatsu K. Novel type V staphylococcal cassette chromosome *mec* driven by a novel cassette chromosome recombinase, *ccrC*. *Antimicrob Agents Chemother* 2004; 48:2637–2651.

25. Ito T, Katayama Y, Asada K, Mori N, Tsutsumimoto K, Tiensasitorn C, Hiramatsu K. Structural comparison of three types of staphylococcal cassette chromosome *mec* integrated in the chromosome in methicillin-resistant *Staphylococcus aureus*. *Antimicrob Agents Chemother* 2001; 45:1323–1336.

26. Ito T, Katayama Y, Hiramatsu K. Cloning and nucleotide sequence determination of the entire *mec* DNA of pre-methicillin-resistant *Staphylococcus aureus* N315. *Antimicrob Agents Chemother* 1999; 43:1449–1458.

27. Ma XX, Ito T, Tiensasitorn C, Jamklang M, Chongtrakool P, Boyle-Vavra S, Daum RS, Hiramatsu K. Novel type of staphylococcal cassette chromosome *mec* identified in community-acquired methicillin-resistant *Staphylococcus aureus* strains. *Antimicrob Agents Chemother* 2002; 46:1147–1152.

28. Sousa MA, de Lencastre H. Evolution of sporadic isolates of methicillin-resistant *Staphylococcus aureus* (MRSA) in hospitals and their similarities to isolates of community-acquired MRSA. *J Clin Microbiol* 2003; 41:3806–3815.

29. Kobayashi N, Wu H, Kojima K, Taniguchi K, Urasawa S, Uehara N, Omizu Y, Kishi Y, Yagihashi A, Kurokawa I. Detection of *mecA*, *femA*, and *femB* genes in clinical strains of staphylococci using polymerase chain reaction. *Epidemiol Infect* 1994; 113:259–266.

30. Ryffel C, Tesch W, Birch-Machin I, Reynolds PE, Barberis-Maino L, Kayser FH, Berger-Bachi B. Sequence comparison of *mecA* genes isolated from methicillin-resistant *Staphylococcus aureus* and *Staphylococcus epidermidis*. *Gene* 1990; 94:137–138.

31. Suzuki E, Hiramatsu K, Yokota T. Survey of methicillin-resistant clinical strains of coagulase-negative staphylococci for *mecA* gene distribution. *Antimicrob Agents Chemother* 1992; 36:429–434.

32. Ubukata K, Nonoguchi R, Song MD, Matsuhashi M, Konno M. Homology of *mecA* gene in methicillin-resistant *Staphylococcus haemolyticus* and *Staphylococcus simulans* to that of *Staphylococcus aureus*. *Antimicrob Agents Chemother* 1990; 34370–172.

33. Song MD, Wachi M, Doi M, Ishino F, Matsuhashi M. Evolution of an inducible penicillin-target protein in methicillin-resistant *Staphylococcus aureus* by gene fusion. *FEBS Lett* 1987; 221:167–171.

34. Georgopapadakou NH, Dix BA, Mauriz YR. Possible physiological functions of penicillin-binding proteins in *Staphylococcus aureus*. *Antimicrob Agents Chemother* 1986; 29:333–336.

35. Reynolds PE. The essential nature of staphylococcal penicillin-binding proteins. In: Actor P, Daneo-Moore L, Higgins ML, Salton MR, Shockman GD, eds. *Antibiotic Inhibition of Bacterial Cell Surface Assembly and Function*. Washington, DC: American Society for Microbiology, 1988:343–351.

36. Giesbrecht P, Kersten T, Maidhof H, Wecke J. Staphylococcal cell wall: morphogenesis and fatal variations in the presence of penicillin. *Microbiol Mol Biol Rev* 1998; 62:1371–1414.

37. Pinho MG, de Lencastre H, Tomasz A. An acquired and a native penicillin-binding protein cooperate in building the cell wall of drug-resistant staphylococci. *Proc Natl Acad Science USA* 2001; 98:10886–10891.

38. de Lencastre H, de Jonge BLM, Matthews PR, Tomasz A. Molecular aspects of methicillin resistance in *Staphylococcus aureus*. *J Antimicrob Chemother* 1994; 33:7–24.

39. Pinho MG, Filipe SR, de Lencastre H, Tomasz A. Complementation of the essential peptidoglycan transpeptidase function of penicillin-binding protein 2 (PBP2) by the drug resistance protein PBP2A in *Staphylococcus aureus*. *J Bacteriol* 2001; 183:6525–6531.

40. Hackbarth CJ, Chambers HF. *balI* and *blaRI* regulate β-lactamase and PBP2a production in methicillin-resistant *Staphylococcus aureus*. *Antimicrob Agents Chemother* 1993; 37:1144–1149.

41. Hiramatsu K, Asada K, Suzuki E, Okonogi K, Yokota T. Molecular cloning and nucleotide sequence determination of the regulator region of *mecA* gene in methicillin-resistant *Staphylococcus aureus* (MRSA). *FEBS Lett* 1992; 298: 133–136.

42. Ryffel C, Kayser FH, Berger-Bächi B. Correlation between regulation of *mecA* transcription and expression of methicillin resistance in staphylococci. *Antimicrob Agents Chemother* 1992; 36:25–31.

43. Lewis RA, Dyke KGH. MecI represses synthesis from the β-lactamase operon of *Staphylococcus aureus*. *J Antimicrob Chemother* 2000; 45:139–144.

44. Berger-Bächi B, Rohrer S. Factors influencing methicillin resistance in staphylococci. *Arch Microbiol* 2002; 178:165–171.

45. Boyce JM, Medeiros AA, Papa EF, O'Gara CJ. Induction of β-lactamase and methicillin resistance in unusual strains of methicillin-resistant *Staphylococcus aureus*. *J Antimicrob Chemother* 1990; 25:73–81.

46. Clarke SR, Dyke KGH. The signal transducer (BlaR1) and the repressor (Bla1) of the *Staphylococcus aureus* β-lactamase operon are inducible. *Microbiology* 2001; 147:803–810.

47. Zhang HZ, Hackbarth CJ, Chansky KM, Chambers HF. A proteolytic transmembrane signaling pathway and resistance to β-lactams in staphylococci. *Science* 2001; 291:1962–1965.

48. Archer GL, Niemeyer DM, Thanassi JA, Pucco MJ. Dissemination among staphylococci of DNA sequences associated with methicillin resistance. *Antimicrob Agents Chemother* 1994; 38:447–454.

49. Suzuki E, Kuwahara-Arai K, Richardson JF, Hiramatsu K. Distribution of *mec* regulator genes in methicillin-resistant *Staphylococcus* clinical strains. *Antimicrob Agents Chemother* 1993; 37:1219–1226.

50. Weller TMA. The distribution of *mecA*, *mecRI* and *mecI* and sequence analysis of *mecI* and the *mec* promoter region in staphylococci expressing resistance to methicillin. *J Antimicrob Chemother* 1999; 43:15–22.

51. Kobayashi N, Taniguchi K, Kojima K, Urasawa S, Uehara N, Omizu Y, Kishi Y, Yagihashi A, Kurokawa I, Watanabe N. Genomic diversity of *mec* regulator genes in methicillin-resistant *Staphylococcus aureus* and *Staphylococcus epidermidis*. *Epidemiol Infect* 1996; 117:289–295.

52. Kobayashi N, Taniguchi K, Urasawa S. Analysis of diversity of mutations in the *mecI* gene and *mecA* promoter/operator region of methicillin-resistant *Staphylococcus aureus* and *Stuphylococcus epidermidis*. *Antimicrob Agents Chemother* 1998; 42:717–720.

53. Schentag JJ, Hyatt JM, Carr JR, Paladino JA, Birmingham MC, Zimmer GS, Cumbo TJ. Genesis of methicillin-resistant *Staphylococcus aureus* (MRSA), how treatment of MRSA infections has selected for vancomycin-resistant *Enterococcus faecium* and the importance of antibiotic management and infection control. *Clin Infect Dis* 1998; 26:1204–1214.

54. Kuwahara-Arai K, Kondo N, Hori S, Tateda-Suzuki E, Hiramatsu K. Suppression of methicillin resistance in a *mecA*-containing pre-methicillin-resistant *Staphylococcus aureus* strain is caused by the *mecI*-mediated repression of PBP 2′ production. *Antimicrob Agents Chemother* 1996; 40:2680–2685.

55. Hackbarth CJ, Miick C, Chambers HF. Altered production of penicillin-binding protein 2a can affect phenotypic expression of methicillin resistance in *Staphylococcus aureus*. *Antimicrob Agents Chemother* 1994; 38:2568–2571.

56. Berger-Bächi B, Strassle A, Kayser FH. Characterization of an isogenic set of methicillin-resistant and susceptible mutants of *Staphylococcus aureus*. *Eur J Clin Microbiol* 1986; 5:697–701.

57. Kornblum J, Hartman BJ, Novick RP, Tomasz A. Conversion of a homogeneously methicillin-resistant strain of *Staphylococcus aureus* to heterogeneous resistance by Tn*551*-mediated insertional inactivation. *Eur J Clin Microbiol* 1986; 5:714–718.

58. Murakami K, Tomasz A. Involvement of multiple genetic determinants in high-level methicillin resistance in *Staphylococcns aureus*. *J Bacteriol* 1989; 171:874–879.

59. Berger-Bächi B, Barberis-Maino L, Strassle A, Kayser FH. FemA, a host-mediated factor essential for methicillin resistance in *Staphylococcus aureus:* molecular cloning and characterization. *Mol Gen Genet* 1989; 219:263–269.

60. Tomasz A. Auxiliary genes assisting in the expression of methicillin resistance in *Staphylococcus aureus*. In: Novick RP, ed. *Molecular Biology of the Staphylococci*. New York: VCH, 1990:565–583.

61. Chambers HE. Methicillin resistance in Staphylococci: molecular and biochemical basis and clinical implications. *Clin Microbiol Rev* 1997; 10:781–791.

62. Berger-Bächi B. Insertional inactivation of staphylococcal methicillin resistance by Tn*551*. *J Bacteriol* 1983; 154:479–487.

63. de Lencastre H, Tomasz A. Reassessment of the number of auxiliary genes essential for expression of high-level methicillin resistance in *Staphylococcus aureus*. *Antimicrob Agents Chemother* 1994; 38:2590–2598.

64. Wu SW, de Lencastre H. Mrp-A new auxiliary gene essential for optimal expression of methicillin resistance in *Staphylococcus aureus*. *Microbial Drug Resist* 1999; 5:9–18.

65. de Lencastre H, Wu SW, Pinho MG, Ludovice AM, Filipe S, Gardete S, Sobral R, Gill S, Chung M, Tomasz A. Antibiotic resistance as a stress response: Complete sequencing of a large number of chromosomal loci in *Staphylococcus aureus* strain COL that impact on the expression of resistance to methicillin. *Microb Drug Res* 1999; 5:163–175.

66. de Jonge BL, Sidow T, Chang YS, Labischinski H, Berger-Bächi B, Gage DA, Tomasz A. Altered muropeptide composition in *Staphylococcus aureus* strains with an inactivated *femA* locus. *J Bacteriol* 1993; 175:2779–2782.

67. Roher S, Ehlert K, Tschierske M, Labischinski H, Berger-Bächi B. The essential *Staphylococcus aureus* gene *fmhB* is involved in the first step of peptidoglycan pentaglycine interpeptide formation. *Proc Natl Acad Sci USA* 1999; 96:9351–9356.

68. Henze U, Sidow T, Wecke J, Labischinski H, Berger-Bächi B. Influence of *femB* on methicillin resistance and peptidoglycan metabolism in *Staphylcoccus aureus*. *J Bacteriol* 1993; 175:1612–1620.

69. Berger-Bächi B, Strassle A, Gustafson JE, Kayser FH. Mapping and characterization of multiple chromosomal factors involved in methicillin resistance in *Staphylococcus aureus*. *Antimicrob Agents Chemother* 1992; 36:1367–1373.

70. Glanzmann P, Gustafson J, Komatsuzawa H, Ohta K, Berger-Bächi B. *glmM* operon and methicillin-resistant *glmM* supressors mutants in *Staphylococcus aureus*. *Antimicrob Agents Chemother* 1999;43:240–245.

71. Berger- Bächi B. Genetic basis of methicillin resistance in *Staphylococcus aureus*. *Cell Mol Life Sci* 1999; 56:764–770.

72. Mulligan ME, Murray-Leisure KA, Ribner BS, Standiford HC, John JF, Korvick JA, Kaufman CA, Yu VL. Methicillin-resistant *Staphylococcus aureus:* a consensus review of the microbiology, pathogenesis, and epidemiology with implications for prevention and management. *Am J Med* 1993; 94:313–328.

73. Jorgensen JH. Laboratory and epidemiologic experience with methicillin-resistant *Staphylococcus aureus* in the USA. *Eur J Clin Microbiol* 1986; 5:693–696.

74. Sabeth LD. Mechanisms of resistance to β-lactam antibiotics in strains of *Staphylococcus aureus*. *Ann Intern Med* 1982; 97:339–344.

75. Tomasz A, Nachman S, Leaf H. Stable classes of phenotypic expression in methicillin-resistant clinical isolates of staphylococci. *Antimicrob Agents Chemother* 1991; 35:124–129.

76. Chambers HF, Hackbarth CJ. Effect of NaCl and nafcillin on penicillin-binding protein 2a and heterogeneous expression of methicillin resistance in *Staphylococcus aureus*. *Antimicrob Agents Chemother* 1987; 31:1982–1988.

77. de Lencastre H, Figueiredo AM, Tomasz A. Genetic control of population structure in heterogeneous strains of methicillin resistant *Staphylococcus aureus*. *Eur J Clin Microbiol Infect Dis* 1993; 12:S13–S18.

78. Jorgensen JH, Redding JS, Maher LA, Ramirez PE. Salt-supplemented medium for testing methicillin-resistant staphylococci with newer β-lactams. *J Clin Microbiol* 1988; 26:1675–1678.

79. Wu S, Piscitelli C, de Lencastre H, Tomasz A. Tracking the evolutionary origin of the methicillin resistance gene: cloning and sequencing of a homologue of *mecA* from a methicillin susceptible strain of *Staphylococcus sciuri*. *Microb Drug Resist* 1996; 2:435–441.

80. Couto I, de Lencastre H, Severina E, Kloos W, Webster JA, Hubner RJ, Sanches IS, Tomasz A. Ubiquitous presence of a *mecA* homologue in natural isolates of *Staphylococcus sciuri*. *Microb Drug Resist* 1996; 2:377–391.

81. Wu SW, de Lencastre H, Tomasz A. Recruitment of the *mecA* gene homologue of *Staphylococcus sciuri* into a resistance determinant and expression of the resistant phenotype in *Staphylococcus aureus*. *J Bacteriol* 2001; 183:2417–2424.

82. Lacey RW, Grinsted J. Genetic analysis of methicillin-resistant strains of *Staphylococcus aureus:* evidence for their evolution from a single clone. *J Med Microbiol* 1973; 6:511–526.

83. Boyce JM, Landry M. Deetz TR, DuPont HL. Epidemiologic studies of an outbreak of nosocomial methicillin-resistant *Staphylococcus aureus* infections. *Infect Control* 1981; 2:110–116.

84. Fitzgerald JR, Sturdeant DE, Mackie SM, Gill SR, Musser JM. Evolutionary genomics of *Staphylococcus aureus*: insights into the origin of methicillin-resistant strains and the toxic shock syndrome epidemic. *Proc Natl Acad Sci USA* 2001; 98:8821–8826.

85. Dennesen PJW, Bonten MJM, Weinstein RA. Multiresistant bacteria as a hospital epidemic problem. *Ann Med* 1998; 30:176–185.

86. Haley RW, Hightower AW, Khabbaz RF, Thornsberry C, Martone JW, Allen JR, Hughes JM. The emergence of methicillin-resistant *Staphylococcus aureus* infections in United States hospitals. Possible role of the house staff-patient transfer circuit. *Ann Intern Med* 1982; 97:297–308.

87. Edmond MB, Wenzel RP, Pasculle AW. Vancomycin-resistant *Staphylococcus aureus:* perspectives on measures needed for control. *Ann Intern Med* 1996; 124:329–334.

88. Casewell MW. The nose: an underestimated source of *Staphylococcus aureus* causing wound infection. *J Hosp Infect* 1998; 40:S4–S11.

89. Williams REO. Healthy carriage of *Staphylococcus aureus:* its prevalence and importance. *Bacteriol Rev* 1963; 27:56–71.

90. Centers for Disease Control and Prevention. National Nosocomial Infection Surveillance System report: data summary from October 1986–April 1996. Atlanta: US Department of Health and Human Services, 1996.

91. Rubin RJ, Harrington CA, Poon A, Dietrich K, Greene JA, Moiduddin A. The economic impact of *Staphylococcus aureus* infection in New York City hospitals. *Emerg Infect Dis* 1999; 5:9–17.

92. Storch GA, Radcliff JL, Meyer PL, Hinrichs JH. Methicillin-resistant *Staphylococcus aureus* in a nursing home. *Infect Control* 1987; 8:24–29.

93. Kaufmann CA, Bradley SF, Terpenning MS. Methicillin-resistant *Staphylococcus aureus* in long-term care facilities. *Infect Control Hosp Epidemiol* 1990; 11: 600–603.

94. Levine DP, Cushing RD, Jui J, Brown WJ. Community-acquired methicillin-resistant *Staphylococcus aureus* endocarditis in the Detroit Medical Center. *Ann Intern Med* 1982; 97:330–338.

95. Craven DE, Reed C, Kollisch N, DeMaria A, Lichtenberg D, Shen K, McCabe WR. A large outbreak of infections caused by a strain of *Staphylococcus aureus* resistant to oxacillin and aminoglycosides. *Am J Med* 1981; 71:53–58.

96. Peacock JE Jr, Marsik FJ, Wenzel RP. Methicillin-resistant *Staphylococcus aureus*: introduction and spread within a hospital. *Ann Intern Med* 1980; 93:526–532.

97. Ward TT, Winn RE, Hartstein AL, Sewell DL. Observations relating to an interhospital outbreak of methicillin-resistant *Staphylococcus aureus*: role of antimicrobial therapy in infection control. *Infect Control* 1981; 2:453–459.

98. Crossley K, Landesman B, Zaske D. An outbreak of infections caused by strains of *Staphylococcus aureus* resistant to methicillin and aminoglycosides. Epidemiologic Studies. *J Infect Dis* 1979; 139:280–287.

99. Boyce JM, Opal SM, Potter-Bynoe G, Medeiros AA. Spread of methicillin-resistant *Staphylococcus aureus* in a hospital after exposure to a health care worker with chronic sinusitis. *Clin Infect Dis* 1993; 17:496–504.

100. Boyce JM, White RL, Causey WA, Lockwood WR. Burn units as a source of methicillin-resistant *Staphylococcus aureus* infections. *JAMA* 1983; 249:2803–2807.

101. Rutala WA, Katz EB, Sherertz RJ, Sarubbi FA Jr. Environmental study of a methicillin-resistant *Staphylococcus aureus* epidemic in a burn unit. *J Clin Microbiol* 1983; 18:683–688.

102. Farrington M, Ling J, Ling T, French GL. Outbreaks of infection with methicillin-resistant *Staphylococcus aureus* on neonatal and burn units of a new hospital. *Epidemiol Infect* 1990; 105:215–228.

103. Monnet DL. Methicillin-resistant *Staphylococcus aureus* and its relationship to antimicrobial use: possible implications for control. *Infect Control Hosp Epidemiol* 1998; 19:552–559.

104. Layton MC, Hierholzer WJ Jr, Patterson JE. The evolving epidemiology of methicillin-resistant *Staphylococcus aureus* at a university hospital. *Infect Control Hosp Epidemiol* 1995; 16:12–17.

105. Asensio A, Guerrero A, Quereda C, Lizan M, Martinez-Ferrer M. Colonization and infection with methicillin-resistant *Staphylococcus aureus*: associated factors and eradication. *Infect Control Hosp Epidemiol* 1996; 17:20–28.

106. Humphreys H, Duckworth G. Methicillin-resistant *Staphylococcus aureus* (MRSA) a reappraisal of control measures in the light of changing circumstances. *J Hosp Infect* 1997; 36:167–170.

107. Thompson RL, Cabezudo I, Wenzel RP. Epidemiology of nosocomial infections caused by methicillin-resistant *Staphylococcus aureus*. *Ann Intern Med* 1982; 97:309–317.

108. Sahm DF, Marsilio MK, Piazza G. Antimicrobial resistance in key bloodstream bacterial isolates: electronic surveillance with The Surveillance Network Database—USA. *Clin Infect Dis* 1999; 29:259–263.

109. Jones ME, Mayfield DC, Thornsberry C, Karlowaky JA, Sahm DF, Peterson D. Prevalence of oxacillin resistance in *Staphylococcus aureus* among inpatients and outpatients in the United States during 2000. *Antimicrob Agents Chemother* 2002; 46:3104–3105.

110. Styers D, Sheehan DJ, Hogan P, Sahm DF. Laboratory-based surveillance of current antinibrobial resistance patterns and trends among *Staphylococcus aureus*: 2005 status in the United States. *Annal Clin Microb Antimicrobial* 2006; 5:2(1–9).

111. Panlilio AL, Culver DH, Gaynes RP, Banerjee S, Henderson TS, Tolson JS, Martone WJ. Methicillin-resistant *Staphylococcus aureus* in U.S. hospitals, 1975–1991. *Infect Control Hosp Epidemiol* 1992; 13:582–586.

112. Tenover FC, McDonald LC. Vancomycin-resistant staphylococci and enterococci: epidemiology and control. *Curr Opin Infect Dis* 2005; 18(4):300–305.

113. Jevitt LA, Smith AJ, Williams PP, Raney PM, McGowan JE Jr, Tenover FC. *In vitro* activities of Daptomycin, Linezolid, and Quinupristin-Dalfopristin against a challenge panel of staphylococci and enterococci, including vancomycin-intermediate *Staphylococcus aureus* and vancomycin-resistant *Enterococcus faecium*. *Microb Drug Resist* 2003; 9:389–393.

114. Boyce JM, Causey WA. Increasing occurrence of methicillin-resistant *Staphylococcus aureus* in the United States. *Infect Control* 1982; 3:377–383.

115. Musser JM, Kapur V. Clonal analysis of methicillin-resistant *Staphylococcus aureus* strains from intercontinental sources: association of the *mec* gene with divergent phylogenetic lineages implies dissemination by horizontal transfer and recombination. *J Clin Microbiol* 1992; 30:2058–2063.

116. Locksley RM, Cohen ML, Quinn TC, Thompkins LS, Coyle MB, Kirihara JM, Counts GW. Multiply antibiotic-resistant *Staphylococcus aureus*: introduction, transmission, and evolution of nosocomial infection. *Ann Intern Med* 1982; 97:317–324.

117. Rimland D. Nosocomial infections with methicillin and tobramycin-resistant *Staphylococcus aureus*, implication of physiotherapy in hospital-wide dissemination. *Am J Med Sci* 1985; 290:91–97.

118. Gaynes RP, Culver DH. National Nosocomial Infection Surveillance (NNIS) System. Nosocomial methicillin-resistant *Staphylococcus aureus* (MRSA) in the United States, 1975–1996. In: *Proceedings of the Annual Meeting of the Infectious Disease Society of America (San Francisco)*. Alexandria, VA: IDSA, 1997:206.

119. Moreno F, Crisp C, Jorgensen JH, Patterson JE. Methicillin-resistant *Staphylococcus aureus* as a community organism. *Clin Infect Dis* 1995; 21:1308–1312.

120. Herold BC, Irnrnergluck LC, Maranan MC, Lauderdale DS, Gaskin RE, Boyle-Vavra S, Leitch CD, Daum RS. Community-acquired methicillin-resistant *Staphylococcus aureus* in children with no identified predisposing risk. *JAMA* 1998; 279:593–598.

121. Salgado CD, Farr BM, Calfee DP. Community-acquired methicillin-resistant *Staphylococcus aureus*: a meta-analysis of prevalence and risk factors. *Clin Infect Dis* 2003; 36:131–139.

122. Saravolatz LD, Markowitz N, Arking L, Pohlod D, Fisher E. Methicillin-resistant *Staphylococcus aureus* epidemiologic observations during a community-acquired outbreak. *Ann Intern Med* 1982; 96:11–16.

123. Saravolatz LD, Pohlod DJ, Arking LM. Community-acquired MRSA infections: a new source of nosocomial outbreaks. *Ann Intern Med* 1982; 97:325–329.

124. Saravolatz LD, Markowitz N, Arking L, Pohlod D, Fisher E. MRSA. Epidemiologic observations during a community-acquired outbreak. *Ann Intern Med* 1982; 96:11–16.

125. Craven DE, Rixinger AI, Goularte TA, McCabe WR. Methicillin-resistant *Staphylococcus aureus* bacteremia linked to intravenous drug abusers using a "shooting gallery." *Am J Med* 1986; 80:770–776.

126. Berman DS, Schafler S, Simberkoff MS, Rahal JJ. *Staphylococcus aureus* colonization in intravenous drug abusers, dialysis patients, and diabetics. *J Infect Dis* 1987; 155929–831.

127. Embil J, Ramotar K, Romance L, Alfa M, Conly J, Cronk S, Taylor G, Sutherland B, Louie T, Henderson E, et al. Methicillin-resistant *Staphylococcus aureus* in tertiary care institutions on the Canadian prairies 1990–1992. *Infect Control Hosp Epidemiol* 1994; 15:646–651.

128. Frank AL, Marcinak JF, Mamgat PD, Schreckenberger PC. Community-acquired and clindamycin-susceptible methicillin-resistant *Staphylococcus aureus* in children. *Pediatr Infect Dis J* 1999; 18:993–1000.
129. Charlebois ED, Perdreau-Remington F, Kreiswirth B, Bangsberg DR, Ciccarone D, Diep BA, Ng VL, Chansky K, Edlin B, Chambers HF. Origins of community strains of methicillin-resistant *Staphylococcus aureus*. *Clin Infect Dis* 2004; 39:47–54.
130. Zetola N, Francis JS, Nuermberger EL, Bishai WR. Community-acquired methicillin-resistant *Staphylococcus aureus*: an emerging threat. *Lancet Infect Dis* 2005; 5:275–286.
131. Lina G, Piemont Y, Godail-Gamot F, Bes M, Peter MO, Gauduchon V, Vandenesch F, Etienne J. Involvement of Panton-Valentine leukocidin-producing *Staphylococcus aureus* in primary skin infections and pneumonia. *Clin Infect Dis* 1999; 29:1128–1132.
132. Gillet Y Issartel B, Vanhems P, Fouret JC, Lina G, Bes M, Vandenesch F, Piemont Y, Brousse N Floret D, Etienne J. Association between *Staphylococcus aureus* strains carrying gene for Panton-Valentine leukocidin and highly lethal necrotizing pneumonia in young immunocompetent patients. *Lancet* 2002; 359:7753–759.
133. Pittet D, Tarara D, Wenzel RP. Nosocomial bloodstream infection in critically ill patients. Excess length of stay, extra costs, and attributable mortality. *JAMA* 1994; 271:1598–1601.
134. Abramsom MA, Sexton DJ. Nosocomial methicillin-resistant and methicillin-susceptible *Staphylococcus aureus* primary bacteremia: at what costs? *Infect Control Hosp Epidemiol* 1999; 20:408–411.
135. Neu HC. The crisis in antibiotic resistance. *Science* 1992; 257:1064–1073.
136. Lugeon C, Blanc DS, Wenger A, Francioli P. Molecular epidemiology of methicillin-resistant *Staphylococcus aureus* at a low-incidence hospital over a 4-year period. *Infect Control Hosp Epidemiol* 1995; 16:260–267.
137. Chambers HF. Community-associated MRSA resistance and virulence converge. *N Engl J Med* 2005; 352:1485–1486.
138. Quale J, Landman D, Saurina G, Atwood E, DiTore V, Patel K. Manipulation of a hospital antimicrobial formulary to control an outbreak of vancomycin-resistant enterococci. *Clin Infect Dis* 1996; 23:1020–1025.
139. Rice LB, Wiley SH, Papanicolaou GA, Medeiros AA, Eliopoulos GM, Moellering RC Jr, Jacoby GA. Outbreak of ceftazidime resistance caused by extended-spectrum β-lactamases at a Massachusetts chronic-care facility. *Antimicrob Agents Chemother* 1990; 34:2193–2199.
140. Meyer KS, Urban C, Eagan JA, Berger BJ, Rahal JJ. Nosocomial outbreak of *Klebsiella* infection resistant to late-generation cephalosporins. *Ann Intern Med* 1993; 119:353–358.
141. Kollef MH, Vlasnik J, Sharpless L, Pasque C, Murphy D, Fraser V. Scheduled change of antibiotic classes: a strategy to decrease the incidence of ventilator-associated pneumonia. *Am J Respir Crit Care Med* 1997; 156:1040–1048.
142. Gould D. Nurses' hand decontamination practice: results of a local study. *J Hosp Infect* 1994; 28:15–30.
143. Doebbeling BN, Stanley GL, Sheetz CT, Pfaller MA, Houston AK, Annis L, Li N, Wenzel RP. Comparative efficacy of alternative hand-washing agents in reducing nosocomial infections in intensive care units. *N Engl J Med* 1992; 327:88–93
144. Albert RK, Condie F. Hand-washing patterns in medical intensive-care units. *N Engl J Med* 1981; 304:1465–1466.
145. Simmons B, Bryant J, Neiman K, Spencer L, Arheart K. The role of hand washing in prevention of endemic intensive care unit infections. *Infect Control Hosp Epidemiol* 1990; 11:589–594.

146. Lu WP, Sun Y, Bauer MD, Paule S, Koenigs PM, Kraft WG. Penicillin-binding protein 2a from methicillin-resistant *Staphylococcus aureus*: kinetic characterization of its interactions with β-lactams using electrospray mass spectrometry. *Biochem* 1999; 38:6537–6546.
147. Hecker SJ, Cho IS, Glinka TW, et al. Discovery of MC-02,331, a new cephalosporin exhibiting potent activity against methicillin-resistant *Staphylococcus aureus*. *J Antibiot* 1998; 51:722–734.

13 Mechanism of Drug Resistance in *Mycobacterium tuberculosis*

Alex S. Pym and Stewart T. Cole

CONTENTS

INTRODUCTION

Tuberculosis is the leading cause of death from a curable infectious disease and there were an estimated 8.9 million new cases of tuberculosis in 2004 [1]. Tackling the global burden of tuberculosis is a major public health challenge that has become more daunting with the realization that strains of multiply drug-resistant *Mycobacterium tuberculosis* (MDR-TB) are increasing rapidly. Between 2000 and 2004, global estimates of MDR-TB incidence increased from 270,000 to over 400,000 new cases annually [2]. These strains are difficult to cure because they usually require two years of therapy with costly and poorly tolerated regimens usually comprising a minimum of five drugs. The ultimate control of MDR-TB will require multiple interventions, but a complete understanding of the mechanisms of drug resistance in *M. tuberculosis* is essential for devising rapid diagnostics and developing new drugs. In this chapter, we review what is currently known about the genetics of resistance to the most important antimycobacterial drugs.

DEVELOPMENT OF TUBERCULOSIS CHEMOTHERAPY

The problem of resistance to antimycobacterial drugs is an old one. Within a decade of the development of the first effective agents against tuberculosis, drug resistance had been described and treatment strategies devised to prevent it. The first two drugs to enter formal clinical trials were developed in the early 1940s: streptomycin (SM), isolated from *Streptomyces griseus* [3], and *para*-aminosalicylic (PAS) acid [4], a synthetic derivative of salicylic acid. These compounds were rapidly shown to be effective in animal models and early reports of their clinical use suggested they were effective against human tuberculosis [4,5]. However, it was soon noted that patients with advanced forms of disease had less chance of responding, and that early response to treatment in others was rapidly followed by deterioration and the emergence of drug-resistant strains. For example, in 1947, the MRC trial of SM versus bed rest for patients with pulmonary tuberculosis showed that after six months therapy, there was reduced mortality and clinical improvement in the SM treated group [6]. Unfortunately, 35 of the 41 SM-treated patients were found to be excreting drug-resistant bacilli, and after five years of follow-up, the mortality in the streptomycin group was only slightly better than in the controls (53% vs. 63%) [7].

The priority of investigators then switched rapidly to investigating ways of preventing the emergence of resistance. It was soon shown that by combining SM with PAS the emergence of resistance to SM could be reduced from 70% to 9% [8]. The discovery of a new more potent antimycobacterial agent, isoniazid (INH) [9], soon followed, and regimens combining this agent with SM and PAS were also found to be highly effective at preventing the emergence of drug resistance [10]. Thus, in the space of little more than a decade, the first principle of modern tuberculosis chemotherapy had been established, namely the necessity of combination drug therapy to combat the emergence of resistance.

The biological basis of the need for combination therapy is thought to be due to the heavy pulmonary bacillary burden that exists prior to therapy, sufficiently large to contain spontaneous mutants resistant to a single anti-tuberculosis drug, which will be rapidly selected for if treatment commences with only a single agent. Canetti quantified the number of bacteria found in surgically resected cavities from patients failing to respond to therapy, and found this to be at least 10^8 [11]. Subsequent estimates of the spontaneous mutation rates for drug resistance to an individual drug have been of the order of one in 10^6 for INH and SM [12,13].

Once the principle of combination therapy had been established, the research agenda switched to defining the minimal duration of therapy [14]. Combinations of INH, SM, and PAS required up to 18 months to obtain adequate results [14]. However, the observation that pyrazinamide (PZA) [15] and the newer agent rifampin (RMP) [16] were uniquely capable of sterilizing organs in animal models of tuberculosis, led to trials of shorter courses of therapy. In a series of painstaking and meticulous studies carried out through the 1970s [14], it was established that treatment regimens that contained either PZA or RMP could be reduced to six months (short course therapy) with cure rates (defined as patients free of tuberculosis after two years of follow-up) in excess of 95%. This is the basis for the standard six-month course of treatment for tuberculosis, which involves an initial intensive treatment for two months with INH, RMP, PZA, and ethambutol (EMB) followed by a continuation phase of four months of RMP and INH.

DRUG-RESISTANT TUBERCULOSIS

DEFINITION

Drug resistance is classified into two types: primary drug resistance occurs in individuals who are infected *de novo* with a drug-resistant strain and secondary (acquired) resistance, which arises in an individual initially infected with a drug-sensitive strain, from which resistant mutants emerge as a result of inadequate therapy. MDR-TB is defined as resistance to at least RMP and INH. There is a certain redundancy built into the standard six-month regimen, which ensures it will be effective in individuals infected with tuberculosis resistant to a single drug, and probably to two drugs except for the combination of RMP and INH resistance [17]. This is the basis for the definition of MDR-TB, as individuals infected with INH/RMP-resistant strains will not respond to short-course therapy and will develop resistance to the accompanying drugs PZA and EMB. Treatment of MDR-TB is complex and requires two years of therapy with a combination of four or five second-line drugs, including a fluoroquinolone and one of three injectable agents: kanamycin, amikacin, or capreomycin. Recent reports have described the emergence of *M. tuberculosis* strains resistant to multiple second-line drugs, which represent a major public health threat as they are potentially untreatable [18,19]. A new category has therefore been proposed, extensively drug-resistant tuberculosis (XDR-TB) defined as MDR (resistance to least INH and RMP) in combination with resistance to fluoroquinolones and at least one of the second-line injectable agents.

EPIDEMIOLOGY

Until the 1980s, MDR-TB was not perceived as a threat to tuberculosis control, and surveys from this period suggested that MDR strains were rare [20], and tended to occur only in the context of multiple courses of inadequate treatment. However, a new phenomenon appeared in the form of micro-epidemics of MDR-TB, described in health care settings in the United States [21–24], but also documented elsewhere [25–27]. These were the result of infectious cases excreting and transmitting MDR strains of tuberculosis to numerous contacts, and occurred particularly among groups of HIV-infected individuals. Population-based molecular epidemiology confirmed the ongoing transmission of MDR-TB within communities and also highlighted the potential of individual strains of drug-resistant *M. tuberculosis* to rapidly spread through vulnerable populations [28]. 22% of all MDR-TB strains isolated in New York in one year represented a single clone (strain W) [29], which has also been identified in Europe and Africa as well as throughout the United States [30], revealing the potential for global dissemination of drug-resistant strains.

Attempts to ascertain the current global burden of drug-resistant tuberculosis are limited by incomplete data from many countries. Drug susceptibility testing requires significant laboratory infrastructure, which is not present in many endemic countries. Data are available from a WHO survey conducted in 77 settings between 1999 and 2002, which detected resistant *M. tuberculosis* strains at 74 sites [31]. Prevalence of MDR in new cases of tuberculosis ranged from 0% in eight countries to exceptionally high levels in most of the former Soviet republics (9.3% in Latvia to 14.2% in Kazakhstan). China and Ecuador were other countries with high prevalences of MDR. However, most of these surveys were conducted at a limited number of centers and the distribution of resistant strains is likely to be highly heterogeneous within a population and may change rapidly. For example, the WHO survey reported an MDR prevalence of <2% for South Africa in 2002, but KwaZulu-Natal has recently experienced an alarming increase in numbers of new MDR and XDR cases that was not predictable from the results of the WHO surveillance conducted in the province [19].

MOLECULAR BASIS OF DRUG RESISTANCE

AMINOGLYCOSIDES

Streptomycin

Streptomycin is an aminoglycosidic aminocyclitol, the first of this class of antibiotics to be identified. Its clinical introduction marked the beginning of the chemotherapy era of tuberculosis control and most of the clinical trials that defined the principles of anti-tuberculosis chemotherapy used SM. An adverse toxicity profile and the need for parenteral administration have resulted in it being replaced by other agents in first-line therapy. However, the emergence of drug resistance has meant that SM is still an important antimycobacterial agent. Other aminoglycoside antibiotics are also being used to treat multidrug-resistant cases of tuberculosis and therefore an understanding of the mechanism of resistance to SM has found a new relevance.

Aminoglycosides are broad spectrum antibiotics and their mode of action has been extensively studied in other organisms [32]. SM binds to the 30S ribosomal subunit leading to inhibition of translational initiation and misreading of messenger RNA [32]. Resistance to SM and other aminoglycosides in Gram-negative organisms is principally due to the acquisition of aminoglycoside-modifying enzymes.

However, no plasmids or transposons bearing drug resistance genes have been detected in *M. tuberculosis,* and the chromosomally encoded aminoglycoside 2′-*N*-acetyltransferase (*aac(2′)-Ic*) is apparently unable to acetylate aminoglycosides, thus differing from its counterpart in *Mycobacterium smegmatis* [33]. An *aph(3″)-Ic* gene, encoding an aminoglycoside 3″-*O*-phosphotransferase which can inactivate SM, has been characterized in *Mycobacterium fortuitum,* but is absent from the genome of *M. tuberculosis* [34].

Streptomycin-resistant strains of *Escherichia coli* have been isolated with mutations in two ribosomal components, the S12 protein and the 16S ribosomal RNA. Analysis of these genes in *M. tuberculosis* was therefore a logical first approach to investigating resistance to SM, since mutations at these loci would produce a dominant phenotype, as *M. tuberculosis* possesses only a single copy of these genes [35]. Various groups in parallel established that mutations in *rpsL,* the gene coding for the S12 ribosomal protein, were associated with resistance [36–39]. These mutations were found to occur at codon 43 (Lys43Arg or rarely Lys43Thr), less frequently at codon 88 (Lys88Arg or Lys88Gln), and in isolates exhibiting high-level resistance to SM (MIC of greater than 500 μg/mL).

Sequence analysis of the *rrs* gene that codes for the 16S rRNA, from SM-resistant clinical isolates also identified a series of single nucleotide substitutions that were in general associated with an intermediate level of resistance and confined to two restricted regions. One group, represented by substitutions at positions 491, 512, 513, or 516 mapped to the 530 loop region of the *E. coli* 16S rRNA, and the other at positions 798, 865, 877, 904, and 906 mapped to the *E. coli* 912 region [40]. The 530 loop region is one of the most highly conserved regions of the 16S ribosomal RNA, both in sequence and secondary structure, reflecting its importance for some translational function. The 912 region has been implicated in translational fidelity and is located at the junction of the three major domains of the 16S rRNA. Chemical footprinting experiments using *E. coli* ribosomes have demonstrated that this region is protected by the binding of SM [41] and mutations here reduce drug binding, [42,43] suggesting it may be the primary site of action of SM.

Mutations in *rpsL* and *rrs* mutations can only be detected in approximately 50% and 10% of clinical isolates, respectively. The substantial proportion of strains with no detectable ribosomal subunit mutations indicates there are other mechanisms of resistance, although these strains are in general only resistant to low levels of SM (MIC < 50 μg/mL). Characterization of the active efflux systems detected in the *M. tuberculosis* genome is not yet very advanced [44], but several have been shown to transport aminoglycosides. The *M. tuberculosis* gene Rv1258c is a homolog of the gene coding for the *M. fortuitum* Tap efflux pump, and a member of the major facilitator superfamily (MFS) of efflux systems [45]. Overexpression of Rv1258c in *Mycobacterium bovis* BCG resulted in a modest increase in the MIC for SM, and gene inactivation increased susceptibility, thus suggesting a role in the intrinsic resistance to aminoglycosides. Similiar results were obtained with Rv2333c and Rv1410c, two other members of the 16 MFS genes described in the genome of *M. tuberculosis* [46,47]. Whether these or other transport systems are actually responsible for resistance in clinical isolates requires further investigation, but a recent study has provided evidence that mutations in *gidB*, a methyltransferase specific for the 16S rRNA, may be responsible for low-level SM resistance in *M. tuberculosis* [48].

Other Aminoglycosides

The emergence of strains resistant to SM and other drugs has necessitated the use of other aminoglycosides to treat individuals infected with such strains. These aminoglycosides have a range of MICs for *M. tuberculosis*, and unlike SM are made up of a 2-deoxystreptamine ring rather than a streptidine ring. Given this structural difference it is not surprising that these 2-deoxystreptamine aminoglycosides bind to different sites on the ribosome, and appear to be fully active against *M. tuberculosis* strains harboring SM resistance-associated mutations in their *rrs* and *rpsL* genes [49]. A structural study of a paromomycin-rRNA complex [50] indicates that these antibiotics bind to a region encompassing the 30S subunit A site including position 1408, which has been demonstrated to be important for resistance to 2-deoxystreptamine aminoglycosides in *E. coli* [51]. In *M. tuberculosis*, position 1400 of the *rrs* gene (the equivalent of position 1408 of the *E. coli rrs*) was also found to be important in mediating resistance to 2-deoxystreptamine aminoglycosides. Studies have found an A to G substitution at position 1400 of the *rrs* gene in 60% to 76% of kanamycin-resistant clinical isolates of *M. tuberculosis* analyzed [52–54]. Other *rrs* mutations at positions 1401 and 1483 have also been described but these occur rarely, indicating 1400 is the principal site involved in mediating resistance to kanamycin. The equivalent *rrs* gene position 1400 can also confer kanamycin resistance in *M. smegmatis* and other mycobacteria [55,56]. As with SM resistance, low-level resistance to kanamycin was not found to be accompanied by mutations in the *rrs* gene, suggesting other pathways to resistance may be involved.

The degree of cross resistance to other aminoglycosides conferred by the *rrs* substitution at position 1400 is not completely known, but appears to confer at least resistance to amikacin [52–54,56]. In various fast growing mycobacteria, a mutation at this position has been shown to convey resistance to five different 2-deoxystreptamine aminoglycosides [55], so the A to G substitution at position 1400 is probably a pan-deoxystreptamine aminoglycoside resistance conferring mutation in *M. tuberculosis* as well. While mutations at position 1401 appear to cause resistance to both amikacin and kanamycin, the substitutions at position 1483 may result in strains that retain susceptibility to amikacin. Other uncharacterized mutations outside of the *rrs* gene also exist that can confer unique resistance to kanamycin [57].

Capreomycin is a macrocyclic peptide antibiotic produced by *Saccharothrix mutabolis* subspecies that was originally discovered in the 1960s [58]. Renewed attention has focused on the mechanism of resistance to this agent to determine if it can be successfully deployed against organisms resistant to other aminoglycosides. Using transposon mutagenesis and insertion site mapping in *M. tuberculosis*, it was recently shown that inactivation of the *tlyA* gene (Rv1694) resulted in resistance to capreomycin, and mutations at this locus have been identified in resistant clinical isolates [59,60]. *tlyA* was subsequently found to encode a 2′-*O*-methyltransferase that modifies nucleotide C1409 in helix 44 of 16S rRNA and nucleotide C1920 in helix 69 of 23S rRNA, and the positioning of these methylations suggested a site of action for capreomycin [61]. The identification of this new resistance mechanism, which can occur in combination with *rrs* mutations, has allowed a rational description of the overlapping resistance profiles of capreomycin and other aminoglycosides [59].

RIFAMYCINS

Development of Rifamycins

During a systematic search for new antibiotic compounds in the 1950s, workers at the Dow-Lepetit Research Laboratories in Milan observed that crude extracts from the fermentation broths of *Nocardia mediterranei* contained a mixture of microbiologically active agents. These were found to be a group of closely related compounds with an ansa structure (*a*romatic *n*ucleus *s*panned by an *a*liphatic *bri*dge), and were named rifamycins [16]. Chemical modification of the rifamycins led to the isolation of rifampin or rifampicin [62], which proved to be a potent anti-tuberculous agent whose clinical introduction enabled the duration of chemotherapy for tuberculosis to be reduced to six months [63]. Other rifamycin derivatives, rifapentine (RPE) and rifabutin (RBU), have been licensed for the treatment of mycobacterial infections.

Rifampin

As was the case with SM, the mode of action and resistance mechanism of RMP had been well characterized in *E. coli*. In *E. coli*, RMP inhibits transcription by targeting DNA-dependent-RNA polymerase [64], and mutations in several restricted and highly conserved regions of the β subunit, coded for by the *rpoB* gene, lead to drug resistance [65]. The availability of the *Mycobacterium leprae*, *rpoB* sequence [66] made it possible for two groups to isolate and characterize the *rpoB* gene from RMP-resistant strains of *M. tuberculosis* [67] and *M. leprae* [68]. These studies demonstrated that missense mutations and short in-frame deletions, exclusively associated with RMP resistance, occurred in a central region of the *rpoB* gene, which corresponded to the region most commonly altered in RMP-resistant *E. coli* strains. Numerous subsequent studies of RMP-resistant strains of *M. tuberculosis* from globally dispersed sources have confirmed these findings, and in the vast majority of strains analyzed a point mutation or in-frame deletion/insertion can be found within an 81bp region corresponding to codons 507 to 533 of the *rpoB* gene [69]. A large number of mutations have been described, though substitutions at two positions, Ser531 and His526, were found to occur in the majority of strains studied. Mutations at these two positions (Ser531Leu, His526Tyr) and an Asp516Val mutation have been shown to confer resistance when episomal vectors carrying an appropriately mutated *rpoB* gene were transformed into *M. tuberculosis* [70,71]. The realization that RMP resistance-conferring mutations are confined to a small genetic region has meant that molecular techniques for diagnosing RMP resistance have been relatively straightforward to develop, and numerous different molecular strategies have been successfully employed for detecting them.

Many of these surveys did not sequence the whole 3516 bp *rpoB* gene, so in the approximately 5% of RMP-resistant *M. tuberculosis* strains that have a wild-type 81bp RMP resistance determining region (RRDR), it is not clear to what extent mutations in other regions of this gene are also involved in mediating RMP resistance.

While several mutations have been reported outside of the RRDR, RMP-resistant strains with a wild-type *rpoB* gene have also been identified, indicating that RMP resistance can also arise through an *rpoB* gene independent mechanism [72]. *M. smegmatis* and many strains of the *Mycobacterium avium* complex are innately resistant to RMP despite a drug-sensitive RRDR sequence [73,74]. In *M. smegmatis*, there is evidence for ribosylative inactivation of the drug [74], but this drug resistance mechanism is absent in *M. tuberculosis*.

ISONIAZID

Historical Studies

Unlike SM and RMP, isonicotinic acid hydrazide (isoniazid or INH) is a highly specific antimycobacterial agent, being exquisitely potent against *M. tuberculosis* (and the other members of the *M. tuberculosis* complex: *M. bovis*, *Mycobacterium microti*, and *Mycobacterium africanum*), but possessing no or little activity against *M. leprae*, atypical mycobacteria, or other organisms. Investigations into its mode of action were therefore restricted to the inherently difficult to manipulate members of the *M. tuberculosis* complex. Despite this, early studies were able to establish that INH-resistant strains were commonly catalase negative [75] with reduced virulence in animal models [76] and that the principal mode of action was likely to be through disruption of the cell wall, probably through inhibition of mycolic acid synthesis (a major cell wall component) [77]. Unifying these observations had to wait, though, until the development of the necessary tools to genetically manipulate *M. tuberculosis*.

Drug Activation

While the role of catalase in the activation of INH to its active form was first proposed in 1958 [78], it was not until 1992 that the first genetic evidence for this hypothesis became available with the cloning of *katG*, the gene encoding the catalase-peroxidase enzyme of *M. tuberculosis* [79]. It was shown that overexpression of *katG* in catalase negative strains of *E. coli* or in an INH-resistant strain of *M. smegmatis* could render these organisms relatively sensitive to INH. It was further demonstrated that two clinical isolates of *M. tuberculosis* with high-level resistance to INH (MIC > 50 μg mL^{-1}) had a chromosomal deletion spanning the *katG* gene, and that transformation of these strains with *katG* could restore their INH sensitivity [80].

Further characterization of *katG* has demonstrated that it encodes a dimeric, heme-containing enzyme with catalase and peroxidase activity, in keeping with its structural similarity to other eubacterial hydroperoxidase 1 (HPI) enzymes [81–83]. Confirmation that INH is a prodrug requiring activation by KatG was provided by demonstrating that InhA (a target for INH discussed below) is only rapidly inactivated by INH in the presence of KatG [82]. A mechanism for the oxidation of INH to its bioactive form has also been proposed, in which the drug is converted into a number of highly reactive species capable of either oxidizing or acylating macromolecules [84,85].

A plethora of studies characterized *M. tuberculosis* strains from diverse geographic locations and have shown the majority of clinical isolates resistant to INH have alterations in their *katG* gene [86–88]. While large scale deletions of *katG* have been detected infrequently, missense mutations and small intragenic deletions are the commonest genetic modification associated with INH resistance. A large number of these mutations have been described, though the serine to threonine mutation at codon 315 is the commonest. The possible explanation for the apparent bias in selection by INH for this mutation over other resistance-conferring mutations in *katG* has been provided by several studies, which have characterized the enzymatic properties of a recombinant KatG protein harboring a Ser315Thr mutation [89–91].

In contrast to other INH resistance-conferring mutations, such as the Thr275Pro mutation, which results in concomitant loss of peroxidatic activity and capacity to activate INH, the Ser315Thr mutation results in an enzyme unable to activate INH, leaving at least 50% of its peroxidase and catalase activities intact. Catalase and peroxidase activities are important for protecting *M. tuberculosis* against reactive oxygen species encountered within macrophages [92], and are essential for a fully virulent phenotype. Strains with reduced or absent activities are less virulent when assayed in animal models [92–94], and more susceptible to H_2O_2 intracellularly [95]. A direct comparison of the pathogenicity of isogenic strains harboring either the Ser315Thr substitution or other *katG* mutations has also been made, and only the 315 mutant retained near normal virulence in mice [96]. Recently, INH-resistant strains of *M. tuberculosis* with the Ser315Thr substitution have been shown to be more transmissible than strains with other *katG* mutations [86]. It is unfortunate for the control of MDR-TB that *M. tuberculosis* can so successfully balance the competing demands of resistance to oxidative stress and resistance to INH.

The recent elucidation of the crystal structure of KatGs from *M. tuberculosis* [84] and two other organisms [97,98] has provided insights into how the S315T substitution could lead to loss of activation of INH but retention of catalase-peroxidase activity. In one model, Ser-315 is located at the periphery of the INH binding pocket, where a threonine substitution would increase the steric bulk at this position, leading to reduced affinity for the drug without fully blocking access to the substrate binding site [84,85].

Drug Targets

InhA

Population genetics of INH-resistant clinical isolates have consistently found that a significant proportion of strains possess a wild-type *katG* gene, demonstrating that there are mechanisms of resistance independent of KatG-mediated INH activation. This suggested that the intracellular targets of activated INH might be involved in mediating resistance. By expressing genomic libraries from two INH-resistant strains of mycobacteria isolated *in vitro* in the fast-growing *M. smegmatis*, it was possible to identify a two-gene operon with homology to proteins involved in fatty acid biosynthesis that could confer INH resistance [99]. Characterization of this operon revealed that only the second gene *inhA* was required for resistance [100], and sequence analysis of the *inhA* gene from the resistant strains revealed a serine to alanine substitution at position 94 relative to the wild-type gene. Transfer of this mutation to *M. tuberculosis* results in INH resistance [101].

Biochemical studies revealed that *inhA* codes for a fatty acid enoyl-acyl carrier protein reductase, which is part of a type II dissociated fatty acid biosynthesis pathway (FASII). *inhA* enzymic activity is nicotinamide adenine dinucleotide (NADH) dependent, and reduces the double bond at position two of a growing fatty acid chain linked to an acyl carrier protein ACP, an activity common to all known fatty acid biosynthetic pathways. InhA has a marked preference for long-chain substrates (which are the precursors of very long alpha-branched fatty acids (C_{40} to C_{60}) known as mycolic acids, a major structural element of the mycobacterial cell wall [102].

Structural analysis of the InhA protein has characterized the nature of the interaction between INH and InhA [103]. The observation that InhA inhibition by INH requires the presence of NADH and that the Ser94Ala mutant protein has a lower affinity for NADH and requires higher concentrations of this cofactor before inhibition occurs, suggested that INH may interact with NADH rather than directly with InhA. This was elegantly demonstrated by co-crystallization of InhA with NADH and INH since the structure showed the activated form of INH covalently linked to NADH within the active site of the enzyme [104]. The INH-NAD adduct has been purified and shown to be a tight-binding inhibitor of InhA [105].

The exact mode of interaction of adduct and enzyme has not been defined, but it is now thought that the adduct is formed in solution before binding to InhA [105, 106]. This model is consistent with the observation that NADH dehydrogenase defects in *M. smegmatis*, leading to a higher than normal NADH/NAD$^+$ ratio are associated with a degree of INH resistance [107]. More recently, *M. bovis* BCG *ndh* (encoding a type II NADH dehydrogenase) mutants have been characterized and found to have aberrant NADH/NAD$^+$ ratios and reduced INH susceptibility as well [108]. The unraveling of the interactions of INH and InhA has helped spawn the development of new classes of InhA inhibitors with promising activity against *M. tuberculosis* and other organisms [109–111]. This is a good example of how understanding the mechanisms of resistance can assist in rational drug design.

Since the original description of the Ser94Ala mutation, other InhA gene substitutions have been identified through sequence analysis of INH-resistant clinical isolates [88], confirming that InhA substitutions could also be selected for *in vivo*. Interestingly, the affinity of the INH-NAD adduct for InhA appears not to be affected by some of the InhA mutations described in INH-resistant clinical isolates, suggesting the mechanism of inhibition by the adduct may be through disruption of InhA interaction with other components of the FAS II pathway [105]. Mutations in the promoter region for *inhA* expression have also been described in INH-resistant clinical isolates, and several of these have been shown, using a gene fusion reporter construct, to confer expression levels from four- to eight-fold greater than wild-type sequences [112], thus demonstrating that up-regulation of *inhA* is also a resistance mechanism. Population-based surveys have found that *inhA* promoter mutations are the second most frequent INH resistance mutation after the Ser315Thr mutation in KatG, and structural mutations in InhA are rare [86–88].

KasA

A second successful approach used to determine the targets of activated INH has been to examine the early adaptive response of *M. tuberculosis* following a challenge with INH. By examining two-dimensional gel electrophoretic protein profiles after treatment with INH, Mdluli [113] and others identified two up-regulated proteins: an acyl carrier protein (AcpM) and a covalent complex of AcpM and KasA (β-ketoacyl-ACP synthase). Labeling after treatment with [^{14}C] INH indicated covalent attachment of INH to this protein complex. These two proteins form part of the FAS type II system. Sequencing of *kasA* in clinical isolates identified several mutations restricted to INH-resistant strains, but these were rare and often occurred in tandem with other INH resistance-conferring mutations [114,115]. These two studies suggest that *kasA* mutations are not clinically important, and recent functional studies have suggested KasA may only have a more peripheral role in INH action and resistance than originally proposed [101,116,117].

Other INH Resistance Mechanisms

The research focus on InhA as the principal target for INH has distracted researchers from investigating more thoroughly the drug's interactions with other intracellular processes. INH is oxidized to a number of highly reactive species capable of either oxidizing or acylating macromolecules [84,85], and can be expected to damage multiple cellular structures. Similar to the interaction with InhA, INH has also been shown to form an INH-NADP adduct that inhibits the *M. tuberculosis* dihydrofolate reductase (DHFR), an enzyme essential for nucleic acid synthesis. It is not yet clear whether there is a resistance mechanism to counter this inhibition, but the expression of *M. tuberculosis* DHFR in *M. smegmatis* can protect against growth inhibition caused by INH [118]. Using affinity chromatography, the same group was able to demonstrate high-affinity binding of the INH-NAD/NADP adducts to 16 other proteins in addition to InhA and DHFR, broadening the range of putative INH targets still further.

One approach to identifying resistance genes has been to study the transcriptional response to a drug challenge. Using this approach, Alland and coworkers identified a three-gene operon *iniABC* (for INH inducible), which is up-regulated following treatment of *M. tuberculosis* with INH and EMB [119,120]. Overexpression of one of these genes, *iniA*, in *M. bovis* BCG resulted in increased survival following an INH challenge, and a *M. tuberculosis* strain containing an *iniA* deletion showed increased susceptibility to INH. The INH-resistant phenotype was abolished by the pump inhibitor reserpine suggesting *iniA* could be involved in transport, compatible with its multimeric pore-like secondary structure. However, studies with radio-labeled drug did not show an effect on intracellular INH levels, indicating its mode of action was not through antibiotic efflux. There is more evidence that one of the resistance-nodulation-cell division (RND) family of transporters encoded in the genome of *M. tuberculosis*, *mmpL7*, is involved directly in INH efflux, as its heterologous expression in *M. smegmatis* is reported to cause INH resistance [44,120].

Compensatory Mutations

Studies using model systems have shown that, although there is a range, most resistance-conferring mutations convey some cost to bacteria in terms of their fitness [122]. However, in some cases, these costs can at least be partially ameliorated by secondary or compensatory mutations that restore levels of fitness toward those of sensitive strains [123–125]. Given the diverse mechanisms of resistance in *M. tuberculosis*, it is likely that compensatory mutations will be important in maintaining the fitness of MDR and XDR strains.

The only compensatory mutations described in *M. tuberculosis* are mutations in the *ahpC* promoter region of *M. tuberculosis*. Initial interest in this gene derived from the observation that an *oxyR* null mutant of *E. coli* (the wild type of this organism is highly resistant to INH) was susceptible to INH [126] due to down-regulation of *ahpC-ahpF*. This led to characterization of the *oxyR-ahpC* locus from different mycobacterial species [127–129]. Members of the *M. tuberculosis* complex were found to possess an *oxyR* pseudogene in conjunction with a functional but feebly expressed *ahpC* gene. This led to speculation that the INH susceptibility of *M. tuberculosis* was due to low levels of *ahpC*. However, the overexpression of *ahpC* in *M. tuberculosis* did not markedly affect INH sensitivity, but it did confer resistance to hydrogen and cumene peroxide [128,129]. Subsequent analysis of INH-resistant isolates identified mutations in the upstream region of *ahpC*, which were associated with enhanced promoter activity and up-regulation of *aphC* [128]. These mutations occur almost exclusively in conjunction with *katG* deficiency [86] and probably enable strains to compensate for the additional oxidative stress imposed by loss of catalase [93,128,130,131]. There is now evidence that compensatory mutations may also be important in restoring the fitness of RMP-resistant strains, although the mutations have yet to be identified [132].

Pyrazinamide

The introduction to the anti-tuberculous pharmacopeia of the nicotinamide analog, PZA, had far-reaching consequences, since it enabled the duration of treatment to be shortened from more than a year to six months. Used during the initial intensive phase of short course chemotherapy, PZA, which is more potent at acidic pH, is believed to be particularly active on intracellular *M. tuberculosis*. For instance, the MIC of the drug for a strain grown *in vitro* at pH 7 is >250 μg/mL, but can be reduced to 15 μg/mL by lowering the pH to 5. It was initially thought that this potentiation effect could be explained by tubercle bacilli residing in acidified phagosomes that concentrated the drug [133]. However, it was subsequently shown that the pH within these vesicles was neutral [134]. Considerable insight into PZA uptake and resistance mechanisms was provided recently by Zhang et al. in an elegant series of publications addressing the issues of the remarkable specificity of the drug for *M. tuberculosis* and its relationship with a broad-spectrum amidase [135–141].

PZA resistance has long been associated with the loss of activity of pyrazinamidase (PZase), a cytosolic enzyme of 186 amino acids that hydrolyses both PZA and nicotinamide, and which may play a role in pyridine nucleotide metabolism. In a seminal study, Scorpio and Zhang cloned the *pncA* gene encoding PZase from

M. tuberculosis and demonstrated restoration of drug susceptibility upon its expression in the naturally resistant organism BCG [137]. On further examination, *pncA* mutations that lowered or abolished PZase activity were found in PZA-resistant isolates of *M. tuberculosis* and also in *M. bovis* and BCG [136]. Subsequently, other workers surveyed their strain collections for altered *pncA* genes and found a near-perfect association between the presence of mutant alleles and PZA resistance [138, 142–144]. Indeed, detection of resistance by molecular techniques is now preferable to susceptibility testing by microbiological methods, as these are notoriously error-prone due to the pH effects discussed above.

Although many mycobacteria contain *pncA* genes and elicit PZase, PZA is most active against *M. tuberculosis*, and its close relatives *M. africanum* and *M. microti*. The natural PZA resistance of *M. bovis*, the other member of the *M. tuberculosis* complex, and its descendants stems from the amino acid substitution His57Asp in the PncA protein. The *M. tuberculosis* PZase shows 69.9% and 67.7% identity to those of *Mycobacterium kansasii* and *M. avium*, respectively, and 35.5% identity with the nicotinamidase of *E. coli*. Both of these mycobacterial PZases restored drug susceptibility to levels similar to those conferred by the *M. tuberculosis* enzyme when expressed in a resistant host, such as BCG, although the *M. tuberculosis* protein probably has higher nicotinamidase and PZase activities [139,140]. Nevertheless, both *M. kansasii* and *M. avium* are naturally resistant to the drug and the likely reason for this will be explained below. *M. smegmatis* also produces a highly active PZase, PzaA, with an apparent molecular weight of 50 kDa. It is quite distinct from the other enzymes, yet hydrolyzes both PZA and nicotinamide. PzaA probably corresponds to a broad-spectrum amidase capable of hydrolyzing a wide range of substrates [145]. Furthermore, *M. smegmatis* also contains a PncA homolog (Y. Zhang, personal communication). The potential contribution of other amidases to PZA and nicotinamide metabolism has been recognized and discussed by others [146].

The mutations present in *pncA* from a large number of drug-resistant isolates of *M. tuberculosis* are known. The majority of these (69%) correspond to missense mutations, although frameshifts, insertions, deletions, and nonsense mutations (31%) also occur [138,142–144]. The higher frequency of missense mutants suggests that the corresponding proteins may retain some residual activity that could confer a competitive advantage. It is of some interest to examine the distribution of the amino acid substitutions resulting from missense mutations. These occur throughout the PncA protein, but are generally located in positions that are conserved in all four enzymes. This suggests that these amino acid residues play critical roles in substrate binding and catalysis, but confirmation by biochemical characterization of well-defined PZase variants is now required.

The toxicity of PZA results from its conversion to pyrazinoic acid [147] and the PZase enzyme from resistant organisms, such as BCG, is unable to catalyze this reaction, which occurs at alkaline, neutral, and acidic pH. However, pyrazinoic acid only accumulates in susceptible tubercle bacilli when the external pH is acidic. In naturally resistant mycobacteria such as *M. smegmatis* or *Mycobacterium vaccae*, efficient conversion of PZA occurs, but pyrazinoic acid is rapidly excreted by a highly active efflux system that can be inhibited by reserpine or valinomycin [141].

M. tuberculosis also appears to have a pyrazinoic acid efflux system, but as this is many orders of magnitude less efficient, copious amounts of the acid build up in the cytoplasm. The natural PZA resistance of *M. kansasii* is attributable to the much lower activity of its PZase rather than to efflux mechanisms, since the introduction of the *M. avium* gene into *M. kansasii* results in highly increased PZA susceptibility and pH-dependent pyrazinoic acid accumulation [140]. By contrast, in *M. avium*, which shows lower levels of PZA resistance than *M. kansasii*, an efflux mechanism has been reported whose efficiency is intermediate between those of *M. smegmatis* and *M. tuberculosis* [140].

Prior to its interaction with PZase, PZA must cross the cell wall and enter the cytoplasm. Daffé et al. demonstrated that the radiolabeled drug diffused passively through the outer envelope of *M. tuberculosis* and was then actively transported by an ATP-dependent uptake system. This transporter, which was inhibited by arsenate (albeit at very high concentrations), appears to transport nicotinamide [146]. Similar transport systems were also detected in *M. avium* and *M. kansasii*, but not in *M. bovis* BCG. The latter observation was also made by Zhang et al., who found that the drug did not accumulate in PZA-resistant strains belonging to the *M. tuberculosis* complex [141]. However, upon introduction of a functional *pncA* gene, PZA uptake and pyrazinoic acid production were restored. This could suggest that the transport system is only expressed when PZase is present. Similar observations regarding nicotinamide uptake and the presence of nicotinamidase have been made in *E. coli* mutants lacking the amidase, and might reflect the existence of regulatory pathways for pyridine nucleotide metabolism [148]. Alternatively, it is conceivable that PZA also diffuses across the cytoplasmic membrane of mycobacteria in a passive manner and is then converted to pyrazinoic acid by PZase, which is trapped in the cell at low pH in *M. tuberculosis* but excreted by the naturally resistant species. The inhibition of PZA uptake observed in the presence of nicotinamide may simply reflect the fact that as nicotinamide is the preferred substrate, PZase hydrolyses the drug at much lower levels.

To summarize, three components appear to be involved in mediating PZA susceptibility in pathogenic mycobacteria: the putative uptake system, PZase, and an efflux pump. The relative contributions of these three factors determine the level of drug susceptibility. Clinically relevant mutations have been described in the *M. tuberculosis* complex that result in loss of PZase activity and concomitantly the absence of pyrazinoic acid. To date, no genes or mutations affecting the PZA transporter (if this exists) or the efflux system have been reported in tubercle bacilli. Overexpression of the putative efflux system would lead to increased excretion of pyrazinoic acid. Such mutants might exist and could explain the resistance observed in a small number of clinical isolates with wild-type *pncA* genes.

Since pyrazinoic acid is the active agent, it was logical to test this compound directly, but its bactericidal activity for tubercle bacilli in infected mice was found to be much lower than that of the acid generated endogenously from PZA [147]. To identify the target for the bioactive form of PZA, attempts were made at isolating laboratory mutants of *M. tuberculosis* that show increased resistance to pyrazinoic acid. These have been repeatedly unsuccessful [136]. Moreover, in well-characterized clinical isolates that display reproducible PZA resistance most strains harbored defective *pncA* genes [136]. These observations suggest that the drug target must be an essential enzyme and current thinking is centered on fatty acid synthase I.

ETHAMBUTOL

Since its introduction in the early 1960s, EMB was thought to act on the mycobacterial envelope, as treatment perturbed the biogenesis of several cell wall components. Through biochemical studies, performed mainly with *M. smegmatis*, it was found that the primary site of action was arabinan synthesis [149] which, in turn, impacts on arabinogalactan and lipoarabinomannan production [150]. It is now clear from work with EMB-resistant mutants and molecular genetics that the main drug target is the arabinosyltransferase(s) involved in the polymerization of arabinan. Using complementary approaches with *M. tuberculosis* [151] and *M. avium* [152], these enzymes were shown to be encoded by linked genes that have evolved by a gene duplication mechanism, controlled by the regulatory gene *embR*. In *M. avium, embR* is transcribed divergently from the adjacent *embAB* genes, whereas in *M. tuberculosis*, the *embCAB* operon is situated 1.87 Mb distal to *embR*. *embR* regulation is dependent on phosphorylation by PknH [152].

The arabinosyltransferases are membrane-bound enzymes [152] with 12 predicted membrane-spanning segments and a large extracytoplasmic domain at the COOH-terminus. Missense mutations located in a tetrapeptide at positions 303 to 306 of a putative cytoplasmic loop of EmbB have been shown to be responsible for acquired drug resistance in the majority of clinical isolates of *M. tuberculosis* and in laboratory mutants of *M. smegmatis* [154–156]. Mutations affecting Met306 are predominant, and replacement by Leu or Val is associated with higher resistance levels (40 µg/mL) than the substitution Met306Ile [156]. High-level resistance has also been described for *M. tuberculosis* strains harboring *embB* genes with Thr630Ile or Phe330Val mutations although causality has not yet been demonstrated. The natural Emb resistance of a variety of non-tuberculous mycobacteria, such as *M. leprae*, *M. fortuitum*, and *Mycobacterium abscessus*, results from the presence of one or more alterations to the otherwise well-conserved motif at positions 303 to 306 [157]. Yet again, as in the case of INH and PZA, serendipity has played a major role in the susceptibility of the tubercle bacillus to a frontline drug.

In roughly 70% of EMB-resistant clinical isolates of *M. tuberculosis*, drug resistance can be explained by alterations to *embB*. However, nothing is known of the mechanism operational in the remaining 30% and these generally display lower levels of resistance (15 µg/mL) [156]. Telenti has proposed that high-level resistance results from a stepwise mutational process in *M. smegmatis* although other workers provide evidence in favor of a single event [154]. There is a clear indication from heterologous expression of *emb* genes in this host that increased gene dosage confers higher resistance levels [151] and it is conceivable that unlinked mutations leading to overexpression of the *M. tuberculosis embCAB* operon may occur in clinical isolates displaying low-level EMB resistance. A strong candidate locus is *embR*, which is required for transcription of the operon [152] although increased efflux of the drug cannot be excluded.

FLUOROQUINOLONES

The discovery that the broad-spectrum bacteriocidal activity of fluoroquinolones extended to mycobacteria led to their rapid clinical deployment [158,159]. They have now been used extensively and have established themselves as a key element of therapy for cases of MDR-TB [160,161]. Their tolerability and oral route of administration make them particularly useful drugs. Most clinical experience has been with ofloxacin and ciprofloxacin, but recently newer fluoroquinolones with lower MICs have become available, of which moxifloxacin and gatifloxacin are currently the most promising. Moxifloxacin has been the most extensively studied in animal models of tuberculosis therapy, and in combination with PZA and RMP has been shown to sterilize mouse lungs more rapidly than conventional anti-tuberculosis regimens [162,163]. This has raised the possibility that including a potent fluoroquinolone in first-line treatment could shorten the duration of therapy from six months, and phase III clinical trials are under way to investigate this possibility.

Fluoroquinolones target bacterial topoisomerases, so it was expected that mutations in DNA gyrase, a type II DNA topoisomerase, would be the principal mechanism of resistance, as a type IV topoisomerase was not identified in the genome of *M. tuberculosis* [164]. DNA gyrase is made up of two A and two B subunits encoded by *gyrA* and *gyrB*, and catalyses negative supercoiling of DNA [165]. Sequencing of the equivalent of the *E. coli* quinolone resistance-determining region (QRDR) of *gyrA* [166] in fluoroquinolone-resistant clinical isolates showed that a number of distinct amino acid substitutions are associated with resistance and those at codon 94 and 90 occur most frequently [167–170]. Recently, functional analysis of mutant gyrase complexes has shown that fluoroquinolone inhibition of DNA supercoiling is reduced by the missense mutations Ala90Val, Ala94Gly, and Ala94His [171,172], with larger effects seen with a double mutant. This mirrors the observation that double *gyrA* mutations in clinical isolates are associated with the highest MICs. This study also characterized a *gyrB* mutant confirming that substitutions at this site, which are only rarely encountered clinically [171,173], can also cause resistance.

There are other pathways to resistance independent of the mutations in the QRDR of *gyrA* and *gyrB*, as fluoroquinolone-resistant clinical isolates that are wild type at these loci have been identified [170,171,174]. Attention has focused on characterizing the active efflux systems of *M. tuberculosis* because of their proven role in conferring fluoroquinolone resistance in other organisms. *M. tuberculosis* possesses representatives of all of the main classes of efflux systems [44,175], and members of the ATP-binding cassette (ABC) superfamily and the MFS have been shown to transport fluoroquinolones in mycobacteria [176,177]. LfrA, an MFS member identified in the non-pathogenic *M. smegmatis*, can extrude quinolones [178, 179]. However, none of these transporters has yet been implicated in clinical resistance.

Recently, work has suggested that efflux pumps and structural changes to DNA gyrase may not be the only ways *M. tuberculosis* handles the toxicity of fluoroquinolones. Using a genetic screen for fluoroquinolone resistance, Takiff and colleagues were able to identify a gene, *mfpA*, that resulted in low-level resistance when overexpressed in *M. smegmatis* [180]. *M. tuberculosis* contains a 183-amino acid homolog of *mfpA* that encodes a member of the pentapeptide repeat family of proteins [181] in which every fifth residue is a leucine or phenylalanine. Structural characterization found that MfpA forms a rod-shaped dimer similar electrostatically and in size to B-form DNA, suggesting the binding and inhibition of DNA gyrase by MfpA was through DNA mimicry [182]. This would also explain how up-regulation of MfpA can block the bacteriocidal activity of fluoroquinolones, which is dependent on binding irreversibly to the gyrase-DNA complex.

New Drugs

The scale of the global tuberculosis problem and the continuing expansion of drug-resistant strains of *M. tuberculosis* have finally galvanized the search for new anti-tuberculosis drugs. These drug discovery efforts have already produced several new compounds that have entered clinical trials. However, even though these agents have novel modes of action, they are not immune from drug resistance.

R207910 is a diarylquinoline (DARC) with potent antimycobacterial activity that was identified by screening prototypes of different chemical series for growth inhibition of *M. smegmatis* [183]. It has a low MIC against *M. tuberculosis*, is active against MDR strains including DNA gyrase mutants, and in combination with selected first-line tuberculosis drugs can sterilize lungs of *M. tuberculosis*-infected mice more rapidly that standard regimens, making it an agent with the potential to shorten the duration of chemotherapy [184–185]. Its mode of action and resistance-conferring mutations were ingeniously elucidated by comparing the near-complete genome sequences of R207910-resistant *M. tuberculosis* and *M. smegmatis* strains selected *in vitro* with their wild-type progenitors. This approach identified single-point mutations in *atpE*, which encodes part of the F_0 subunit of ATP synthase. The mutations Asp32Val in *M. smegmatis* and Ala63Pro in *M. tuberculosis* were both located in the membrane-spanning domain of the protein and could confer resistance to R207910 when transformed on appropriate constructs into wild-type organisms. Further work has established that the region involved in resistance is conserved across a wide range of mycobacteria except for *M. xenopi*, which unlike most myco-bacteria, is naturally resistant to R207910 and has a Met63 genotype rather than the highly conserved Ala63 [186].

The observation that nitroimidazoles, such as metronidazole, have some antimycobacterial toxicity, prompted researchers to identify related compounds with enhanced bacteriocidal activity against *M. tuberculosis*. Two of these compounds, OPC-67683 (a nitro-dihydro-imidazole) [187] and PA-824 (a nitroimidazole-oxazine), [188] have already advanced into clinical trials. To date, only the resistance mechanisms of PA-824 have been studied extensively. The intracellular targets have not yet been identified, but PA-824 inhibits protein and cell wall synthesis after reductive activation, which requires a F_{420}-dependent glucose-6-phosphate dehydrogenase (FGD1) encoded by Rv0407. Therefore, it was not surprising that mutations in FGD1 and the genes required for synthesis of cofactor F_{420} (*fbiA, fbiB*, and *fbiC*) result in resistance to PA-824 [189,190]. However, a proportion of PA-824-resistant strains can produce FGD1 and F_{420} but are still unable to activate the drug. Barry and colleagues sequenced the genome of these strains and identified mutations in an additional gene (Rv3547), which they propose encodes an essential part of the F_{420}-dependent nitroreductase [191]. The involvement of at least five genes in PA-824 activation means that the *in vitro* selection rate for resistant mutants is relatively high [191], and it will be important to determine if resistance to PA-824 emerges rapidly during therapy.

CONCLUSIONS

The impact of genomics on mycobacteriology has been immense, and has been driven by an expanding catalog of complete genome sequences of *M. tuberculosis* strains and closely related members of the *M. tuberculosis* complex [192]. In addition to providing a fast-track method of determining resistance mutations to new drugs [183] genomics has also confirmed a clonal population structure for *M. tuberculosis* [193] and the infrequency of recombination [194,195]. Unlike other human pathogens in which the horizontal acquisition of genetic material has been instrumental in the dissemination of resistance, *M. tuberculosis* lacks plasmids and other mobile genetic elements that can transfer resistance genes and all resistance mechanisms described in this chapter are chromosomally encoded. Despite this, *M. tuberculosis* has proved remarkably adept at developing resistance to all antimycobacterial drugs, even though their targets are diverse. A key objective of future drug development will therefore be the discovery of agents with higher genetic barriers to resistance, which will come from an enhanced knowledge of drug resistance mechanisms.

ACKNOWLEDGMENTS

The authors wish to acknowledge the Institut Pasteur, the Association Française Raoul Follereau, and the Wellcome Trust.

REFERENCES

1. Dye, C. 2006. Global epidemiology of tuberculosis. *Lancet* 367:938–40.
2. Zignol, M., M. S. Hosseini, A. Wright, C. L. Weezenbeek, P. Nunn, C. J. Watt, B. G. Williams, and C. Dye. 2006. Global incidence of multidrug-resistant tuberculosis. *J Infect Dis* 194:479–85.
3. Schatz, A., E. Bugie, and S. A. Waksman. 1944. Streptomycin. Substance exhibiting activity against Gram-positive and Gram-negative bacteria. *Proc Soc Exp Biol Med* 55:66–9.
4. Lehman, J. 1946. Para-aminosalicylic acid in the treatment of tuberculosis. *Lancet* 250:15–6.
5. Pfuetze, K. H., M. M. Pyle, and H. C. Hinshaw. 1955. The first clinical trial of streptomycin in human tuberculosis. *Am Rev Tuberc* 71:752–4.
6. Council, M. R. 1948. Medical Research Council streptomycin treatment of pulmonary tuberculosis. *Br Med J* 2:769–82.
7. Fox, W., I. Sutherland, and M. Daniels. 1954. A five year assessment of patients in a controlled trial of streptomycin in pulmonary tuberculosis. Quart J Med 23:347–366.
8. Council, M. R. 1950. Medical Research Council treatment of pulmonary tuberculosis with streptomycin and para-aminosalicylic acid. *Br Med J* 2:1073–85.
9. Robitzek, E. H., and I. J. Selikoff. 1952. Hydrazine derivatives of isonicotinic acid (Rimifon, Marsilid) in the treatment of active progressive caseous-pneumonic tuberculosis. A preliminary report. *Am Rev Tuberc* 65:402–28.

10. Council, M. R. 1955. Various combinations of isoniazid with streptomycin or with PAS in the treatment of pulmonary tuberculosis. *Br Med J* 1:435–45.

11. Canetti, G. 1959. Modifications des populations des foyers tuberculeux au cours de la chimiothérapie antibacillaire. *Ann Inst Pasteur* 97:53–79.

12. Canetti, G., and J. Grosset. 1961. Teneur des souches sauvages de *Mycobacterium tuberculosis* en variants résistants à la streptomycine sur milieu de Lowenstein-Jensen. *Ann Inst Pasteur* 101:28–46.

13. David, H. L. 1970. Probability distribution of drug-resistant mutants in unselected populations of *Mycobacterium tuberculosis*. *Appl Microbiol* 20:810–4.

14. Fox, W., G. A. Ellard, and D. A. Mitchison. 1999. Studies on the treatment of tuberculosis undertaken by the British Medical Research Council tuberculosis units, 1946–1986, with relevant subsequent publications. *Int J Tuberc Lung Dis* 3:S231–79.

15. Yeager, R., W. E. Monroe, and F. I. Dessau. 1952. Pyrazinamide in the treatment of pulmonary tuberculosis. *Am Rev Tuberc* 65:523–546.

16. Sensi, P., A. M. Greco, and R. Ballotta. 1960. Rifomycin. I. Isolation and properties of rifomycin B and rifomycin complex. *Antibiotics Annual* 1959:262–270.

17. Mitchison, D. A., and A. J. Nunn. 1986. Influence of initial drug resistance on the response to short-course chemotherapy of pulmonary tuberculosis. *Am Rev Respir Dis* 133:423–30.

18. 2006. Emergence of *Mycobacterium tuberculosis* with extensive resistance to second-line drugs—worldwide, 2000–2004. *MMWR Morb Mortal Wkly Rep* 55:301–5.

19. Gandhi, N. R., A. Moll, A. W. Sturm, R. Pawinski, T. Govender, U. Lalloo, K. Zeller, J. Andrews, and G. Friedland. 2006. Extensively drug-resistant tuberculosis as a cause of death in patients co-infected with tuberculosis and HIV in a rural area of South Africa. *Lancet* 368:1575–80.

20. Kopanoff, D. E., J. O. Kilburn, J. L. Glassroth, D. J. Snider, L. S. Farer, and R. C. Good. 1978. A continuing survey of tuberculosis primary drug resistance in the United States: March 1975 to November 1977. A United States Public Health Service cooperative study. *Am Rev Respir Dis* 118:835–42.

21. Beck, S. C., S. W. Dooley, M. D. Hutton, J. Otten, A. Breeden, J. T. Crawford, A. E. Pitchenik, C. Woodley, G. Cauthen, and W. R. Jarvis. 1992. Hospital outbreak of multi-drug-resistant *Mycobacterium tuberculosis* infections. Factors in transmission to staff and HIV-infected patients. *JAMA* 268:1280–6.

22. Edlin, B. R., J. I. Tokars, M. H. Grieco, J. T. Crawford, J. Williams, E. M. Sordillo, K. R. Ong, J. O. Kilburn, S. W. Dooley, K. G. Castro, et al. 1992. An outbreak of multidrug-resistant tuberculosis among hospitalized patients with the acquired immuno-deficiency syndrome. *N Engl J Med* 326:1514–21.

23. Small, P. M., R. W. Shafer, P. C. Hopewell, S. P. Singh, M. J. Murphy, E. Desmond, M. F. Sierra, and G. K. Schoolnik. 1993. Exogenous reinfection with multidrug-resistant *Mycobacterium tuberculosis* in patients with advanced HIV infection. *N Engl J Med* 328:1137–44.

24. Valway, S. E., S. B. Richards, J. Kovacovich, R. B. Greifinger, J. T. Crawford, and S. W. Dooley. 1994. Outbreak of multi-drug-resistant tuberculosis in a New York State prison, 1991. *Am J Epidemiol* 140:113–22.

25. Morb Mortal Wkly Rep. 1996. Multidrug-resistant tuberculosis outbreak on an HIV ward—Madrid, Spain, 1991–1995. *Morb Mortal Wkly Rep* 45:330–3.

26. Portugal, I., M. J. Covas, L. Brum, M. Viveiros, P. Ferrinho, P. J. Moniz, and H. David. 1999. Outbreak of multiple drug-resistant tuberculosis in Lisbon: detection by restriction fragment length polymorphism analysis. *Int J Tuberc Lung Dis* 3:207–13.

27. Ritacco, V., L. M. Di, A. Reniero, M. Ambroggi, L. Barrera, A. Dambrosi, B. Lopez, N. Isola, and K. I. de. 1997. Nosocomial spread of human immunodeficiency virus-related multidrug-resistant tuberculosis in Buenos Aires. *J Infect Dis* 176:637–42.

28. Frieden, T. R., T. Sterling, M. A. Pablos, J. O. Kilburn, G. M. Cauthen, and S. W. Dooley. 1993. The emergence of drug-resistant tuberculosis in New York City. *N Engl J Med* 328:521–6.

29. Moss, A. R., D. Alland, E. Telzak, D. J. Hewlett, V. Sharp, P. Chiliade, V. LaBombardi, D. Kabus, B. Hanna, L. Palumbo, K. Brudney, A. Weltman, K. Stoeckle, K. Chirgwin, M. Simberkoff, S. Moghazeh, W. Eisner, M. Lutfey, and B. Kreiswirth. 1997. A city-wide outbreak of a multiple-drug-resistant strain of *Mycobacterium tuberculosis* in New York. *Int J Tuberc Lung Dis* 1:115–21.

30. Agerton, T. B., S. E. Valway, R. J. Blinkhorn, K. L. Shilkret, R. Reves, W. W. Schluter, B. Gore, C. J. Pozsik, B. B. Plikaytis, C. Woodley, and I. M. Onorato. 1999. Spread of strain W, a highly drug-resistant strain of *Mycobacterium tuberculosis*, across the United States. *Clin Infect Dis* 29:85–92.

31. WHO. 2004. Anti-tuberculosis Drug Resistance in the World—Third Global Report. World Health Organization: Geneva, Switzerland WHO/HTM/TB/2004.343.

32. Noller, H. F. 1984. Structure of ribosomal RNA. *Annu Rev Biochem* 53:119–62.

33. Ainsa, J. A., E. Perez, V. Pelicic, F. X. Berthet, B. Gicquel, and C. Martin. 1997. Aminoglycoside 2′-N-acetyltransferase genes are universally present in mycobacteria: characterization of the *aac(2′)-Ic* gene from *Mycobacterium tuberculosis* and the *aac(2′)-Id* gene from *Mycobacterium smegmatis*. *Mol Microbiol* 24:431–41.

34. Ramon-Garcia, S., I. Otal, C. Martin, R. Gomez-Lus, and J. A. Ainsa. 2006. Novel streptomycin resistance gene from *Mycobacterium fortuitum*. *Antimicrob Agents Chemother* 50:3920–2.

35. Cole, S. T., R. Brosch, J. Parkhill, T. Garnier, C. Churcher, D. Harris, S. V. Gordon, K. Eiglmeier, S. Gas, C. E. Barry III, F. Tekaia, K. Badcock, D. Basham, D. Brown, T. Chillingworth, R. Connor, R. Davies, K. Devlin, T. Feltwell, S. Gentles, N. Hamlin, S. Holroyd, T. Hornsby, K. Jagels, A. Krogh, A. McLean, S. Moule, L. Murphy, K. Oliver, J. Osborne, M. A. Quail, M.-A. Rajandream, J. Rogers, S. Rutter, K. Seeger, J. Skelton, R. Squares, S. Squares, J. E. Sulston, K. Taylor, S. Whitehead, and B. G. Barrell. 1998. Deciphering the biology of *Mycobacterium tuberculosis* from the complete genome sequence. *Nature* 393:537–44.

36. Cooksey, R. C., G. P. Morlock, A. McQueen, S. E. Glickman, and J. T. Crawford. 1996. Characterization of streptomycin resistance mechanisms among *Mycobacterium tuberculosis* isolates from patients in New York City. *Antimicrob Agents Chemother* 40:1186–8.

37. Finken, M., P. Kirschner, A. Meier, A. Wrede, and E. C. Bottger. 1993. Molecular basis of streptomycin resistance in *Mycobacterium tuberculosis*: alterations of the ribosomal protein S12 gene and point mutations within a functional 16S ribosomal RNA pseudoknot. *Mol Microbiol* 9:1239–46.

38. Honore, N., and S. T. Cole. 1994. Streptomycin resistance in mycobacteria. *Antimicrob Agents Chemother* 38:238–42.

39. Nair, J., D. A. Rouse, G. H. Bai, and S. L. Morris. 1993. The *rpsL* gene and streptomycin resistance in single and multiple drug-resistant strains of *Mycobacterium tuberculosis*. *Mol Microbiol* 10:521–7.

40. Sreevatsan, S., X. Pan, K. E. Stockbauer, D. L. Williams, B. N. Kreiswirth, and J. M. Musser. 1996. Characterization of *rpsL* and *rrs* mutations in streptomycin-resistant *Mycobacterium tuberculosis* isolates from diverse geographic localities. *Antimicrob Agents Chemother* 40:1024–6.

41. Moazed, D., and H. F. Noller. 1987. Interaction of antibiotics with functional sites in 16S ribosomal RNA. *Nature* 327:389–94.

42. Leclerc, D., P. Melancon, and G. L. Brakier. 1991. Mutations in the 915 region of *Escherichia coli* 16S ribosomal RNA reduce the binding of streptomycin to the ribosome. *Nucleic Acids Res* 19:3973–7.

43. Powers, T., and H. F. Noller. 1991. A functional pseudoknot in 16S ribosomal RNA. *EMBO J* 10:2203–14.

44. De Rossi, E., J. A. Ainsa, and G. Riccardi. 2006. Role of mycobacterial efflux transporters in drug resistance: an unresolved question. *FEMS Microbiol Rev* 30:36–52.

45. Ainsa, J. A., M. C. J. Blokpoel, I. otal, D. B. Young, K. A. L. de Smet, and C. Martin. 1998. Molecular cloning and characterization of Tap, a putative multidrug efflux pump present in *Mycobacterium fortuitum* and *Mycobacterium tuberculosis*. *J Bacteriol* 180:5836–43.

46. Ramon-Garcia, S., C. Martin, E. De Rossi, and J. A. Ainsa. 2007. Contribution of the Rv2333c efflux pump (the Stp protein) from *Mycobacterium tuberculosis* to intrinsic antibiotic resistance in *Mycobacterium bovis* BCG. *J Antimicrob Chemother* 59:544–7.

47. Silva, P. E., F. Bigi, M. P. Santangelo, M. I. Romano, C. Martin, A. Cataldi, and J. A. Ainsa. 2001. Characterization of P55, a multidrug efflux pump in *Mycobacterium bovis* and *Mycobacterium tuberculosis*. *Antimicrob Agents Chemother* 45:800–4.

48. Okamoto, S., A. Tamaru, C. Nakajima, K. Nishimura, Y. Tanaka, S. Tokuyama, Y. Suzuki, and K. Ochi. 2007. Loss of a conserved 7-methylguanosine modification in 16S rRNA confers low-level streptomycin resistance in bacteria. *Mol Microbiol* 63:1096–106.

49. Meier, A., P. Sander, K. J. Schaper, M. Scholz, and E. C. Bottger. 1996. Correlation of molecular resistance mechanisms and phenotypic resistance levels in streptomycin-resistant *Mycobacterium tuberculosis*. *Antimicrob Agents Chemother* 40:2452–4.

50. Fourmy, D., M. I. Recht, S. C. Blanchard, and J. D. Puglisi. 1996. Structure of the A site of *Escherichia coli* 16S ribosomal RNA complexed with an aminoglycoside antibiotic. *Science* 274:1367–71.

51. Beauclerk, A. A., and E. Cundliffe. 1987. Sites of action of two ribosomal RNA methylases responsible for resistance to aminoglycosides. *J Mol Biol* 193:661–71.

52. Alangaden, G. J., B. N. Kreiswirth, A. Aouad, M. Khetarpal, F. R. Igno, S. L. Moghazeh, E. K. Manavathu, and S. A. Lerner. 1998. Mechanism of resistance to amikacin and kanamycin in *Mycobacterium tuberculosis*. *Antimicrob Agents Chemother* 42:1295–7.

53. Suzuki, Y., C. Katsukawa, A. Tamaru, C. Abe, M. Makino, Y. Mizuguchi, and H. Taniguchi. 1998. Detection of kanamycin-resistant *Mycobacterium tuberculosis* by identifying mutations in the 16S rRNA gene. *J Clin Microbiol* 36:1220–5.

54. Taniguchi, H., B. Chang, C. Abe, Y. Nikaido, Y. Mizuguchi, and S. I. Yoshida. 1997. Molecular analysis of kanamycin and viomycin resistance in *Mycobacterium smegmatis* by use of the conjugation system. *J Bacteriol* 179:4795–801.

55. Prammananan, T., P. Sander, B. A. Brown, K. Frischkorn, G. O. Onyi, Y. Zhang, E. C. Bottger, and R. J. Wallace, Jr. 1998. A single 16S ribosomal RNA substitution is responsible for resistance to amikacin and other 2-deoxystreptamine aminoglycosides in *Mycobacterium abscessus* and *Mycobacterium chelonae*. *J Infect Dis* 177:1573–81.

56. Sander, P., T. Prammananan, and E. C. Bottger. 1996. Introducing mutations into a chromosomal rRNA gene using a genetically modified eubacterial host with a single rRNA operon. *Mol Microbiol* 22:841–8.

57. Kruuner, A., P. Jureen, K. Levina, S. Ghebremichael, and S. Hoffner. 2003. Discordant resistance to kanamycin and amikacin in drug-resistant *Mycobacterium tuberculosis*. *Antimicrob Agents Chemother* 47:2971–3.

58. Herr, E. B., Jr., and M. O. Redstone. 1966. Chemical and physical characterization of capreomycin. *Ann NY Acad Sci* 135:940–6.

59. Maus, C. E., B. B. Plikaytis, and T. M. Shinnick. 2005. Molecular analysis of cross-resistance to capreomycin, kanamycin, amikacin, and viomycin in *Mycobacterium tuberculosis*. *Antimicrob Agents Chemother* 49:3192–7.

60. Maus, C. E., B. B. Plikaytis, and T. M. Shinnick. 2005. Mutation of *tlyA* confers capreomycin resistance in *Mycobacterium tuberculosis*. *Antimicrob Agents Chemother* 49:571–7.

61. Johansen, S. K., C. E. Maus, B. B. Plikaytis, and S. Douthwaite. 2006. Capreomycin binds across the ribosomal subunit interface using tlyA-encoded 2'-*O*-methylations in 16S and 23S rRNAs. *Mol Cell* 23:173–82.

62. Maggi, N., C. R. Pasqualucci, R. Ballota, and P. Sensi. 1966. Rifampicin: a new orally active rifamycin. *Chemotherapia* 11:285–92.

63. East Africa, B. M. R. C. 1972. Controlled clinical trial of short course (6 months) regimens of chemotherapy for treatment of pulmonary tuberculosis. *Lancet* 1:1072–85.

64. Hartmann, G., K. O. Honikel, F. Knusel, and J. Nuesch. 1967. The specific inhibition of the DNA-directed RNA synthesis by rifamycin. *Biochim Biophys Acta* 145:843–4.

65. Ovchinnikov, Y. A., G. S. Monastyrskaya, S. O. Guriev, N. F. Kalinina, E. D. Sverdlov, A. I. Gragerov, I. A. Bass, I. F. Kiver, E. P. Moiseyeva, V. N. Igumnov, S. Z. Mindlin, V. G. Nikiforov, and R. B. Khesin. 1983. RNA polymerase rifampicin resistance mutations in *Escherichia coli*: sequence changes and dominance. *Mol Gen Genet* 190:344–8.

66. Honore, N., S. Bergh, S. Chanteau, P. F. Doucet, K. Eiglmeier, T. Garnier, C. Georges, P. Launois, T. Limpaiboon, S. Newton, and a. l. et. 1993. Nucleotide sequence of the first cosmid from the *Mycobacterium leprae* genome project: structure and function of the Rif-Str regions. *Mol Microbiol* 7:207–14.

67. Telenti, A., P. Imboden, F. Marchesi, D. Lowrie, S. Cole, M. J. Colston, L. Matter, K. Schopfer, and T. Bodmer. 1993. Detection of rifampicin-resistance mutations in *Mycobacterium tuberculosis*. *Lancet* 341:647–50.

68. Honore, N., and S. T. Cole. 1993. Molecular basis of rifampin resistance in *Mycobacterium leprae*. *Antimicrob Agents Chemother* 37:414–8.

69. Ramaswamy, S., and J. M. Musser. 1998. Molecular genetic basis of antimicrobial agent resistance in *Mycobacterium tuberculosis*: 1998 update. *Tuber Lung Dis* 79:3–29.

70. Miller, L. P., J. T. Crawford, and T. M. Shinnick. 1994. The *rpoB* gene of *Mycobacterium tuberculosis*. *Antimicrob Agents Chemother* 38:805–11.

71. Williams, D. L., L. Spring, L. Collins, L. P. Miller, L. B. Heifets, P. R. Gangadharam, and T. P. Gillis. 1998. Contribution of *rpoB* mutations to development of rifamycin cross-resistance in *Mycobacterium tuberculosis*. *Antimicrob Agents Chemother* 42:1853–7.

72. Musser, J. M. 1995. Antimicrobial agent resistance in mycobacteria: molecular genetic insights. *Clin Microbiol Rev* 8:496–514.

73. Guerrero, C., L. Stockman, F. Marchesi, T. Bodmer, G. D. Roberts, and A. Telenti. 1994. Evaluation of the *rpoB* gene in rifampicin-susceptible and -resistant *Mycobacterium avium* and *Mycobacterium intracellulare*. *J Antimicrob Chemother* 33:661–3.

74. Quan, S., H. Venter, and E. R. Dabbs. 1997. Ribosylative inactivation of rifampin by *Mycobacterium smegmatis* is a principal contributor to its low susceptibility to this antibiotic. *Antimicrob Agents Chemother* 41:2456–60.

75. Middlebrook, G. 1954. Isoniazid-resistance and catalase activities of tubercle bacilli. A preliminary report. *Am Rev Tuberc* 69:471–2.

76. Middlebrook, G., and M. L. Cohn. 1953. Some observations on the pathogenicity of isoniazid-resistant variants of tubercle bacilli. *Science* 118:297–9.

77. Winder, F. G. 1982. Mode of action of the antimycobacterial agents and associated aspects of the molecular biology of mycobacteria. In: Ratledge C, Stanford J, eds. *The Biology of Mycobacteria*. New York, Academic Press:353–438.

78. Kruger-Thiemer, E. 1958. Isonicotinic acid hypothesis of the antituberculous action of isoniazid. *Am Rev Tuberc* 77:364–7.

79. Zhang, Y., B. Heym, B. Allen, D. Young, and S. Cole. 1992. The catalase-peroxidase gene and isoniazid resistance of *Mycobacterium tuberculosis*. *Nature* 358:591–3.
80. Zhang, Y., T. Garbe, and D. Young. 1993. Transformation with *katG* restores isoniazid-sensitivity in *Mycobacterium tuberculosis* isolates resistant to a range of drug concentrations. *Mol Microbiol* 8:521–4.
81. Heym, B., P. M. Alzari, N. Honore, and S. T. Cole. 1995. Missense mutations in the catalase-peroxidase gene, *katG*, are associated with isoniazid resistance in *Mycobacterium tuberculosis*. *Mol Microbiol* 15:235–45.
82. Johnsson, K., W. A. Froland, and P. G. Schultz. 1997. Overexpression, purification, and characterization of the catalase-peroxidase KatG from *Mycobacterium tuberculosis*. *J Biol Chem* 272:2834–40.
83. Johnsson, K., and P. G. Schultz. 1994. Mechanistic studies of the oxidation of isoniazid by the catalase peroxidase from *Mycobacterium tuberculosis*. *J Am Chem Soc* 116:7425–26.
84. Bertrand, T., N. A. Eady, J. N. Jones, Jesmin, J. M. Nagy, B. Jamart–Gregoire, E. L. Raven, and K. A. Brown. 2004. Crystal structure of *Mycobacterium tuberculosis* catalase-peroxidase. *J Biol Chem* 279:38991–9.
85. Pierattelli, R., L. Banci, N. A. Eady, J. Bodiguel, J. N. Jones, P. C. Moody, E. L. Raven, B. Jamart-Gregoire, and K. A. Brown. 2004. Enzyme-catalyzed mechanism of isoniazid activation in class I and class III peroxidases. *J Biol Chem* 279:39000–9.
86. Gagneux, S., M. V. Burgos, K. DeRiemer, A. Encisco, S. Munoz, P. C. Hopewell, P. M. Small, and A. S. Pym. 2006. Impact of bacterial genetics on the transmission of isoniazid-resistant *Mycobacterium tuberculosis*. *PLoS Pathog* 2:e61.
87. Hazbon, M. H., M. Brimacombe, M. Bobadilla del Valle, M. Cavatore, M. I. Guerrero, M. Varma-Basil, H. Billman-Jacobe, C. Lavender, J. Fyfe, L. Garcia-Garcia, C. I. Leon, M. Bose, F. Chaves, M. Murray, K. D. Eisenach, J. Sifuentes-Osornio, M. D. Cave, A. Ponce de Leon, and D. Alland. 2006. Population genetics study of isoniazid resistance mutations and evolution of multidrug-resistant *Mycobacterium tuberculosis*. *Antimicrob Agents Chemother* 50:2640–9.
88. Ramaswamy, S. V., R. Reich, S. J. Dou, L. Jasperse, X. Pan, A. Wanger, T. Quitugua, and E. A. Graviss. 2003. Single nucleotide polymorphisms in genes associated with isoniazid resistance in *Mycobacterium tuberculosis*. *Antimicrob Agents Chemother* 47:1241–50.
89. Saint-Joanis, B., H. Souchon, M. Wilming, K. Johnsson, P. M. Alzari, and S. T. Cole. 1999. Use of site-directed mutagenesis to probe the structure, function and isoniazid activation of the catalase/peroxidase, KatG, from *Mycobacterium tuberculosis*. *Biochem J* 338:753–60.
90. Wengenack, N. L., S. Todorovic, L. Yu, and F. Rusnak. 1998. Evidence for differential binding of isoniazid by *Mycobacterium tuberculosis* KatG and the isoniazid-resistant mutant KatG(S315T). *Biochemistry* 37:15825–34.
91. Wengenack, N. L., J. R. Uhl, A. A. St, A. J. Tomlinson, L. M. Benson, S. Naylor, B. C. Kline, F. r. Cockerill, and F. Rusnak. 1997. Recombinant *Mycobacterium tuberculosis* KatG(S315T) is a competent catalase-peroxidase with reduced activity toward isoniazid. *J Infect Dis* 176:722–7.
92. Ng, V. H., J. S. Cox, A. O. Sousa, J. D. MacMicking, and J. D. McKinney. 2004. Role of KatG catalase-peroxidase in mycobacterial pathogenesis: countering the phagocyte oxidative burst. *Mol Microbiol* 52:1291–302.
93. Heym, B., E. Stavropoulos, N. Honore, P. Domenech, J. B. Saint, T. M. Wilson, D. M. Collins, M. J. Colston, and S. T. Cole. 1997. Effects of overexpression of the alkyl hydroperoxide reductase AhpC on the virulence and isoniazid resistance of *Mycobacterium tuberculosis*. *Infect Immun* 65:1395–401.

94. Jackett, P. S., V. R. Aber, and D. B. Lowrie. 1978. Virulence and resistance to super-oxide, low pH and hydrogen peroxide among strains of *Mycobacterium tuberculosis*. *J Gen Microbiol* 104:37–45.
95. Manca, C., S. Paul, C. R. Barry, V. H. Freedman, and G. Kaplan. 1999. *Mycobacterium tuberculosis* catalase and peroxidase activities and resistance to oxidative killing in human monocytes *in vitro*. *Infect Immun* 67:74–9.
96. Pym, A. S., B. Saint-Joanis, and S. T. Cole. 2002. Effect of *katG* mutations on the virulence of *Mycobacterium tuberculosis* and the implication for transmission in humans. *Infect Immun* 70:4955–60.
97. Carpena, X., S. Loprasert, S. Mongkolsuk, J. Switala, P. C. Loewen, and I. Fita. 2003. Catalase-peroxidase KatG of *Burkholderia pseudomallei* at 1.7A resolution. *J Mol Biol* 327:475–89.
98. Yamada, Y., T. Fujiwara, T. Sato, N. Igarashi, and N. Tanaka. 2002. The 2.0 A crystal structure of catalase-peroxidase from *Haloarcula marismortui*. *Nat Struct Biol* 9:691–5.
99. Banerjee, A., E. Dubnau, A. Quemard, V. Balasubramanian, K. S. Um, T. Wilson, D. Collins, L. G. de, and W. J. Jacobs. 1994. *inhA*, a gene encoding a target for isoniazid and ethionamide in *Mycobacterium tuberculosis*. *Science* 263:227–30.
100. Banerjee, A., M. Sugantino, J. C. Sacchettini, and W. J. Jacobs. 1998. The mabA gene from the *inhA* operon of *Mycobacterium tuberculosis* encodes a 3-ketoacyl reductase that fails to confer isoniazid resistance. *Microbiology* 144:2697–704.
101. Vilcheze, C., F. Wang, M. Arai, M. H. Hazbon, R. Colangeli, L. Kremer, T. R. Weisbrod, D. Alland, J. C. Sacchettini, and W. R. Jacobs, Jr. 2006. Transfer of a point mutation in *Mycobacterium tuberculosis inhA* resolves the target of isoniazid. *Nat Med* 12:1027–9.
102. Quemard, A., J. C. Sacchettini, A. Dessen, C. Vilcheze, R. Bittman, W. J. Jacobs, and J. S. Blanchard. 1995. Enzymatic characterization of the target for isoniazid in *Mycobacterium tuberculosis*. *Biochemistry* 34:8235–41.
103. Dessen, A., A. Quemard, J. S. Blanchard, W. R. Jacobs, Jr., and J. C. Sacchettini. 1995. Crystal structure and function of the isoniazid target of *Mycobacterium tuberculosis*. *Science* 267:1638–41.
104. Rozwarski, D. A., G. A. Grant, D. Barton, W. J. Jacobs, and J. C. Sacchettini. 1998. Modification of the NADH of the isoniazid target (InhA) from *Mycobacterium tuberculosis*. *Science* 279:98–102.
105. Rawat, R., A. Whitty, and P. J. Tonge. 2003. The isoniazid-NAD adduct is a slow, tight-binding inhibitor of InhA, the *Mycobacterium tuberculosis* enoyl reductase: adduct affinity and drug resistance. *Proc Natl Acad Sci USA* 100:13881–6.
106. Wilming, M., and K. Johnsson. 1999. Spontaneous formation of the bioactive form of the tuberculosis drug isoniazid. *Angew Chem Int Ed Engl* 38:2588–90.
107. Miesel, L., T. R. Weisbrod, J. A. Marcinkeviciene, R. Bittman, and W. J. Jacobs. 1998. NADH dehydrogenase defects confer isoniazid resistance and conditional lethality in *Mycobacterium smegmatis*. *J Bacteriol* 180:2459–67.
108. Vilcheze, C., T. R. Weisbrod, B. Chen, L. Kremer, M. H. Hazbon, F. Wang, D. Alland, J. C. Sacchettini, and W. R. Jacobs, Jr. 2005. Altered NADH/NAD+ ratio mediates coresistance to isoniazid and ethionamide in mycobacteria. *Antimicrob Agents Chemother* 49:708–20.
109. Kuo, M. R., H. R. Morbidoni, D. Alland, S. F. Sneddon, B. B. Gourlie, M. M. Staveski, M. Leonard, J. S. Gregory, A. D. Janjigian, C. Yee, J. M. Musser, B. Kreiswirth, H. Iwamoto, R. Perozzo, W. R. Jacobs, Jr., J. C. Sacchettini, and D. A. Fidock. 2003. Targeting tuberculosis and malaria through inhibition of Enoyl reductase: compound activity and structural data. *J Biol Chem* 278:20851–9.

110. Parikh, S. L., G. Xiao, and P. J. Tonge. 2000. Inhibition of InhA, the enoyl reductase from *Mycobacterium tuberculosis*, by triclosan and isoniazid. *Biochemistry* 39:7645–50.
111. Sullivan, T. J., J. J. Truglio, M. E. Boyne, P. Novichenok, X. Zhang, C. F. Stratton, H. J. Li, T. Kaur, A. Amin, F. Johnson, R. A. Slayden, C. Kisker, and P. J. Tonge. 2006. High affinity InhA inhibitors with activity against drug-resistant strains of *Mycobacterium tuberculosis*. *ACS Chem Biol* 1:43–53.
112. Mdluli, K., D. R. Sherman, M. J. Hickey, B. N. Kreiswirth, S. Morris, C. K. Stover, and C. E. Barry. 1996. Biochemical and genetic data suggest that InhA is not the primary target for activated isoniazid in *Mycobacterium tuberculosis*. *J Infect Dis* 174:1085–90.
113. Mdluli, K., R. A. Slayden, Y. Zhu, S. Ramaswamy, X. Pan, D. Mead, D. D. Crane, J. M. Musser, and C. r. Barry. 1998. Inhibition of a *Mycobacterium tuberculosis* beta-ketoacyl ACP synthase by isoniazid. *Science* 280:1607–10.
114. Lee, A. S., I. H. Lim, L. L. Tang, A. Telenti, and S. Y. Wong. 1999. Contribution of *kasA* analysis to detection of isoniazid-resistant *Mycobacterium tuberculosis* in Singapore. *Antimicrob Agents Chemother* 43:2087–9.
115. Piatek, A. S., A. Telenti, M. R. Murray, H. H. El, W. J. Jacobs, F. R. Kramer, and D. Alland. 2000. Genotypic analysis of *Mycobacterium tuberculosis* in two distinct populations using molecular beacons: implications for rapid susceptibility testing. *Antimicrob Agents Chemother* 44:103–10.
116. Kremer, L., L. G. Dover, H. R. Morbidoni, C. Vilcheze, W. N. Maughan, A. Baulard, S. C. Tu, N. Honore, V. Deretic, J. C. Sacchettini, C. Locht, W. R. Jacobs, Jr., and G. S. Besra. 2003. Inhibition of InhA activity, but not KasA activity, induces formation of a KasA-containing complex in mycobacteria. *J Biol Chem* 278:20547–54.
117. Larsen, M. H., C. Vilcheze, L. Kremer, G. S. Besra, L. Parsons, M. Salfinger, L. Heifets, M. H. Hazbon, D. Alland, J. C. Sacchettini, and W. R. Jacobs, Jr. 2002. Overexpression of *inhA*, but not *kasA*, confers resistance to isoniazid and ethionamide in *Mycobacterium smegmatis*, *M. bovis* BCG and *M. tuberculosis*. *Mol Microbiol* 46:453–66.
118. Argyrou, A., M. W. Vetting, B. Aladegbami, and J. S. Blanchard. 2006. *Mycobacterium tuberculosis* dihydrofolate reductase is a target for isoniazid. *Nat Struct Mol Biol* 13:408–13.
119. Alland, D., A. J. Steyn, T. Weisbrod, K. Aldrich, and W. J. Jacobs. 2000. Characterization of the *Mycobacterium tuberculosis iniBAC* promoter, a promoter that responds to cell wall biosynthesis inhibition. *J Bacteriol* 182:1802–11.
120. Colangeli, R., D. Helb, S. Sridharan, J. Sun, M. Varma-Basil, M. H. Hazbon, R. Harbacheuski, N. J. Megjugorac, W. R. Jacobs, Jr., A. Holzenburg, J. C. Sacchettini, and D. Alland. 2005. The *Mycobacterium tuberculosis iniA* gene is essential for activity of an efflux pump that confers drug tolerance to both isoniazid and ethambutol. *Mol Microbiol* 55:1829–40.
121. Pasca, M. R., P. Guglierame, E. De Rossi, F. Zara, and G. Riccardi. 2005. mmpL7 gene of *Mycobacterium tuberculosis* is responsible for isoniazid efflux in *Mycobacterium smegmatis*. *Antimicrob Agents Chemother* 49:4775–7.
122. Andersson, D. I., and B. R. Levin. 1999. The biological cost of antibiotic resistance. *Curr Opin Microbiol* 2:489–93.
123. Bjorkman, J., D. Hughes, and D. I. Andersson. 1998. Virulence of antibiotic-resistant *Salmonella typhimurium*. *Proc Natl Acad Sci USA* 95:3949–53.
124. Maisnier-Patin, S., O. G. Berg, L. Liljas, and D. I. Andersson. 2002. Compensatory adaptation to the deleterious effect of antibiotic resistance in *Salmonella typhimurium*. *Mol Microbiol* 46:355–66.
125. Schrag, S. J., and V. Perrot. 1996. Reducing antibiotic resistance. *Nature* 381:120–1.

126. Rosner, J. L. 1993. Susceptibilities of *oxyR* regulon mutants of *Escherichia coli* and *Salmonella typhimurium* to isoniazid. *Antimicrob Agents Chemother* 37:2251–3.
127. Deretic, V., W. Philipp, S. Dhandayuthapani, M. H. Mudd, R. Curcic, T. Garbe, B. Heym, L. E. Via, and S. T. Cole. 1995. *Mycobacterium tuberculosis* is a natural mutant with an inactivated oxidative-stress regulatory gene: implications for sensitivity to isoniazid. *Mol Microbiol* 17:889–900.
128. Sherman, D. R., K. Mdluli, M. J. Hickey, T. M. Arain, S. L. Morris, C. r. Barry, and C. K. Stover. 1996. Compensatory *ahpC* gene expression in isoniazid-resistant *Mycobacterium tuberculosis*. *Science* 272:1641–3.
129. Sherman, D. R., P. J. Sabo, M. J. Hickey, T. M. Arain, G. G. Mahairas, Y. Yuan, C. r. Barry, and C. K. Stover. 1995. Disparate responses to oxidative stress in saprophytic and pathogenic mycobacteria. *Proc Natl Acad Sci USA* 92:6625–9.
130. Wilson, T., G. W. de Lisle, J. A. Marcinkeviciene, J. S. Blanchard, and D. M. Collins. 1998. Antisense RNA to *ahpC*, an oxidative stress defence gene involved in isoniazid resistance, indicates that AhpC of *Mycobacterium bovis* has virulence properties. *Microbiology* 144:2687–95.
131. Wilson, T. M., L. G. de, and D. M. Collins. 1995. Effect of *inhA* and *katG* on isoniazid resistance and virulence of *Mycobacterium bovis*. *Mol Microbiol* 15:1009–15.
132. Gagneux, S., C. D. Long, P. M. Small, T. Van, G. K. Schoolnik, and B. J. Bohannan. 2006. The competitive cost of antibiotic resistance in *Mycobacterium tuberculosis*. *Science* 312:1944–6.
133. Mackaness, G. B. 1956. The intracellular activation of pyrazinamide and nicotinamide. *Am Rev Tubercul Pul Dis* 74:718–28.
134. Sturgill-Koszycki, S., P. H. Schlesinger, P. Chakraborty, P. L. Haddix, H. L. Collins, A. K. Fok, R. D. Allen, S. L. Gluck, J. Heuser, and D. G. Russell. 1994. Lack of acidification in Mycobacterium phagosomes produced by exclusion of the vesicular proton-ATPase. *Science* 263:678–81.
135. Scorpio, A., D. Collins, D. Whipple, D. Cave, J. Bates, and Y. Zhang. 1997. Rapid differentiation of bovine and human tubercle bacilli based on a characteristic mutation in the bovine pyrazinamidase gene. *J Clin Microbiol* 35:106–10.
136. Scorpio, A., L. P. Lindholm, L. Heifets, R. Gilman, S. Siddiqi, M. Cynamon, and Y. Zhang. 1997. Characterization of *pncA* mutations in pyrazinamide-resistant *Mycobacterium tuberculosis*. *Antimicrob Agents Chemother* 41:540–3.
137. Scorpio, A., and Y. Zhang. 1996. Mutations in *pncA*, a gene encoding pyrazinamidase/nicotinamidase, cause resistance to the antituberculous drug pyrazinamide in tubercle bacillus. *Nat Med* 2:662–7.
138. Sreevatsan, S., X. Pan, Y. Zhang, B. N. Kreiswirth, and J. M. Musser. 1997. Mutations associated with pyrazinamide resistance in *pncA* of *Mycobacterium tuberculosis* complex organisms. *Antimicrob Agents Chemother* 41:636–40.
139. Sun, Z., A. Scorpio, and Y. Zhang. 1997. The *pncA* gene from naturally pyrazinamide-resistant *Mycobacterium avium* encodes pyrazinamidase and confers pyrazinamide susceptibility to resistant *M. tuberculosis* complex organisms. *Microbiology* 143:3367–73.
140. Sun, Z., and Y. Zhang. 1999. Reduced pyrazinamidase activity and the natural resistance of *Mycobacterium kansasii* to the antituberculosis drug pyrazinamide. *Antimicrob Agents Chemother* 43:537–42.
141. Zhang, Y., A. Scorpio, H. Nikaido, and Z. Sun. 1999. Role of acid pH and deficient efflux of pyrazinoic acid in unique susceptibility of *Mycobacterium tuberculosis* to pyrazinamide. *J Bacteriol* 181:2044–9.
142. Hirano, K., M. Takahashi, Y. Kazumi, Y. Fukasawa, and C. Abe. 1998. Mutation in *pncA* is a major mechanism of pyrazinamide resistance in *Mycobacterium tuberculosis*. *Tubercl Lung Dis* 78:117–22.

143. Lemaitre, N., W. Sougakoff, C. Truffot-Pernot, and V. Jarlier. 1999. Characterization of new mutations in pyrazinamide-resistant strains of *Mycobacterium tuberculosis* and identification of conserved regions important for the catalytic activity of the pyrazinamidase PncA. *Antimicrob Agents Chemother* 43:1761–63.

144. Marttila, H. J., M. Marjamäki, E. Vyshnevskaya, B. I. Vyshnevskiy, T. F. Otten, A. V. Vasilyef, and M. K. Viljanen. 1999. *pncA* mutations in pyrazinamide-resistant *Mycobacterium tuberculosis* isolates from northwestern Russia. *Antimicrob Agents Chemother* 43:1764–66.

145. Boshoff, H. I., and V. Mizrahi. 1998. Purification, gene cloning, targeted knockout, overexpression, and biochemical characterization of the major pyrazinamidase from *Mycobacterium smegmatis*. *J Bacteriol* 180:5809–14.

146. Raynaud, C., M.-A. Lanéelle, R. H. Senaratne, P. Draper, G. Lanéelle, and M. Daffé. 1999. Mechanisms of pyrazinamide resistance in mycobacteria: importance of lack of uptake in addition to lack of pyrazinamidase activity. *Microbiology* 145:1359–67.

147. Konno, K., F. M. Feldman, and W. McDermott. 1967. Pyrazinamide susceptibility and amidase activity of tubercle bacilli. *Am Rev Respir Dis* 95:461–9.

148. McLaren, J., D. T. C. Ngo, and M. Olivera. 1973. Pyridine nucleotide metabolism in *Escherichia coli*. *J Biol Chem* 248:5144–9.

149. Takayama, K., and J. O. Kilburn. 1989. Inhibition of synthesis of arabinogalactan by ethambutol in *Mycobacterium smegmatis*. *Antimicrob Agents Chemother* 33:1493–9.

150. Khoo, K. H., E. Douglas, P. Azadi, J. M. Inamine, G. S. Besra, K. Mikusova, P. J. Brennan, and D. Chatterjee. 1996. Truncated structural variants of lipoarabinomannan in ethambutol drug-resistant strains of *Mycobacterium smegmatis*. Inhibition of arabinan biosynthesis by ethambutol. *J Bacteriol Chem* 271:28682–90.

151. Telenti, A., W. J. Philipp, S. Sreevatsan, C. Bernasconi, K. E. Stockbauer, B. Wieles, J. M. Musser, and W. J. Jacobs. 1997. The emb operon, a gene cluster of *Mycobacterium tuberculosis* involved in resistance to ethambutol. *Nat Med* 3:567–70.

152. Belanger, A. E., G. S. Besra, M. E. Ford, K. Mikusova, J. T. Belisle, P. J. Brennan, and J. M. Inamine. 1996. The *embAB* genes of *Mycobacterium avium* encode an arabinosyl transferase involved in cell wall arabinan biosynthesis that is the target for the antimycobacterial drug ethambutol. *Proc Natl Acad Sci USA* 93:11919–24.

153. Sharma, K., M. Gupta, M. Pathak, N. Gupta, A. Koul, S. Sarangi, R. Baweja, and Y. Singh. 2006. Transcriptional control of the mycobacterial *embCAB* operon by PknH through a regulatory protein, EmbR, *in vivo*. *J Bacteriol* 188:2936–44.

154. Lety, M. A., S. Nair, P. Berche, and V. Escuyer. 1997. A single point mutation in the *embB* gene is responsible for resistance to ethambutol in *Mycobacterium smegmatis*. *Antimicrob Agents Chemother* 41:2629–33.

155. Plinke, C., S. Rusch-Gerdes, and S. Niemann. 2006. Significance of mutations in embB codon 306 for prediction of ethambutol resistance in clinical *Mycobacterium tuberculosis* isolates. *Antimicrob Agents Chemother* 50:1900–2.

156. Sreevatsan, S., K. E. Stockbauer, X. Pan, B. N. Kreiswirth, S. L. Moghazeh, W. J. Jacobs, A. Telenti, and J. M. Musser. 1997. Ethambutol resistance in *Mycobacterium tuberculosis*: critical role of *embB* mutations. *Antimicrob Agents Chemother* 41:1677–81.

157. Alcaide, F., G. E. Pfyffer, and A. Telenti. 1997. Role of *embB* in natural and acquired resistance to ethambutol in mycobacteria. *Antimicrob Agents Chemother* 41:2270–3.

158. Tsukamura, M. 1985. *In vitro* antituberculosis activity of a new antibacterial substance ofloxacin (DL8280). *Am Rev Respir Dis* 131:348–51.

159. Tsukamura, M., E. Nakamura, S. Yoshii, and H. Amano. 1985. Therapeutic effect of a new antibacterial substance ofloxacin (DL8280) on pulmonary tuberculosis. *Am Rev Respir Dis* 131:352–6.

160. Iseman, M. D. 1999. Management of multidrug-resistant tuberculosis. *Chemotherapy* 2:3–11.
161. Iseman, M. D. 1993. Treatment of multidrug-resistant tuberculosis. *N Engl J Med* 329:784–91.
162. Nuermberger, E. L., T. Yoshimatsu, S. Tyagi, R. J. O'Brien, A. N. Vernon, R. E. Chaisson, W. R. Bishai, and J. H. Grosset. 2004. Moxifloxacin-containing regimen greatly reduces time to culture conversion in murine tuberculosis. *Am J Respir Crit Care Med* 169:421–6.
163. Nuermberger, E. L., T. Yoshimatsu, S. Tyagi, K. Williams, I. Rosenthal, R. J. O'Brien, A. A. Vernon, R. E. Chaisson, W. R. Bishai, and J. H. Grosset. 2004. Moxifloxacin-containing regimens of reduced duration produce a stable cure in murine tuberculosis. *Am J Respir Crit Care Med* 170:1131–4.
164. Camus, J. C., M. J. Pryor, C. Medigue, and S. T. Cole. 2002. Re-annotation of the genome sequence of *Mycobacterium tuberculosis* H37Rv. *Microbiology* 148: 2967–73.
165. Wang, J. C. 1985. DNA topoisomerases. *Ann Rev Biochem* 54:665–97.
166. Yoshida, H., M. Bogaki, M. Nakamura, and S. Nakamura. 1990. Quinolone resistance-determining region in the DNA gyrase *gyrA* gene of *Escherichia coli. Antimicrob Agents Chemother* 34:1271–2.
167. Alangaden, G. J., E. K. Manavathu, S. B. Vakulenko, N. M. Zvonok, and S. A. Lerner. 1995. Characterization of fluoroquinolone-resistant mutant strains of *Mycobacterium tuberculosis* selected in the laboratory and isolated from patients. *Antimicrob Agents Chemother* 39:1700–3.
168. Cambau, E., W. Sougakoff, M. Besson, P. C. Truffot, J. Grosset, and V. Jarlier. 1994. Selection of a *gyrA* mutant of *Mycobacterium tuberculosis* resistant to fluoroquinolones during treatment with ofloxacin. *J Infect Dis* 170:479–83.
169. Takiff, H. E., L. Salazar, C. Guerrero, W. Philipp, W. M. Huang, B. Kreiswirth, S. T. Cole, W. J. Jacobs, and A. Telenti. 1994. Cloning and nucleotide sequence of *Mycobacterium tuberculosis gyrA* and *gyrB* genes and detection of quinolone resistance mutations. *Antimicrob Agents Chemother* 38:773–80.
170. Xu, C., B. N. Kreiswirth, S. Sreevatsan, J. M. Musser, and K. Drlica. 1996. Fluoro-quinolone resistance associated with specific gyrase mutations in clinical isolates of multidrug-resistant *Mycobacterium tuberculosis. J Infect Dis* 174:1127–30.
171. Aubry, A., N. Veziris, E. Cambau, C. Truffot-Pernot, V. Jarlier, and L. M. Fisher. 2006. Novel gyrase mutations in quinolone-resistant and -hypersusceptible clinical isolates of *Mycobacterium tuberculosis*: functional analysis of mutant enzymes. *Antimicrob Agents Chemother* 50:104–12.
172. Matrat, S., N. Veziris, C. Mayer, V. Jarlier, C. Truffot-Pernot, J. Camuset, E. Bouvet, E. Cambau, and A. Aubry. 2006. Functional analysis of DNA gyrase mutant enzymes carrying mutations at position 88 in the A subunit found in clinical strains of *Mycobacterium tuberculosis* resistant to fluoroquinolones. *Antimicrob Agents Chemother* 50:4170–3.
173. Pitaksajjakul, P., W. Wongwit, W. Punprasit, B. Eampokalap, S. Peacock, and P. Ramasoota. 2005. Mutations in the gyrA and gyrB genes of fluoroquinolone-resistant *Mycobacterium tuberculosis* from TB patients in Thailand. *Southeast Asian J Trop Med Public Health* 36 Suppl 4:228–37.
174. Sullivan, E. A., B. N. Kreiswirth, L. Palumbo, V. Kapur, J. M. Musser, A. Ebrahimza-deh, and T. R. Frieden. 1995. Emergence of fluoroquinolone-resistant tuberculosis in New York City. *Lancet* 345:1148–50.
175. De Rossi, E., P. Arrigo, M. Bellinzoni, P. A. Silva, C. Martin, J. A. Ainsa, P. Guglierame, and G. Riccardi. 2002. The multidrug transporters belonging to major facilitator superfamily in *Mycobacterium tuberculosis. Mol Med* 8:714–24.

176. Pasca, M. R., P. Guglierame, F. Arcesi, M. Bellinzoni, E. De Rossi, and G. Riccardi. 2004. Rv2686c-Rv2687c-Rv2688c, an ABC fluoroquinolone efflux pump in *Mycobacterium tuberculosis. Antimicrob Agents Chemother* 48:3175–8.

177. Takiff, H. E., M. Cimino, M. C. Musso, T. Weisbrod, M. B. Delgado, L. Salazar, B. R. Bloom, and J. Jacobs, W.R. 1996. Efflux pump of the proton aniporter family confers low-level fluoroquinolone resistance in *Mycobacterium smegmatis. Proc Natl Acad Sci USA* 93:362–366.

178. Buroni, S., G. Manina, P. Guglierame, M. R. Pasca, G. Riccardi, and E. De Rossi. 2006. LfrR is a repressor that regulates expression of the efflux pump LfrA in *Mycobacterium smegmatis. Antimicrob Agents Chemother* 50:4044 52.

179. Sander, P., E. De Rossi, B. Boddinghaus, R. Cantoni, M. Branzoni, E. C. Bottger, H. Takiff, R. Rodriquez, G. Lopez, and G. Riccardi. 2000. Contribution of the multi-drug efflux pump LfrA to innate mycobacterial drug resistance. *FEMS Microbiol Lett* 193:19–23.

180. Montero, C., G. Mateu, R. Rodriguez, and H. Takiff. 2001. Intrinsic resistance of *Mycobacterium smegmatis* to fluoroquinolones may be influenced by new pentapeptide protein MfpA. *Antimicrob Agents Chemother* 45:3387–92.

181. Vetting, M. W., S. S. Hegde, J. E. Fajardo, A. Fiser, S. L. Roderick, H. E. Takiff, and J. S. Blanchard. 2006. Pentapeptide repeat proteins. *Biochemistry* 45:1–10.

182. Hegde, S. S., M. W. Vetting, S. L. Roderick, L. A. Mitchenall, A. Maxwell, H. E. Takiff, and J. S. Blanchard. 2005. A fluoroquinolone resistance protein from *Mycobacterium tuberculosis* that mimics DNA. *Science* 308:1480–3.

183. Andries, K., P. Verhasselt, J. Guillemont, H. W. Gohlmann, J. M. Neefs, H. Winkler, J. Van Gestel, P. Timmerman, M. Zhu, E. Lee, P. Williams, D. de Chaffoy, E. Huitric, S. Hoffner, E. Cambau, C. Truffot-Pernot, N. Lounis, and V. Jarlier. 2005. A diarylquinoline drug active on the ATP synthase of *Mycobacterium tuberculosis. Science* 307:223–7.

184. Ibrahim, M., K. Andries, N. Lounis, A. Chauffour, C. Truffot-Pernot, V. Jarlier, and N. Veziris. 2007. Synergistic activity of R207910 combined with pyrazinamide against murine tuberculosis. *Antimicrob Agents Chemother* 51:1011–5.

185. Lounis, N., N. Veziris, A. Chauffour, C. Truffot-Pernot, K. Andries, and V. Jarlier. 2006. Combinations of R207910 with drugs used to treat multidrug-resistant tuberculosis have the potential to shorten treatment duration. *Antimicrob Agents Chemother* 50:3543–7.

186. Petrella, S., E. Cambau, A. Chauffour, K. Andries, V. Jarlier, and W. Sougakoff. 2006. Genetic basis for natural and acquired resistance to the diarylquinoline R207910 in mycobacteria. *Antimicrob Agents Chemother* 50:2853–6.

187. Matsumoto, M., H. Hashizume, T. Tomishige, M. Kawasaki, H. Tsubouchi, H. Sasaki, Y. Shimokawa, and M. Komatsu. 2006. OPC-67683, a nitro-dihydro-imidazooxazole derivative with promising action against tuberculosis *in vitro* and in mice. *PLoS Med* 3:e466.

188. Stover, C. K., P. Warrener, D. R. VanDevanter, D. R. Sherman, T. M. Arain, M. H. Langhorne, S. W. Anderson, J. A. Towell, Y. Yuan, D. N. McMurray, B. N. Kreiswirth, C. E. Barry, and W. R. Baker. 2000. A small-molecule nitroimidazopyran drug candidate for the treatment of tuberculosis. *Nature* 405:962–6.

189. Choi, K. P., T. B. Bair, Y. M. Bae, and L. Daniels. 2001. Use of transposon *Tn5367* mutagenesis and a nitroimidazopyran-based selection system to demonstrate a requirement for *fbiA* and *fbiB* in coenzyme F(420) biosynthesis by *Mycobacterium bovis* BCG. *J Bacteriol* 183:7058–66.

190. Choi, K. P., N. Kendrick, and L. Daniels. 2002. Demonstration that *fbiC* is required by *Mycobacterium bovis* BCG for coenzyme F(420) and FO biosynthesis. *J Bacteriol* 184:2420–8.

191. Manjunatha, U. H., H. Boshoff, C. S. Dowd, L. Zhang, T. J. Albert, J. E. Norton, L. Daniels, T. Dick, S. S. Pang, and C. E. Barry, 3rd. 2006. Identification of a nitroimidazo-oxazine-specific protein involved in PA-824 resistance in *Mycobacterium tuberculosis*. *Proc Natl Acad Sci USA* 103:431–6.
192. Cole, S. T. 2002. Comparative and functional genomics of the *Mycobacterium tuberculosis* complex. *Microbiology* 148:2919–28.
193. Baker, L., T. Brown, M. C. Maiden, and F. Drobniewski. 2004. Silent nucleotide polymorphisms and a phylogeny for *Mycobacterium tuberculosis*. *Emerg Infect Dis* 10:1568–77.
194. Hirsh, A. E., A. G. Tsolaki, K. DeRiemer, M. W. Feldman, and P. M. Small. 2004. Stable association between strains of *Mycobacterium tuberculosis* and their human host populations. *Proc Natl Acad Sci USA* 101:4871–6.
195. Liu, X., M. M. Gutacker, J. M. Musser, and Y. X. Fu. 2006. Evidence for recombination in *Mycobacterium tuberculosis*. *J Bacteriol* 188:8169–77.

14 Antibiotic Resistance in Enterobacteria

Nafsika H. Georgopapadakou

CONTENTS

Enterobacteria cause a variety of nosocomial and community-acquired (including foodborne) infections and include some of the most deadly pathogens. As a result, their resistance to antibiotics has profound clinical implications. The major antibiotic classes currently in use for enterobacterial infections are the β-lactams, quinolones, aminoglycosides, tetracyclines, and sulfonamides. Resistance to β-lactams is relatively common and involves mainly serine β-lactamases: inducible, typically chromosomal (class C) as well as constitutive, typically plasmid-mediated, extended spectrum (classes A and D). There have been recent reports of class B metallo-β-lactamases in *Klebsiella*. Integron-borne β-lactamases (classes A, B, and D) occur in Enterobacteria species together with non-β-lactam resistance genes, giving rise to multidrug-resistant bacteria. They pose a threat, particularly in the hospital environment, as non-β-lactam agents may select potent β-lactamases through integron-mediated resistance. Resistance to quinolones is associated with changes in the target DNA gyrase (chromosome encoded) or target protection by the proteins QnrA, QnrB, and QnrS (plasmid encoded) and affects quinolones in use as well as in clinical development. Reduced accumulation in the cell, due to active efflux through the cytoplasmic membrane and decreased influx through the outer membrane, may facilitate the emergence of quinolone resistance. Resistance to aminoglycosides is predominantly due to enzymatic inactivation in the periplasmic space, the exact nature of the modification depending on the particular aminoglycoside. The major mechanism for tetracycline resistance involves an active efflux system; ribosomal protection is not a clinically important mechanism in Enterobacteria. Sulfonamide resistance is due to an additional, plasmid-mediated, sulfonamide-resistant, dihydropteroate synthase target. Overall, the biggest clinical concern is multidrug resistance, particularly the ongoing erosion of the effectiveness of β-lactams and quinolones, two bactericidal and generally safe antibacterial classes.

INTRODUCTION

The family Enterobacteriaceae is the widest and most heterogeneous group of medically important Gram-negative bacteria [1]. It includes many species found in the gastrointestinal tract of humans and animals, as well in soil, water, and plants. About one-third of the 30 genera known contain human pathogens, causing a variety of diseases ranging from mild intestinal infections to urinary tract infections, nosocomial respiratory tract infections, wound infections, and septic shock (Table 14.1). Individual species have been associated with specific epidemics that continue to the present: infamously *Yersinia pestis* [2], responsible for plague (the Black Death of the Middle Ages) and currently considered a potential bioterrorism (category A) agent [3–6]; more modestly *Escherichia coli* O157:H7, responsible for 70,000 cases of infection and 60 deaths in the United States yearly (http://www.cdc.gov/ncidod/dbmd/diseaseinfo/escherichiacoli_g.htm) [7]).

The pathogenicity of Enterobacteria is associated with clusters of genes (plasmid- or chromosomal-borne) that play critical roles in bacterial colonization and virulence [8–13]. Pathogenic strains often live in a sea of avirulent strains that colonize people and animals; the latter strains represent a reservoir of resistance genes and of potential hosts for the mobile genetic elements that encode virulence factors.

Antibiotic resistance is a direct consequence of antibiotic use both in humans and in animals [14–16]. For example, quinolone resistance in *Salmonella*, a food-borne pathogen causing perhaps a million cases of—mostly self-limited—infection and a thousand deaths in the United States yearly, almost certainly originated from animals [17–21]. The over-reliance on antibiotics, and insufficient application of infection control measures and improved hygiene, has eroded the effectiveness of older, inexpensive agents and threatens the efficacy of recently introduced ones [22].

Antibiotic resistance is commonly detected by susceptibility testing, which describes the resistance phenotype of an organism and has practical implications for patient treatment. For epidemiological/surveillance purposes, strain typing is often performed by serologic methods or, more precisely, by pulsed-field gel electrophoresis (PFGE) of digested DNA [23–24]. Resistance is further characterized by biochemical and molecular biology techniques. The former include function assays; the latter DNA restriction analysis, DNA probes, and nucleic acid amplification by the polymerase chain reaction (PCR) [25]. Plasmid DNAs can be isolated and analyzed by digestion with restriction endonucleases to estimate size and type; DNA

TABLE 14.1
Pathogenic Enterobacteria and the Diseases They Produce

Genus	Species	Infection/Disease	Comments
Citrobacter	freundii	UT[a], RT, wound, blood	Nosocomial
Enterobacter	aerogenes	UT, RT, wound, blood	Nosocomial
	cloacae	UT, RT, wound, blood	Nosocomial
Escherichia	coli	GIT[b], UT[c], RT, wound, blood	Food borne, nosocomial
Klebsiella	oxytoca	UT, RT, wound, blood	Nosocomial
	pneumoniae	UT, RT, wound, blood	Nosocomial
Morganella	morganii	UT, RT, wound, blood	Nosocomial
Proteus	mirabilis	UT, RT, wound, blood	Nosocomial
	vulgaris	UT, RT, wound, blood	Nosocomial
Providencia	rettgeri	UT, wound, pneumonia, blood	Nosocomial
	stuartii	UT, wound, pneumonia, blood	Nosocomial
Salmonella	enteritidis	GIT	Food-borne
	typhi	GIT, typhoid fever	Food-borne
	typhimurium	GIT	Food-borne
Serratia	marcescens	UTI, wound, pneumonia, blood	Nosocomial
Shigella[d]	dysenteriae	GIT (shigellosis)	Food-borne
	flexneri	GIT (shigellosis)	Food-borne
	sonnei	GIT (shigellosis)	Food-borne
Yersinia	enterocolitica	GIT	Food/water-borne
	pseudotuberculosis	GIT	Zoonotic, food-borne
	pestis	Lymph nodes (bubonic plague)	Zoonotic (rodents/fleas)

[a]*Abbreviations*: GIT, gastrointestinal tract, UT, urinary tract, RT, respiratory tract.
[b]Most common cause of UT infection.
[c]Infections caused by particular virulent *E. coli* (EC) strains: enterotoxigenic (ETEC), diarrhea; entero-pathogenic (EPEC), infantile diarrhea; enteroinvasive (EIEC), dysentery; enterohemorrhagic (EHEC), such as serotype O157:H7, hemorrhagic colitis.
[d]In the U.S., *Shigella sonnei* (group D *Shigella),* accounts for over two-thirds of the shigellosis, while *Shigella flexneri* (group B *Shigella*), accounts for most of the rest. *Shigella dysenteriae* type 1 is found in the developing world, where it causes deadly epidemics.

of fragments can be sequenced to identify specific resistance genes [26–28]. Resistance mechanisms may operate synergistically—for example, transport-associated resistance and antibiotic inactivation—and the contribution of each to the overall resistance may be difficult to assess [29–30].

RESISTANCE MECHANISMS IN ENTEROBACTERIA

GENETIC MECHANISMS

Resistance in Enterobacteria can result from gene mutations or transfer of resistance determinants (R-determinants) between strains or species. Clinically, gene transfer is the most common mechanism of transferring resistance [31–32]. R-determinants are typically on plasmids, but may also be part of mobile genetic elements (transposons, integrons, and gene cassettes), which can move between plasmids or chromosomes in the same species or to a new species [33–37].

Plasmids (4 to 400 kb) are self-replicating, extrachromosomal elements that contain genes for resistance, virulence, and other functions and are dispensable under certain conditions. Some larger plasmids are conjugative (R-plasmids) and can transfer between organisms, spreading resistance genes. For example, conjugative plasmids were responsible for the spread of sulfonamide resistance to *Shigella dysenteriae* in the 1950s. Resistance genes can thus disseminate independently of the host organism (horizontal transfer) in addition to disseminating along with the host (clonal spread).

Transposons (2 to 20 kb) are mobile genetic elements that contain insertion sequences (0.2 to 6 kb) and one or more resistance genes. They are not capable of autonomous self-replication, but can move (transpose) from one site on the chromosome to another site on the same or different chromosome or plasmid and replicate along with it. Transposition is made possible by short inverted repeats of DNA.

Integrons are mobile genetic elements of specific structure that consist of two conserved segments flanking a central region in which resistance gene cassettes are inserted [32–34]. On the 5′-conserved segment is an *int* gene that encodes a site-specific recombinase capable of capturing DNA, including resistance genes. Although the probability of capture of a resistance gene is low, it can confer a selective advantage to its host. Adjacent to it are a suitably oriented promoter for expression of the cassette genes and the receptor site for the gene cassettes (*attI* site). On the 3′-conserved segment, which is of variable length, is typically the *sulI* gene that encodes a sulfonamide-resistant dihydropteroate synthase [36]. Additional resistance genes can also be present, their distance from the promoter determining their level of expression.

Alarmingly, as resistance genes are inserted into mobile genetic elements, they sometimes link with other resistance genes in resistance clusters, whose transfer can then result in simultaneous acquisition of resistance to several unrelated drugs (multidrug resistance) [32–33,35,37–38].

BIOCHEMICAL MECHANISMS

Biochemical mechanisms of antibiotic resistance include altered transport (influx or efflux) and thereby reduced intracellular accumulation [30] (see Chapter 8, this volume); enzymatic inactivation (hydrolysis or derivatization); altered or additional resistant target; bypassed target; and compensatory changes downstream of target. Table 14.2 summarizes resistance mechanisms for specific antibacterial classes used against Enterobacteria.

RESISTANCE TO MAJOR ANTIBIOTIC CLASSES IN ENTEROBACTERIA

β-LACTAMS

β-Lactam antibiotics constitute the most enduring and widely used class of antibacterials, encompassing a large number of mostly semisynthetic compounds.

TABLE 14.2

Biochemical Resistance Mechanisms in Enterobacteria

Antibiotic Class	New/Altered Enzyme Protein/Gene	Gene Location	Comments	References
Antibiotic Inactivation				
Hydrolysis				
β-Lactams	β-Lactamase	Ch, P	Most common resistance mechanism	44 46, 56–58
Modification				
Aminoglycosides	N-Acetyltransferase, O-adenylyl- transferase, O-phosphotransferase	Ch, P		104–105 111, 116
Chloramphenicol	O-Acetyltransferase			153
Altered/Additional Target				
Decreased binding				
Quinolones	DNA gyrase, topoIV	Ch, P	Most common resistance mechanism	95–97
Quinolones	Target protection (*qnrA, qnrB, qnrS*)	P	Associated with multidrug resistance	80–92
Aminoglycosides	RNA, ribosomal protein S12			104–105
Tetracyclines	Ribosomal protection(*tetM, O*)			124–125
Sulfonamides	Dihydropteroate synthetase	P		36, 145
Trimethoprim	Dihydrofolate reductase (DHFR)	P		145, 150
Overproduced Target				
Trimethoprim	DHFR (type IV)	P		145, 150
Decreased Intracellular Accumulation				
Decreased uptake				
β-Lactams	Porins	Ch		78, 82
Quinolones	Porins	Ch		95, 98
Aminoglycosides	Altered active transport			122
Increased efflux				
β-Lactams	AcrAB-TolC	Ch		163–164
Quinolones	AcrAB-TolC, EmrAB	Ch		95, 98
Tetracyclines	tet A,B,C,D,E,K,L (I)	Ch, P	Most common resistance mechanism	133

Abbreviations: Ch, Chromosomal; I, inducible; P, plasmid-mediated; R, resistance.

The clinically useful β-lactams are divided on the basis of structure into penams, penems and cephems (Figure 14.1) [39]. Their targets are peptidoglycan transpeptidases, cell wall synthesizing enzymes located on the outer face of the cytoplasmic membrane [40]. These enzymes are ubiquitous in bacteria and are commonly detected by their ability to bind covalently and specifically penicillin and other β-lactam antibiotics (hence the name, penicillin-binding proteins [PBPs]). Not all PBPs are peptidoglycan transpeptidases, or essential; in Enterobacteria, only three of the eight PBPs are essential.

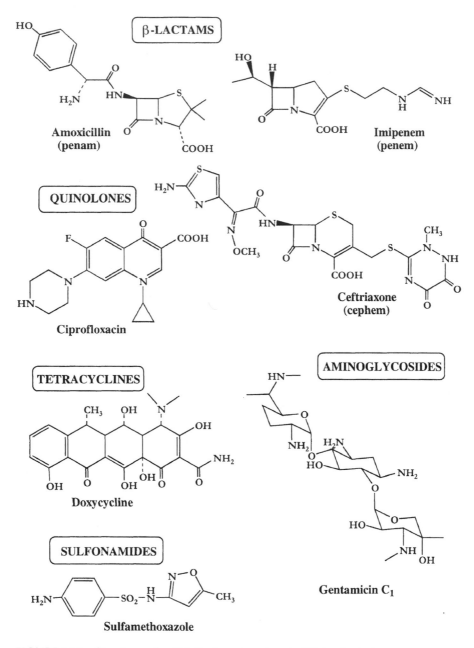

FIGURE 14.1   Structures of antibiotic classes used against Enterobacteria.

Clinically, the most important mechanism of resistance in Enterobacteria is hydrolysis by β-lactamases, common bacterial enzymes related to the cell wall targets [41]. β-lactamases are divided into four groups based on amino acid sequence homology (Ambler classification) [42–43], though other classification schemes also exist [44]. In Enterobacteria, all four classes are represented (Table 14.3), but classes

TABLE 14.3
Common β-Lactamases in Enterobacteria

Enzyme	Original Host	Substrate Profile	Inhibitor Profile	Comments
Class A: Serine Enzymes (~30 kDa), Mostly Plasmid-Mediated, Constitutively Expressed, ca. 250				
TEM-1	E. coli	pen/ceph	clox/clav/sulb/taz	Most common type
TEM-2	E. coli	pen/ceph	clox/clav/sulb/taz	Differs from TEM-1 by one amino acid
TEM-3 to TEM-70	K. pneumoniae	pen/ceph	clox/clav/sulb/taz	ESBL variants of TEM-1, in nosocomial outbreaks (see www.lahey.org/lcinternet/studies/webt.htm)
SHV-1	K. pneumoniae	pen/ceph	clav/sulb/taz	a.k.a. PIT-2; ESBL, often chromosomal; 24 variants to-date (www.lahey.org/lcinternet/studies/webt.htm)
CTX-M	E. coli	pen/ceph		Plasmid, extended spectrum
K-1	K. oxytoca	ceph		Chromosomal, extended spectrum
Class B: Metalloenzymes (~22 kDa), Mostly Chromosomal, ca. 30				
IMP-1	Stenotrophomonas maltophilia	pen/ceph/cpen		Uncommon; reported in S. marcescens, S. flexneri (integron-borne)
VIM-1			pen/ceph/cpen	Uncommon, reported in E. cloacae, K. pneumoniae (integron-borne)
Class C: Serine Enzymes (~40 kDa), Mostly Chromosomal, Inducible, ca. 30				
AmpC	E. coli	ceph	taz	
P99	E. cloacae	ceph	taz	
MIR-1	K. pneumoniae	ceph		Plasmid-mediated
CMY-1 to CMY-5	K. pneumoniae	ceph		Plasmid-mediated
Class D: Serine Enzymes (~12 kDa), Plasmid Mediated, ca. 15				
OXA-1	E. coli	pen/ceph		Related, less common enzymes: OXA-3 to OXA-7

Abbreviations: ceph, cephems, clav; clavulanic acid; clox, cloxacillin; cpen, carbapenem; ESBL, extended spectrum β-lactamases; pen, penams; taz, tazobactam.

A and C are of the greatest clinical significance [45–48]. Both penams and cephems are affected, though rarely penems. Surprisingly, *Y. pestis* appears to have decreased susceptibility to penems, but not to penams or cephems [49–50].

In class A β-lactamases, the most important enzymes are TEM-1, SHV-1, CTX-M. TEM-1, which originated in *E. coli*, is now very common in *Klebsiella* and other

Enterobacteria, while SHV-1 is commonly found in *Klebsiella pneumoniae* and can be plasmid mediated or chromosomal [51–54]. CTX-M enzymes were first recognized in 1986 in Japan, in 1989 in Argentina, and thereafter worldwide [55–56]. Of great clinical concern is the emergence of extended-spectrum variants of TEM, SHV, and CTX-M enzymes in the 1980s that continues to the present [57–64]. To date, over 250 variants of these enzymes have been identified (http://www.lahey.org/Studies). These extended-spectrum β-lactamases (ESBLs) are particularly problematic because they can hydrolyze oxyimino β-lactams (cefotaxime, ceftriaxone, and ceftazidime) and can easily spread to other species. They are generally sensitive to inhibition by clavulanic acid, though resistant variants have been reported [60–62]. Clinical isolates that produce ESBLs are frequently associated with nosocomial outbreaks [63–65]. Class D and class B [57–58] enzymes, though uncommon, are also ESBLs.

Many clinical laboratories lack the necessary technology, and thus ESBL detection in the clinical microbiology laboratory is often problematic. For example, in a survey by Tenover et al. [65], the percentage of labs in Connecticut that failed to detect resistance in the ESBL- or ampC-producing isolates ranged from 24% to 32%. A 1998 survey of 369 microbiology labs participating in the CDC Emerging Infections Network found that only 32% tested for ESBL producers, and of that subset, only 17% used adequate methods to confirm ESBL presence [66].

Ambler class C β-lactamases are produced by most Enterobacteria, but are particularly important in clinical isolates of *Enterobacter cloacae* and *Enterobacter aerogenes*, *Citrobacter freundii*, and *Serratia marcescens* [67–71]. They hydrolyze both penicillins and cephalosporins, including cephamycins, such as cefoxitin, and are resistant to clavulanic acid. They are normally inducible, regulation of expression being linked to the cell wall synthesis and recycling [72–73]. Relatively recent developments are the evolution and spread of class A enzymes, the mobilization of class C enzymes MIR-1, CMY, MOX into plasmids and the expansion of the activity spectrum to include carbapenems, cephamycins, and oxyiminocephalosporins.

A major factor contributing to β-lactam resistance is decreased outer-membrane permeability [74–82]. Because they live in the gut, Enterobacteria have developed a particularly "finicky" outer membrane. This is a protein-rich, asymmetric lipid bilayer that contains lipopolysaccharide in the outer leaflet and envelops the peptidoglycan. It functions as a molecular sieve, with water-filled channels (pores) formed by 35 to 40 kDa protein trimers (porins). It is through these channels that non-specific transport of small hydrophilic molecules, such as β-lactams, occurs. There are at least two porin species in *E. coli*, OmpC and OmpF, which form channels of 11 and 12 A diameter, respectively, with an exclusion limit of 600 to 800 Da. Other Enterobacteria also have two to three porins, homologous to those of *E. coli* [78–79]. Porin-deficient mutants of Enterobacteria have reduced β-lactam susceptibility relative to isogenic wild-type strains [80–82].

A key aspect to the susceptibility of Enterobacteria to β-lactams is the interplay of outer-membrane permeability, affinity/turnover for β-lactamases in the periplasmic space, and affinity for target PBPs. Although β-lactamases constitute the major form of β-lactam resistance in Enterobacteria, it is the combined presence of β-lactamases and reduced outer membrane permeability that affects resistance. This cooperative action effectively reduces the concentration of β-lactams in the periplasm. Decreased target affinity has not been reported in clinical isolates of Enterobacteria, perhaps because of the fitness cost it entails [83].

QUINOLONES

Quinolones are broad-spectrum, bactericidal antibiotics whose potent activity, including activity even against intracellular pathogens, and ease of administration (oral and parenteral) have firmly established them both in the hospital and the community (see Chapter 7, this volume). Quinolones enter bacterial cells through the porins in the outer membrane and by diffusion through the cytoplasmic membrane [84,85]. They then complex immediately, selectively, and reversibly with DNA gyrase and the related topoisomerase IV, bacterial enzymes essential for transcription, replication, and chromosome decatenation. They trap a covalent enzyme-DNA complex (cleavable complex), in which the enzyme has broken the phosphodiester backbone of the DNA, and thereby inhibit the subsequent religation of DNA [86–87]. The result is inhibition of supercoiling (DNA gyrase), chromosome decatenation (topoisomerase IV) and, most importantly, the induction of DNA lesions, which trigger the SOS response and ultimately lead to cell death.

With the exception of Qnr proteins, which are plasmid mediated [88–92], quinolone resistance is exclusively chromosomal, spreading with the resistant organism. The Qnr proteins are capable of protecting DNA gyrase from quinolones and have homologs in water-dwelling bacteria. They seem to have been in circulation for some time, having achieved global distribution in a variety of plasmid environments and bacterial genera [90]. Though *qnr* genes provide low-level quinolone resistance, they facilitate the emergence of higher-level, clinical resistance. Of further concern is the rapid, horizontal spread of the *qnr* genes and their co-selection with other resistance elements. AAC(6′)-Ib-cr, a variant aminoglycoside acetyltransferase capable of modifying ciprofloxacin and modestly reducing its activity, seems to have emerged more recently [93], but might be even more prevalent than the Qnr proteins.

The most common mechanism of high-level, clinical resistance in Enterobacteria is associated with mutations in *gyrA*, which encodes subunit A of DNA gyrase [94–96]. Resistance mutations tend to cluster between residues 67 and 106 (quinolone resistance-determining region [QRDR]) [96]. Mutations in *parC*, which encodes the homologous subunit A of topoisomerase IV, are also associated with resistance [97]. Reduced accumulation in the cell, due to active efflux through the cytoplasmic membrane combined with decreased influx through the outer membrane, appears to cause only low levels of resistance, but can facilitate the emergence of fluoroquinolone-resistant strains [98–102].

Clinical resistance is not yet very common in Enterobacteria [102], despite the widespread use of quinolones and the emergence of plasmid-associated quinolone resistance. The only exception is the foodborne pathogen *Salmonella*, where resistance in some specific phage types has been found in Europe, most likely due to the extensive use of quinolones in food animals [103]. Nevertheless, because resistance affects not only quinolones in use, but also in clinical development and—in the case of *qnr*-associated resistance—is linked to other resistance genes, it is a cause for concern.

AMINOGLYCOSIDES

Aminoglycosides are bactericidal, broad-spectrum antibiotics discovered in the 1940s (see Chapter 5, this volume). They are still widely used (gentamicin and amikacin), usually in combination with β-lactam agents, against problem pathogens despite their ototoxicity and nephrotoxicity (104). Structurally, aminoglycosides are polycationic amino sugars, the amino groups being protonated in biological media. A number of subclasses have been identified and semisynthetic derivatives less prone to enzymatic inactivation have been developed [105–106].

The mechanism of aminoglycoside action involves binding to the A site of the 16S ribosomal RNA and thereby inhibition of protein synthesis [107–108], after entering through the outer membrane via a porin-independent, "self-promoted" pathway and through the cytoplasmic membrane via an energy-dependent pathway [109–111].

Clinically, the most significant mechanism of aminoglycoside resistance is enzymatic modification [112], the exact profile depending on the aminoglycoside being used [113–114]. The modifying enzymes, N-acetyltransferases, O-phosphoryltransferases, and O-adenyltransferases, have broad substrate specificity and can catalyze more than one reaction [115–119]. Their origin is diverse [120]. Impaired uptake may also contribute to resistance [121–123].

TETRACYCLINES

Tetracyclines are broad-spectrum, bacteriostatic agents that also act by inhibiting protein synthesis [124–125]. They bind reversibly to a single, high-affinity site on the 30S ribosomal subunit and disrupt the codon–anticodon interaction between aminoacyl-tRNA and mRNA, thereby inhibiting the binding of aminoacyl-tRNA to the acceptor site on the ribosome. Their selective antibacterial toxicity may be due, at least in part, to selective, concentrative uptake by bacteria [126–127].

The major mechanism for tetracycline resistance involves an inducible active efflux system, whereby the intracellular concentration of these compounds is reduced [128–132]. Several genes encoding for components of this system (tetA-E in Gram-negative bacteria) [133–136], located mostly on plasmids, have been identified. Different tetracyclines are not equally recognized by transport proteins; for example, TetA does not recognize minocycline and doxycycline. Ribosomal protection by a soluble, usually plasmid-encoded, 72-kDa protein homologous to the elongation factor G that is involved in protein synthesis [137–138], is not a significant resistance mechanism in Enterobacteria. A notable development in the field was the discovery of glycylcyclines, minocycline derivatives, one of which, tigecycline (formerly known as GAR 936), has received FDA approval and is now available as Tygacil® [139–143]. Tigecycline is active against most tetracycline-resistant strains. Another recent development is the potentiation of the antibacterial activity of tetracyclines by inhibitors of Tet efflux proteins [144].

ANTIFOLATES: SULFONAMIDES AND TRIMETHOPRIM

Sulfonamides, the oldest, totally synthetic antibacterial agents, are competitive inhibitors of dihydropteroate synthetase by virtue of their active (sulfone) form being a structural analog of the *p*-aminobenzoic acid substrate. Clinically, the most common and important mechanism of resistance to these bacteriostatic agents is altered, usually plasmid-mediated, target enzyme [36,145]. Two distinct types of altered dihydropteroate synthetase have been characterized in Gram-negative bacteria, I and II, encoded by *sulI* and *sulII*, respectively [146–148]. They have reduced binding to inhibitors, but—remarkably—retain normal binding to the *p*-aminobenzoic acid substrate. The *sulI* gene is often located in transposons related to Tn*21* or on large R-plasmids with a resistance region similar to Tn*21* [36,148]. The *sulII* gene is carried mainly on small non-conjugative plasmids.

Trimethoprim, also totally synthetic and commonly used in combination with sulfonamides, is a bactericidal agent. It is a selective, potent, competitive inhibitor of the bacterial dihydrofolate reductase (DHFR). The resulting tetrahydrofolate depletion affects methyl transfer reactions, particularly the one involved in thymine biosynthesis, thereby causing thymineless death. The most common mechanism of trimethoprim resistance is altered, usually plasmid-mediated, target enzyme [149–151]. Mutant forms of the normal, chromosomal DHFR are far less common in clinical isolates. Seven major types of plasmid-encoded, trimethoprim-resistant DHFRs (I to VII) have been found in Gram-negative bacteria. They share variable homology with each other and with the normal, chromosomal enzyme, suggesting both divergent and convergent evolution.

CHLORAMPHENICOL

Chloramphenicol, still an important bacteriostatic agent, owes its selective antibacterial activity to inhibition of the peptidyltransferase reaction of protein syntesis via binding to the 50S ribosomal subunit. The major mechanism of clinical resistance to chloramphenicol is its inactivation by acetylation [152]. Three genetically distinct groups of chloramphenicol acetyltransferases have been found so far, some inducible and others constitutive, but all sharing sequence homology [153]. As previously stated, multidrug-resistant *S. typhimurium* (DT104), which represent approximately 10% of the Salmonella isolates in the United States, is resistant to chloramphenicol [154].

CONCLUSIONS AND FUTURE DIRECTIONS

Enterobacteria cause a variety of nosocomial and community-acquired (including foodborne) infections and thus their resistance to antibiotics has profound clinical implications. Of the five major antibiotic classes currently used for enterobacterial infections (β-lactams, quinolones, aminoglycosides, tetracyclines, and sulfonamides), quinolones (ciprofloxacin) and β-lactams (amoxicillin/clavulanic acid combination and third-generation cephalosporins) are the least affected by resistance. The biggest threat for the future is the gradual erosion of the effectiveness of these two bactericidal and generally safe antibacterial classes [155,156].

The fact that enterobacterial infections are treatable with existing antibiotics has kept them out of the limelight. *Enterobacter* and *Klebsiella*, though often resistant to several antibiotics, are simply not in the *Pseudomonas/Acinetobacter/Enterococcus* resistance league. Neither are as spectacularly invasive as some *Streptococcus* strains, yet they are very common pathogens and their emerging resistance in institutional settings should be a cause for concern [157–161].

Since the first edition of this book was written, several new antibacterials have entered the clinic. Unfortunately, all but one (tigecycline) target mostly Gram-positive bacteria, which are admittedly easier to kill since they lack an outer membrane, a cell structure particularly effective in restricting entry of antibiotics in Enterobacteria. Resistance to β-lactams has alarmingly increased in *Klebsiella* and plasmid-mediated quinolone resistance appears to be more and more common.

The plea mentioned in the first edition of this book has therefore assumed new urgency. Drug development is a long, tortuous process; we need to be more proactive and start targeting Enterobacteria now, with new agents that do not cross-react with existing ones. The possibility of covering also *Pseudomonas* and *Acinetobacter* would be an added bonus. In this context, it would be valuable to draw on the recent advances in our understanding of efflux mechanisms [162–164], with the goal of perhaps avoiding them rather than targeting them. The hydra-like nature of transport proteins, whereby suppression of one unmasks another, argues for caution. Nevertheless, recent work has shown that it is possible to potentiate antibacterial activity by inhibiting drug efflux [165–166], just as earlier work had shown that it was possible to potentiate antibacterial action by promoting drug influx through the outer membrane [167].

REFERENCES

1. Holt JG, Krieg NR, Sneath PHA, Staley JT, Williams ST (eds) *Bergey's Manual of Determinative Bacteriology*, 9th ed. Baltimore: Williams & Wilkins, 1993.
2. Dykhuizen DE. *Yersinia pestis*: an instant species? *Trends Microbiol* 2000; 8:296–8.
3. Achtman M, Zurth K, Morelli G, Torrea G, Guiyoule A, Carniel E. *Yersinia pestis*, the cause of plague, is a recently emerged clone of *Yersinia pseudotuberculosis*. *Proc Natl Acad Sci USA* 1999; 96:14043–8.
4. Skurnik M, Peippo A, Ervela E. Characterization of the O-antigen gene clusters *of Yersinia pseudotuberculosis* and the cryptic O-antigen gene cluster of *Yersinia pestis* shows that the plague bacillus is most closely related to and has evolved from *Y. pseudotuberculosis* serotype O:1b. *Mol Microbiol* 2000; 37:316–30.
5. Derbise A, Chenal-Francisque V, Pouillot F, Fayolle C, Prevost MC, Medigue C, Hinnebusch BJ, Carniel E. A horizontally acquired filamentous phage contributes to the pathogenicity of the plague bacillus. *Mol Microbiol* 2007; 63:1145–57.
6. Ligon BL. Plague: a review of its history and potential as a biological weapon. *Semin Pediatr Infect Dis* 2006; 17:161–70.
7. Armstrong GL, Hollingsworth J, Morris JG Jr. Emerging foodborne pathogens: *Escherichia coli* O157:H7 as a model of entry of a new pathogen into the food supply of the developed world. *Epidemiol Rev* 1996; 18, 29–51.
8. Hacker J, Blum-Oehler G, Muhldorfer I, Tschape H. Pathogenicity islands of virulent bacteria: structure, function and impact on microbial evolution. *Mol Microbiol* 1997; 23:1089–97.
9. Wood MW, Jones MA, Watson PR, Hedges S, Wallis TS, Galyov EE. Identification of a pathogenicity island required for *Salmonella* enteropathogenicity. *Mol Microbiol* 1998; 29:883–91.
10. Hensel M, Nikolaus T, Egelseer C. Molecular and functional analysis indicates a mosaic structure of *Salmonella* pathogenicity island 2. *Mol Microbiol* 1999; 31:489–98.

11. Vokes SA, Reeves SA, Torres AG, Payne SM. The aerobactin iron transport system genes in *Shigella flexneri* are present within a pathogenicity island. *Mol Microbiol* 1999; 33:63–73.

12. Hacker J, Hochhut B, Middendorf B, Schneider G, Buchrieser C, Gottschalk G, Dobrindt U. Pathogenomics of mobile genetic elements of toxigenic bacteria. *Int J Med Microbiol* 2004; 293:453–61.

13. Deng W, Puente JL, Gruenheid S, Li Y, Vallance BA, Vazquez A, Barba J, Ibarra JA, O'Donnell P, Metalnikov P, Ashman K, Lee S, Goode D, Pawson T, Finlay BB. Dissecting virulence: systematic and functional analyses of a pathogenicity island. *Proc Natl Acad Sci USA* 2004; 101:3597–602.

14. Tenover FC, McGowan JE Jr. Reasons for the emergence of antibiotic resistance. *Am J Med Sci* 1996; 311:9–16.

15. Barbosa TM, Levy SB. The impact of antibiotic use on resistance development and persistence. *Drug Resist Updates* 2000; 3:303–311.

16. Stohr K, Wegener HC. Animal use of antimicrobials: impact on resistance. *Drug Resist Updates* 2000; 3:207–209.

17. Angulo FJ, Johnson KR, Tauxe RV, Cohen ML. Origins and consequences of antimicrobial-resistant nontyphoidal *Salmonella*: implications for the use of fluoroquinolones in food animals. *Microb Drug Resist* 2000; 6:77–83.

18. Glynn MK, Bopp C, Dewitt W, Dabney P, Mokhtar M, Angulo FJ. Emergence of multidrug-resistant *Salmonella enterica* serotype Typhimurium DT104 infections in the United States. *N Engl J Med* 1998; 338:1333–8.

19. Akkina JE, Hogue AT, Angulo FJ, Johnson R, Petersen KE, Saini PK, Fedorka-Cray PJ, Schlosser WD. Epidemiologic aspects, control, and importance of multiple-drug resistant *Salmonella typhimurium* DT104 in the United States. *J Am Vet Med Assoc* 1999; 214:790–8.

20. Briggs CE, Fratamico PM. Molecular characterization of an antibiotic resistance gene cluster of *Salmonella typhimurium* DT104. *Antimicrob Agents Chemother* 1999; 43:846–9.

21. Angulo FJ, Griffin PM. Changes in antimicrobial resistance in *Salmonella enterica* serovar Typhimurium [letter]. *Emerg Infect Dis* 2000; 6:436–8.

22. Hughes JM, Tenover FC. Approaches to limiting emergence of antimicrobial resistance in bacteria in human populations. *Clin Infect Dis* 1997; 24 (Suppl 1):S131–5.

23. Podzorski RP, Persing DH. Molecular detection and identification of microorganisms. In: Murray PR, Baron JE, Pfaller MA, Tenover FC, Yolken RH (eds). *Manual of Clinical Microbiology*, 6th ed. Washington DC: ASM Press, 1995:130–157.

24. Tenover FC, Arbeit RD, Goering RV, Mickelsen PA, Murray BE, Persing DH, Swaminathan B. Interpreting chromosomal DNA restriction patterns produced by pulsed-field gel electrophoresis: criteria for bacterial strain typing. *J Clin Microbiol* 1995; 33:2233–9.

25. Tenover FC, Arbeit RD, Goering RV. How to select and interpret molecular strain typing methods for epidemiological studies of bacterial infections: a review for healthcare epidemiologists. Molecular TypingWorking Group of the Society for Healthcare Epidemiology of America. *Infect Control Hosp Epidemiol* 1997; 18:6,426–39.

26. Courvalin P. Impact of molecular biology on antibiotic susceptibility: testing and therapy. *Am J Med* 1995; 99 (6A):21S-25S.

27. Tenover FC. Rapid detection and identification of bacterial pathogens using novel molecular technologies: infection control and beyond. *Clin Infect Dis* 2007; 44:418–23.

28. Giles WP, Benson AK, Olson ME, Hutkins RW, Whichard JM, Winokur PL, Fey PD. DNA sequence analysis of regions surrounding bla_{CMY-2} from multiple *Salmonella* plasmid backbones. *Antimicrob Agents Chemother* 2004; 48:2845–52.

29. Hiraoka M, Okamoto R, Inoue M, Mitsuhashi S. Effects of β-lactamases and *omp* mutation on susceptibility to β-lactam antibiotics in *Escherichia coli. Antimicrob Agents Chemother* 1989; 33:382–386.

30. Bauernfeind A, Georgopapadakou NH. Clinical significance of antibacterial transport. In Georgopapadakou NH (ed). *Drug transport in antimicrobial and anticancer chemotherapy.* New York: Marcel Dekker, 1995:1–19.

31. Courvalin P. Transfer of antibiotic resistance genes between Gram-positive and Gram-negative bacteria. *Antimicrob Agents Chemother* 1994; 38:1447–51.

32. Depardieu F, Podglajen I, Leclercq R, Collatz E, Courvalin P. Modes and modulations of antibiotic resistance gene expression. *Clin Microbiol Rev* 2007; 20:79–114.

33. Hall RM, Collis CM. Antibiotic resistance in Gram-negative bacteria: the role of gene cassettes and integrons. *Drug Resist Updates* 1998; 1:109–119.

34. Hall RM. Mobile gene cassettes and integrons: moving antibiotic resistance genes in Gram-negative bacteria. *Ciba Found Symp* 1997; 207:192–202; discussion 202–5.

35. Liebert CA, Hall RM, Summers AO. Transposon Tn*21*, flagship of the floating genome. *Microbiol Mol Biol Rev* 1999; 63:507–22.

36. Skold O. Sulfonamide resistance: mechanisms and trends. *Drug Resist Updates* 2000; 3:155–160.

37. Poirel L, Cattoir V, Soares A, Soussy CJ, Nordmann P. Novel Ambler class A β-lactamase LAP-1 and its association with the plasmid-mediated quinolone resistance determinant QnrS1. *Antimicrob Agents Chemother* 2007; 51:631–7.

38. Pai H, Seo MR, Choi TY. Association of QnrB determinants and production of extended-spectrum β-lactamases or plasmid-mediated AmpC β-lactamases in clinical isolates of *Klebsiella pneumoniae. Antimicrob Agents Chemother* 2007; 51:366–8

39. Neuhaus FC, Georgopapadakou NH. Strategies in β-lactam design. In Sutcliffe JA, Georgopapadakou NH (eds). *Emerging targets in antibacterial and antifungal chemotherapy.* New York: Chapman & Hall, 1992:204–273.

40. Georgopapadakou NH. Penicillin-binding proteins and bacterial resistance to β-lactams. *Antimicrob Agents Chemother* 1993; 37:2045–53.

41. Ghuysen JM. Serine β-lactamases and penicillin-binding proteins. *Annu Rev Microbiol* 1991;45:37–67.

42. Ambler RP. The structure of β-lactamases. *Philos Trans R Soc Lond B Biol Sci* 1980; 289:321–31.

43. Ledent P, Raquet X, Joris B, Van Beeumen J, Frere JM. A comparative study of class-D β-lactamases. *Biochem J* 1993; 292:555–62.

44. Bush K, Jacoby GA, Medeiros AA. A functional classification scheme for β-lactamases and its correlation with molecular structure. *Antimicrob Agents Chemother* 1995; 39:1211–33.

45. Rice LB, Bonomo RA. β-Lactamases: which ones are clinically important? *Drug Resist Updates* 2000; 3:178–89.

46. Jacoby GA, Munoz-Price LS. The new β-lactamases. *New Engl J Med* 2005; 352:380–91.

47. Walther-Rasmussen J, Hoiby N. OXA-type carbapenemases. *J Antimicrob Chemother* 2006 Mar;57(3):373–83.

48. Deshpande LM, Jones RN, Fritsche TR, Sader HS. Occurrence and characterization of carbapenemase-producing Enterobacteriaceae: report from the SENTRY Antimicrobial Surveillance Program (2000–2004). *Microb Drug Resist* 2006; 12:223–30.

49. Wong JD, Barash JR, Sandfort RF, Janda JM. Susceptibilities of *Yersinia pestis* strains to 12 antimicrobial agents. *Antimicrob Agents Chemother* 2000; 44:1995–6.

50. Galimand M, Guiyoule A, Gerbaud G, Rasoamanana B, Chanteau S, Carniel E, Courvalin P. Multidrug resistance in *Yersinia pestis* mediated by a transferable plasmid. *N Engl J Med* 1997; 337:10, 677–80.

51. Rice LB, Carias LL, Hujer AM, Bonafede M, et al. High-level expression of chromosomally encoded SHV-1 β-lactamase and an outer membrane protein change confer resistance to ceftazidime and piperacillin/tazobactam in a clinical isolate of *Klebsiella pneumoniae*. *Antimicrob Agents Chemother* 2000; 44:362–367.

52. Rasheed JK, Jay C, Metchock B, Berkowitz F, Weigel L, Crellin J, Steward C, Hill B, Medeiros AA, Tenover FC. Evolution of extended-spectrum β-lactam resistance (SHV-8) in a strain of *Escherichia coli* during multiple episodes of bacteremia. *Antimicrob Agents Chemother* 1997; 41:647–53.

53. Rasheed JK, Anderson GJ, Yigit H, Queenan AM, Domenech-Sanchez A, Swenson JM, Biddle JW, Ferraro MJ, Jacoby GA, Tenover FC. Characterization of the extended-spectrum β-lactamase reference strain, *Klebsiella pneumoniae* K6 (ATCC 700603), which produces the novel enzyme SHV-18. *Antimicrob Agents Chemother* 2000; 44:9, 2382–8.

54. Coque TM, Oliver A, Perez-Diaz JC, Baquero F, Canton R. Genes encoding TEM-4, SHV-2, and CTX-M-10 extended-spectrum β-lactamases are carried by multiple *Klebsiella pneumoniae* clones in a single hospital (Madrid, 1989 to 2000). *Antimicrob Agents Chemother* 2002; 46:500–10.

55. Bonnet R. Growing group of extended-spectrum β-lactamases: the CTX-M enzymes. *Antimicrob Agents Chemother* 2004; 48:1–14.

56. Canton R, Coque TM. The CTX-M β-lactamase pandemic. *Curr Opin Microbiol* 2006; 9:466–75.

57. Jacoby GA. Extended-spectrum β-lactamases and other enzymes providing resistance to oxyimino-β-lactams. *Infect Dis Clin North Am* 1997; 11: 4, 875–87.

58. Walsh TR, Toleman MA, Poirel L, Nordmann P. Metallo-β-lactamases: the quiet before the storm? *Clin Microbiol Rev* 2005; 18(2):306–25.

59. Franklin C, Liolios L, Peleg AY. Phenotypic detection of carbapenem-susceptible metallo-β-lactamase-producing Gram-negative bacilli in the clinical laboratory. *J Clin Microbiol* 2006; 44:3139–44.

60. Georgopapadakou NH. 2004. β-Lactamase inhibitors: evolving compounds for evolving resistance targets. *Expert Opin Investig Drugs* 2004; 13:1307–1318.

61. MGP Page. β-Lactamase inhibitors. *Drug Resist Updates* 2000; 3:109–125.

62. Stapleton PD, Shannon KP, French GL. Construction and characterization of mutants of the TEM β-lactamase containing amino acid substitutions associated with both extended-spectrum resistance and resistance to β-lactamase inhibitors. *Antimicrob Agents Chemother* 1999; 43:1881–1887.

63. Rahal JJ, Urban C, Horn D, et al. Class restriction of cephalosporin use to control total cephalosporin resistance in nosocomial *Klebsiella*. *J Am Med Assoc* 1998; 280:1233–37.

64. Lucet JC, Decre D, Fichelle A, et al. Control of a prolonged outbreak of extended-spectrum β-lactamase-producing Enterobacteriaceae in a university hospital. *Clin Infect Dis* 1999; 29:1411–1418.

65. Monnet DL, Biddle JW, Edwards JR, Culver DH, Tolson JS, Martone WJ, Tenover FC, Gaynes RP. Evidence of interhospital transmission of extended-spectrum β-lactam-resistant *Klebsiella pneumoniae* in the United States, 1986 to 1993. The National Nosocomial Infections Surveillance System. *Infect Control Hosp Epidemiol* 1997; 18: 7, 492–8

66. Tenover FC, Mohammed MJ, Gorton TS, Dembek ZE. Detection and reporting of organisms producing extended-spectrum β-lactamases: survey of laboratories in Connecticut. *J Clin Microbiol* 1999; 37:4065–70.

67. Laboratory capacity to detect antimicrobial resistance. *MMWR Morb Mortal Wkly Rep* 2000; 48:1167–71.

68. Martinez-Martinez L, Conejo MC, Pascual A, Hernandez-Alles S, Ballesta S, Ramirez De Arellano-Ramos E, Benedi VJ, Perea EJ. Activities of imipenem and cephalosporins against clonally related strains of *Escherichia coli* hyperproducing chromosomal β-lactamase and showing altered porin profiles. *Antimicrob Agents Chemother* 2000; 44:2534–6.

69. Joris B, De Meester F, Galleni M, Frere JM, Van Beeumen J. The K1 β-lactamase of *Klebsiella pneumoniae*. *Biochem J* 1987; 243:561–7.
70. Martinez-Martinez L, Pascual A, Hernandez-Alles S, Alvarez-Diaz D, Suarez AI, Tran J, Benedi VJ, Jacoby GA. Roles of β-lactamases and porins in activities of carbapenems and cephalosporins against *Klebsiella pneumoniae*. *Antimicrob Agents Chemother* 1999; 43:1669–73.
71. Weindorf H, Schmidt H, Martin HH. Contribution of overproduced chromosomal β-lactamase and defective outer membrane porins to resistance to extended-spectrum β-lactam antibiotics in *Serratia marcescens*. *J Antimicrob Chemother* 1998; 41:189–95.
72. Wiedemann B, Pfeiffle D, Wiegand I, Janas E. β-Lactamase induction and cell wall recycling in Gram-negative bacteria. *Drug Resist Updates* 1998; 1:223–226.
73. Jacobs C. Life in the balance: cell walls and antibiotic resistance. *Science* 1997; 278:1731–1732.
74. Nikaido H, Vaara M. Molecular basis of bacterial outer membrane permeability. *Microbiol Rev* 1985; 49:1–32.
75. Georgopapadakou NH. Antibiotic permeation through the bacterial outer membrane. *J Chemother* 1990; 2:275–9.
76. Nikaido H. Molecular basis of bacterial outer membrane permeability revisited. *Microbiol Mol Biol Rev* 2003; 67:593–656.
77. Alcantar-Curiel MD, Garcia-Latorre E, Santos JI. *Klebsiella pneumoniae* 35 and 36 kDa porins are common antigens in different serotypes and induce opsonizing antibodies. *Arch Med Res* 2000; 31:28–36.
78. Hernandez-Alles S, Alberti S, Alvarez D, Domenech-Sanchez A, Martinez-Martinez L, Gil J, Tomas JM, Benedi VJ. Porin expression in clinical isolates of *Klebsiella pneumoniae*. *Microbiology* 1999; 145: 673–9.
79. Hutsul JA, Worobec E. Molecular characterization of the *Serratia marcescens* OmpF porin, and analysis of *S. marcescens* OmpF and OmpC osmoregulation. *Microbiology* 1997; 143: 2797–806.
80. Chevalier J, Pages JM, Mallea M. *In vivo* modification of porin activity conferring antibiotic resistance to *Enterobacter aerogenes*. *Biochem Biophys Res Commun* 1999; 266:248–51.
81. Mallea M, Chevalier J, Bornet C, Eyraud A, Davin-Regli A, Bollet C, Pages JM. Porin alteration and active efflux: two *in vivo* drug resistance strategies used by *Enterobacter aerogenes*. *Microbiology* 1998; 144:3003–9.
82. Domenech-Sanchez A, Hernandez-Alles S, Martinez-Martinez L, Benedi VJ, Alberti S. Identification and characterization of a new porin gene of *Klebsiella pneumoniae*: its role in β-lactam antibiotic resistance. *J Bacteriol* 1999; 181:2726–32.
83. Bjorkman J, Andersson DI. The cost of antibiotic resistance from a bacterial perspective. *Drug Resist Updates* 2000; 3:237–245.
84. Chapman JS, Georgopapadakou NH. Routes of quinolone permeation in *Escherichia coli*. *Antimicrob Agents Chemother* 1988; 32:438–42.
85. McCaffrey C, Bertasso A, Pace J, Georgopapadakou NH. Quinolone accumulation in *Escherichia coli*, *Pseudomonas aeruginosa*, and *Staphylococcus aureus*. *Antimicrob Agents Chemother* 1992; 36:1601–5.
86. Hooper DC. Mode of action of fluoroquinolones. *Drugs* 1999; 58 (Suppl 2):6–10.
87. Hooper DC. Mechanisms of action of antimicrobials: focus on fluoroquinolones. *Clin Infect Dis* 2001.
88. Martinez-Martinez L, Pascual A, Jacoby GA. Quinolone resistance from a transferable plasmid. *Lancet* 1998; 351:797–9.
89. Wang M, Sahm DF, Jacoby GA, Hooper DC. Emerging plasmid-mediated quinolone resistance associated with the *qnr* gene in *Klebsiella pneumoniae* clinical isolates in the United States. *Antimicrob Agents Chemother* 2004; 48:1295–9.

90. Nordmann P, Poirel L. Emergence of plasmid-mediated resistance to quinolones in Enterobacteriaceae. *J Antimicrob Chemother* 2005; 56:463–9.
91. Poirel L, Nordmann P, De Champs C., Eloy C. Nosocomial spread of QnrA-mediated quinolone resistance in *Enterobacter sakazakii. Int J Antimicrob Agents* 2007; 29:223–4.
92. Robicsek A, Jacoby GA, Hooper DC. The worldwide emergence of plasmid-mediated quinolone resistance. *Lancet Infect Dis* 2006; 6(10):629–40.
93. Park CH, Robicsek A, Jacoby GA, Sahm D, Hooper DC. Prevalence in the United States of aac(6′)-Ib-cr encoding a ciprofloxacin-modifying enzyme. *Antimicrob Agents Chemother* 2006; 50(11):3953–5.
94. Welgel LM, Steward CD, Tenover FC. *gyrA* mutations associated with fluoroquinolone resistance in eight species of Enterobacteriaceae. *Antimicrob Agents Chemother* 1998; 42:10, 2661–7.
95. Hooper DC. Mechanisms of fluoroquinolone resistance. *Drug Resist Updates* 1999; 2:38–55.
96. Yoshida H, Bogaki M, Nakamura M et al. Quinolone-resistance-determining region in the DNA gyrase *gyrA* gene of *Escherichia coli. Antimicrob Agents Chemother* 1990; 34:1271–2.
97. Breines DM, Ouabdesselam S, Ng EY, Tankovic J, Shah S, Soussy CJ, Hooper DC. Quinolone resistance locus *nfxD* of *Escherichia coli* is a mutant allele of the *parE* gene encoding a subunit of topoisomerase IV. *Antimicrob Agents Chemother* 1997; 41:175–9.
98. Georgopapadakou NH. Quinolone uptake and efflux. In Georgopapadakou NH (ed). *Drug transport in antimicrobial and anticancer chemotherapy.* New York: Marcel Dekker, 1995:245–267.
99. Cohen SP, Hooper DC, Wolfson JS et al. An endogenous active efflux of norfloxacin in *Escherichia coli. Antimicrob Agents Chemother* 1988; 32:1187–91.
100. Chapman JS, Bertasso A, Georgopapadakou NH. Fleroxacin resistance in *Escherichia coli. Antimicrob Agents Chemother* 1989; 33:239–41.
101. Ishii H, Sato K, Hoshino K et al. Active efflux of ofloxacin by a highly quinolone-resistant strain of *Proteus vulgaris. J Antimicrob Chemother* 1991; 28:827–36.
102. O'Hara CM, Steward CD, Wright JL, Tenover FC, Miller JM. Isolation of *Enterobacter intermedium* from the gallbladder of a patient with cholecystitis. *J Clin Microbiol* 1998; 36:3055–6.
103. Dworkin RJ. Aminoglycosides for the treatment of Gram-negative infections: therapeutic use, resistance and future outlook. *Drug Resist Updates* 1999; 2:173–9.
104. Wright GD, Berghuis AM, Mobashery S. Aminoglycoside antibiotics. Structures, functions, and resistance. *Adv Exp Med Biol* 1998; 456:27–69.
105. Kotra LP, Haddad J, Mobashery S. Aminoglycosides: perspectives on mechanisms of action and resistance and strategies to counter resistance. *Antimicrob Agents Chemother* 2000; 44:3249–56.
106. Davies JE. Aminoglycosides: ancient and modern. *J Antibiot* 2006; 59:529–32.
107. Fourmy D, Recht MI, Blanchard SC, Puglisi JD. Structure of the A site of *Escherichia coli* 16S ribosomal RNA complexed with an aminoglycoside antibiotic. *Science* 1996; 274:1367–71.
108. Moazed D, Noller HF. Interaction of antibiotics with functional sites in 16S ribosomal RNA. *Nature* 1987; 327:389–94.
109. Bryan LE. Mechanisms of action of aminoglycoside antibiotics. In Root RK, Sande MA (eds). *New dimensions in antimicrobial therapy.* New York: Churchill-Livingstone, 1984:17–36.
110. Bryan LE, Kwan S. Roles of ribosomal binding, membrane potential, and electron transport in bacterial uptake of streptomycin and gentamicin. *Antimicrob Agents Chemother* 1983; 23:835–45.

111. Davies J, Wright GD. Bacterial resistance to aminoglycoside antibiotics. *Trends Microbiol* 1997; 5:234–40.
112. Gerding DN, Larson TA, Hughes RA et al. Aminoglycoside resistance and aminoglycoside usage: ten years of experience in one hospital. *Antimicrob Agents Chemother* 1991; 35:1284–90.
113. Schmitz FJ, Verhoef J, Fluit AC. Prevalence of aminoglycoside resistance in 20 European university hospitals participating in the European SENTRY Antimicrobial Surveillance Programme. *Eur J Clin Microbiol Infect Dis* 1999; 18:414–21.
114. Busch-Sorensen C, Sonmezoglu M, Frimodt-Moller N, Hojbjerg T, Miller GH, Espersen F. Aminoglycoside resistance mechanisms in Enterobacteriaceae and *Pseudomonas* spp. from two Danish hospitals: correlation with type of aminoglycoside used. *APMIS* 1996; 104:763–8.
115. Shaw KJ, Rather PN, Hare RS, Miller GH. Molecular genetics of aminoglycoside resistance genes and familial relationships of the aminoglycoside-modifying enzymes. *Microbiol Rev* 1993; 57:138–63.
116. Thompson PR, Schwartzenhauer J, Hughes DW, Berghuis AM, Wright GD. The COOH terminus of aminoglycoside phosphotransferase (3′)-IIIa is critical for antibiotic recognition and resistance. *J Biol Chem* 1999; 274:43, 30697–706.
117. Sandvang D, Aarestrup FM. Characterization of aminoglycoside resistance genes and class 1 integrons in porcine and bovine gentamicin-resistant *Escherichia coli*. *Microb Drug Resist* 2000; 6:19–27.
118. Lambert T, Gerbaud G, Courvalin P. Characterization of transposon *Tn1528*, which confers amikacin resistance by synthesis of aminoglycoside 3′-O-phosphotransferase type VI. *Antimicrob Agents Chemother* 1994; 38:702–6.
119. Wu HY, Miller GH, Blanco MG, Hare RS, Shaw KJ. Cloning and characterization of an aminoglycoside 6′-N-acetyltransferase gene from *Citrobacter freundii* which confers an altered resistance profile. *Antimicrob Agents Chemother* 1997; 41:2439–47.
120. Rather PN. Origins of the aminoglycoside-modifying enzymes. *Drug Resist Updates* 1998; 1:285–291.
121. Perlin MH, Lerner SA. High-level amikacin resistance in *E. coli* due to phosphorylation and impaired aminoglycoside uptake. *Antimicrob Agents Chemother* 1986; 29:216–24.
122. Acosta MB, Ferreira RC, Padilla G, Ferreira LC, Costa SO. Altered expression of oligopeptide-binding protein (OppA) and aminoglycoside resistance in laboratory and clinical *Escherichia coli* strains. *J Med Microbiol* 2000; 49:409–13.
123. Kashiwagi K, Tsuhako MH, Sakata K, Saisho T, Igarashi A, da Costa SO, Igarashi K. Relationship between spontaneous aminoglycoside resistance in *Escherichia coli* and a decrease in oligopeptide binding protein. *J Bacteriol* 1998; 180:5484–8.
124. Chopra I, Hawkey PM, Hinton M. Tetracyclines, molecular and clinical aspects. *J Antimicrob Chemother* 1992; 29:245–77.
125. Shlaes DM. An update on tetracyclines. *Curr Opin Investig Drugs* 2006; 7:167–71.
126. Chopra I. Tetracycline uptake and efflux in bacteria. In Georgopapadakou NH (ed). *Drug transport in antimicrobial and anticancer chemotherapy*. New York: Marcel Dekker, 1995:221–243.
127. Nikaido H, Thanassi DG. Penetration of lipophilic agents with multiple protonation sites into bacterial cells: tetracyclines and fluoroquinolones as examples. *Antimicrob Agents Chemother* 1993; 37:1393-9.
128. Speer BS, Shoemaker NB, Salyers AA: Bacterial resistance to tetracycline: mechanisms, transfer, and clinical implications. *Clin Microbiol Rev* 1992; 5:387–99.
129. Levy SB. Active efflux mechanisms for antimicrobial resistance. *Antimicrob Agents Chemother* 1992; 36:695–703.
130. Thanassi DG, Suh GS, Nikaido H. Role of outer membrane barrier in efflux-mediated tetracycline resistance of *Escherichia coli*. *J Bacteriol* 1995; 177:998–1007.

131. Levy SB, McMurry LM, Barbosa TM, Burdett V, Courvalin P, Hillen W, Roberts MC, Rood JI, Taylor DE. Nomenclature for new tetracycline resistance determinants. *Antimicrob Agents Chemother* 1999; 43:6, 1523–4.

132. McMurry LM, Petrucci RE, Levy SB. Active efflux of tetracycline encoded by four genetically different tetracycline resistant determinants in *Escherichia coli. Proc Natl Acad Sci USA* 1980; 77:3974–7.

133. Roberts MC. Tetracycline resistance determinants: mechanisms of action, regulation of expression, genetic mobility, and distribution. *FEMS Microbiol Rev* 1996; 19:1–24.

134. Roberts MC. Genetic mobility and distribution of tetracycline resistance determinants. *Ciba Found Symp* 1997; 207:206–18; discussion 219–22.

135. Makino S, Asakura H, Obayashi T, Shirahata T, Ikeda T, Takeshi K. Molecular epidemiological study on tetracycline resistance R plasmids in enterohaemorrhagic *Escherichia coli* O157:H7. *Epidemiol Infect* 1999; 123:25–30.

136. Magalhaes VD, Schuman W, Castilho BA. A new tetracycline resistance determinant cloned from *Proteus mirabilis. Biochim Biophys Acta* 1998; 1443:262–6.

137. Taylor DE, Chau A. Tetracycline resistance mediated by ribosomal protection. *Antimicrob Agents Chemother* 1996; 40:1–5.

138. Oliva B, Chopra I. *tet* Determinants provide poor protection against some tetracyclines: further evidence for division of tetracyclines into two classes. *Antimicrob Agents Chemother* 1992; 36:876–8.

139. Testa RT, Petersen PJ, Jacobus NV et al. *In vitro* and *in vivo* antibacterial activities of the glycylglycines, a new class of semisynthetic tetracyclines. *Antimicrob Agents Chemother* 1993; 37:2270–7.

140. Tally FT, Ellestad GA, Testa RT. Glycylcyclines: a new generation of tetracyclines. *J Antimicrob Chemother* 1995; 35:449–52.

141. Petersen PJ, Jacobus NV, Weiss WJ, Sum PE, Testa RT. *In vitro* and *in vivo* antibacterial activities of a novel glycylcycline, the 9-t-butylglycylamido derivative of minocycline (GAR-936). *Antimicrob Agents Chemother* 1999; 43:738–44.

142. Doan TL, Fung HB, Mehta D, Riska PF. Tigecycline: a glycylcycline antimicrobial agent. *Clin Ther* 2006; 28(8):1079–106.

143. Stein GE, Craig WA.Tigecycline: a critical analysis. *Clin Infect Dis* 2006; 43(4): 518–24.

144. Nelson ML, Levy SB. Reversal of tetracycline resistance mediated by different bacterial tetracycline resistance determinants by an inhibitor of the Tet(B) antiport protein. *Antimicrob Agents Chemother* 1999; 43:1719–24.

145. Huovinen P, Sundstrom L, Swedberg G, Skold O. Trimethoprim and sulfonamide resistance. *Antimicrob Agents Chemother* 1995; 39:279–89.

146. Swedberg G, Skold O. Characterization of different plasmid-borne dihydropteroate synthases mediating bacterial resistance to sulfonamides. *J Bacteriol* 1980; 142:1–7.

147. Radstrom P, Swedberg G. RSF1010 and a conjugative plasmid contain *sulII,* one of the two known genes for plasmid-borne sulfonamide resistant dihydropteroate synthase. *Antimicrob Agents Chemother* 1988; 32:1684–92.

148. Sundstrom L, Radstrom P, Swedberg G, Skold O. Site-specific recombination promotes linkage between trimethoprim- and sulfonamide resistance genes. Sequence characterization of *dhfrV* and *sulI* and a recombination active locus of *Tn21. Mol Gen Genet* 1988; 213:191–201.

149. Amyes SGB, Towner KJ. Trimethoprim resistance: epidemiology and molecular aspects. *J Med Microbiol* 1990; 31:1–9.

150. Grey D, Hamilton-Miller JMT, Brumfitt W. Incidence and mechanisms of resistance to trimethoprim in clinically isolated Gram-negative bacteria. *Chemotherapy* 1979; 25:147–56.

151. Huovinen P. Increases in rates of resistance to trimethoprim. *Clin Infect Dis* 1997; 24(Suppl 10): S63–6.
152. Shaw WV. Chloramphenicol acetyltransferase: enzymology and molecular biology. *Crit Rev Biochem* 1983; 14:1–46.
153. Parent R, Roy PH. The chloramphenicol acetyltransferase gene of Tn*2424*: a new breed of *cat*. *J Bacteriol* 1992; 174:2891–7.
154. Arcangioli MA, Leroy-Setrin S, Martel JL, Chaslus-Dancla E. Evolution of chloramphenicol resistance, with emergence of cross-resistance to florfenicol, in bovine *Salmonella typhimurium* strains implicates definitive phage type (DT) 104. *J Med Microbiol* 2000; 49:103–10.
155. Schwarz S, Kehrenberg C, Doublet B, Cloeckaert A. Molecular basis of bacterial resistance to chloramphenicol and florfenicol. *FEMS Microbiol Rev* 2004; 28(5): 519–42.
156. Centers for Disease Control and Prevention/National Nosocomial Infections Surveillance System. Semiannual report. December 1999 (corrected March 2000) Atlanta, Ga: CDC/NNIS. 2000 [database online: http://www.cdc.gov/ncidod/hip/NNIS/dec99sar. pdf].
157. Jacoby GA. Antimicrobial-resistant pathogens in the 1990s. *Annu Rev Med* 1996; 47:169–79.
158. Struelens MJ, Byl B, Vincent J-L. Antibiotic policy: a tool for controlling resistance of hospital pathogens. *Clin Microbiol Infect* 1999; 5:S19–S24.
159. Flynn DM, Weinstein RA, Nathan C et al. Patients' endogenous flora as the source of "nosocomial" *Enterobacter* in cardiac surgery. *J Infect Dis* 1987; 156:363–8.
160. Gupta K, Scholes D, Stamm WE. Increasing prevalence of antimicrobial resistance among uropathogens causing acute uncomplicated cystitis in women. *J Am Med Assoc* 1999; 281:736–8.
161. Wiener J, Quinn JP, Bradford PA et al. Multiple antibiotic-resistant *Klebsiella* and *Escherichia coli* in nursing homes. *J Am Med Assoc* 1999; 281:517–23.
162. Guyot A, Barrett SP, Threlfall EJ, Hampton MD, Cheasty T. Molecular epidemiology of multi-resistant *Escherichia coli*. *J Hosp Infect* 1999; 43:39–48.
163. Nikaido H. The role of outer membrane and efflux pumps in the resistance of Gram-negative bacteria. Can we improve access? *Drug Resist Updates* 1998; 1:93–98.
164. Poole K. Efflux-mediated antimicrobial resistance. *J Antimicrob Chemother* 2005; 56(1):20–51.
165. Lee A, Mao W, Warren MS, Mistry A, Hoshino K, Okumura R, Ishida H, Lomovskaya O. Interplay between efflux pumps may provide either additive or multiplicative effects on drug resistance. *J Bacteriol* 2000; 182:3142–50.
166. Hancock REW. Alterations in outer membrane permeability. *Annu Rev Microbiol* 1984; 38:237.
167. Viljanen P, Kayhty H, Vaara M, Vaara T. Susceptibility of Gram-negative bacteria to the synergistic bactericidal action of serum and polymyxin B nonapeptide. *Can J Microbiol* 1986; 32:66–9.

15 Resistance as a Worldwide Problem

Paul Shears

CONTENTS

Bacterial resistance to antimicrobial agents is a global problem, affecting both industrialized and resource-poor countries. Such resistance increases the difficulty of treating both community- and hospital-acquired infections, with a resulting effect on morbidity, mortality, and economic cost. There is a limited group of infections responsible for the main burden of antimicrobial resistance, including acute respiratory infections, diarrheal disease, sexually transmitted infections, tuberculosis (TB), and health care–related (nosocomial) infections.

There are limited data to show the direct effect of antimicrobial resistance on patient outcome, particularly in community-acquired infections, but hospital studies have shown definite increases in morbidity and length of stay resulting from multiply-resistant pathogens, such as methicillin-resistant *Staphylococcus aureus* (MRSA).

Containing antimicrobial resistance requires improved strategies of antimicrobial use, particularly better control of prescribing, public health strategies to reduce the spread of antimicrobial-resistant communicable disease, and improved laboratory monitoring and epidemiological surveillance of resistant bacteria.

INTRODUCTION

Infections and infectious diseases continue to be major causes of morbidity and mortality, particularly in resource-poor and tropical countries. The availability of effective antimicrobial agents was one of the health inputs, along with immunizations and improved public health, that would have contributed to the World Health Organization aim of "Health for All by 2000." The global occurrence and spread of antimicrobial resistance has been one of the major factors in that goal not being achieved.

On a worldwide scale, community acquired infections, notably acute respiratory disease, gastrointestinal infections, sexually transmitted diseases, and tuberculosis, together are responsible for the burden of bacterial infection–related morbidity and mortality. In industrialized countries, while community-acquired infections continue to be of importance, health care–associated (nosocomial) infections have become of major concern. Most nosocomial infections are caused by bacteria resistant to multiple antibiotics, with some, such as vancomycin-resistant enterococci, multiply-resistant Gram-negative bacteria, and MRSA, leaving few options available for treatment. It is now evident from the limited studies available that multiply-resistant nosocomial pathogens are also present in hospitals in low-income countries.

Whether community or hospital acquired, some resistant bacteria have evolved locally, whereas for others, there is evidence for the international spread of specific clones. This chapter describes the global occurrence of resistant bacteria, their impact on clinical outcomes and health care systems, the factors contributing to the emergence and spread of resistance, and strategies that need to be implemented to contain the spread of antimicrobial resistance.

GLOBAL PATTERNS OF RESISTANCE IN MAJOR BACTERIAL PATHOGENS

ACUTE RESPIRATORY INFECTIONS

In both industrialized and less-developed countries, acute respiratory infections continue to be a major cause of morbidity, and particularly among children in low-income countries, a cause of mortality. *Streptococcus pneumoniae* and *Haemophilus influenzae* are by far the two most important pathogens. While penicillin was formerly the drug of choice for *S. pneumoniae*, intermediate (MIC 0.12 to 1.0 mg/L) and high level (MIC > 1 mg/L) resistance to penicillin occurs globally. In Europe and North America, reported rates of high-level resistance range from 10% to 30% [1–3]. Penicillin resistance is of greater concern in low-income countries, where alternative antimicrobials may not be available or are too expensive for poorer communities who may have to pay for medication. Penicillin resistance has now been described in many African countries in addition to South Africa, where resistance was first reported. Studies have shown penicillin resistance (either intermediate or high-level) rates of 45% in Kenya [4], 24% in Côte d'Ivoire [5], and 14% in Zambia [6]. Pneumococcal infection is a particular problem in HIV-infected patients, a study among HIV-infected adults in Kenya showed 77% resistance to penicillin [7].

Penicillin resistance has also been described from most countries in Asia. In a surveillance study of isolates from 11 Asian countries, the ANSORP (Asian Network for Surveillance of Resistant Pathogens) Study [8], 23% of pneumococci showed intermediate resistance and 30% high-level resistance. The highest resistance rates were in Vietnam (72%) and Korea (75%). Studies from China and India showed 45% and 34%, respectively, of pneumococcal isolates to be resistant.

Resistance of pneumococci to macrolides, particularly erythromycin, and to cotrimoxazole, the two other antimicrobials widely recommended for treatment of acute respiratory infections, have also been widely reported [9]. For erythromycin, resistance rates of 78%, 37%, and 92% have been reported in China, India, and Vietnam, respectively [8]. Of considerable concern in developing countries is resistance to cotrimoxazole, which is widely recommended for the symptomatic treatment of acute respiratory infections. Resistance rates of 81% in India, 94% in Kenya, and 12% in Zambia have been reported [4,6,10].

H. influenzae is the second most common cause of acute respiratory infections in the developing world, and may also cause more severe sepsis, including meningitis. Although less frequently isolated in routine laboratory investigations, more detailed prevalence studies have confirmed its importance. Resistance to ampicillin, cotrimoxazole, and chloramphenicol, the principal antimicrobials considered for treatment, have been reported, although there are less extensive data than for pneumococci. In India, a study has shown 23% resistant to ampicillin and 67% resistant to cotrimoxazole [11]. In Kenya, in a pediatric population, the resistance rates of *H. influenzae* to amoxycillin, chloramphenicol, and cotrimoxazole were 66%, 66%, and 38%, respectively [12], and in Bangladesh comparable figures were 32%, 21%, and 49% [13]. In a global study (the SENTRY Surveillance Program), country combined resistance prevalence to ampicillin was 29% in North America, 16% in Latin America, and 16% in Europe. Resistance rates for cotrimoxazole were 24%, 40%, and 24% [14].

With the increasing use of immunization against *H. influenzae* b (Hib), the occurrence of invasive Haemophilus disease has fallen in most industrialized countries but remains widespread in low-income regions.

Diarrheal Diseases

Shigella Infections

Of the four species and many serotypes of Shigella, *Shigella dysenteriae* 1 is responsible for the most severe morbidity in many parts of the world, where inadequate water and sanitation lead to both local outbreaks and occasionally widespread epidemics. Multiple, transferable antimicrobial resistance was described in Shigella in the 1950s, and since then there has been a wide distribution of resistant strains and the development of resistance to an increasing number of antimicrobial agents. Combined resistance to ampicillin, tetracycline, chloramphenicol, and cotrimoxazole has been widely described [15], and resistance to nalidixic acid, the recent preferred treatment of choice, has become increasingly common. Resistance to nalidixic acid is associated with reduced sensitivity to ciprofloxacin and other fluoroquinolones, leaving few treatment options, particularly in low-income countries. Nalidixic acid strains were first reported in Bangladesh [16] and in 1994 in central Africa [17].

Multiple resistance was described in outbreaks in eastern India in 2002 and 2003, resistant to most available oral antimicrobials and with reduced sensitivity to ciprofloxacin [18]. Ciprofloxacin-resistant strains have also been described in Bangladesh and Nepal [19].

Cholera

Both endemic and epidemic cholera caused by *Vibrio cholerae* 01 and 0139 continue to occur in many resource-poor countries, particularly where public health infrastructure is limited, and where population displacement and refugee exodus occur. While the mainstay of cholera management is timely and effective rehydration, appropriate antimicrobial management can reduce the period of Vibrio excretion, and the risk of symptomatic disease in household contacts. Despite its widespread use, tetracycline resistance was not reported until 1977 in an outbreak in Tanzania [20]. Since that time, there has been a varying development of resistance in different locations. Outbreaks in different parts of Africa have consistently demonstrated resistance to tetracycline and chloramphenicol [21,22]. Studies in India in both serotypes 01 and 0139 have generally shown a low level of antimicrobial resistance [23,24]. In Southeast Asia, resistance to tetracycline has been demonstrated in outbreaks in Laos [25] and Thailand [26], with variable resistance in Vietnam [27].

Typhoid and Other Salmonella Infections

Multiple antimicrobial resistance (ampicillin, teracycline, and chloramphenicol) in typhoid was described in an increasing number of countries in the 1990s. Initially, such multiply-resistant strains remained susceptible to fluoroquinolones, which became the preferred treatment. There are now increasing reports of strains with reduced sensitivity to fluoroquinolones including ciprofloxacin (MICs 0.25 to 0.5 mg/L). Such strains have been demonstrated in east Africa [28], the Indian subcontinent [29], and Southeast Asia [30]. Strains with reduced fluoroquinolone sensitivity present difficulties in treatment, requiring either longer courses of fluoroquinolone treatment, or alternative, more expensive, and often less available antimicrobials.

Non-typhi Salmonella infections, in addition to causing food-borne gastroenteritis in industrialized countries, have recently been shown to be an important cause of bacteraemia in Africa, particularly in pediatric patients and patients with HIV [31]. Many strains are multiply-resistant, possibly acquiring resistant determinants from bacteria forming part of the bowel flora. The use of antimicrobials in animal husbandry may be an important contribution to resistance in food-associated Salmonellae. Mutiple resistance has been widely documented in *Salmonella typhimurium* and could be a source of resistance in other strains in humans [32].

SEXUALLY TRANSMITTED INFECTIONS

While HIV infection (and its resistance to anti-retroviral therapy) is beyond the scope of this chapter, its association with other sexually transmitted infections is a major public health problem. If bacterial sexually transmitted infections are inadequately treated because of antimicrobial resistance, HIV infections will be more easily transmitted.

The treatment of uncomplicated gonorrhea infections has progressed through various stages as resistance to different antimicrobials has occurred. Resistance to penicillin, resulting from penicillinase production (PPNG), has reached over 80% in many countries in South and Southeast Asia and the western Pacific as shown through surveillance by the WHO GASP (Gonococcal Antimicrobial Susceptibility Program) [33]. Many of these resistant strains have subsequently spread to Europe and the Americas.

Similar figures of penicillin resistance have been reported from different countries in Africa, 85% in a study in Ethiopia [34], 89% in Rwanda [35], and 98% in Nigeria [36].

Tetracycline became an alternative to penicillin, but rapid development of resistance led to the recommendation of ciprofloxacin for the syndromic management of gonococcal infections. Since its introduction, there has been a rapid rise and spread of resistance. The GASP program showed ciprofloxacin resistance rates varying from 10% to 100% in countries of South and Southeast Asia. In a study from Australia [37], 23% of 3640 isolates were ciprofloxacin resistant, and in a study from Sweden, the prevalence of ciprofloxacin resistance was over 50% [38]. Currently there are few reports of ciprofloxacin resistance in Africa, but this may be due to lack of laboratory facilities and under-reporting.

TUBERCULOSIS

The introduction of directly observed therapy (DOTS) has had a major impact in many parts of the world on controlling both the spread of tuberculosis and the development of resistant strains. However, in some areas where there is a large background of tuberculosis infection in the community, and particularly in populations where HIV infection is endemic, multidrug-resistant tuberculosis continues to be a problem. New molecular techniques have assisted in describing the epidemiology of multi-resistant strains [39–42]. There are many endemic areas where laboratory facilities, particularly at the periphery, may be unable to perform culture and sensitivity testing, and hence the prevalence of drug resistance is unknown.

MDR TB is a major burden on poor communities where limited availability and high cost of second-line therapy may limit treatment options.

ANTIBIOTIC RESISTANCE IN HEALTH CARE–ASSOCIATED (NOSOCOMIAL) INFECTIONS

It is in health care–associated infections that the most complex problems of anti-microbial resistance now occur in industrialized countries. The major infections are ventilator-associated pneumonia, intravascular line-related infections, surgical site infections, and urinary tract infections related to indwelling catheters. Important resistant bacteria include MRSA, vancomycin-resistant enterococci (VRE) and multiply-resistant Gram-negative bacteria, particularly Enterobacteriaceae, *Pseudomonas* spp., and *Acinetobacter* spp., with varying resistance to aminoglycosides, carbapenems, other β-lactam antibiotics, and cephalosporins. In European studies, the percentage of *S. aureus* bacteremias due to MRSA have been shown to be 35% to 45% [43,44]. In a study of 200 intensive care units in Europe [45], *S. aureus, Pseudomonas,* and *E. coli* were the major causes of sepsis, with a high proportion of resistant isolates.

Patients infected or colonized with multiply-resistant nosocomial bacteria may be responsible for spread into the community and subsequent widespread geographic spread. A study from Canada [46] described an outbreak of MRSA in a tertiary care hospital in British Columbia introduced by a patient who had been hospitalized in India.

Although there is less information on health care–associated infections in low-income countries, limited studies confirm that they are a global problem [47]. Several studies in Africa have reported MRSA infections including Côte d'Ivoire [48], Nigeria [49], and Sudan [50]. Vancomycin intermediate *S. aureus* has been reported from South Africa [51].

THE CLINICAL AND ECONOMIC IMPACT OF ANTIMICROBIAL RESISTANCE

While there are extensive data on the prevalence of antimicrobial-resistant bacteria, there are limited systematic data on how such resistance influences disease outcome. This is particularly the case for community-acquired infections in resource poor countries, where in terms of total populations affected, the greatest burden occurs. There are many unanswered questions. Do children with pneumonia caused by cotrimoxazole resistant *S. pneumoniae* die? Do they survive but with increased morbidity after so many days of treatment? Do they seek hospital admission and receive treatment with, for example, a third-generation cephalosporin? Is poor outcome not due to antimicrobial resistance, but rather to malnutrition, undiagnosed HIV, or other, unrelated infection? In the majority of cases there will be no pathogen isolated, and no sensitivity known, and so such questions are impossible to answer. However, despite the constraints in research in such situations, it is essential that attempts are made to begin to answer these questions. One strategy to reduce the impact of infections caused by potentially resistant pathogens in resource-poor countries is the Integrated Management of Childhood Ilnessess (IMCI) initiative, where syndromic management is based on anticipated sensitivity patterns [52]. An important negative effect on the lack of information on antimicrobial sensitivity is the over-use of broad-spectrum antimicrobials in the "hope" they will cover whatever resistant organisms may occur.

The most rigorous information on the clinical and economic impact of infections caused by resistant bacteria come from hospital studies in industrialized countries [53]. Several studies have shown increased lengths of stay and increased costs in patients with MRSA infections [54,55]. Multi-resistant organisms, including MRSA, have also been shown to be associated with higher mortality [56].

In addition to the increased morbidity in the index patient, patients infected with resistant organisms will be infected longer and are more likely to transmit infection to others.

LABORATORY AND EPIDEMIOLOGICAL ISSUES IN RELATION TO RESISTANCE DETERMINATION AND SURVEILLANCE

While industrialized countries have competent laboratory facilities for culture and sensitivity testing of bacterial pathogens, many use different methodologies for sensitivity testing, and comparisons between hospitals and between countries need to be interpreted cautiously. It would be preferable if all countries adopted one standard, preferably the Clinical and Laboratory Standards Institute (CLSI) standards, which are most widely used.

The problem is a different order of magnitude in developing countries, where apart from central or university hospitals, many hospitals do not have laboratories that can perform bacterial culture and sensitivity testing, and, where they do exist, limited quality control and availability of reagents may result in data of limited accuracy.

There have been a number of initiatives to set up standardized surveillance systems to provide updated resistance information and to compare rates between different locations [57]. Notably among these are the SENTRY Antimicrobial Surveillance Program [58], the Meropenem Yearly Susceptibility Test Information Collection (MYSTIC) project [59], and the Alexander International Multicentre Longitudinal Surveillance study of antimicrobial susceptibility among respiratory pathogens [60]. More comprehensive projects include the International Network for the Study and Prevention of Emerging Antimicrobial Resistance (INSPEAR) from the Centers for Disease Control and Prevention (CDC) [61] and the World Health Organization Network for Antimicrobial Resistance Surveillance (WHO NET), a surveillance project set up by WHO to improve country-level surveillance programs, with easy-to-use software [62].

While each of these programs provides valuable data on resistance prevalence, it is important that they are linked into programs to plan antimicrobial guidelines and prescribing.

FACTORS CONTRIBUTING TO THE EMERGENCE AND SPREAD OF ANTIMICROBIAL RESISTANCE

The emergence and spread of antimicrobial resistance is dependent on both biological and behavioral/political factors. The ability of bacteria to become resistant to antimicrobials by mutation or the acquisition of resistance genes by plasmid or other transfer are described in other chapters of this book. The selection of resistant strains is maintained by antibiotic pressure, and inappropriate antimicrobial prescribing and use are major factors that contribute to this.

Suboptimal use of antimicrobials is a particular problem in low-income countries, where there is often a lack of laboratory data to guide prescribing, and where antimicrobials are often "prescribed" by non-medical practitioners or are freely available in markets [63]. In such situations, subtherapeutic doses are often taken, increasing the risk of selecting resistant strains. Resistant strains once evolved may spread between patients in the community, in health care settings, and globally by the increase in international travel and migration.

Transfer of resistant strains in the community is particularly a problem in areas characterized inadequate water and sanitation facilities. Enteric carriage of resistant organisms has been shown to contaminate water storage vessels, with subsequent transfer to other family members, including children, who may have had little exposure to antibiotics [64].

In hospitals, the use of multiple antimicrobial agents, patients susceptible to complex infections because of the advanced medical procedures undertaken, and the spread between patients of inherently resistant nosocomial bacteria, all contribute to a major problem of difficult-to-treat infections.

STRATEGIES FOR THE CONTAINMENT OF THE GLOBAL OCCURRENCE AND SPREAD OF RESISTANCE

IMPROVED STRATEGIES FOR ANTIMICROBIAL USE

In many developing countries, non-medical practitioners, while they may be responsible for inappropriate antimicrobial use, do play an important role in the health care system, where qualified staff are limited particularly in remote areas. Thus education of these practitioners in appropriate prescribing is an important strategy, encouraging standardized treatment regimes for the syndromic management of infections [65]. While it is difficult to demonstrate a containment of resistance by such strategies, evidence from the DOTS TB control strategy is that there is a lower development of resistant TB. At the national level, control of antimicrobials available in the country, with registration and control of manufacture, should limit the range of antimicrobials available to the prescriber.

In industrialized countries where generally antimicrobials are only available through medical prescription, antimicrobial use is more controlled. However, in hospital practice, where the greatest intensity of antimicrobial use occurs, antibiotic guidelines are often not followed, despite the input of pharmacists and microbiologists. Where strict antibiotic policies have been followed, there is evidence that the occurrence of resistant nosocomial infections is reduced [66]. Failure to adhere to antimicrobial policies may lead to multiply-resistant bacteria, and overuse of inappropriate antibiotics may alter the normal flora of the bowel. This may in turn lead to overgrowth by bacteria, such as *Clostridium difficile*, a cause of antibiotic-related colitis.

PUBLIC HEALTH STRATEGIES

Public health interventions can contribute to reducing the occurrence and spread of resistant infections in both industrialized and low-income countries.

Immunization programs for *S. pneumoniae* and *H. influenzae* would greatly reduce the burden of these causes of acute respiratory infections, and hence the problem of treating resistant strains. There is a need for effective and low-cost vaccines against diarrheal diseases, particularly Shigella, cholera, and typhoid, where few oral antimicrobials are currently effective.

Improved water supply and sanitation in resource-poor countries will reduce the spread of both resistant enteric pathogens and non-pathogenic but resistant enteric bacteria that may act as a reservoir of transferable antimicrobial resistance.

Health education can play a major role in people's understanding of infection and the importance of using antibiotics appropriately [67]. Such programs can include both formal teaching situations and the wider use of publicly available media.

Control of Health Care–Associated Infections

In industrialized countries, the greatest problem of antimicrobial resistance is in nosocomial pathogens, particularly MRSA, VRE, and multiply-resistant Gram-negative bacteria. Controlling the development and spread of these pathogens is a priority if the specter of untreatable infections is to be avoided. New, comprehensive strategies, such as the Saving Lives campaign in the United Kingdom [68] and the Making Hospitals Safer initiative supported by the World Health Organization [69], are aimed at reducing health care–associated infections. The programs include an emphasis on hygiene and hand washing to reduce the path of transmission between patients and health care workers. The WHO program is not restricted to industrialized countries, and should be a catalyst to initiate nosocomial infection control in low-income countries.

Improved Laboratory Monitoring and Epidemiological Surveillance of Resistant Bacteria

Timely and accurate information on the prevalence of antimicrobial resistance is essential for planning resistance containment programs and rational antimicrobial policies. In resource-poor countries, the development of functioning laboratories at central and district level is a priority. There have been numerous attempts to do this in particular countries, but an internationally supported program is required to build sustainable and quality-controlled laboratory networks. Resistance/susceptibility data should form the basis of informing clinicians of results relevant to individual patients, provide local resistance data to form a "community antibiogram" [70], and link in to national and international surveillance systems.

Development of New Antimicrobial Agents

While the priority strategy at all levels must be to contain antimicrobial resistance using the strategies described above, the development of new antimicrobial agents is also a priority in the treatment of multiply-resistant infections [71]. New understanding of bacterial genomics and molecular biology techniques provide an understanding of new, selective, targets for antimicrobial action [72]. Such techniques may also provide agents that can combat bacterial resistance strategies.

CONCLUSION

The control of the emergence and spread of antimicrobial resistance is a public health priority on a global scale. Failure to do so will have a major and continuing impact on the morbidity and mortality caused by infection, and be an economic burden in both industrialized and low-income countries. A major collaborative effort will be required by national governments and their health ministries, by pharmaceutical companies, and international agencies if containment of antimicrobial resistance is to be achieved [73].

REFERENCES

1. Goldsmith CE, Moore JE, Murphy PG. Pneumococcal resistance in the UK. *J Antimicrob Chemother* 1997; 40: Suppl A:11–18.
2. Picazzo JJ, Betriu C, Rodriguez-Avial I, et al. Surveillance of antimicrobial resistance: VIRA study 2004. *Enferm Infec Microbiol Clin* 2004; 22: 517–523.
3. Thornsberry C, Jones ME, Hickey ML, et al. Resistance surveillance of *Streptococcus pneumoniae, Haemophilus influenzae* and *Moraxella catarrhalis* isolated in the United States, 1997–1998. *J Antimicrob Chemother* 1999; 44: 749–759.
4. Kariuki S, Muyodi J, Mirza B, et al. Antimicrobial susceptibility in community acquired bacterial pneumonia in adults. *East Afr Med J* 2003; 80: 213–217.
5. Kacou-N'douba A, Guessennd-Kouadio N, Kouassi-M'bengue A, Dosso M. Evolution of *Streptococcus pneumoniae* antibiotic resistance in Abidjan: update on nasopharyngeal carriage, from 1997 to 2001. *Med Mal Infect* 2004; 34: 83–85.
6. Woolfson A, Huebner R, Wasas A, et al. Nasopharyngeal carriage of community acquired antibiotic resistant *Streptococcus pneumoniae* in a Zambian paediatric population. *Bull World Health Organ* 1997; 75: 453–462.
7. Medina MJ, Greene CM, Gertz RE, et al. Novel antibiotic-resistant pneumococcal strains recovered from the upper respiratory tracts of HIV infected adults and their children in Kisumu, Kenya. *Microb Drug Resist* 2005; 11: 9–17.
8. Song JH, Jung SI, Ko KS, et al. High prevalence of antimicrobial resistance among clinical *Streptococcus pneumoniae* isolates in Asia (an ANSORP study). *Antimicrob Agents Chemother* 2004; 48: 2101–2107.
9. Klugman KP, Lonks JR. Hidden epidemic of macrolide resistant pneumococci. *Emerg Infect Dis* 2005; 11: 802–807.
10. Coles CL, Rahmathullah L, Kanungo R, et al. Nasopharyngeal carriage of resistant pneumococci in young South Indian infants. *Epidemiol Infect* 2002; 129: 491–497.
11. Jain A, Kumar P, Awasthi S. High nasopharyngeal carriage of drug resistant *Streptococcus pneumoniae* and *Haemophilus influenzae* in North Indian school children. *Trop Med Int Health* 2005; 10: 234–239.
12. Scott JA, Mwarumba S, Ngetsa C, et al. Progressive increase in antimicrobial resistance among invasive isolates of *Haemophilus influenzae* obtained from children admitted to a hospital in Kilifi, Kenya, from 1994 to 2002. *Antimicrob Agents Chemother* 2005; 49: 3021–3024.
13. Saha SK, Baqui AH, Darmstadt GL, et al. Invasive Haemophilus type b disease in Bangladesh with increased resistance to antibiotics. *J Pediatr* 2005; 146: 227–233.
14. Johnson DM, Sader HS, Fritsche TR, et al. Susceptibility trends of *Haemophilus influenzae* and *Moraxella catarrhalis* against orally administered antimicrobial agents: five year report from the SENTRY Antimicrobial Surveillance Programme. *Diagn Microbiol Infect Dis* 2003; 47: 373–376.
15. Shears P. Shigella infections. *Ann Trop Med Parasitol* 1996; 90: 105–114.

16. Munshi MH, Sack DA, Haider K, et al. Nalidixic acid resistance to *Shigella dysenteriae* type 1. *Lancet* 1987; ii: 419–421.

17. Ries AA, Wells JG, Olivola D, et al. Epidemic *Shigella dysenteriae* type 1 in Burundi: panresistance and implications for prevention. *J Infect Dis* 1994; 169: 1035–1041.

18. Pazhani GP, Sarkar B, Ramamurthy T, et al. Clonal multi-drug resistant *Shigella dysenteriae* type 1 strains associated with epidemic and sporadic dysenteries in eastern India. *Antimicrob Agents Chemother* 2004; 48: 681–684.

19. Talukder KA, Khajanchi BK, Islam MA, et al. Genetic relatedness of ciprofloxacin resistant *Shigella dysenteriae* type 1 strains isolated in south Asia. *J Antimicrob Chemother* 2004; 54; 730–734.

20. Mhalu FS, Muari PW, Ijumba J. Rapid emergence of El Tor *Vibrio cholerae* resistant to antimicrobials during the first six months of the fourth cholera epidemic in Tanzania. *Lancet* 1979; i: 345–347.

21. Dalsgaard A, Forslund A, Sandvang D, et al. *Vibrio cholerae* 01 outbreak isolates in Mozambique and South Africa in 1998 are multiple drug resistant, contain the SXT element and the aadA2 gene located on class 1 integrons. *J Antimicrob Chemother* 2001; 48: 827–838.

22. Urassa WK, Mhando YB, Mhalu FS, Mjonga SJ. Antimicrobial susceptibility pattern of *Vibrio cholerae* 01 strains during two cholera outbreaks in Dar es Salaam, Tanzania. *East Afr Med J* 2000; 77: 350–353.

23. Sundaram SP, Revathi J, Sarkar BL, Battacharya SK. Bacteriological profile of cholera in Tamil Nadu (1980–2001). *Indian J Med Res* 2002; 116: 258–263.

24. Sur D, Dutta P, Nair GB, Battacharya SK. Severe cholera outbreak following floods in a northern district of West Bengal. *Indian J Med Res* 2000; 112: 178–182.

25. Iwanaga M, Toma C, Miyazato T, et al. Antibiotic resistance conferred by a class 1 integron in *Vibrio cholerae* 01 strains isolated in Laos. *Antimicrob Agents Chemother* 2004; 48: 2364–2369.

26. Tabtieng R, Wattanasri S, Echeverria P, et al. An epidemic of *Vibrio cholerae* el tor Inaba resistant to several antibiotics with a conjugative group C plasmid coding for type II dihydrofolate reductase in Thailand. *Am J Trop Med Hyg* 1982; 41: 680–686.

27. Ehara M, Nguyen BM, Nguyen DT, et al. Drug susceptibility and its genetic basis in epidemic *Vibrio cholerae* in Vietnam. *Epidemiol Infect* 2004; 132: 595–600.

28. Kariuki S, Revathi G, Muyodi J, et al. Characterisation of multi drug resistant typhoid outbreaks in Kenya. *J Clin Microbiol* 2004; 42 1477–1482.

29. Rahman MM, Haq JA, Morshed MA, Rahman MA. *Salmonella enterica* serovar Typhi with decreased susceptibility to ciprofloxacin—an emerging problem in Bangladesh. *Int J Antimicrob Agents* 2005; 25: 345–346.

30. Parry CM. The treatment of multi drug resistant and nalidixic acid resistant typhoid in Vietnam. *Trans R Soc Trop Med Hyg* 2004; 98: 413–422.

31. Gordon MA, Walsh AL, Chaponda M, et al. Bacteraemia and mortality among adult medical admissions in Malawi—predominance of non-typhi salmonellae and *Streptococcus pneumoniae*. *J Infect* 2001; 42: 44–49.

32. Weinberger M, Keller N. Recent trends in the epidemiology of non typhoid Salmonella and antimicrobial resistance: a worldwide review. *Curr Opin Infect Dis* 2005; 18: 513–521.

33. Ray K, Bala M, Kumari S, Narain JP. Antimicrobial resistance of *Neisseria gonorrhoeae* in selected World Health Organisation Southeast Asia Region countries: an overview. *Sex Transm Dis* 2005; 32: 178–184.

34. Tadesse A, Mekonnen A, Kassu A, Amelash T. Antimicrobial sensitivity of *Neisseria gonorrhoea* in Gondar, Ethiopia. *East Afr Med J* 2001; 78: 259–261.

35. Van Dyck E, Karita E, Abdellati S, et al. Antimicrobial susceptibilities of *Neisseria gonorrhoeae* in Kigali, Rwanda, and trends of resistance between 1986 and 2000. *Sex Transm Dis* 2001; 28: 539–545.

36. Bakare RA, Oni AA, Arowojolu AO, et al. Penicillinase producing *Neisseria gonorrhoeae*: The review of the present situation in Ibadan, Nigeria. *Niger Postgrad Med J* 2002; 9: 59–62.
37. Australian Gonococcal Surveillance Programme. Annual report of the Australian Gonococcal Surveillance Programme, 2004. *Commun Dis Intell* 2005; 29: 137–142.
38. Berglund T, Colucci B, Lund B, et al. Increasing incidence of ciprofloxacin resistant gonorrhoeae in Sweden. *Lakartidningen* 2004; 101: 2332–2335.
39. Herrera L, Valverde A, Saiz P, et al. Molecular characterisation of isoniazid resistant *Mycobacterium tuberculosis* clinical strains isolated in the Philippines. *Int J Antimicrob Agents* 2004; 23: 572–576.
40. Harris KA, Mukundan U, Musser JM, et al. Genetic diversity and evidence for acquired antimicrobial resistance in *Mycobacterium tuberculosis* at a large hospital in South India. *Int J Infect Dis* 2000; 4: 140–147.
41. Tudo G, Gonzalez J, Obama R, et al. Study of resistance to anti-tuberculous drugs in five districts of Equatorial Guinea: rates, risk factors, genotyping of gene mutations and molecular epidemiology. *Int J Tuberc Lung Dis* 2004; 8: 15–22.
42. Sharaf-Eldin,GS, Saeed NS, Hamid ME, et al. Molecular analysis of clinical isolates of *Mycobacterium tuberculosis* collected from patients with persistent disease in the Khartoum region of Sudan. *J Infect* 2002; 44: 244–251.
43. Bertrand X, Costa Y, Pina P. Surveillance of antimicrobial resistance of bacteria isolated from blood stream infections: data of the French National Observatory for Epidemiology of Bacterial Resistance to Antibiotics, 1998–2003. *Med Mal Infect* 2005; 35:329–334.
44. Pan A, Carnevale G, Catenazzi P, et al. Trends in methicillin resistant *Staphylococcus aureus* (MRSA) blood stream infections: effect of the MRSA "search and isolate" strategy in a hospital in Italy with hyperendemic MRSA. *Infect Control Hosp Epidemiol* 2005; 26: 127–135.
45. Vincent JL, Sakr Y, Sprung CL, et al. Sepsis in European intensive care units: results of the SOAP study. *Crit Care Med* 2006; 34: 344–353.
46. Roman S, Smith J, Walker M, et al. Rapid geographic spread of a methicillin resistant *Staphylococcus aureus* strain. *Clin Infect Dis* 1997; 25: 698–705.
47. Blomberg B, Mwakagile DS, Urassa WK, et al. Surveillance of antimicrobial resistance at a tertiary hospital in Tanzania. *BMC Public Health* 2004; 4: 45–49.
48. Akoua Koffi C, Dje K, Toure R, et al. Nasal carriage of methicillin resistant *Staphylococcus aureus* among health care personnel in Abidjan (Cote d'Ivoire). *Dakar Med* 2004; 49: 70–74.
49. Taiwo SS, Bamidele M, Omonigbehin EA, et al. Molecular epidemiology of methicillin resistant *Staphylococcus aureus* in Ilorin, Nigeria. *West Afr Med* 2005; 24: 100–106.
50. Musa HA, Shears P, Khagali A. First report of methicillin resistant *Staphylococcus aureus* in Sudan. *J Hosp Infect* 1999; 42: 74.
51. Amod F, Moodley I, Peer AK, et al. Ventriculitis due to a hetero strain of vancomycin intermediate *Staphylococcus aureus* (hVISA); successful treatment with linezolid in combination with intraventricular vancomycin. *J Infect* 2005; 50: 252–257.
52. Gouws E, Bryce J, Habicht JP, et al. Improving antimicrobial use among health workers in first-level facilities: results from the multi country evaluation of the Integrated Management of Childhood Illness Strategy. *Bull World Health Organ* 2004; 82: 509–515.
53. Cosgrove SE, Carmeli Y. The impact of antimicrobial resistance on health and economic outcomes. *Clin Infect Dis* 2003; 36: 1433–1437.
54. Lodise TP, McKinnon PS. Clinical and economic impact of methicillin resistance in patients with *Staphylococcal aureus* bacteraemia. *Diagn Microbiol Infect Dis* 2005; 52: 113–122.

55. Cosgrove SE, Qi Y, Kaye KS, et al. The impact of methicillin resistance in *Staphylococcus aureus* on patient outcomes: mortality, length of stay, and hospital charges. *Infect Control Hosp Epidemiol* 2005; 26 166–173.

56. Melzer M, Eykyn SJ, Gransden WR, Chinn S. Is methicillin resistant *Staphylococcus aureus* more virulent than methicillin susceptible *S. aureus*? A comparative cohort study of British patients with nosocomial infection and bacteraemia. *Clin Infect Dis* 2003; 37: 1453–1460.

57. Masterton RG. Surveillance studies: can they help the management of infection? *J Antimicrob Chemother* 2000; 46: T2, 53–58.

58. Deshpande LM, Fritsche TR, Jones RM. Molecular epidemiology of selected multi drug resistant bacteria: a global report from the SENTRY Antimicrobial Surveillance Programme. *Diagn Microbiol Infect Dis* 2004; 49: 231–236.

59. Mutnick AH, Rhomberg PR, Sader HS, Jones RN. Antimicrobial useage and resistance trend relationships from the MYSTIC Programme in North America (1999–2001). *J Antimicrob Chemother* 2004; 53: 290–296.

60. Felmingham D, White AR, Jacobs MR, et al. The Alexander Project: the benefits from a decade of surveillance. *J Antimicrob Chemother* 2005; 56: Suppl 2: ii3–ii21.

61. Richet HM, Mohammed J, Mcdonald LC, et al. Building Communication Networks: International Network for the Study and Prevention of Emerging Antimicrobial Resistance (INSPEAR). *Emerg Infect Dis* 2001; 7: 319–322.

62. O'Brien TF, Stelling JM. WHONET: removing obstacles to the full use of information about antibiotic resistance. *Diag Microbiol Infect Dis* 1996; 25: 163–168.

63. Mamun KZ, Tabassum S, Shears P, Hart CA. A survey of antimicrobial prescribing and dispensing practices in rural Bangladesh. *Mymensingh Med J* 2006; 15: 81–84.

64. Shears P, Hussein MA, Chowdhury AH, Mamun KZ. Water sources and environmental transmission of multiply resistant enteric bacteria in rural Bangladesh. *Ann Trop Med Parasitol* 1995; 89: 297–303.

65. Bexell A, Lwando E, von Hofsten B, et al. Improving drug use through continuing education: a randomised controlled trial in Zambia. *J Clin Epidemiol* 1996; 49: 355–357.

66. Martin C, Ofotokun I, Rapp R, et al. Results of an antimicrobial control programme at a university hospital. *Am J Health Syst Pharm* 2005; 62: 732–738.

67. Suttajit S, Wagner AK, Tantipidoke R, et al. Patterns, appropriateness, and predictors of antimicrobial prescribing for adults with upper respiratory tract infections in urban slum communities of Bangkok. *Southeast Asian J Trop Med Public Health* 2005; 36: 489–497.

68. Department of Health, UK. Saving Lives: a delivery programme to reduce Health Care Associated Infections including MRSA. Department of Health, UK, 2005.

69. Lazzari S, Allegranzi B, Concia E. Making hospitals safer: the need for a global strategy for infection control in health care settings. *World Hosp Health Serv* 2004; 40: 36–42.

70. Halstead DC, Gomez N, McCarter YS. Reality of developing a community wide antibiogram. *J Clin Microbiol* 2004; 42: 1–6.

71. Paine K, Flower DR. Bacterial bioinformatics: pathogenesis and the genome. *J Mol Microbiol Biotechnol* 2002; 4: 357–365.

72. Rogers BL. Bacterial targets to antimicrobial leads and development candidates. *Curr Opin Drug Discov Devel* 2004; 7: 211–222.

73. Simonsen GS, Tapsall JW, Allegranzi B, et al. The antimicrobial resistance containment and surveillance approach—a public health tool. *Bull World Health Organ* 2004; 82: 928–934.

16 Public Health Responses to Antimicrobial Resistance in Outpatient and Inpatient Settings

Cindy R. Friedman and Arjun Srinivasan

CONTENTS

In 1998, the Centers for Disease Control and Prevention issued a report, Preventing Emerging Infectious Diseases, which outlined a plan designed to address key emerging infectious disease issues [1]. Antimicrobial resistance was seen as one of the major infectious disease issues facing the world as we entered the new millennium. In 2003, an Institute of Medicine report entitled Microbial Threats to Health reiterated this issue [2]. In this chapter, we review some of the approaches that have been taken to combat antimicrobial resistance in outpatient and inpatient settings.

In the outpatient setting, we focus on issues related to pneumococcal resistance and the role that inappropriate use of antimicrobials has played in promoting the development of resistant bacteria in the community. We address the importance of surveillance and epidemiologic investigation in measuring the magnitude of the resistance and for assessing the impact of interventions designed to reduce resistance and to promote judicious use of antibiotics. Two interventions we address are vaccination and educational campaigns.

In the inpatient setting, we address the unique nature of the hospital environment, which makes this aspect of health care delivery a focus for the emergence and spread of many antimicrobial-resistant pathogens. We review surveillance data that have shown increasing rates of resistance for most pathogens associated with nosocomial infections. We also describe many opportunities to prevent the emergence and spread of these resistant pathogens through a systematic review of surveillance data on antimicrobial resistance and antimicrobial prescribing and improved use of established infection-control measures.

EMERGENCE OF DRUG RESISTANCE IN THE COMMUNITY

The appearance of bacterial strains that are increasingly resistant to antimicrobial agents is a significant problem facing all sectors of the health care community. The relationship between the development of resistance and the use of antibiotics has been shown in both community and hospital settings [3,4]. The emergence of drug-resistant *Streptococcus pneumoniae* (DRSP) in the United States in the early 1990s serves as a good case study of antimicrobial resistance in the community.

In the United States, *S. pneumoniae* is an important cause of community-acquired infections, such as pneumonia and otitis media, and serious invasive infections, such as meningitis and bacteremia. The precise incidence of non-invasive infections is difficult to ascertain because of the lack of routine diagnostic testing. In recent years, the incidence of invasive disease has been on the decline because of the use of the pneumococcal conjugate vaccine [5,6]. Numerous studies have documented the association between recent antibiotic use and both carriage of and invasive disease caused by DRSP [7]. In one study, children who were recently given an antibiotic were two to seven times more likely colonized with DRSP than were children who did not have recent antibiotic use [7]. In addition, children with invasive disease caused by DRSP were significantly more likely to report recent antibiotic use than were children with invasive disease due to drug susceptible pneumococci.

S. pneumoniae is not the only community-acquired organism where antibiotic use and emergence of drug resistance are linked. In Finland from 1988 to 1990, there was an 8% increase in erythromycin resistance among group A *Streptococcus* (GAS) isolates, although the rates declined shortly after [8]. In the United States, high-level macrolide resistance among GAS pharyngeal isolates has been documented. The prevalence is estimated to be 5% [9].

The widespread use of antimicrobial agents, whether used appropriately or inappropriately, has led to the emergence of resistant organisms. If antibiotics were used exclusively for conditions for which they are known to be clinically effective, this increased risk would be viewed as one of the necessary but unavoidable consequences of therapy. However, this is not the case. In an analysis of data from the National Ambulatory Medical Care Survey, McCaig and Hughes documented that over 30% of all antibiotics are prescribed for colds, upper respiratory infections (URI), and bronchitis [10]. These conditions are largely viral in etiology and would not be expected to improve with antibiotic therapy [11]. An analysis of the same data by Gonzales et al. demonstrated that antibiotics were prescribed to 51% of adults with colds, 52% with URIs, and 66% with bronchitis [11]. In children, antibiotics were prescribed to 44% with colds, 46% with URIs, and 75% with bronchitis [12]. These are all conditions for which prescribing could be reduced or eliminated without adversely affecting patient care.

Physicians prescribe unnecessary antibiotics for many reasons; however, studies indicate that physicians believe they overprescribe because of patient/parent demand. Focus groups conducted with pediatricians and family practitioners in Atlanta identified parental expectations for antibiotics as the primary reason for this inappropriate use [13]. In a national survey of pediatricians, 48% reported that parents pressure them to prescribe antibiotics [14]. Seventy-eight percent of surveyed pediatricians felt that the single most important thing that could be done to promote judicious antibiotic use would be to educate parents about the proper indications for antibiotic use. This is supported by a study addressing the relationship between parental expectations and pediatrician antimicrobial prescribing. Physician perception of parental expectation was the only significant predictor of prescribing for conditions of viral origin [15]. When pediatricians believed that a parent wanted an antibiotic, they prescribed them in 62% of cases as compared with 7% of cases when they believed the parent did not want an antibiotic. Interestingly, there was no association between parents' true expectations and physician prescribing. Hamm found very similar findings in a study of antibiotic prescribing for respiratory tract infections in adults [16]. Although prescribing was related to physician-perceived patient expectations, patient satisfaction was not related to receipt of an antibiotic. One could conclude from these studies that if physicians and parents were able to communicate more openly, inappropriate prescribing might be reduced.

Although more difficult to document, it is likely that physicians' lack of understanding of the wide variety of presentations and natural history of viral illnesses plays a role in overprescribing. For example, although green nasal discharge is normal in a child with a cold [17], many physicians use this finding as an indication for antibiotic prescribing [13]. This lack of understanding, combined with the diagnostic uncertainty inherent in most clinical encounters and the time pressures of outpatient practice, contributes to the problem of inappropriate antibiotic prescribing.

PUBLIC HEALTH RESPONSE TO EMERGENCE OF DRUG RESISTANCE IN THE COMMUNITY

In 1994, in response to the increasing prevalence of DRSP, the Centers for Disease Control and Prevention (CDC) convened a working group of public health practitioners, clinical laboratorians, health care providers, and representatives of key professional societies. To minimize the impact of DRSP, the working group developed a strategy, which included three areas to be addressed in the public health response to antimicrobial resistance: (*i*) surveillance, (*ii*) epidemiologic investigation, and (*iii*) prevention and control strategies.

The primary goals of establishing surveillance included monitoring the prevalence and the geographic distribution of DRSP and rapid recognition of new patterns of resistance. The primary goal of epidemiologic investigation was to use the surveillance infrastructure and data to perform special studies to help ascertain information such as the risk factors for DRSP. The goal of prevention and control of DRSP was to minimize complications and reduce the number of antimicrobial-resistant infections. Nationwide surveillance data provide necessary information for fulfilling the objectives identified for prevention and control. Area-specific data can be used by clinicians to heighten their awareness and guide their selection of antimicrobial drugs for treatment of infections caused by resistant organisms.

SURVEILLANCE

In response to DRSP, officials in some state health departments began conducting surveillance, and in 1994 the Council of State and Territorial Epidemiologists (CSTE) recommended that DRSP become one of the nationally notifiable diseases. In addition, the CDC began supporting active, population-based surveillance for invasive pneumococcal infections in selected areas. In 1995, the surveillance program was named Active Bacterial Core surveillance (ABCs) and is a core component of the CDC's Emerging Infections Programs Network (EIP), which is a collaboration between the CDC, state health departments, and universities. ABCs is an active laboratory- and population-based surveillance system for invasive bacterial pathogens of public health importance [18]. For each case of invasive disease in the surveillance population, a case report with basic demographic information is completed and bacterial isolates are sent to CDC and other reference laboratories for additional laboratory evaluation. ABCs also provides an infrastructure for further public health research, including special studies aiming at identifying risk factors for disease, post-licensure evaluation of vaccine efficacy and monitoring effectiveness of prevention policies.

ABCs was initially established in four states. It currently operates among 10 EIP sites across the United States, representing a population of over 38 million persons. (www.cdc.gov/abcs) The sites conducting *S. pneumoniae* surveillance include all counties in the states of Connecticut, Minnesota, and New Mexico, and multiple counties in California, Colorado, Georgia, Maryland, New York, Oregon, and Tennessee. Currently, ABCs conducts surveillance for six pathogens: group A and group B *Streptococcus* (GAS and GBS), *Haemophilus influenzae*, *Neisseria meningitidis*, *S. pneumoniae*, and methicillin-resistant *Staphylococcus aureus* (MRSA).

The current population under surveillance for *S. pneumoniae* is 27,816,794. Antimicrobial susceptibility testing is performed on all *S. pneumoniae* isolates. Nonsusceptibility and resistance are defined by using Clinical Laboratory Standards Institute (formerly NCCLS) criteria [19]. Genotyping is performed on certain groups of *S. pneumoniae* isolates. Techniques include pulsed field gel electrophoresis, multilocus sequence typing, the penicillin-binding protein gene *dhf* specific restriction fragment length polymorphism/sequence analysis, and molecular analysis of resistance mechanisms. Antimicrobial susceptibility testing is performed on a subset of GAS isolates. Serotyping and genotyping techniques based on the M protein molecule and the *emm* gene encoding the molecule are performed on all GAS isolates. Opacity factor sequence typing is done on a subset of GAS isolates. Medical record review is performed by surveillance personnel. Data from case report forms, isolate forms, and special study forms are entered into a computerized database at each surveillance site. The data are transmitted to the CDC where they are verified and aggregated. Every month, summary tables and laboratory testing results are sent to the surveillance sites from the CDC.

Public health officials can use this information to improve vaccination use, targeting areas most likely to benefit from intervention (e.g., regions with high levels of resistance to antimicrobial drugs or communities of persons at highest risk for infection), to promote the judicious use of antimicrobial agents in areas with high levels of antimicrobial resistance, and to publish national and regional trends in pneumococcal antimicrobial resistance [6,20,21].

EPIDEMIOLOGIC INVESTIGATIONS

ABCs provides a framework for special epidemiologic investigations of DRSP to be performed. A study using DRSP population-based surveillance data for the eight-county metropolitan Atlanta area in 1994 identified 27% and 24% of invasive *S. pneumoniae* isolates as penicillin nonsusceptible among children and adults, respectively. Higher proportions of whites than blacks had infections caused by penicillin-resistant or multidrug-resistant strains of *S. pneumoniae*. Higher proportions of white children less than six years of age had infections caused by penicillin-resistant, cefotaxime-resistant, or multidrug-resistant strains of *S. pneumoniae* than black children of the same age. Suburban residence was also associated with an increased risk of infections with an antimicrobial-resistant organism [22]. The high rate of antimicrobial-resistant infections found in adults suggested that antimicrobial resistance is not a problem limited to pediatric patients. Because of concern about antimicrobial-resistant infections in children, recommendations for empirical therapy for children with suspected life-threatening pneumococcal infections were developed [23]. The results of this study provided evidence for the necessity of recommendations for empirical therapy of pneumococcal infections in adults.

A case-control study to identify risk factors for invasive pediatric pneumococcal disease was performed within the ABC surveillance sites. Along with risk factors identified for invasive pediatric pneumococcal disease, recent antimicrobial use was associated with increased risk for invasive infection with penicillin-resistant *S. pneumoniae* [24].

Two additional epidemiologic studies resulting from ABCs surveillance data looked at resistant *S. pneumoniae*. Surveillance data initially showed that disease caused by macrolide-resistant *S. pneumoniae* was a rapidly increasing problem in the 1990s [25]. A follow-up study showed a dramatic decline in macrolide resistance following the introduction of the pneumococcal conjugate vaccine in metropolitan Atlanta. The decrease was due to both direct and herd immunity effects of the vaccine to decrease infections [26]. In a more recent study of ABCs data, rates of resistant invasive disease caused by vaccine serotypes fell dramatically [27]. These results support the strategy of using vaccines to prevent resistant infections.

PREVENTION AND CONTROL

VACCINATION

The CDC is working with state health departments to promote the use of pneumococcal vaccines. Currently, there are two types of pneumococcal vaccines available for use, the 23-valent pneumococcal capsular polysaccharide vaccine and the 7-valent pneumococcal protein polysaccharide conjugate vaccine. The 23-valent pneumococcal vaccine contains 23 purified pneumococcal capsular polysaccharide antigens, which comprise at least 85% to 90% of the serotypes that cause invasive disease in the United States and include serotypes of drug-resistant strains that most frequently cause invasive disease [20,28]. This vaccine is not recommended for children under two years old because antibody response to most pneumococcal capsular types is generally poor or inconsistent in this age group [29].

In 1998, an estimated 3400 adults aged ≥65 years died as a result of invasive pneumococcal disease [20]. According to data from the National Health Interview Survey (NHIS), pneumococcal vaccination coverage increased by 32% (from 42.6% to 56.3%) among persons aged ≥65 years from 1997 to 2005, but coverage has remained nearly unchanged since 2002 (56.2%) (available at http://www.cdc.gov/nchs/about/major/nhis/released200609.htm#4). In 2004, the overall proportion of respondents of the Behavioral Risk Factor Surveillance System (BRFSS) surveys aged ≥65 years reporting ever having received pneumococcal vaccine was 63.4% (CI = 62.7% to 64.1%). Vaccination coverage ranged from 32.7% (Puerto Rico) to 71.6% (Montana), with a median of 64.6%. In 2005, the overall proportion of respondents aged ≥65 years reporting ever having received pneumococcal vaccine was 63.7% (CI = 63.1% to 64.4%). Vaccination coverage ranged from 28.3% (Puerto Rico) to 71.7% (North Dakota), with a median of 65.7% [30]. However, these vaccination rates are still far from the national health objective for year 2010, which is to achieve 90% coverage of non-institutionalized adults aged ≥65 years for pneumococcal vaccination (objective 14 to 29 available at http://www.health.gov/healthypeople) [31].

Underutilization of the 23-valent polysaccharide vaccine may be partly due to some ongoing controversy regarding vaccine efficacy, duration of protection, side effects, and adverse reactions. Nonetheless, pneumococcal polysaccharide vaccine has been shown to be effective against bacteremic disease caused by organisms whose serotypes are contained in the vaccine [32]. Effectiveness in case-control studies has ranged from 56% to 81%. Vaccine effectiveness of 65% to 84% has been demonstrated among immunocompetent persons aged 65 years or older and among persons with diabetes mellitus, coronary vascular disease, congestive heart failure, chronic obstructive pulmonary disease and anatomic asplenia [33–36]. In addition, a meta-analysis of nine randomized controlled trials concluded that pneumococcal vaccine is efficacious in reducing the frequency of bacteremic pneumococcal pneumonia among low-risk adults [29].

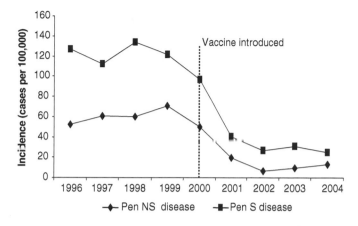

FIGURE 16.1 Annual incidence (cases per 100,000 persons) of invasive disease caused by penicillin-susceptible (Pen S) and penicillin-nonsusceptible pneumococci (Pen NS) in children <2 years of age, 1996–2004 (66). (From Kyaw, M.H., Lynfield, R., Schaffner, W., Craig, A.S., Hadler, J., Reingold, A., et al. *N. Engl. J. Med.*, 354, 14, 2006. With permission.)

In February, 2000, a 7-valent pneumococcal protein polysaccharide conjugate vaccine was licensed for use in young children and has been recommended for use by the Advisory Committee on Immunization Practices [37]. Clinical trials found a high efficacy for the prevention of meningitis and bacteremia, with efficacy against clinically defined otitis media and known to be caused by *S. pneumoniae* of vaccine serotypes [38,39]. Because conjugate vaccines reduce carriage of pneumococci of vaccine serotypes, and approximately 80% of DRSP occurs within serotypes included in the heptavalent vaccine, their impact on transmission of DRSP has been considerable. In a study of ABCs data by Kyaw et al., rates of resistant invasive disease caused by vaccine serotypes fell 87% [27]. In children under two years of age disease caused by penicillin non-susceptible *S. pneumoniae* of all serotypes declined 81%, and in those 65 years and older, the decline was 49% (Figures 16.1 and 16.2).

APPROPRIATE USE EDUCATION CAMPAIGNS

In 1995, the CDC launched the Campaign for Appropriate Antibiotic Use in the Community [40]. The Campaign targets both health care consumers and providers to promote the appropriate use of antibiotics in the community for respiratory infections. The Campaign targets the five respiratory conditions that in 1992 accounted for more than 75% of all office-based prescribing for all ages combined: otitis media, sinusitis, pharyngitis, bronchitis, and the common cold [10]. The goal of the Campaign is to reduce the spread of antibiotic resistance by: (*i*) promoting adherence to appropriate prescribing guidelines among providers, (*ii*) decreasing demand for antibiotics for viral upper respiratory infections among healthy adults and parents of young children, and (*iii*) increasing adherence to prescribed antibiotics for upper respiratory infections. The Campaign focused on four areas: (*i*) forming a coalition of partnerships, (*ii*) developing educational materials for providers and consumers,

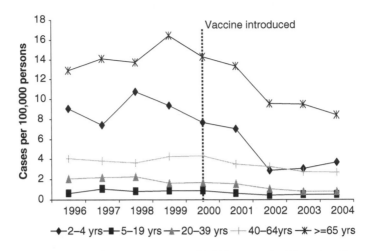

FIGURE 16.2 Annual incidence (cases per 100,000 persons) of invasive disease caused by penicillin-nonsusceptible pneumococci in persons ≥2 years, 1996 to 2004. (From Kyaw, M.H., Lynfield, R., Schaffner, W., Craig, A.S., Hadler, J., Reingold, A., et al. *N. Engl. J. Med.*, 354, 14, 2006. With permission.)

(*iii*) developing and implementing specific interventions, and (*iv*) assessing the impact on antibiotic use and physician–patient satisfaction.

In 2003, the Campaign for Appropriate Antibiotic Use in the Community was expanded into a national media campaign to provide a coordinated message on appropriate antibiotic use and to create a foundation for local efforts across the country. The Campaign was renamed Get Smart: Know When Antibiotics Work. The appropriate use messages were disseminated via print, television, radio, and billboards. Educational materials and media toolkits were distributed to CDC-funded sites for use in conjunction with their local campaigns.

A social ecological framework has been applied to the CDC's educational campaign. This approach considers the contributions of individual influences and social-environmental influences on health behavior. The framework is useful to ensure efforts are targeted at all levels of influence. The factors of influence may be further divided into five levels: individual or interpersonal factors, groups or social networks, organizational factors, community factors, and public policy. The Campaign focuses its activities in the first three levels, attempting to change the behaviors of patients, consumers, and health care providers, to modify group norms, and to promote organizational adoption of policies and tools that support appropriate antibiotic use techniques and philosophies [41].

PARTNERSHIPS

Over the last decade the CDC's partnerships for the Campaign have included state and local health departments, federal agencies, managed care organizations, pharmacy benefits management companies, consumer advocacy groups, community organizations, pharmaceutical companies, health care purchasers, professional associations, and medical schools. The various partnerships the CDC has established help support programs and target a variety of audiences.

The CDC funds state and local health departments for the development, implementation, and evaluation of local Campaigns to promote appropriate antibiotic use. The number of funded sites increased from 8 in 1998 to 30 in 2007. In 2004, a multicultural outreach program was also developed to create and distribute educational materials on appropriate antibiotic use to diverse audiences. In partnership with the Indian Health Service and various Latino organizations culturally appropriate educational materials have been designed, tested, and distributed for Native American and Spanish-speaking consumers.

EDUCATIONAL MATERIALS, GUIDELINES, CURRICULA

Campaign activities over the last decade have included developing, distributing, and promoting the adoption of prescribing guidelines and educational materials for health care providers. One of the initial products of the CDC's Campaign was the development and distribution of principles for appropriate antibiotic use for pediatric upper respiratory tract infections [42]. Developed in collaboration by the CDC, the American Academy of Pediatrics, and members of the American Academy of Family Physicians, these guidelines provide a definition of appropriate prescribing and have been distributed to numerous state and local health departments, health plans, and physician groups. The guidelines serve to change knowledge and attitudes of providers in ways that favor appropriate prescribing.

Following the development of the pediatric principles, the CDC produced a series of health education materials for both parents and providers to promote appropriate antibiotic use. These include brochures, posters, question-and-answer fact sheets for parents on runny nose and otitis media, instructional or "detailing" sheets for small-group physician education modeled after materials used by the pharmaceutical industry, a day care letter, and a viral prescription pad. The prescription pad in particular has been extremely popular and useful as a communication tool. Health care providers can use this tool to recommend strategies for symptomatic relief of viral illnesses, thereby acknowledging the patient's discomfort and suggesting solutions without prescribing an antibiotic unnecessarily. These health education materials, as well as the pediatric guidelines, are available at the CDC website—www.cdc.gov/getsmart—and may be downloaded and copied free of charge. Materials may be purchased in bulk from the Public Health Foundation—www.phf.org.

Approximately 15 million prescriptions are written each year in the United States for acute otitis media (AOM). AOM has a high rate of spontaneous resolution without antibiotic treatment. Many studies have shown similar outcomes and complication rates in children with AOM who were not treated with antibiotics. As a result of this, the American Academy of Pediatrics issued guidelines in 2004 proposing an observation (or wait-and-see) option for children with AOM. This option refers to managing selected patients diagnosed with uncomplicated AOM with symptomatic relief only for up to 72 h [43]. The wait-and-see approach has been shown to reduce the unnecessary use of antibiotics in children with AOM [44].

In 2003, the CDC developed an appropriate antibiotic use curriculum for medical students in collaboration with Westat (Rockville, Maryland), the University of California, San Diego, and the American Association of Medical Colleges. The curriculum was pilot tested and distributed to 25 medical schools. It includes didactic lectures, small-group interactive sessions, and case studies. The American Association of Medical Colleges is leading the effort to distribute and promote to all U.S. medical schools in 2007. Two additional curricula are in development for residency programs. The first was modeled after the above medical school curriculum and is being developed for use at the Oregon Health Science Center for family practice and internal medicine residents. The second is being developed by the Children's Hospital of Pittsburgh and is designed to improve pediatric residents' proficiency in diagnosing acute otitis media. The curriculum uses video otoscopes and other innovative teaching methods.

INTERVENTIONS TO PROMOTE JUDICIOUS USE

In June 1999, the CDC convened a panel of investigators to evaluate the impact of selected intervention studies designed to promote judicious antimicrobial use. The projects ranged from managed-care and community-level interventions to large-scale, statewide interventions, all focusing on educating medical-care providers, parents, and patients about appropriate indications for, and use of, antibiotics [45,46]. Projects used a variety of strategies and materials to improve communication between physicians and patients and to promote the use of symptomatic therapy as an alternative to antibiotics.

The panel reached a number of conclusions. First, interventions to promote judicious use of antibiotics may be effective in reducing inappropriate prescribing. Successful projects all combined physician and patient education, acknowledging the role that both groups play in the promotion of appropriate prescribing. Second, to be successful, projects must present their messages via multiple vehicles to make use of the variety of means by which people learn. Third, the problem of inappropriate prescribing is not uniform across the country. For instance, some investigators documented significant antibiotic overprescribing for the common cold, while other investigators found that this was rarely occurring. In sites where this was not occurring, physicians were often offended when told not to prescribe for this condition. Successful programs must tailor the prevention messages to local conditions.

A few projects highlight these findings, as shown in Table 16.1. Gonzales et al. were able to reduce antibiotic prescribing for adults with bronchitis by 40% through an intervention targeting clinicians and patients in a managed-care setting [45]. This reduction was achieved without any increase in adverse events or decrease in patient satisfaction. This controlled intervention compared a full intervention that provided education to physicians and patients to either no intervention or an intervention directed only at physicians. These researchers demonstrated that an intervention directed solely at physicians was ineffective. Physicians will reduce their antibiotic prescribing only when they feel that their patients are receptive to this change. In Wisconsin, in addition to educating providers and patients, Belongia et al. conducted an education campaign in the community (i.e., child care providers, schools, parenting groups, etc.) [46]. There was a significant decline in the prescription rate, particularly for liquid prescriptions (i.e., those mainly prescribed for pediatric patients).

TABLE 16.1

Summary of Published Controlled Trials Promoting Appropriate Antibiotic Use

Location of Study (Year)	Setting	Provider Education	Patient Education	Public Education	Scope	Prescription Rate Decline Intervention	Control
Denver (1997–1998) [45]	HMO	Prescribing rate feedback, small group presentations, and practice tips for withholding antibiotics (full intervention sites only)	Office-based educational materials including posters for exam rooms and patient information sheets (limited and full intervention sites); household mailing to patients including pamphlets, refrigerator magnets, and letter from clinic director (full intervention sites only)	No additional efforts aimed at general public	Adult bronchitis only	35% (for full intervention site)	3%
Boston/ Seattle (1997–1998) [67]	HMO	Small group office presentations led by pediatrician "peer leaders;" prescribing rate feedback	Brochures mailed to patients at home; posters and pamphlets in waiting rooms and exam rooms	No additional efforts aimed at general public	Respiratory conditions among children less than 6 years old	18.6% (3 to <36 mo. old) 15% (36 to <72 mo. old)	11.5% (3 to <36 mo. old) 9.8% (36 to <72 mo. old)
Tennessee (1997–1998) [68]	Metropolitan area, Medicaid managed care enrollees	Lectures to targeted providers in multiple settings (primary care clinics, grand rounds, hospital staff meetings); guideline distribution; articles in county health journals	Brochures distributed to parents of newborns, children in day care and K–3rd grade; patient education materials distributed to targeted providers	Brochures distributed to hospitals, clinics, dentists, persons receiving flu vaccine, and pharmacy clients; TV, radio, and newspaper coverage; public service announcements	Respiratory conditions among children less than 15 years old	19%	8%

TABLE 16.1 (continued)
Summary of Published Controlled Trials Promoting Appropriate Antibiotic Use

Location of Study (Year)	Setting	Provider Education	Patient Education	Public Education	Scope	Prescription Rate Decline Intervention	Control
Wisconsin (1997) [46]	Rural communities	Grand rounds presentations; practice-based small group meetings led by physician educators; guideline distribution; CDC fact sheets, patient education materials, viral prescription pad	Information and educational materials presented to clinics by project nurses; "cold kits" provided to clinics for distribution to adults and adolescents	Education for child care providers, public health agencies, parent groups and community organizations; educational materials (brochures and posters) distributed to clinics, pharmacies, child care facilities and schools	Respiratory conditions among children	11% (liquids) 19% (solids)	(+12%) (liquids) 8% (solids)
Alaska (1998–2000) [47]	Rural villages	Workshops for community health aides and physicians		Presentations at village-wide meetings; health newsletters mailed to homes; high school education; health fairs	Respiratory conditions among children and adults	31%	9.5%

Source: From McCaig, L.F., Besser, R.E., Hughes, J.M., *Emerg. Infect. Dis.,* 2003; 9(4): 432–7. With permission.

The CDC and the National Committee on Quality Assurance have written measures for the Health Plan Employer Data and Information Set (HEDIS®), the performance measurement tool used by over 90% of the nation's health plans. In 2004, the first two measures were written to assess the appropriate treatment of children with pharyngitis and with upper respiratory infections. The first measure attempts to increase the proportion of children who are tested for GAS before receiving antibiotics for sore throats. The second measure aims to decrease the proportion of children who receive an antibiotic for the common cold. In 2006, two additional HEDIS measures were pilot tested, one to measure the inappropriate antibiotic treatment of acute bronchitis and the other to measure antibiotic utilization (http://www.ncqa.org/PROGRAMS/HEDIS/index.htm).

EVALUATION AND FUTURE DIRECTIONS

From 1992 to 2001, data from the National Ambulatory Medical Care Survey (NAMCS) and the National Hospital Ambulatory Medical Care Survey (NHAMCS) showed that the overall antimicrobial prescribing population and visit-based rates declined by 23% and 28%, respectively [48]. The decline in antimicrobial prescribing is still being observed in physician offices through 2004 (Figure 16.3) [49]; however, for the first time since 1992, the overall population-based antimicrobial prescribing rate and the visit-based rate for persons 15 to 24 years of age are no longer decreasing. In addition, rates continue to increase for persons 15 years of age and over seen in the outpatient department. In the emergency department (ED), the prescribing rate for adults is rising, most prominently in persons 65 years of age and older. During the same time period, increasing rates of use were seen for broad-spectrum antibiotics, such as azithromycin and clarithromycin (up 260%) and fluoroquinolones (up 98%, Figure 16.4). Since data from NAMCS and NAHMCS do not link prescription and diagnosis data, additional studies are needed to determine the conditions for

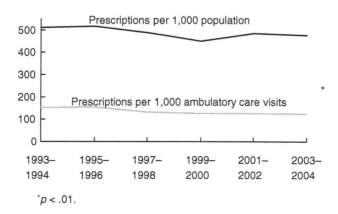

FIGURE 16.3 Trends in antimicrobial prescription rates in ambulatory care settings, United States, 1993–2004. (From McCaig, L., Friedman, C.R., *Annual Conference on Antimicrobial Resistance*. Bethesda, MD: National Foundation for Infectious Diseases, 2006.)

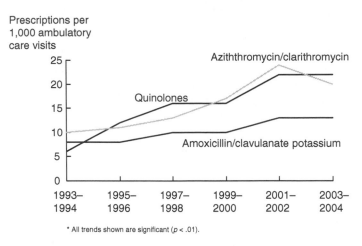

FIGURE 16.4   Trends in antimicrobial prescription rates in ambulatory care settings by drug class, United States, 1993–2004. (From McCaig, L., Friedman, C.R., *Annual Conference on Antimicrobial Resistance*. Bethesda, MD: National Foundation for Infectious Diseases, 2006.)

which these drugs are being prescribed. Future education efforts may need to be tailored toward health care providers in the ED, the elderly and their caregivers, and toward appropriate antimicrobial choice to help decrease the use of broad-spectrum drugs when not indicated.

Lessons learned from the education intervention projects that have been critically evaluated are being used to guide the development of many new projects. Through the use of surveillance for antimicrobial resistance, and the systematic analysis of prescriber databases we should be able to determine whether these efforts are effective in slowing the rise or reversing the trends in antimicrobial resistance in the community.

ANTIMICROBIAL RESISTANCE IN THE INPATIENT SETTING

Antimicrobial resistance remains a significant and growing problem in inpatient settings. Data from the Centers for Disease Control and Prevention's (CDC) National Nosocomial Infection Surveillance System (NNIS) from 2003 continue to show increases in resistance rates among several organism/antimicrobial pairs that have been monitored since 1998 (Figure 16.1). Though the problem of antimicrobial resistance in intensive care units (ICUs) remains well recognized, there are now reports demonstrating that this problem is not limited to the ICU, but is present in a variety of health care settings including outpatient, hemodialysis, long-term acute care, and long-term care settings [50].

There are several factors that make health care environments a focus for the emergence and spread of antimicrobial-resistant pathogens. The sometimes-urgent nature of acute care might not allow for necessary aseptic technique or hand hygiene in all instances, particularly in ICU settings. The large number and variety of types of health care workers attending to patients' needs mean that training and compliance with infection-control measures such as hand hygiene, gloving, and gowning may be inconsistent. These challenges are greater in teaching environments where the presence of trainees leads to increases in the number of people caring for patients and in staff turnover due to rotations and graduations. Evidence suggests that antimicrobial-resistant pathogens are commonly carried from patient to patient (exogenous flora) by way of the unwashed hands of health care workers [51]. The introduction of antimicrobial-resistant bacteria into health care settings may also occur upon transfer of patients unknowingly colonized or infected with such bacteria from other facilities. Normal defenses among hospitalized patients are also compromised by the presence of invasive devices such as intravenous and urinary catheters and mechanical ventilators. These devices have consistently been shown to be significant risk factors for infections, often with resistant pathogens. The prolonged lengths of stay in the long-term acute and long-term care settings also provide increased opportunities for both the development and dissemination of antimicrobial-resistant pathogens. Finally, and perhaps most importantly, there are the issues of antimicrobial use, which contribute significantly to antimicrobial resistance in health care settings [4,52,53]. Of studies involving hospital-acquired pathogens, 22 reviewed by McGowan have shown a fairly consistent association between use and resistance [4]. Unfortunately, nearly all of these studies were reports from single hospitals, which may not be representative of other hospitals; however, a previous multicenter study in the 1970s demonstrated that changes in aminoglycoside use paralleled changes in aminoglycoside-resistant Gram-negative bacilli [54]. One other multicenter study demonstrated this type of relationship among several antimicrobials and the corresponding resistant pathogens, including ceftazidime use and ceftazidime-resistant *Enterobacter cloacae* [55].

The issue of antimicrobial resistance is well known to clinicians in multiple-practice settings and at all levels of experience. In a survey of 490 internal medicine physicians from a variety of practice settings, including public, university, community, and long-term care hospitals, 87% of respondents reported that antimicrobial resistance was a very important national problem [56]. Nearly all respondents to this survey [97%] reported that widespread and inappropriate antibiotic use were important causes of resistance. In another survey of 117 clinicians from a variety of hospital types, 95% of respondents agreed that antimicrobial resistance was a national problem [57]. Finally, in a survey of 179 medical house staff in a variety of specialties, 96% agreed that hospitals in general face serious problems with antibiotic resistance and 97% agreed that better use of antibiotics would reduce resistance [58].

Public health organizations like the World Health Organization (WHO) and the CDC have also long recognized that antimicrobial resistance in health care settings presents a significant threat to public health. Historically, the public health presence in this area has been most visible in surveillance activities. For example, in 1995 the WHO developed and made available at no charge the WHONET software, which allows users to track rates of antimicrobial resistance and to share that surveillance data (http://www.who.int/drugresistance/whonetsoftware/en/). In this country, the CDC has monitored and reported on rates of antimicrobial resistance, primarily in the intensive care unit setting, through the NNIS. The NNIS was established in 1974 as a sentinel surveillance program for health care–associated infections. Approximately 300 hospitals from around the country have volunteered to systematically collect these data and report them to the CDC. Since 1986, surveillance efforts have been targeted in ICUs, where the majority of health care–associated infections and antimicrobial resistance occur. In addition to collecting information on the rates of infections, the NNIS also allows participants to enter susceptibility data for infecting organisms. These data are published in the NNIS annual report and have provided excellent national information on susceptibility trends among inpatients in the ICU setting. In 1999, the NNIS added the "antimicrobial use and resistance component" that specifically tracks antimicrobial resistance for a select group of pathogen/antimicrobial pairs in both the inpatient and outpatient settings and also records data on the rates of antimicrobial use among inpatients.

BACKGROUND ON THE CAMPAIGN

In 2002, the public health response to antimicrobial resistance in health care facilities entered a new phase with the advent of the CDC Campaign to Prevent Antimicrobial Resistance in Health care Settings. Surveillance activities had characterized the growing problems of antimicrobial resistance in health care settings. The Campaign was created to complement these surveillance activities by presenting a framework that would help address the causes of antimicrobial resistance. It was envisioned as a nationwide effort to facilitate the implementation of educational and behavioral interventions that would assist clinicians in preventing antimicrobial resistance in health care settings. The Campaign was centered on the premise that there are a number of simple, evidence-based practices that could be implemented in health care settings, which would not only impact antimicrobial resistance, but also improve patient care. The goal of the Campaign was not to provide new recommendations, but rather to combine existing, evidence-based strategies into a comprehensive and cohesive framework. Based on discussions with clinicians, it was determined that the optimal format for the Campaign would be "12-step" programs containing evidence-based recommendations.

Research on preventing antimicrobial resistance was reviewed, and four core strategies emerged that formed the foundation of the Campaign: Prevent Infection, Diagnose and Treat Infection Effectively, Use Antimicrobials Wisely, and Prevent Transmission. However, because clinicians who treat different patient populations face different challenges, it was recognized early on that there could not be a "one-size-fits-all" solution. Hence, it was decided that there should be not one, but five, 12-step programs within the Campaign to target clinicians caring for a variety of patient populations, including hospitalized adults, dialysis patients, surgical patients, hospitalized children, and long-term care residents. Though the core strategies and some of the individual recommendations are common to all the programs, each also contains evidence-based recommendations unique to that patient population.

DEVELOPMENT AND DISSEMINATION OF THE CAMPAIGN PROGRAMS

Patient populations were identified based on surveillance data and other research demonstrating significant problems caused by antimicrobial resistance. Because the

Campaign programs are intended for clinicians, it was decided that each should be developed under the leadership of experienced clinicians practicing in the chosen field. The CDC sought the assistance of specialty professional societies to identify recognized clinical leaders in the various fields who were interested in leading this effort. Once selected, these clinicians formed the expert panel that participated in several sessions during which specific steps for the program were discussed. Potential steps were identified based on both the evidence basis and the potential impact of the recommendation on antimicrobial resistance. Candidate steps were then assembled and prioritized to yield the preliminary 12 step program for the Campaign. To solicit broader input, each program was then presented to multiple clinician focus groups for review and comment. The goal was that each program would be discussed by at least two nationally representative focus groups of practicing clinicians. Comments from the focus group discussions were carefully reviewed and the steps were modified accordingly to yield the final program. The steps of each program of the Campaign are summarized in Table 16.2.

Following development of each program of the Campaign, the next critical step was disseminating information to clinicians. A first step in that process was the publication of a manuscript, which highlighted the evidence basis for the steps of the Campaign program. The goal was that this paper be published in a peer-reviewed journal that is widely read by the target audience of that program. Publication of the paper not only served to alert clinicians to the presence of the Campaign program, but also provided peer-reviewed evidence to support the efficacy for the steps that were chosen—an issue that has been shown to be vitally important in clinician acceptance of recommendations [59]. To date, such papers have been published for the programs directed at clinicians caring for hospitalized adults, dialysis patients [60], surgical patients [61], and long-term care patients [62]. To further heighten both the awareness and acceptance of the Campaign, endorsement of the various programs was sought from professional societies that represent practitioners of the specialty. The overall Campaign approach has been endorsed by several societies representing practitioners of internal medicine, infectious diseases and health care epidemiology and infection control. Furthermore, each of the specific programs has been endorsed by one or more specialty societies (Table 16.3). In order to alert clinicians about the availability of the Campaign programs, well-publicized "launches" have been conducted for each of the Campaigns at national meetings of many of these societies. Finally, the CDC staff work both with clinicians and experts in health communication to develop educational materials about the Campaign program, including slide sets, fact sheets, posters, and pocket cards as well as informational sheets for patients. These materials are made available at no or minimal cost through the CDC Campaign Web site (http://www.cdc.gov/drugresistance/health care/default.htm).

ASSESSING THE CAMPAIGN

Soon after the idea of the Campaign was developed, investigators began conducting interviews, focus groups and surveys to examine clinicians' perceptions about issues surrounding implementation of Campaign recommendations. These evaluations have demonstrated that different measures are given different priorities in various practice settings. For example, "Prevent Transmission" was listed as the most important

TABLE 16.2

Comparison of the Steps of the CDC Campaigns to Prevent Antimicrobial Resistance in Health Care Settings

Hospitalized Adults	Dialysis Patients	Surgical Patients	Hospitalized Children	Long-Term Care Residents
		Prevent Infection		
Step 1. Vaccinate • Give influenza/pneumococcal vaccine to at-risk patients before discharge • Get influenza vaccine annually	**Step 1. Vaccinate staff and patients** • Get influenza vaccine • Give influenza and pneumococcal vaccine to patients in addition to routine vaccines (e.g., hepatitis B)	**Step 1. Prevent surgical site infections** • Monitor and maintain normal glycemia • Maintain normothermia • Perform proper skin preparation using appropriate aseptic agent and, when necessary, hair-removal techniques • Think outside the wound to stop surgical-site infections	**Step 1. Vaccinate hospitalized children and staff** • Vaccinate according to AAP/ACIP/AAFP recommendations • Review immunization records and catch-up with routine vaccinations • Give influenza vaccine to at-risk infants and children • Give influenza vaccine to all health workers	**Step 1. Vaccinate** • Give influenza and pneumococcal vaccinations to residents • Promote vaccination among all staff
Step 2. Get the catheters out • Use catheters only when essential • Use the correct catheter • Use proper insertion and catheter-care protocols • Remove catheters when they are no longer essential	**Step 2. Get the catheters out** *Hemodialysis* • Use catheters only when essential • Maximize use of fistulas/grafts • Remove catheters when they are no longer essential *Peritoneal Dialysis* • Remove/replace infected catheters	**Step 2. Prevent device-related infections: get the devices out** • Use catheters only when essential • Use proper insertion and catheter-care protocols • Use drains appropriately • Remove catheters and drains when they are no longer essential	**Step 2. Get the devices out** • Insert catheters and devices only when essential • Use proper insertion techniques and follow guidelines for catheter care • Remove catheters and other devices when no longer essential	**Step 2. Prevent conditions that lead to infection** • Prevent aspiration • Prevent pressure ulcers • Maintain hydration
	Step 3. Optimize access care • Follow established KDOQI and CDC guidelines for access care	**Step 3. Prevent hospital-acquired pneumonia** • Wean from the ventilator when appropriate		**Step 3. Get the unnecessary devices out** • Insert catheters and devices only when

essential and minimize duration of exposure
- Use proper insertion and catheter-care protocols
- Reassess catheters regularly
- Remove catheters and other devices when no longer essential

Step 4. Use established criteria for diagnosis of infection
- Target empiric therapy to likely pathogens
- Target definitive therapy to known pathogens
- Obtain appropriate cultures and interpret results with care
- Consider *C. difficile* in patients with diarrhea and antibiotic exposure

Step 5. Use local resources
- Consult the infectious disease experts for complicated infections and potential outbreaks

- Elevate head of bed to 30°
- Drain circuit/tubing condensate away from patient
- Prevent contamination of respiratory therapy equipment, ventilator circuits, and respiratory medications

Diagnose and Treat Infection Effectively

Step 4. Target the pathogen
- Target empiric antimicrobial therapy to likely pathogens
- Obtain appropriate cultures
- Target definitive antimicrobial therapy to known pathogens
- Optimize timing, regimen, dose, route, and duration of antimicrobial therapy
- Practice safe source control (e.g., debridement, or open wound as indicated)

Step 5. Access the experts
- Consult the appropriate expert for complicated infections: surgeons; infectious disease experts; clinical pharmacists

Step 3. Use appropriate methods for diagnosis
- Order appropriate laboratory tests
- Obtain appropriate specimens

Step 4. Target the pathogen
- Target empiric antimicrobial therapy to likely pathogens
- Target definitive antimicrobial therapy to known pathogens

- Use proper insertion and catheter-care protocols
- Remove access device when infected
- Use the correct catheter

Step 4. Target the pathogen
- Obtain appropriate cultures
- Target empiric therapy to likely pathogens
- Target definitive antimicrobial therapy to known pathogens
- Optimize timing, regimen, dose, route, and duration

Step 5. Access the experts
- Consult the appropriate expert for complicated infections

Step 3. Target the pathogen
- Culture the patient
- Target empiric therapy to likely pathogens and local antibiogram
- Target definitive therapy to known pathogens and antimicrobial susceptibility test results

Step 4. Access the experts
- Consult infectious disease experts for patients with serious infections

continued

TABLE 16.2 (continued)
Comparison of the Steps of the CDC Campaigns to Prevent Antimicrobial Resistance in Health Care Settings

Hospitalized Adults	Dialysis Patients	Surgical Patients	Hospitalized Children	Long-Term Care Residents
Step 5. Practice antimicrobial control • Engage in local antimicrobial control efforts	**Step 6. Use local data** • Know your local antibiogram • Get previous microbiology results when patients transfer to your facility	**Use Antimicrobials Wisely** **Step 6. Start prophylactic antimicrobials promptly** • Give the initial dose within one hour preceding incision • Use the appropriate antimicrobial and dosing • Repeat the dose during surgery as needed to maintain blood levels	**Step 5. Access the experts** • Consult infectious disease experts for complicated infections **Step 6. Practice antimicrobial control** • Optimize timing, regimen, dose, route, and duration of antimicrobial treatment and prophylaxis • Follow policies and protocols in your institution	• Know your local and/or regional data • Get previous microbiology data for transfer residents **Step 6. Know when to say "no"** • Minimize use of broad-spectrum antibiotics • Avoid chronic or long-term antimicrobial prophylaxis • Develop a system to monitor antibiotic use and provide feedback to appropriate personnel **Step 7. Treat infection, not colonization or contamination** • Perform proper antisepsis with culture collection • Re-evaluate the need for continued therapy after 48–72 hours • Do not treat asymptomatic bacteriuria

Public Health Responses to Antimicrobial Resistance 397

Step 6. Use local data • Know your antibiogram • Know your patient population **Step 7. Treat infection, not contamination** • Use proper antisepsis for blood and other cultures • Culture the blood, not the skin or catheter hub • Use proper methods to obtain and process all cultures **Step 8. Treat infection, not colonization** • Treat pneumonia, not the tracheal aspirate • Treat bacteremia, not the catheter tip or hub • Treat urinary tract infection, not the indwelling catheter	**Step 7. Know when to say "no" to vanco** • Follow CDC guidelines for vancomycin use • Consider 1st generation cephalosporins instead of vancomycin **Step 8. Treat infection, not contamination or colonization** • Use proper antisepsis for drawing blood cultures • Get one peripheral vein blood culture, if possible • Avoid culturing vascular catheter tips • Treat bacteremia, not the catheter tip **Step 9. Stop antimicrobial treatment** • When infection is treated • When infection is not diagnosed	**Step 7. Stop prophylactic antimicrobials within 24 hours** • Discontinue use even with catheters or drains still in place **Step 8. Use local data** • Know your antibiogram • Know your formulary • Know your patient population **Step 9. Know when to say "no" to vanco** • Vanco should be used to treat known infections, not for routine prophylaxis • Treat staphylococcal infection, not contaminants or colonization • Consider other antimicrobials in treating MRSA	**Step 7. Use local data** • Know your regional, institutional, and high-risk, unit-specific antibiograms • Know your formulary • Know your patient population (birthweight, age, and setting) **Step 8. Treat infection, not contamination or colonization** • Use proper antisepsis for drawing blood cultures • Avoid routine culturing of catheter tips • Treat bacteremia, not catheter colonization or contamination **Step 9. Know when to say "no"** • Avoid routine use of vancomycin, extended-spectrum cephalosporins (third [cefotaxime] and fourth [cefepime] generation cephalosporins) carbapenems oral quinolones, and linezolid • Follow antimicrobial prescribing guidelines from CDC, AAP, and other professional societies	**Step 8. Stop antimicrobial treatment** • When cultures are negative and infection is unlikely • When infection has resolved **Step 9. Isolate the pathogen** • Use Standard Precautions • Contain infectious body fluids (use approved Droplet and Contact isolation precautions) **Step 10. Break the chain of contagion** • Follow CDC recommendations for work restrictions and stay home when sick • Cover your mouth when you cough or sneeze • Educate staff, residents, and families • Promote wellness in staff and residents

continued

TABLE 16.2 (Continued)

Comparison of the Steps of the CDC Campaigns to Prevent Antimicrobial Resistance in Health Care Settings

Hospitalized Adults	Dialysis Patients	Surgical Patients	Hospitalized Children	Long-Term Care Residents
Step 9. Know when to say "no" to vanco • Treat infection, not contaminants or colonization • Fever in a patient with an intravenous catheter is not a routine indication for vancomycin **Step 10. Stop antimicrobial treatment** • When infection is cured • When cultures are negative and infection is unlikely • When infection is not diagnosed		**Step 10. Treat infection, not contamination or colonization** • Use proper antisepsis for drawing blood cultures • Get at least one peripheral vein blood culture, if possible • Avoid culturing vascular catheter tips • Treat bacteremia, not the catheter tip	**Step 10. Stop treatment** • When infection is unlikely • When culture results indicate no clinical need for antimicrobials • When infection is cured	
Step 11. Isolate the pathogen • Use standard infection control precautions • Contain infectious body fluids	**Step 10. Follow infection control precautions** • Use standard infection-control precautions for dialysis centers	**Prevent Transmission** **Step 11. Contain your contaminant and contagion** • Follow infection-control precautions	**Step 11. Practice infection control** • Be familiar with recommended infection-control precautions	**Step 11. Perform hand hygiene** • Use alcohol-based handrubs or wash your hands

(use approved airborne/droplet/contact isolation precautions) • When in doubt, consult infection-control experts	• Consult local infection-control experts	• Consult infection-control teams	• Consult infection-control teams • Stay home when you are sick • Restrict visitors with symptoms of respiratory or gastrointestinal tract infections from contact with your patients	• Encourage staff and visitors
Step 12. Break the chain of contagion • Stay home when you are sick • Keep your hands clean • Set an example	**Step 11. Practice hand hygiene** • Wash your hands or use an alcohol-based handrub • Set an example	**Step 12. Practice hand hygiene** • Set an example • Wash your hands or use an alcohol-based handrub • Do not operate with open sores on hands • Do not operate with artificial nails • Promote good habits for the entire surgical team	**Step 12. Practice hand hygiene** • Wash your hands or use an alcohol-based handrub before and after patient contact • Set an example	**Step 12. Identify residents with multidrug resistant organisms (MDROs)** • Identify both new admissions and existing residents with MDROs • Follow standard recommendations for MDRO case management
	Step 12. Partner with your patients • Educate on access care and infection-control measures • Re-educate regularly			

TABLE 16.3

Professional Societies That Have Endorsed the CDC's Campaign to Prevent Antimicrobial Resistance in Health Care Settings

The following organizations have endorsed the entire Campaign:
   American College of Physicians / American Society of Internal Medicine (ACP/ASIM)
   Association for Professionals in Infection Control and Epidemiology (APIC)
   Infectious Diseases Society of America (IDSA)
   National Foundation for Infectious Diseases (NFID)
   Society for Health care Epidemiology of America (SHEA)
   American Academy of Physician Assistants (AAPA)
The following organizations have endorsed specific programs:

Hospitalized Adults
Society of Hospital Medicine (SHM)

Dialysis Patients
American Society of Nephrology (ASN)
National Kidney Foundation (NKF)
Renal Physicians Association (RPA)

Surgical Patients
American College of Surgeons (ACS)
Surgical Infection Society (SIS)

Hospitalized Children
American Academy of Pediatrics (AAP)
Pediatric Infectious Diseases Society (PIDS)
National Association of Pediatric Nurse Practitioners (NAPNAP)

Long-Term Care Residents
American Association of Homes and Services for the Aging (AAHSA)
American Society on Aging (ASA)
National Association of Directors of Nursing Administration in Long-Term Care (NADONA/LTC)

strategy by long-term care practitioners, while "Diagnose and Treat Infection Effectively" was most important to physicians caring for hospitalized adults. These differences reflect the differing importance of risk factors for antimicrobial resistance in various health care settings and support the Campaign approach of targeting specific programs at different patient populations. The differing opinions on priority areas also underscore one of the important advantages of the Campaign format. The fact that recommendations are divided into four strategies and 12 steps allows facilities flexibility with respect to implementation because they can be implemented as individual strategies or steps to combat specific problems or together as the full program.

One of these assessments also attempted to determine specific barriers to gaining acceptance of Campaign steps. This survey of a variety of clinicians in several practice settings found that aspects of the "health care culture" were reported as the single most important obstacle to implementing Campaign steps [63]. Specifically, respondents pointed to a tendency to overtreat infections, a lack of accountability among health care workers, and non-compliance with infection control precautions as major health care cultural barriers that would have to be overcome in implementing Campaign steps. A secondary issue that was raised was a general lack of knowledge and education among both health care workers and the public about issues of antimicrobial resistance. These types of evaluations are especially important in helping guide implementation strategies in various health care settings. Campaign steps are likely to be adopted by clinicians only if they are perceived as addressing an important problem for the patient population, and thus an understanding of the issues confronted by clinicians in the field is critical in designing successful implementation strategies. Likewise, it is important to understand in advance what barriers to implementation might exist so that that they can be anticipated and addressed.

The clinical relevance of individual steps of a Campaign program has been assessed in one study that examined the steps pertaining to antimicrobial use in the 12-step Program to Prevent Antimicrobial Resistance Among Hospitalized Adults. In a study looking at post-prescription reviews of antimicrobial therapy, investigators reviewed 179 antimicrobial interventions to determine how many fell into one of the steps addressing antimicrobial use [64]. Overall, 90% of the interventions fell into one of the Campaign steps addressing antimicrobial use, demonstrating that these steps were indeed well chosen to address common problems with antimicrobial use among hospitalized adults.

IMPLEMENTING THE CAMPAIGN

Several health care facilities have now implemented the Campaign, either using individual steps and strategies or in its entirety. These implementation projects have permitted an assessment not only of the logistics of Campaign implementation, but also of the impact of the Campaign itself. Recently, three facilities published their experiences with implementation projects [65].

The first project implemented the eight Campaign steps in the two strategies pertaining to antimicrobial use, "Diagnose and Treat Infection Effectively" and "Use Antimicrobials Wisely," as part of a new antimicrobial management program at a 410-bed, non-teaching hospital. The new program, a joint effort of pharmacy and infectious diseases, was undertaken with the support of the administration and was spearheaded by a well-respected infectious diseases specialist. An antimicrobial management team of pharmacists and the infectious diseases specialist was created to review antimicrobial choices and make recommendations to improve therapy in accordance with Campaign steps. Compliance with the team's recommendations was more than 90%, a finding that validates the clinical importance of these steps in the Campaign, since clinicians altered their patient management to comply with the recommendations. In addition to improving antimicrobial use at their hospital, it was estimated that the interventions are saving the institution $136,000 annually as a result of shortened lengths of stay for patients with infections.

A project at another facility, a 500-bed, academic, tertiary care center, involved implementation of step 1 of the Campaign, "Vaccinate" as the initial project of a broader plan to implement all 12 steps. Specifically, the facility wanted to improve influenza vaccination rates among eligible hospital inpatients as this had historically been a problem area. The year before the Campaign was implemented only 95 doses of influenza vaccine had been given to hospital inpatients. The facility assembled a multi-disciplinary team, supported by the facility administration, chaired by a well-respected clinician, and consisting of representatives from infection control, nursing, medicine, pharmacy, quality improvement, and information technology. The team reviewed available literature on improving compliance with influenza vaccination and developed a variety of approaches targeted at overcoming specific obstacles to vaccination at their facility. One of their most important approaches was to identify well-recognized "project champions" from all practice areas, including physicians, nurses, pharmacists, and administrators who could promote the importance of the project to specific practice groups. The year they implemented the project, the number of inpatients vaccinated for influenza increased by more than six-fold to just over 600 patients.

The third project was a broad implementation of the entire Campaign that was undertaken as a facility-wide quality improvement project. Again, a team approach was used to develop the implementation strategy for this 500-bed, non-teaching hospital and again the effort was fully supported by the facility administration. The team was led by a well-respected infectious diseases specialist and had representatives from several practice areas including pharmacy and infection control. Extensive educational efforts formed one aspect of the implementation with a series of continuing medical education programs for physicians and a facility-wide Campaign education fair for all staff. Along with these educational efforts, the implementation team undertook a variety of separate projects to overcome barriers to complying with individual steps. These included projects to remove catheters and improve vaccination rates, antimicrobial use, and hand-hygiene. The facility monitored outcomes data for two years following implementation of the Campaign with impressive results. For strategy one ("Prevent Infection"), an influenza vaccination program led to significant increases in compliance, and a program to optimize catheter use led to decreases in catheter-associated infections. For strategies two ("Diagnose and Treat Infection Effectively") and three ("Use Antimicrobials Wisely") facility-specific guidelines for the use of antimicrobials were implemented based on Campaign steps. This initiative led to a 40% improvement in antimicrobial use and also led to decreases in the rates of recovery of MRSA and vancomycin-resistant *Enterococcus*. As with the first project, the antimicrobial-use interventions also led to cost savings through decreased antimicrobial expenditures and shorter lengths of stay for patients with infections. Finally, for strategy four ("Prevent Transmission") the facility implemented an aggressive Campaign to improve compliance with hand hygiene, which led to significant and sustained improvements throughout the hospital.

Data from a fourth project implementing steps in the 12-step Program to Prevent Antimicrobial Resistance Among Surgical Patients was recently presented at a national meeting. This project focused on improving compliance with recommendations on surgical antimicrobial prophylaxis using a multidisciplinary, team-based approach to identify and overcome barriers to compliance. Pre- and post-implementation comparisons demonstrated significant improvements in the number of patients who received the appropriate agent and dose and in compliance with the recommendation that prophylactic antimicrobials be discontinued within 24 hours of surgery.

These projects provide important information on the CDC's Campaign to Prevent Antimicrobial Resistance in Health care Settings. First, they demonstrate some of the critical factors in successfully implementing the Campaign. All four project leaders commented on how critical it was not only to have well-respected clinicians leading the projects, to appeal to other clinicians, but also to have strong administrative backing to demonstrate both a multidisciplinary approach and the importance of the project to the facility. Second, they provide real-world examples of the flexibility of the Campaign. The Campaign was designed so that it could be implemented as any combination of steps and strategies or in its entirety and these projects provide objective evidence that this theoretical flexibility indeed applies in the real-world as they ran the spectrum from implementing a single step to the full Campaign. Finally, they provide strong data in support of the efficacy of the Campaign, as each project was successful in attaining positive outcomes. Most importantly, project three, which looked at antimicrobial resistance as an outcome and had the longest follow-up, did indeed show a reduction in the rates of recovery of antimicrobial resistant pathogens following implementation of the Campaign.

NEXT STEPS FOR THE CAMPAIGN

The Campaign to Prevent Antimicrobial Resistance in Health care Settings provides a unified and comprehensive approach to the challenge of antimicrobial resistance. The recommendations provided in each of the 12-step programs are supported by published evidence and the Campaign combines these measures in a format directed at clinicians. Each program has been carefully reviewed and vetted by clinicians and assessments of implementation challenges have been examined. Finally, initial implementation studies have now demonstrated the effectiveness of the Campaign. What is needed now are on-going and expanded efforts at implementing the Campaign in health care facilities and systems around the country. These efforts are under way and will provide more information that will help refine and improve both the Campaign and other implementation efforts. More recently, the CDC has begun working with health departments in several states that are attempting to develop plans for local implementation of Campaign measures. These efforts are especially attractive in that they leverage the profile and support of the public health authorities to bring together a variety of facilities in a specific area, who can then work together to combat significant regional issues in antimicrobial resistance.

REFERENCES

1. Centers for Disease Control and Prevention. Preventing Emerging Infectious Diseases: A Strategy for the 21st Century. Atlanta, GA: Centers for Disease Control and Prevention; 1998.
2. Smolinski M. Microbial Threats to Health: Emergence Detection and Response. Washington, D.C.: Institute of Medicine; 2003 March 18.
3. Reichler M, Allphin, AA, Breiman, RF, Schreiber, JR, Arnold, JE, et al. The spread of multiply resistant *Streptococcus pneumoniae* at a day care center in Ohio. *J Infect Dis.* 1992;166(6):1346–53.
4. McGowan JE, Jr. Antimicrobial resistance in hospital organisms and its relation to antibiotic use. *Rev Infect Dis.* 1983;5(6):1033–48.

5. Whitney CG, Farley MM, Hadler J, Harrison KH, Bennett NM, Lynfield R, et al. Decline in invasive pneumococcal disease following the introduction of protein-polysaccharide conjugate vaccine. *N Engl J Med*. 2003;348:1737–46.

6. Centers for Disease Control and Prevention. Direct and indirect effects of routine vaccination of children with 7–valent pneumoccocal conjugate vaccine on incidence of invasive pneumococcal disease—United States, 1998–2003. *MMWR*. 2005. p. 893–97.

7. Dowell S SB. Resistant pneumococci: Protecting patients through judicious use of antibiotics. *Am Fam Physcian*. 1997;55:1647–54.

8. Seppala H, Klaukka T, Vuopio-Varkila J, Muotiala A, Helenius H, Lager K, et al. The effect of changes in the consumption of macrolide antibiotics on erythromycin resistance in group A streptococci in Finland. Finnish Study Group for Antimicrobial Resistance. *N Engl J Med*. 1997;337(7):441–6.

9. Green MD, Beall B, Marcon MJ, Allen CH, Bradley JS, Dashefsky B, et al. Multicentre surveillance of the prevalence and molecular epidemiology of macrolide resistance among pharyngeal isolates of group A streptococci in the USA. *J Antimicrob Chemother*. 2006;57(6):1240–3.

10. McCaig LF, Hughes JM. Trends in antimicrobial drug prescribing among office-based physicians in the United States. *JAMA*. 1995;273:214–9.

11. Gonzales R, Steiner JF, Sande MA. Antibiotic prescribing for adults with colds, upper respiratory tract infections, and bronchitis by ambulatory care physicians. *JAMA*. 1997;278:901–4.

12. Nyquist AC, Gonzales R, Steiner JF, Sande MA. Antibiotic prescribing for children with colds, upper respiratory tract infections, and bronchitis. *JAMA*. 1998;279:875–7.

13. Barden LS, Dowell SF, Schwartz B, Lackey C. Current attitudes regarding use of antimicrobial agents: results from physicians' and parents' focus group discussions. *Clin Pediatr (Phila)*. 1998;37(11):665–71.

14. Bauchner H, Pelton SI, Klein JO. Parents, physicians, and antibiotic use. *Pediatrics*. 1999;103(2):395–401.

15. Mangione-Smith R, McGlynn EA, Elliott MN, Krogstad P, Brook RH. The relationship between perceived parental expectations and pediatrician antimicrobial prescribing behavior. *Pediatrics*. 1999;103(4 Pt 1):711–8.

16. Hamm RM, Hicks RJ, Bemben DA. Antibiotics and respiratory infections: are patients more satisfied when expectations are met? *J Fam Pract*. 1996;43(1):56–62.

17. Todd JK. Bacteriology and clinical relevance of nasopharyngeal and oropharyngeal cultures. *Pediatr Infect Dis*. 1984;3(2):159–63.

18. Schuchat A, Hilger T, Zell E, et al. Active Bacterial Core Surveillance of the Emerging Infections Program Network. *Emerg Infect Dis*. 2001;7:1–8.

19. NCCLS. Performance Standards for Antimicrobial Susceptibility Testing; Sixteenth Informational Supplement (M100-S16). NCCLS document M100-S16. Wayne, Pa.: National Committee for Clinical Laboratory Standards; 2006.

20. Robinson KA, Baughman W, Rothrock G, Barrett NL, Pass M, Lexau C, et al. Epidemiology of invasive *Streptococcus pneumoniae* infections in the United States, 1995–1998: opportunities for prevention in the conjugate vaccine era. *JAMA*. 2001;285:1729–35.

21. Centers for Disease Control and Prevention. Geographic variation in penicillin resistance in *Streptococcus pneumoniae*—selected sites, United States, 1997. *MMWR*. 1999;48:656–61.

22. Hofmann J, Cetron MS, Farley MM, Baughman WS, Facklam RR, Elliott JA, et al. The prevalence of drug-resistant *Streptococcus pneumoniae* in Atlanta. *N Engl J Med*. 1995;333:481–6.

23. American Academy of Pediatrics Committee on Infectious Diseases. Therapy for children with invasive pneumococcal disease. *Pediatrics*. 1997;99:289–99.

24. Levine OS, Farley M, Harrison LH, Lefkowitz L, McGeer A, Schwartz B. Risk factors for invasive pneumococcal disease in children: a population-based case-control study in North America. *Pediatrics.* 1999;103(3):E28.

25. Hyde TB, Gay K, Stephens DS, Vugia DJ, Pass M, Johnson S, et al. Macrolide resistance among invasive *Streptococcus pneumoniae* isolates. *JAMA.* 2001;286:1857–62.

26. Stephens DS, Zughaier SM, Whitney CG, Baughman WS, Barker L, Gay K, et al. Incidence of macrolide resistance in *Streptococcus pneumoniae* after introduction of the pneumococcal conjugate vaccine: population-based assessment. *Lancet.* 2005:855–63.

27. Kyaw MH, Lynfield R, Schaffner W, Craig AS, Hadler J, Reingold A, et al. Effect of introduction of the pneumococcal conjugate vaccine on drug resistant *Streptococcus pneumoniae. N Engl J Med.* 2006;354(14):1455–63.

28. Breiman R, Spika JS, Navarro VJ, et al. Pneumococcal bacteremia in Charleston County, South Carolina: a decade later. *Arch Intern Med.* 1990;150:1401–5.

29. Fine MJ, Smith MA, Carson CA, et al. Efficacy of pneumococcal vaccination in adults: a meta-analysis of randomized controlled trials. *Arch Intern Med.* 1994;154:2666–77.

30. Centers for Disease Control and Prevention. Influenza and Pneumococcal Vaccination Coverage among Persons Aged >65 Years—United States, 2004–2005. *MMWR.* 2006 October 6, 2006 55(39):1065–8.

31. U.S. Department of Health and Human Services. Healthy people 2010. Conference ed. Washington, DC: US Department of Health and Human Services; 2000.

32. Butler JC, Breiman RF, Campbell JF, Lipman HB, Broome CV, Facklam RR. Polysaccharide pneumococcal vaccine efficacy: an evaluation of current recommendations. *JAMA.* 1993;270:1826–31.

33. Sims RV, Steinmann WC, McConville JH, King LR, Zwick WC, Schwartz JS. The clinical effectiveness of pneumococcal vaccine in the elderly. *Ann Intern Med.* 1988;108:653–7.

34. Shapiro ED, Berg AT, Austrian R, Schroeder D, Parcells V, Margolis A, et al. The protective efficacy of polyvalent pneumococcal polysaccharide vaccine. *N Engl J Med.* 1991;325:1453–60.

35. Farr BM, Johnston BL, Cobb DK, Fisch MJ, Germanson TP, Adal KA, et al. Preventing pneumococcal bacteremia in patients at risk: results of a matched case-control study. *Arch Intern Med.* 1995;155:2336–40.

36. Centers for Disease Control and Prevention. Prevention of pneumococcal disease: recommendations of the Advisory Committee on Immunization Practices (ACIP). *MMWR.* 1997; 46(No. RR-8):1–24.

37. Centers for Disease Control and Prevention. Prevention of pneumococcal disease among infants and young children: recommendations of the Advisory Committee on Immunization Practices. *MMWR.* 2000;49 (No. RR-9):1–35.

38. Black S, Shinefield H, Fireman B, Lewis E, Ray P, Hansen JR, et al. Efficacy, safety and immunogenicity of heptavalent pneumococcal conjugate vaccine in children. Northern California Kaiser Permanente Vaccine Study Center Group. *Pediatr Infect Dis J.* 2000; 19:187–95.

39. Eskola J, Kilpi T, Palmu A, Jokinen J, Haapakoski J, Herva E, et al. Efficacy of a pneumococcal conjugate vaccine against acute otitis media. *N Engl J Med.* 2001; 344: 403–9.

40. Emmer C, Besser RE. Combating antimicrobial resistance:intervention programs to promote appropriate antibiotic use. *Infect Med.* 2002;19:160–73.

41. Weissman J, Besser RE. Promoting appropriate antibiotic use for pediatric patients: a social ecological framework. *Semin Pediatr Infect Dis.* 2004;15(1):41–51.

42. Dowell SF, Marcy SM, Phillips WR, Gerber MA, Schwartz B. Principles of judicious use of antimicrobial agents for pediatric upper respiratory tract infections. *Pediatrics.* 1998;101:163–5.

43. Subcommittee on Management of Acute Otitis Media. Diagnosis and management of acute otitis media. *Pediatrics*. 2004;113:1451–65.

44. Spiro DM, Tay KY, Arnold DH, Dziura JD, Baker MD, Shapiro ED. Wait-and-see prescription for the treatment of acute otitis media: a randomized controlled trial. *JAMA*. 2006;296(10):1235–41.

45. Gonzales R, Steiner JF, Lum A, Barrett Jr PH. Decreasing antibiotic use in ambulatory practice: impact of a multidimensional intervention on the treatment of uncomplicated acute bronchitis in adults. *JAMA*. 1999;281:1512–9.

46. Belongia EA, Sullivan BJ, Chyou PH, Madagame E, Reed KD, Schwartz B. A community intervention trial to promote judicious antibiotic use and reduce penicillin-resistant *Streptococcus pneumoniae* carriage in children. *Pediatrics*. 2001;108(3):575–83.

47. Hennessy TW, Petersen KM, Bruden D, Parkinson AJ, Hurlburt D, Getty M, et al. Changes in antibiotic-prescribing practices and carriage of penicillin-resistant *Streptococcus pneumoniae*: A controlled intervention trial in rural Alaska. *Clin Infect Dis*. 2002 Jun 15;34(12):1543–50.

48. McCaig LF, Besser RE, Hughes JM. Antimicrobial drug prescription in ambulatory care settings, United States, 1992–2000. *Emerg Infect Dis*. 2003;9(4):432–7.

49. McCaig L, Friedman CR. Trends in antimicrobial prescribing in ambulatory care settings in the United States, 1993–2004. Annual Conference on Antimicrobial Resistance. Bethesda, MD: National Foundation for Infectious Diseases; 2006.

50. Gould CV, Rothenberg R, Steinberg JP. Antibiotic resistance in long-term acute care hospitals: the perfect storm. *Infect Control Hosp Epidemiol*. 2006;27(9):920–5.

51. Centers for Disease Control and Prevention. Guideline for handwashing and hospital environmental control, 1985. MMWR. 1988;36.

52. Shlaes DM, Gerding DN, John JF, Jr., Craig WA, Bornstein DL, Duncan RA, et al. Society for Health care Epidemiology of America and Infectious Diseases Society of America Joint Committee on the Prevention of Antimicrobial Resistance: guidelines for the prevention of antimicrobial resistance in hospitals. *Clin Infect Dis*. 1997;25(3): 584–99.

53. Gold HS, Moellering RC Jr. Antimicrobial-drug resistance. *N Engl J Med*. 1996;335: 1445–53.

54. Gerding DN, Larson TA. Resistance surveillance programs and the incidence of Gram-negative bacillary resistance to amikacin from 1967–1985. *Am J Med* 1986; 80(6B):22–8.

55. Ballow CH, Schentag JJ. Trends in antibiotic utilization and bacterial resistance: Report of the National Nosocomial Resistance Surveillance Group. *Diagn Microbiol Infect Dis*. 1992;15:37S–42S.

56. Wester CW, Durairaj L, Evans AT, Schwartz DN, Husain S, Martinez E. Resistance: a survey of physician perceptions. *Arch Intern Med*. 2002;162(19):2210–6.

57. Giblin TB, Sinkowitz-Cochran RL, Harris PL, Jacobs S, Liberatore K, Palfreyman MA, Harrison EI, Cardo DM. CDC Campaign to Prevent Antimicrobial Resistance Team. Clinicians' perceptions of the problem of antimicrobial resistance in health care facilities. *Arch Intern Med*. 2004;164(15):1662–8.

58. Srinivasan A, Song X, Richards A, Sinkowitz-Cochran R, Cardo D, Rand C. A survey of knowledge, attitudes, and beliefs of house staff physicians from various specialties concerning antimicrobial use and resistance. *Arch Intern Med*. 2004;164(13):1451–6.

59. Brinsley KJ, Sinkowitz-Cochran RL, Cardo DM. Assessing the motivation for physicians to prevent antimicrobial resistance in hospitalized children using the Health Belief Model as a framework. *Am J Infect Control*. 2005;33:175–81.

60. Berns JS, Tokars J. Preventing bacterial infections and antimicrobial resistance in dialysis patients. *Am J Kidney Dis*. 2002;40:886–98.

61. Raymond L, Kingsbury D, Duncan JF. Penicillin resistance of *D. pneumoniae* in upper respiratory infections associated with diving. *Aerospace Medicine*. 1971;42(2):196–8.

62. Kupronis BA, Richards CL, Whitney CG. Active Bacterial Core Surveillance Team. Invasive pneumococcal disease in older adults residing in long-term care facilities and in the community. *J Am Geriatr Soc.* 2003;51:1520–5.
63. Brinsley K, Sinkowitz-Cochran R, Cardo D. An assessment of issues surrounding implementation of the Campaign to Prevent Antimicrobial Resistance in Health Care Settings. *Am J Infect Control* 2005;33:402–9.
64. Srinivasan A, Patel A, Sinkowitz-Cochran R, Cardo D, Miller R. The Johns Hopkins Hospital Antibiotic Management Program (AMP) Reporter: Automated post-prescription antibiotic review to improve compliance with the CDC's Campaign to Prevent Antimicrobial Resistance in Health Care Settings. 13th Annual Scientific Meeting of the Society for Health care Epidemiology of America. Arlington, VA; 2003.
65. Brinsley K, Srinivasan A, Sinkowitz-Cochran R, Lawton R, McIntyre R, Kravitz G, Burke B, Shadowen R, Cardo D. Implementation of the Campaign to Prevent Antimicrobial Resistance in Health care Settings: 12 Steps to Prevent Antimicrobial Resistance Among Hospitalized Adults—experiences from 3 institutions. *Am J Infect Control.* 2005;33(53–54).
66. National Nosocomial Infections Surveillance System. National Nosocomial Infections Surveillance (NNIS) System report, data summary from January 1992 through June 2004. *Am J Infect Control* 2004;32:470–85.
67. Finkelstein JA, Davis RL, Dowell SF, Metlay JP, Soumerai SB, Rifas-Shiman SL, et al. Reducing antibiotic use in children: a randomized trial in 12 practices. *Pediatrics.* 2001;108(1):1–7.
68. Perz JF, Craig AS, Coffey CS, Jorgensen DM, Mitchel E, Hall S, et al. Changes in antibiotic prescribing for children after a community-wide Campaign. *JAMA.* 2002; 287(23):3103–9.

17 Antibacterial Drug Discovery in the 21st Century

Steven J. Projan

CONTENTS

DO WE NEED NEW ANTIBACTERIAL DRUGS?

Antimicrobial drug resistance constitutes a major problem in both developing and developed countries, yet, despite stories in the media that gain the attention of the public for one or two news cycles each year, resistance as a public health problem is vastly underappreciated. In fact, patients today are contracting multidrug-resistant infections and are suffering higher levels of morbidity and mortality. While the connection between multidrug resistance and these poorer patient outcomes remains controversial, the increased expense and stress to healthcare systems are undeniable. Most importantly, the number of therapeutic options for serious, life-threatening bacterial infections are becoming more limited specifically because of multidrug resistance. It is a common misconception that virtually all bacterial infections remain treatable with our current armamentarium of antibacterial drugs. In fact, untreatable, pan-resistant bacterial infections do occur and are becoming increasingly common. Furthermore, as stated above, deaths following bacterial infections are on the rise even in patients who are infected with bacteria that are ostensibly treatable with currently available drugs. This is undoubtedly because the sickest patients are those most at risk for multidrug-resistant infections, yet it is also a fact that otherwise healthy people are being infected with multidrug-resistant bacteria and these infections are now becoming increasingly prevalent in the community, whereas in the past they were almost solely a hospital problem. We have seen multiple reports of community-associated, multidrug-resistant *Staphylococcus aureus* infections and these are increasing in frequency. In the past, physicians could reliably treat such infections with penicillin-like drugs, but this is no longer the case and has already led to changes in prescribing patterns that can be predicted to lead to even more multidrug-resistant strains proliferating in the community setting.

Many important advances in the practice of medicine are actually at serious risk. Multidrug-resistant bacteria are compromising our ability to perform what are now considered routine surgical procedures, such as hip replacements or coronary artery bypass grafts (CABG). A ubiquitous phrase encountered in obituaries is "died from complications following surgery," but what is not well understood is that these "complications" are quite frequently multidrug-resistant infections. We are also now seeing an increase in all manner of immunosuppressive therapy, not just in transplants, but in oncology, rheumatoid arthritis, and asthma, to name but a few. Clearly, these novel, in many cases breakthrough, therapies may become untenable because of untreatable, multidrug-resistant infections.

Antiinfective drugs are unique in that, unlike almost all other therapies, the use of drugs has a clear societal impact, not just in the communicability of infections, but in the fact that increased use leads almost inexorably to increased levels of resistance and loss of utility. Successfully treating an infected patient today may lead to clinical failure in treating an infected patient tomorrow. Therefore, there is a clear conflict between the needs of society to have effective antibacterial therapies in the future and the needs of today's patients. Antibacterial drugs are scarce resources, but they are resources we dare not ration where a patient may benefit from their use today. Such rationing, however, has actually become commonplace.

An important underlying assumption is that because of the very nature of microbial evolution a continuous flow of novel therapeutic and preventative strategies will be necessary merely to maintain our capabilities in warding off infections. However, coupled with this clear and growing unmet medical need has been a marked decline in the pursuit of novel antibacterial drugs in the biopharmaceutical industry [1]. The forces behind this reduced level of activity are many. While the market for antibacterial drugs remains large, it holds the promise of only modest growth (at best) in the near term. There are many antibacterial products currently available, and the market place is crowded and confusing. Together with this complex market is the fact that recent industrial and academic productivity in the discovery of novel antibacterial strategies has been poor. The lack of new agents entering the pipeline has actually not been for lack of effort. Some of this work appears to have borne fruit, but it may well be that the recent rounds of industrial retrenchment and consolidations will lead to a squandering of research that appears to be on the verge of delivering novel agents. Another important reason for the dry pipeline and decline in industrial interest is that the time it takes to develop a novel antiinfective agent after it has been discovered has greatly increased. This is mainly because the drug approval process has become increasingly complicated, time-consuming, contentious, and costly, with less assurance of favorable outcomes than in the past (this is true in all areas of research, not only antiinfectives, but it has particularly affected antibacterial research). Indeed, among the regulatory authorities there are those who believe that there really is not a need for novel antibacterial agents. This view is even more commonly held in Europe than in the United States, where there is barely concealed contempt when pharmaceutical companies bring forward a novel drug for approval. In some ways this is not surprising, since in Europe the governments are both responsible for ultimately approving new drug applications and then paying for those drugs when they are marketed. A new (and relatively expensive) drug on the market may actually represent an economic threat to national healthcare systems.

The decline in industrial research aimed at providing these novel approaches just at a time when they are needed represents a serious and immediate public health crisis, and is a problem begging for aggressive and imaginative solutions. Yet from whence will these solutions come? It is a sad truth that amateurs hardly ever discover drugs and never have the wherewithal to develop them, given the time and expense of clinical trials, so with the number of professional drug hunters decreasing who is going to discover and develop the next generation of antibacterial drugs? The hope here is that an increased academic and governmental focus on understanding resistance and identifying novel agents will serve to improve our overall knowledge base, which is clearly necessary for future successes. But how to actually discover these needed drugs?

WHAT ANTIBACTERIAL DRUGS DO WE NEED?

Interestingly, there is not a consensus as what types of new drugs are needed to treat antibacterial infections. Neither major regulatory authorities (the U.S. Food and Drug Administration, the European Agency for the Evaluation of Medicinal Products) nor competent authorities (e.g., the Centers for Disease Control and Prevention) have come forward with even a draft proposal as to which organisms causing what kinds of infections should be the subjects of future investigations. However, the Infectious Diseases Society of America, in its monograph "Bad Bugs. No Drugs," has summarized both the issues and the problem organisms [2]. Likewise, Dr. Louis Rice has explicitly stated what the problem organisms are in a recent publication, and these include what could be called the usual suspects [3]. Included among the Gram-positive pathogens are methicillin-resistant *Staphylococcus aureus* and *S. epidermidis*, vancomycin-resistant *Enterococcus faecium and E. faecalis,* and the rapidly growing mycobacteria. In the past five years, however, no fewer than four novel agents have been approved that have clinical activity versus these bacteria. It is really among the multidrug-resistant Gram-negative bacteria that we find growing unmet medical need, and only a single new agent has been approved in well over a decade. These include pan-resistant strains of *Pseudomonas aeruginosa* and *Klebsiella pneumoniae*, *Stenotrophomonas maltophilia*, and *Acinetobacter baumanii*. Compromising the utility of many workhorse β-lactam drugs have been extended spectrum β-lactamase (ESBL)-producing organisms among the Gram negatives, and novel approaches are particularly needed for these organisms.

DO WE REALLY UNDERSTAND RESISTANCE?

Complicating the search for new antibacterial agents is the fact that much of what we believe to know about resistance is either wrong or only partially correct. Resistance is not simply a matter of "select for them and they will come." We rarely observe the emergence of multidrug-resistant phenotypes among small genome organisms (e.g., *Streptococcus pyogenes*), but we routinely see the rapid emergence of pan resistance in large genome organisms like *P. aeruginosa*. None of the current dogma on resistance accounts for this observation. Even the basics, like mistakenly calling for "hand washing" among healthcare workers when what is really called for is hand disinfection, are routinely misreported to the public. Perhaps the most interesting conundrum lacking a valid scientific explanation is that using a combination of piperacillin and the β-lactamase inhibitor tazobactam as a workhorse antibiotic in healthcare settings leads to lower levels of resistance than using a third-generation cephalosporin [4,5]. Both have about the same potency, spectrum of activity, and even molecular target. There is nothing in the current dogma that accounts for this observation. Therefore, the question must be posed: "If we do not really understand resistance, then how do we prevent it?" Perhaps, the most vexing unanswered question is "What is the actual clinical impact of resistance?" Surprisingly, this has not been well addressed, with the number of valid studies specifically focusing on the clinical impact of resistance totaling fewer than 10 with no clear consensus emerging. Unless and until adequate resources (read "public funding") are brought to bear on the fundamental questions of resistance, we will be fighting an uphill battle.

THE DRUG DISCOVERY AND DEVELOPMENT PROCESS

The goal of the antibacterial drug discovery process is simple in concept: (*i*) Find agents that inhibit bacterial growth and/or kill bacteria. (*ii*) Determine whether these agents work in animal models of infection without outright toxicity. (*iii*) Test them in humans with defined and documented bacterial infections compared to existing therapies. But how does one approach this process, that is, find those specific, non-toxic inhibitors of bacterial growth?

THE OLD-FASHIONED WAY

The time-honored (and validated) method is to screen a series of chemical compounds (or natural product extracts) for antibacterial activity initially *in vitro* (and then in animal models of disease). All of the antibacterial drugs currently used therapeutically were identified by variations of this method or by synthetic modifications of agents identified by this method. Once such agents have been identified, it is typical to make, via medicinal chemistry, variants of the initial "hit," and this is true whether one is working with a synthetic chemical starting point or a natural product. In almost all cases, new medical entities (NMEs) can be generated with activities superior to the original compound, and increases in potency or improvements in pharmacokinetics of 1000-fold are not unusual. Therefore, any antibacterial drug discovery program that lacks a synthetic chemistry component will, at best, have limited success.

THE NEW-FANGLED APPROACH— TARGET-BASED DRUG DISCOVERY

Using Post-Genomic Technologies for Target-Based Drug Discovery [6]

One critical point that cannot be repeated often enough is that any target-based assay must be more sensitive than is achieved by merely screening compounds for frank growth inhibition ("it's the assay, stupid"). Otherwise this assay has virtually zero chance of discovering a novel antibacterial agent (because such phenotypic screening has already taken place). A corollary of this law of drug discovery is that if one does succeed in opening up new regions of chemical space, then the most logical starting point is not a target-based approach, but use of the tried-and-true method of screening for growth inhibition. Another key consideration is the effect of permeability barriers, especially if one starts with a biochemical screen/assay. Such permeability barriers are more profound for the Gram-negative bacteria, with their inner and outer membranes, than for Gram-positive bacteria (and this explains the greater success at identifying novel Gram-positive agents).

THE FAILURE OF TARGET-BASED DRUG DISCOVERY: "PAYNE'S LAW"

Perhaps the area of greatest frustration for those who have labored in this field has been the realization that screening for antibacterial agents has become progressively more difficult. David Payne at Glaxo SmithKline has observed that an automated, antibacterial ("high throughput") screen is about five times less likely to succeed than a screen for any other kind of therapeutic target. This appears to be true whether one is using whole cell screening methods or biochemical assays. However, there have been some recent successes, although none of the new chemical entities identified to date has made its way into clinical testing. There is no obvious reason for this lack of efficacy in that the assays that have been developed and used have, in many cases, been authored by the same scientists who have developed very successful approaches for other, non-bacterial targets. So as tempting as it is to blame incompetence for this futility, the data do not bear that out. Rather, one may suggest that bacteria (and their potential molecular targets) are the oldest of living organisms and thus have been subject to three billion years of evolution in harsh environments and therefore have been selected to withstand chemical assault.

CHEMICAL SPACE—THE FINAL FRONTIER?

It may also be true that if we keep screening the same old stuff, we are not going to find anything new. This means that new chemistries are needed to open other quadrants of chemical space. The good news is that this is not as outlandish as it sounds.

On the natural products front, there are also prospects for mining seawater for pharmaceutical gold in the newfound ability to cultivate previously uncultivable bacteria [7]. The open question is whether these organisms take other approaches to the synthesis of secondary metabolites than do those bacteria that we have been able to grow and that have been found to produce antimicrobial agents.

TARGET-BASED SCREENING IN THE POST-GENOMIC ERA

New antiinfectives will frequently be given to critically ill people, to treat conditions that may be life threatening within a matter of hours. As stated above, these drugs must be exceptionally effective and must pose no serious toxicity. Attacking unique aspects of microbial biochemistry has produced drugs with these characteristics in the past and this appears unlikely to change. Our ability to perform comprehensive comparative genomics, with virtually every clinically important bacterium sequenced, has given a new impetus to this approach, but has not really changed it. The identification of novel targets has moved from the characterization of new enzymological activities to the examination of new open reading frames encoding previously undiscovered essential functions.

In the most general of applications, genome sequence information has been used to select target genes based on the concept that open reading frames that are well conserved among bacteria (especially pathogenic bacteria), but lacking close human homologs, represent the best choices for finding valid ("essential"?) targets whose inhibitors will have a lower potential for toxicity. One should note that once this bioinformatic exercise is completed then the "wet work" must begin in earnest, validating the putative targets, configuring assays, running those assays and identifying and validating inhibitors of the targets. However, there may be a flaw in this logic. We now know that the target of trimethoprim, the enzyme dihydrofolate reductase, is actually quite well conserved, yet inhibitors of the bacterial version of this enzyme have sufficient specificity to be relatively non-toxic to humans. Likewise, there is great similarity between bacterial and mitochondrial ribosomes, but the natural product inhibitors of protein synthesis routinely have greater than 1000-fold specificity, preferentially targeting the bacterial over the mitochondrial version. Perhaps we need to sharpen our bioinformatic algorithms to capture these targets as well.

WHAT IS "ESSENTIAL"?

The principal drug discovery use of bacterial genomics has been viewed as a means of identifying open reading frames that may represent "essential genes" in target bacteria. Exactly what is meant by essential can vary depending on one's frame of reference. From the geneticist's point of view an essential gene is one whose inactivation results in a loss of bacterial viability. Genomics enables several techniques to be more effectively used in the determination of such essential genes (see below). However, this definition of essentiality may constitute a misleading, if not dangerous, oversimplification. In *Escherichia coli*, either *uvrD* or the highly similar *rep* gene can be inactivated with little effect on bacterial growth. However, the double mutant is not viable. In Gram-positive bacteria, for example, *S. aureus*, the *uvrD/rep* homolog is the *pcrA* gene [8], which is a gene that cannot be similarly inactivated (i.e., there is no redundancy in the Gram-positive bacteria for this essential DNA helicase). The relative similarity between *rep* and *uvrD* (and *pcrA*) suggests that an inhibitor could be identified, which successfully targets all three enzymes (and therefore inhibit bacterial growth in a broad spectrum manner). Thus, despite the fact that neither *rep* nor *uvrA* is "essential" in *E. coli* (and other Gram negatives) these probably do represent valid targets for antibacterial drug discovery. Conversely, there have been several reports describing "essential" two-component regulatory systems (aka TCRS) and significantly more effort has been devoted to discovering inhibitors of these TCRS than for bacterial helicases. Yet the precise essential function of these TCRS have yet to be elucidated, and may merely reflect the toxic derepression of genes negatively regulated by the TCRS, rather than some function necessary for growth carried out by either the histidine autokinase (HAK) or response regulator (RR). The lesson one may learn for the above examples is that, despite powerful genomic and genomic-enabled technologies, there is simply no substitute for an intimate understanding of microbial physiology.

WHAT MAKES A GOOD TARGET?

What is meant by a good target is one that can be targeted by a small molecule to elicit either a static or a cidal activity. Indeed, it has been argued that antibacterial drugs that kill bacteria, as opposed to merely inhibiting bacterial growth, are superior ("cidal trumps static"). This has simply not been borne out by the clinical data, however, even in patients with compromised immune systems. Targeting an essential activity, helicase activity in the above example, defines a good target, even though the definition applies to a single activity as opposed to a single gene. In antiinfectives, this kind of example is becoming the rule, as we gain an increased mechanistic understanding of how effective antibacterial agents function, particularly ones that are less prone to the emergence of resistance. It is the opinion of this author that the "best" targets are in biosynthetic pathways. We have seen that by combining two "mediocre" inhibitors, both in the folic acid biosynthetic pathway—a powerful and effective drug (trimethoprim-sulfa), resistant to the rapid emergence of resistance, can be created. Given that most of the genes for these pathways are well conserved among almost all pathogenic bacteria, this would appear to be an approach whose time has come. The downside is that moving not one, but two, novel agents through the approval process more than doubles the regulatory complexity; however, it can and has been done. One such biosynthetic pathway is fatty acid biosynthesis [9], for which several synthetic and natural product inhibitors of specific steps in this pathway have been discovered, but relatively few have been used clinically. However, there has been considerable recent progress in this area [10].

HOW MANY TARGETS ARE THERE?

There have been many estimates as to the number of essential genes/targets there are in a given bacterium/pathogenic bacteria. A number in excess of 300 has been suggested (based on the so-called "minimal essential gene set," [11]) and this number has the potential to expand further if one views inhibition of virulence as a viable strategy for antibacterial chemotherapy. However, if the goal is to identify a broad-spectrum inhibitor (even one limited to the Gram-positive bacteria as an example) the number of potential targets drops precipitously. It is the view here that, for the near future, narrowly targeted, pathogen-specific drugs will find little clinical utility and even smaller markets (but are likely to be quite expensive, if not prohibitively so). This, however, is the subject for discussion elsewhere (although it clearly affects how many view antibacterial drug discovery, that is, taking a pathogen-specific route).

So why hasn't target-based screening led to a plethora of novel antibacterial agents? The fact is that progress is being made and novel, target-based agents are moving, albeit slowly, into clinical development. But, unlike other fields of therapeutic endeavor (e.g., Alzheimer's disease), there are already scores of safe and effective antibacterial agents that set the bar quite high for new drugs. Also given the decades of empiric screening for antibacterial activity, the law of diminishing returns dictates that it will require increasing resources to find the next novel class of antibiotics and that, indeed, has been the experience.

NOVEL APPROACHES FOR THE 21ST CENTURY

TARGETING RESISTANCE

Probably the most successful (commercially and probably clinically) class of antibacterial agents has been a β-lactam antibiotic coupled with an inhibitor of β-lactamases. It is, therefore, not just theoretically possible to target a resistance mechanism and develop a clinical and commercially viable combination therapy, it has been an extremely successful approach. Perhaps the most common resistance determinants are efflux pumps [12]. While these pumps are much more diverse in genetics and structure than the β-lactamases, they do share a fair amount of structural similarity. And because not all agents are effluxed by all pumps, but rather a smaller subset of them, it may be possible to at least enhance the antibacterial activity of several classes of agents with specific efflux pump inhibitors. Given the success of the β-lactamase inhibitors, it is surprising that there has not been a similar focus on efflux pump inhibitors.

TARGETING VIRULENCE [13]

It has been suggested that disabling a bacterium's ability to cause disease may be a "kinder and gentler" approach to antibacterial chemotherapy. Targeting quorum sensing was quite popular in the late 1990s and several researchers are intent on attacking the dedicated virulence secretion systems (e.g., type three secretion). There may well be some valid uses for these agents (assuming they are discovered);

however, the idea that inhibiting virulence will mitigate against the emergence of resistance must follow the axiom that anything that interferes with a bacterium's ability to interact with its host will impose a selective pressure and select for resistant mutants. By not killing the bacterium, the likelihood is that anti-virulence strategies will actually select for resistance more quickly.

IMMUNOTHERAPEUTICS

Prior to the antibiotic era, the state of the art for antibacterial therapy was the use of immune antisera, typically generated from horses [14], and this has prompted the search for novel monoclonal antibodies as therapeutics, especially for "orphan" infectious diseases [15]. It is now commonplace for monoclonal antibodies to be used as therapeutic agents. There is one example in the field of antiinfectives. A monoclonal antibody directed against rotavirus is now used clinically, mainly in low-birth-weight babies. Monoclonal antibodies produced in culture, either of human origin or humanized to resemble human antibodies, should obviate issues like serum sickness caused by the use of animal antisera. While generating specific monoclonal antibodies is now actually quite easy, these are not likely to be the answer to our resistance problems. Even the old polyclonal preparations were only partially effective in reversing the course of a serious infection (efficacy rates appear to be far less than for standard antibacterial chemotherapy), and it is unlikely that a monoclonal antibody would be better than a polyclonal approach. Therefore, these antibodies, usually directed toward a critical, surface-expressed virulence factor, are more likely to be used in combination with classical antibiotics (and at considerable expense).

PHAGE THERAPY

Identifying bacteriophage that kill a specific target bacterium is actually a trivial exercise. Why then has bacteriophage therapy not come to the fore? There are many reasons for this, starting with the fact that as easy as it is to find such a bacteriophage, it is just as easy to select for a bacterium resistant to that bacteriophage. Furthermore, there is often such tight specificity between phage and target bacteria that no single bacteriophage will be able to therapeutically target an entire species of bacteria, thus requiring the use of "cocktails" [16]. Bacteriophage have found utility in novel targeting approaches [17], and, as such, pursuing the century-old dream of using a bacterium's natural predators as a means of therapy is worthy of study.

CONCLUDING REMARKS

In the late 1940s, the successes of Waksman and Schatz (streptomycin [18]) and Duggar (tetracycline) led many to believe that bacterial infections were basically conquered. That conceit led to widespread misuse and outright abuse of antibacterial agents. Nevertheless, we still neither fully understand nor appreciate resistance to antibacterial agents, and this needs to be addressed by assiduous study both in the lab and in the clinic. The powerful tools we now have at hand in this post-genomic century will allow us to discover new and better antibacterial drugs; it is only a matter of time and effort.

This chapter is dedicated to the memory of John F. Barrett, 1954–2006.

REFERENCES

1. Projan SJ, Shlaes DM. Antibacterial drug discovery: is it all downhill from here? *Clin Microbiol Inf* 10 Suppl 4:18–22, 2004.
2. Talbot GH, Bradley J, Edwards JE, Gilbert D, Scheld M, Bartlett JG. Bad bugs need drugs: an update on the development pipeline from the antimicrobial availability task force of the Infectious Diseases Society of America. *Clin Inf Dis* 42: 657–68, 2006.
3. Rice LB. Do we really need new anti-infective drugs? *Curr Opin Phamacol* 3(5):459–63, 2003.
4. Rice LB, Eckstein EC, DeVente J, Shlaes DM. Ceftazidime-resistant *Klebsiella pneumoniae* isolates recovered at the Cleveland Department of Veterans Affairs Medical Center. *Clin Inf Dis* 23(1):118–24, 1996.
5. Smith DW. Decreased antimicrobial resistance following changes in antibiotic use. *Surg Inf* 1:73–8, 2000.
6. Projan SJ. New (and not so new) antibacterial targets—from where and when will the novel drugs come? *Curr Opin Pharmacol* 2(5):513–22, 2002.
7. Kaeberlein T, Lewis K, Epstein SS. Isolating "uncultivable" microorganisms in pure culture in a simulated natural environment. *Science* 296:1127–29, 2002.
8. Petit MA, Dervyn E, Rose M, Entian KD, McGovern S, Ehrlich SD, Bruand C. PcrA is an essential DNA helicase of *Bacillus subtilis* fulfilling functions both in repair and rolling-circle replication. *Mol Microbiol* 29(1):261–73, 1998.
9. Payne DJ. The potential of bacterial fatty acid biosynthetic enzymes as a source of novel antibacterial agents. *Drug News Persp* 17(3):187–94, 2004.
10. Payne DJ, Miller WH, Berry V, Brosky J, Burgess WJ, Chen E, DeWolf WE Jr. Fosberry AP, Greenwood R, Head MS, Heerding DA, Janson CA, Jaworski DD, Keller PM, Manley PJ, Moore TD, Newlander KA, Pearson S, Polizzi BJ, Qiu X. Rittenhouse SF, Slater-Radosti C, Salyers KL, Seefeld MA, Smyth MG, Takata DT, Uzinskas IN, Vaidya K, Wallis NG, Winram SB, Yuan CC, Huffman WF. Discovery of a novel and potent class of FabI-directed antibacterial agents. *Antimicrob Agents Chemother* 46(10):3118–24, 2002.
11. Salama NR, Shepherd B, Falkow S. Global transposon mutagenesis and essential gene analysis of *Helicobacter pylori*. *J Bacteriol* 186(23):7926–35, 2004.
12. Kaatz GW. Inhibition of bacterial efflux pumps: a new strategy to combat increasing antimicrobial agent resistance. *Expert Opin Emerg Drugs* 7(2):223–33, 2002.
13. Alksne LE, Projan SJ. Bacterial virulence as a target for antimicrobial chemotherapy. [erratum appears in Curr Opin Biotechnol 2001 Feb;12(1):112]. *Curr Opin Biotechnol* 11(6):625–36, 2000.
14. Casadevall A, Scharff MD. Serum therapy revisited: animal models of infection and development of passive antibody therapy. *Antimicrob Agents Chemother* 38(8):1695–702, 1994.
15. Mukherjee J, Feldmesser M, Scharff MD, Casadevall A. Monoclonal antibodies to *Cryptococcus neoformans* glucuronoxylomannan enhance fluconazole efficacy. *Antimicrob Agents Chemother* 39(7):1398–405, 1995.
16. Projan S. Phage-inspired antibiotics? *Nat Biotechnol* 22(2):167–8, 2004.
17. Liu J, Dehbi M, Moeck G, Arhin F, Bauda P, Bergeron D, Callejo M, Ferretti V, Ha N, Kwan T, McCarty J, Srikumar R, Williams D, Wu JJ, Gros P, Pelletier J, DuBow M. Antimicrobial drug discovery through bacteriophage genomics. *Nat Biotechnol* 22(2):185–91, 2004.
18. Daniel TM. Selman Abraham Waksman and the discovery of streptomycin. *Int J Tuberc Lung Dis* 9(2):120–2, 2005.

Index

A

AAC(2')-Ia
  in *Providencia stuartii*, 33–38
    genetic regulation, 33
    negative regulators, 34–35
    physiological functions, 33
    positive regulators, 36
    quorum sensing, 37
aarA, 34
aarB, 34
aarC, 34
aarD, 35
aarE, 36
aarF, 36
aarG, 35
aarP, 36
ABC, *see* Adenosine triphosphate (ATP) binding
    cassette (ABC) transporters
Absorption, distribution, metabolism, excretion,
    toxicity (ADMET) profile, 56
Acetyltransferases, 72
  intrinsic, 32
Acinetobacter, 77
Acinetobacter baumanii
  aminoglycoside antibiotic efflux, 78
  PER-1 β-lactamase, 113–114
Acinetobacter haemolyticus, 33
Acquired genes modification and protection,
    147–150
Acquired metabolic pathway remodeling, 143
AcrA, 48
AcrAB-TolC tripartite efflux complex, 49
AcrB, 47
  asymmetric crystal, 52
  efflux pump, 46
  peristaltic pump, 53
Actinobacillus actinomycetemcomitans, 216
Active Bacterial Core Surveillance, 380, 381
Acute otitis media (AOM), 385
Acute respiratory infections
  global resistance, 364
Adaptation
  chemical environment, 4
  MDR, 5
Adenosine triphosphate (ATP) binding cassette
    (ABC) transporters, 169

Adenylyltransferases (ANT), 72
ADMET, *see* Absorption, distribution,
    metabolism, excretion, toxicity
    (ADMET) profile
Albomycin, 175
Alexander International Multicentre
    Longitudinal Surveillance, 369
Amikacin, 74
Aminoglycoside
  action mode, 73–75
  bacterial ribosome, 73–75
  bactericidal action, 76–77
  drug-resistant tuberculosis, 316–319
  Gram-negative bacteria
    outer membrane, 173
  modifying enzymes, 79
    circumvention, 87–90
  structure, 74, 348
  uptake, 76
Aminoglycoside antibiotic efflux, 77, 78
Aminoglycoside *O*-nucleotidyltransferases
    (ANT), 85–86
Aminoglycoside *O*-phosphotransferases (APH),
    72, 80–81
Aminoglycoside resistance, 32, 77–86, 148
  altered uptake, 77
  enterobacteria, 347, 352
  mechanisms, 71–102
  modification, 78–86
  to synergistic killing, 265–267
  target modification, 78
Aminoglycoside-resistant mutants, 77
Amoxicillin resistance, 365
Ampicillin resistance
  Enterococcus faecalis, 256–257
  Enterococcus faecium, 257–258
  Haemophilus influenzae, 365
  Shigella dysenteriae, 366
  typhoid, 366
ANT, *see* Aminoglycoside
    O-nucleotidyltransferases (ANT)
Antibacterial drug discovery, 409–418
Antibacterial drugs, 409–411
Antibiogram, 371
Antibiotic permeability, 169–182
  bacterial cell envelopes, 172
  defined, 170